LIFE IN MOVING FLUIDS

STEVEN VOGEL

Life in Moving Fluids

THE PHYSICAL BIOLOGY
OF FLOW

SECOND EDITION
REVISED AND EXPANDED
ILLUSTRATED BY
SUSAN TANNER BEETY
AND THE AUTHOR

PRINCETON UNIVERSITY PRESS
PRINCETON, N.J.

Library of Congress Cataloging-in-Publication Data
Vogel, Steven, 1940–
Life in moving fluids : the philosophical biology of flow /
Steven Vogel.—2nd ed., rev. and expanded.
p.   cm.
Includes bibliographical references and index.
ISBN 0-691-03485-0
ISBN 0-691-02616-5 (pbk.)
1. Fluid mechanics. 2. Biophysics. I. Title
QH505.V63 1994
574.19'1—dc20 93-46149

To the Department of Zoology of Duke University

as it existed from the mid-sixties to the mid-nineties—
a cohesive oasis of intellectual stimulation, academic excellence,
and interpersonal support and sensitivity

Contents

Preface to the New Edition   ix

Acknowledgments   xiii

Chapter 1. Remarks at the Start   3

Chapter 2. What Is a Fluid and How Much So   16

Chapter 3. Neither Hiding nor Crossing Streamlines   32

Chapter 4. Pressure and Momentum   50

Chapter 5. Drag, Scale, and the Reynolds Number   81

Chapter 6. The Drag of Simple Shapes and Sessile Systems   106

Chapter 7. Shape and Drag: Motile Animals   132

Chapter 8. Velocity Gradients and Boundary Layers   156

Chapter 9. Life in Velocity Gradients   174

Chapter 10. Making and Using Vortices   204

Chapter 11. Lift, Airfoils, Gliding, and Soaring   230

Chapter 12. The Thrust of Flying and Swimming   262

Chapter 13. Flows within Pipes and Other Structures   290

Chapter 14. Internal Flows in Organisms   308

Chapter 15. Flow at Very Low Reynolds Numbers   331

Chapter 16. Unsteady Flows   362

Chapter 17. Flow at Fluid-Fluid Interfaces   378

Chapter 18. Do It Yourself   398

List of Symbols   403

Bibliography and Index of Citations   407

Subject Index   441

Preface to the New Edition

ABOUT A dozen years ago, calling up a degree of hubris that I now find quite inexplicable, I wrote a book about the interface between biology and fluid dynamics. I had never deliberately written a book, and I had never taken a proper course in fluids. But I had learned through teaching—both something about the subject and something about the dearth of material that might provide a useful avenue of approach for biologist and engineer. Each seemed dazzled and dismayed by the complexity of the other's domain. The book happened in a hurry, in a kind of race against the impending end of a sabbatical semester, and in a kind of mad fit of passion driven by the simple realization (and astonishment) that it was actually happening.

The reception of *Life in Moving Fluids* turned out to surpass my most self-indulgent fantasies—it reached the people I had hoped to reach, from ecologists and marine biologists to physical and applied scientists of various persuasions, and it seems to have played a catalytic or instigational role in quite a few instances. Quite clearly the book has been the most important thing of a professional sort that I've ever done; certainly that's true if measured by the frequency with which the first punning sentence of its preface is flung back at me. (That my writing has been more important than my research in furthering my area of science suggests that doing hands-on science, which I enjoy, is really just a personal indulgence—quite a curious state of affairs!)

But the book was done quickly, and I was so concerned about keeping my feet and mouth decently distant when talking about physical matters that I barely realized how thin was my coverage of the biology. Some omissions got apologetic mention, others were quietly given the blind eye, and a lot represented simple personal innocence. Mistakes were made, most of which were not accidental, and people wrote to point out (always kindly) the errors of my ways. Mistakes and ambiguities became particularly distressing when I found them contaminating the primary literature—that's not the ideal measure of a book's influence!

Doing a new version is an enormous luxury, one afforded only a small fraction of authors of instructional material. In the present case, correcting errors has turned out to be the smallest part of the task. What I've been able to do is to rewrite the book with what was almost entirely lacking before—a sense of who would use it and what role it would serve. In effect, I now have a criterion by which to judge appropriate content, level, and so forth. I'm still trying to make something that serves a variety of roles. The primary one, as before, is as guide for the biologist who needs to know something about fluids in motion. But I've given more attention to the problem of the biologist or engineer who wants to know a little about what people interested in biological fluid mechanics have been up to. That's what most of the additional material, the near doubling of the number of words, is about. In the earlier version, the three hundred or so references represented the full depth of my plumbing of the relevant literature. In the present one, the seven hundred-odd references are a culling from several thousand sources given some degree of attention. Even so, I'm uncomfortably aware that the job has been a bit superficial and spotty. Part of that reflects limitations of time and energy; part represents the positive conviction that compendia and textbooks are different creatures and that this intends to be one of the latter. But I must forthrightly admit that omissions don't always represent informed or even specific judgments—the book at best gives the flavor of a very heterogeneous area of inquiry and has a story line driven by the physical rather than biological content.

Certain topics have been deliberately omitted to maintain the intended level of presentation, an entry-level work for people with backgrounds typical of biologists. Vorticity, stokeslets, potential flow, the Navier-Stokes equations—none lacks biological relevance, but usefully lacing them into the present discussion didn't seem practical. At least I would have had to violate my first rule for writing—*explain*, don't just *mention*. And some perfectly biological material just got too complicated and specialized, in particular some of the fancier aspects of swimming and flying. For the latter, I've tried to direct the reader toward other sources.

Other topics were omitted as judgments of scope and for simple reasons of space. My focus as a biologist is on organisms, not cells, molecules, habitats, or ecosystems. Fluid mechanics is quite as relevant at levels of organization other than organismal, but the present book simply doesn't worry much about them. Thus for eddy viscosity, for much on gravity waves, for atmospheric, oceanic, lacustrine, and other large-scale motion, for diffusion with drift and flows where the mean free path of molecules is significant, for non-Newtonian and intracellular flows, for convective heat transfer—for these the reader will have to turn elsewhere. Again, I've tried to suggest some appropriate sources. Finally, the appendices on techniques of the earlier edition have disappeared. I just couldn't figure out how to do them justice in a reasonable space, and I think we're now at a stage at which the proper vehicle is some network-accessible bulletin board that permits interaction and continuous alteration.

On the other hand, the new version has gained whole new topics, not just a lot more biological detail and citations. Swimming has surfaced, pumps are now of prime concern, blood flow is no longer dismissed with some sanguine phrases, unsteady flows and the acceleration reaction get a proper start, events at the air-water interface get more than vaporous mention, jet propulsion isn't just recoiled from, Péclét number is present if perhaps peculiarly done, and Froude propulsion efficiency is pushed with dispatch if not great efficiency.

In the preface to the earlier version I offered to send my accumulated teaching material to anyone who wrote to me. Quite a few people took advantage of the offer, and I make it again. The set of problems proved more useful than the other items, so problems are what you'll get if you write (answers, too, unless I get suspicious about your motives).

In going through the bibliography I find citations of no fewer than forty-two people who were part of one or another class in front of which I said my piece. I happily admit the grossest bias in choosing cases and sources, mainly because of my joy in discovering that such large-scale favoritism has been possible. I hope these written words have some similar effect.

Acknowledgments

A LARGE number of people have given suggestions and advice for this rewriting; that so many took so much time is a most pleasant reminder that the book has been useful. The ones I can specifically recall, and to whom I express my gratitude, are Sarah Armstrong, Douglas Craig, Hugh Crenshaw, Mark Denny, Robert Dudley, Olaf Ellers, Charles Ellington, Shelley Etnier, Matthew Healy, Carl Heine, Mimi Koehl, Anne Moore, Charles Pell, Roy Plotnick, John J. Socha, Lloyd Trefethen, Vance Tucker, Jane Vogel, Stephen Wainwright, and Paul Webb. In addition, I appreciate the continuous flow of advice and support that has come from the members of the comparative biomechanics and functional morphology group ("BLIMP") of the Duke Zoology Department. All suggestions were actually taken seriously, and a few suggestions were actually taken. At Princeton University Press, Emily Wilkinson stoked the fires of the project, keeping it off the back burner, and Alice Calaprice steered me around turbulence and stagnation points and dealt deftly with both my allusions and illusions.

The following figures have been redrawn from existing material. The author gratefully acknowledges permission for reuse from the copyright holders (for previously published material) and the original creator (for other material). Figure 3.8: David M. Fields and Jeannette Yen; 4.6, 14.5b: Marine Biological Laboratory (*Biological Bulletin*); 4.11a: Lisa S. Orton; 6.1: Cornell University Press (*Aerodynamics* by T. Von Kármán); 6.5a: Barbara A. Best; 6.5b: Douglas A. Craig; 6.6, 11.12: The Company of Biologists, Ltd. (*Journal of Experimental Biology*); 6.11b, 10.9a, 10.9b: American Society of Limnology and Oceanography (*Limnology and Oceanography*); 7.5: Springer-Verlag (*Journal of Comparative Physiology*); 8.6; American Society of Zoologists (*American Zoologist*); 9.1a, 9.1b: Birkhäuser Verlag (*Swiss Journal of Hydrology*); 9.1e, 15.3: John Wiley and Sons, Inc. (*Fresh Water Invertebrates of the United States* by R. W. Pannak); 9.2: Olaf Ellers (Ph.D. dissertation); 9.5a: Cambridge University Press (*The Invertebrata* by L. A. Borradaile et al.); 10.8: National Research Board of Canada (*Canadian Journal of Zoology*); 11.7, 11.2: The Royal Society (*Philosophical Transactions of the Royal Society*); 11.8: Springer-Verlag (*Oecologia*); 12.5b: Cambridge University Press (*Biological Reviews*); 13.6: United Engineering Trustees, Inc. (*Fluid Mechanics for Hydraulic Engineers* by H. Rouse); 14.2b: Holt, Rinehart, and Winston, and Mrs. Alfred S. Romer (*The Vertebrate Body*, by A. S. Romer); 14.5a: Cambridge University Press (*Journal of the Marine Biological Association of the United Kingdom*); 15.2: Springer-Verlag (*Marine Biology*); 16.1, 16.2, 16.3: Thomas L. Daniel (Ph.D. dissertation).

LIFE IN MOVING FLUIDS

CHAPTER 1

Remarks at the Start

WITH THE easy confidence of long tradition, the biologist measures the effects of temperature on every parameter of life. Lack of sophistication poses no barrier; heat storage and exchange may be ignored or Arrhenius abused; but temperature is, after time, our favorite abscissa. One doesn't have to be a card-carrying thermodynamicist to wield a thermometer.

By contrast, only a few of us measure the rates at which fluids flow, however potent the possible effects of winds and currents on our particular systems. Fluid mechanics is intimidating, with courses and texts designed for practicing engineers and other masters of vector calculus and similar arcane arts. Besides, we've developed no comfortably familiar and appropriate technology with which to produce and measure the flow of fluids under biologically interesting circumstances. So the effects of flow are too commonly either ignored or else relegated to parentheses, speculation, or anecdote.

A life immersed in a fluid—air or water—is, of course, nothing unusual for an organism. Almost as commonly, the organism and fluid move with respect to each other, either through locomotion, as winds or currents across some sedentary creature, or as fluid passes through internal conduits. Clearly, then, fluid motion is something with which many organisms must contend; as clearly, it ought to be a factor to which the design of organisms reflects adaptation.

It is this particular set of phenomena—the adaptations of organisms to moving fluids—that this book is mainly about. Its intended messages are that such adaptations are of considerable interest and that fluid flow need not be viewed with fear or alarm. Flow may indeed be one of the messier aspects of the physical world, but most of the messiness can be explained in words, simple formulas, and graphs. Quantitative rules are available to bring respectable organization to a wide range of phenomena. With a little experience, one's intuition can develop into a reasonably reliable guide to flows and the forces they generate. Even the technology for experimental work on flows proves less formidable than one might anticipate. Indeed, the underlying complexity of fluid mechanics can be something of a boon, since it greatly restricts the possibilities of exact mathematical solutions or trustworthy simulations. Thus the investigator must often resort to the world of direct experimentation and simple physical models, a world in which the biologist can feel quite at home.

Supplying a comprehensive review of what's known about the interrelationships of the movements of fluids and the design of organisms isn't my intention. I will cite no small number of phenomena and investigations; but they're mainly meant to illustrate the diversity of situations to which flow is relevant and the ways in which such situations can be analyzed. Instead, the main objective is an attempt to imbue the reader with some intuitive feeling for the behavior of fluids under biologically interesting circumstances, to supply some of the comfortable familiarity with fluid motion that most of us have for solids. I take the view that with that familiarity the biologist is likely to notice relationships and phenomena and to put hydrodynamic hypotheses to proper experimental scrutiny. And I feel strongly that such investigations should be unhesitatingly pursued by any biologist and should not constitute the peculiar province of some *au courant* priesthood.

One might organize a book such as this with either biology or physics as framework. The physical phenomena, though, flow more easily in an orderly and useful sequence; my attempts to make good order of the biological topics inevitably have a more severe air of artificiality. So the physics of flow will provide the skeleton, fleshed out, in turn, by consideration of the bioportentousness of each item. Where relevance to a particularly large segment of biology wants examination, as when considering velocity gradients or drag, wholly biological chapters will be interjected. While I hope that the arrangement will be effective for both the biologist seeking an easy entry into fluids and the fluid mechanist dazzled by biological diversity, I've opted for the biologist where hard choices had to be made. The reader may be a little vague on the distinctions between work and power or stress and strain but is assumed to be quite sound on vertebrates and invertebrates as well as cucumbers and sea cucumbers. For one thing, I'm very much a biologist myself; for another, the relevant biological details are easy to obtain from textbooks or other references. Since the framework is physical rather than biological, topics such as seed dispersal and suspension feeding will wander in and out; since specific biological topics sometimes involve several applications of fluid mechanics, a little redundancy has been inevitable.

Biological examples will vary from well-established through half-baked to wildly speculative; I'll try to indicate the degree of confidence with which each should be vested. Speculation is the crucial raw material of science, and it seems especially important at this stage of this subject. If some assertion generates either enough antipathy or enthusiasm to provoke a decent investigation, then we'll all be a bit farther ahead. In any case, I claim no proprietary rights to any hypothesis here, whether it's explicit or implied. (But neither does any idea come warranted against post facto silliness.) Incidentally, it's an entirely practical procedure to fix on some physical phenomenon and then go looking to see how organisms have responded to it. While this may sound like shooting at a wall and then drawing targets around the bullet holes, science is, after all, an opportunistic affair rather than a sporting proposition. But whatever the respectability of the approach, it does at least take some advantage of the way things are organized here.

It's probably not the best idea to use this book solely as a reference, with the index for intromission. The reader coming from biology ought to move in sequence through at least the basic material on viscosity, the principle of continuity, Bernoulli's equation, the Reynolds number, and the characteristics of velocity gradients. The book certainly can be used out of order or with parts omitted, but certain pitfalls lurk. Biologists have been all too willing to use equations with no more than a guileless glance to check that the right variables were represented. As a result, much mischief has been perpetrated. Bernoulli's equation doesn't work well in boundary layers and has little direct applicability to circulatory systems. The Hagen-Poiseuille equation presumes that flow is laminar and that one is far from the entrance to a pipe. Stokes' law applies (in general) to small spheres, not large ones, and it usually needs a correction for spheres of gas moving in a liquid medium. Equations certainly abound in these pages, but they can be found as easily elsewhere. More important are the discussions of how and where they apply—when to rush in and when to fear to tread. Especially in this latter matter, the specific sequential treatment of topics is of significance.

INCLUSIONS, EXCLUSIONS, GENERAL SOURCES

At least for physical phenomena, some idea of what's covered in this book is apparent from the table of contents. Naturally, the omissions are less evident, but it's important that they be mentioned. Physical fluid mechanics ramifies in many directions; much of it carries the unmistakable odor of our technological concerns and has little relevance to biology. Thus we can safely ignore such things as high speed (compressible) flows and the flows of rarefied gases. For reasons that flatter neither author nor most readers, the mathematical niceties of fluid mechanics will not loom large here. More attention will be given to phenomena that I judged simple or widespread than to ones that I regarded as more complex or specialized. The focus here is on flow in and around organisms, decent-sized chunks of organisms, and small assemblages of organisms, and this focus has necessitated some omissions. Only a little will be said about the statics of fluids, biometeorology, and the flow of substances that are incompletely fluid, such as the contents of cells and life's various slurries and slimes. Thus I'll largely eschew the treacherous quagmires of mitosis and cyclosis, of surface waves and ocean currents, of the hydraulics of streams and rivers, of atmospheric winds and wind-driven circulation in lakes, and of flow through porous media. In addition, I've made some essentially arbitrary exclusions and will have little to say about the sensory side of responses to flows and the effects of flow on chemical communication in either air or water. Flows either driven by or involving temperature differences—convective heat transfer—are a major omission rationalized only by a lack of space and steam. Most of these topics are treated well elsewhere, and sources of enlightenment will be mentioned at appropriate points.

More extensive and detailed information on fluid dynamics may be obtained from conventional engineering textbooks, such as those by Streeter and Wylie (1985) or Massey (1989). The basic processes that will be of concern here were about as well understood fifty or sixty years ago as they are today, so the age of the source is usually immaterial. Indeed, the earlier generations of fluid dynamicists may have worried more than their successors about low-speed phenomena and other items that turn out to have biological relevance; and inexpensive reprints of several quite useful old texts are currently available. Prandtl and Tietjens' (1934) *Applied Hydro- and Aeromechanics* is particularly good on boundary layers and drag. *Fluid Mechanics for Hydraulic Engineers* (1938) contains a fine treatment of dimensional analysis and the origin of the common dimensionless indices, while *Elementary Mechanics of Fluids* (1946) covers an unusually diverse range of topics; both are by Hunter Rouse. Goldstein's (1938) two volumes, *Modern Developments in Fluid Dynamics*, have nice verbal descriptions of phenomena in between the equations. Mises (1945) gives excellent explanations of how wings and propellers work in *Theory of Flight*. With several of these books at hand one can usually find an understandable and intuitively satisfying explanation of a puzzling aspect of flow. Finally, both aesthetically and technically pleasing, there's *An Album of Fluid Motion* (1982), by Van Dyke.

Trustworthy popular accounts are surprisingly scarce, with no treatment of fluid mechanics coming anywhere near the breadth and elegance that Gordon's (1978) *Structures* brings to solid mechanics. My favorite three are Von Kármán's (1954) *Aerodynamics*, Sutton's (1955) *The Science of Flight*, and Shapiro's (1961) *Shape and Flow*. The first is particularly witty, the second is especially clear and honest, and the third, *mirabile dictu*, focuses on fluid phenomena that biologists should encounter.

Several other general sources deserve mention. *Fluid Behavior in Biological Systems*, by Leyton (1975), is nearest to this book in intent. It gives less attention to drag, boundary layers, and propulsion but more to flow in porous media, heat transfer, non-Newtonian fluids, thermodynamics, and micrometeorology. Ward-Smith's *Biophysical Aerodynamics* (1984) focuses on seed dispersal and animal flight. *Waves and Beaches* (1980) by Bascom, *Biology and the Mechanics of the Wave-Swept Environment* by Denny (1988), and *Air and Water*, also by Denny (1993), are invaluable for anyone working along the edge of the ocean; in effect the Denny picks up in both elegance and focus where the Bascom leaves off. And the privately published compendium of Hoerner (1965), *Fluid-Dynamic Drag*, is filled with information about the behavior of simple objects to which analogous data for organisms can be compared.

TECHNOLOGY

The how-to-do-it aspects of biological fluid mechanics present special snares. High technology is certainly no stranger to the microscopist, molecular biologist, or neurophysiologist. But few of us are facile at making and measuring moving fluids, and there's rarely a lab-down-the-hall where folks are already, so to speak, well immersed in flow. Faced with some upcoming investigation, one's first impulse is to seek out a friendly engineer, who then prescribes a hot-wire or laser anemometer; the problem is thereby reduced to a search for kilobucks. There is, though, a second and less ordinary problem. The technology used by the engineers is a product of, by, and for engineers. The range of magnitude of the flows we usually need to produce and of the forces we typically want to measure is rather different, and engineering technology is often as inappropriate as it is expensive.

In over thirty years of facing problems of flow, I've found that the devices I've had to use were, compared to those of my colleagues in more established fields of biology, rather cheap. On the other hand, they have rarely been available prepackaged and precalibrated, with a factory to phone when all else fails. Flow tanks, wind tunnels, flow meters, anemometers, and force meters have simply been built in my laboratory as needed. I've developed a fair contempt for fancy commercial gear except for items of very general applicability—digital voltmeters, power supplies, potentiometric chart recorders, variable speed motors, gears and pulleys, electronic stroboscopes, analog-to-digital converters, video cameras and recorders, and so forth. The consistently most valuable tools have been lathe, milling machine, and drill press—but I'm a quite unreconstructed primitivist with the perverse passion of the impecunious and impatient. The first edition of this book included appendices on making and measuring flows. I've not retained them, since the earlier edition is still available as a reference and since I'm now convinced that it would take a whole book to do the job properly.

DIMENSIONS AND UNITS

Fluid mechanics makes use of a wide array of different variables, some (density, viscosity, lift, drag) familiar and others (circulation, friction factor, pressure coefficient) out of the biologist's normal menagerie of terms. A list of symbols and quantities used in this book precedes the index; it might be worth putting a protruding label on that page. Matters will be somewhat simplified if the reader pays a little attention to the dimensions that attach to each quantity. By dimensions I don't mean units. Thus velocity always has dimensions of length per unit time, whether data are given in units of meters per second, miles per hour, or furlongs per fortnight. Dimensions take on rather special significance in fluid mechanics not just as a result of the general messiness of the subject but because of a condition that may sound trivial but proves surprisingly potent. *For an equation to have any applicability to the real world, not only must the two sides be numerically equal, but they must be dimensionally equal as well.* The general statement asserts that proper equations must be dimensionally homogeneous—each term must have the same dimensions. When theory, memory, or intuition fails, this injunction can go a long way toward indicating the form of an appropriate equation.

An example should clarify the matter. Assume you want an equation relating the tension in the wall of a sphere or a cylinder to the pressure inside—perhaps you know the surface tension of water and want to know what this does to the internal pressure of a gas bubble (in connection with Chapter 15 or 17). Tension has dimensions of force per unit length (as you can tell, if need be, from the units in which surface tension is quoted). Pressure has dimensions of force per unit area (as in pounds of force per square inch of area). To relate pressure to tension, clearly pressure must be multiplied by some linear dimension of the sphere or cylinder. Thus the equation will be of the form

tension = constant × radius × pressure,

where we know nothing about the constant except that it's dimensionless. Evidently a given tension generates a larger pressure when the radius is small than when the radius is large. Even without further information about the constant we're no longer surprised that (surface tension being constant) tiny bubbles have high internal pressures. And it ought to take a higher pressure to generate a given tension when the radius is small, so we're much less mystified about why it's relatively hard to *start* blowing up a balloon despite the obvious flaccidity of the rubber. We're no longer surprised that a small plant cell can withstand pressure differences of many atmospheres across its thin walls, nor are we puzzled at how arterioles can get by with much thinner walls than that of the aorta when both are subjected to similar internal pressures.

The easiest way to compare the dimensions of different variables is to reduce them to combinations of a few so-called fundamental dimensions. We will require only three such dimensions here: *length, mass,* and *time* (*temperature* is a frequent fourth). This reduction is accomplished by use of definitions or previously memorized functions. For instance, force is mass times acceleration, and therefore force has fundamental dimensions of MLT^{-2}. Pressure or stress is force per area, so both have dimensions of $ML^{-1}T^{-2}$. Incidentally, this latter example emphasizes the fact that just because two quantities have the same dimensions doesn't mean that they are synonymous or equivalent. With such simple manipulations, the fundamental dimensions of each term in almost any equation can be obtained. Considerable use will be made of this sort of dimensional reasoning in forthcoming chapters. More extensive and formal introductions to the subject of dimensional analysis can be found in books by Bridgman (1931) and Langhaar (1951); biological contexts are supplied by McMahon and Bonner (1983) and Pennycuick (1992).

Not only constants but also variables may be dimensionless and still be rich in relevance to the real world. Any number that is the ratio of two quantities measured in the same dimensions will be dimensionless. A fairly simple example is *strain*, as used in solid mechanics. Strain is the ratio of the extension in length of a stressed object to its original, unstressed length. Unstressed length is simple and commonly fixed; things get even more interesting when several of the quantities in a dimensionless index decide to vary. Such indices turn out to be scales that achieve quite useful simplifications of otherwise complex situations and that can lead to remarkable insights into what really matters beneath a confusion of varying quantities. Thus surface-to-volume ratio depends on size as well as shape; it has a dimension of inverse length (L^{-1}). By contrast, surface cubed over volume squared is dimensionless and quite indifferent to the size of an object per se; so, for instance, a cube has the same value whatever its size. It's therefore much more useful as an index of shape. Dimensionless numbers, usually named after their first promulgators, find wide use in fluid mechanics; while initially they seem odd, one rapidly achieves proper contemptuous familiarity even with graphs of one dimensionless number plotted against another. I hardly need mention that dimensionless numbers are automatically unitless and therefore quite indifferent to which system of units is in use.

Units, though, can't be completely ignored. All variables in a dimensionally proper equation ought to be given in a consistent set of units. An earlier generation of biologists, when they meant metric units, usually used something approximating the physicist's centimeter-gram-second (CGS) system, along with a few oddities such as heat as calories and pressure as millimeters of mercury in a column. We're now enjoined to adopt a specific version of the metric system common to all of science, the SI or "Système Internationale," which will be used here. Fundamental units (for the fundamental dimensions) are kilograms (mass), meters (length), and seconds

9

(time). SI allows the common prefixes going up or down by factors of 1000 (mega-, milli-, micro-, and so forth) to be attached to any fundamental or derived unit. Only a single prefix, though, may be used with a single unit, and the prefix must attach to the numerator. Thus meganewtons per square meter is legitimate but newtons per square millimeter is not. I'll only stoop to such frowned-upon units as centimeters, liters, and kilometers as a kind of vernacular where no calculations are contemplated. SI units often get mildly ludicrous in the context of organisms. Thus the drag of a cruising fruit fly (Vogel 1966) is about three micronewtons. And Wainwright et al. (1976) give the strength of spider silk as about 10^9 newtons per square meter of cross section; it would take one hundred billion (U.S.) strands to get that combined area. But consistency is really a sufficiently compensatory blessing.

Table 1.1 gives a list of quantities with their fundamental dimensions and SI units. For further introduction to SI units and conventions, see *Quantities, Units, and Symbols* by the Symbols Committee of the Royal Society (1975) or *The International System of Units* by Goldman and Bell (1986). For the inevitable nuisance of dealing with different systems of units, Pennycuick's (1988) booklet, *Conversion Factors*, is an absolute godsend. It's cheap enough so one might scatter a few copies around home, office, and laboratory.

TABLE 1.1 COMMON QUANTITIES FUNDAMENTAL DIMENSIONS, AND SI UNITS.

Quantities	Dimensions	SI units
Length, distance (1)	L	meter (m)
Area, surface (S)	L^2	square meter (m^2)
Volume (V)	L^3	cubic meter (m^3)
Time (t)	T	second (s)
Velocity, speed (U)	LT^{-1}	meter per second (m s^{-1})
Acceleration (a)	LT^{-2}	meter per second squared (m s^{-2})
Mass (m)	M	kilogram per cubic meter (kg m^{-3})
Force, weight (F, W)	MLT^{-2}	newton (N or kg m s^{-2})
Density (ρ)	ML^{-3}	kilogram per cubic meter (kg m^{-3})
Work, energy (W)	ML^2T^{-2}	joule (J or N m)
Power (P)	ML^2T^{-3}	watt (W or J s^{-1})
Pressure, stress (p, τ)	$ML^{-1}T^{-2}$	pascal (Pa or N m^{-2})
Dynamic viscosity (μ)	$ML^{-1}T^{-1}$	pascal second (Pa s)
Kinematic viscosity (ν)	L^2T^{-1}	square meter per second (m^2 s^{-1})
Tension, surface t. (γ)	MT^{-2}	newton per meter (N m^{-1})

NOTES: For the dimensions, M = mass; L = length; and T = time. Notice that some symbols are shared between quantities and units; since units are never indicated in formulas and text can be made explicit, no ambiguity need result.

If you're decently scrupulous about consistency of dimensions and use of units then you never have to specify units when giving equations, a considerable convenience. The main place where the system falters at all is when equations with variable exponents are fitted to empirical data, as in any statement such as "metabolic rate is proportional to body mass to the 0.75 power." The exponent of proportionality comes out the same in any set of units, but the constant with which it forms an equation does not. Moreover, the rule about dimensional homogeneity is violated unless one tacitly heaps all the unpleasantness on the constant of proportionality. If the expression is written as an equation rather than a proportionality, the usual practice is merely to specify the set of units that must be used. Here the practice among fluid mechanists and biologists is about the same, both being practical people who have scattered data, imperfect theories, and the like.

MEASUREMENTS AND ACCURACY

First, a few words about what's meant by accuracy. As Eisenhart (1968) has pointed out, lack of accuracy reflects two distinct disabilities in data. First, there's imprecision, or lack of repeatability of determinations. And second, there's systematic error, or bias, the gap between the measured value and some actual, "true" value—the tendency to measure something other than what was really intended. In the kinds of problems we'll discuss, unavoidable imprecision is usually pretty horrid, at least by the standards of physical rather than biological sciences. Thus it's rarely worth great attention to fine standards for calibration where these just drive systematic errors well below what proves to be the more intractable problem of imprecision.

Quantities such as density can be measured very precisely. But the inevitable irregularities in flows, the nature of vortices and turbulence, and quite a few other phenomena severely limit the precision with which the behavior and effects of moving fluids can be usefully measured. The drag of an object measured in one wind tunnel will often differ considerably from that measured in another, while a third datum will result from towing the object through otherwise still air. If a figure of, say, 1 m s^{-1} is cited as the transition point from laminar to turbulent flow in some pipe, that figure should not be interpreted as 1 ± 0.01 or often even 1 ± 0.1 m s^{-1}. With extreme care it may be possible to postpone the transition to 10 m s^{-1} or more. Some empirical formulas given in standard works, especially those for convective flows, use constants with three significant figures. My own experience suggests that such numbers should be viewed with enormous skepticism for anything except, perhaps, the very specific experimental conditions under which they were determined. And citing calculations for Reynolds numbers (Chapter 5), for instance, without minimally

11

rounding off to the nearest part in a hundred is at the least a bit self-deceptive.

The development of electronic calculators has had a pernicious effect on our notions of precision. The art estimating just how much precision is truly necessary to resolve the question at hand has suffered from the demise of the slide rule. As a practical matter, the flow of fluids, even without organisms, is a subject that enjoys barely a slide rule level of precision—rarely better than one percent and often much worse.

RELATIVE MOTION

Another matter ought to be set straight at the start, an item that arises with some frequency among biologists taking their first look at fluid flow. Frames of reference can be chosen for convenience, and the surface of the earth upon which we walk has no absolute claim as a "correct" reference frame. One might imagine that a seed carried beneath a parachute of fluffy fibers will trail behind the fibers as the unit is blown across the landscape by a steady wind. In fact the image is quite wrong for anything beyond the initial events of detachment from the plant—when the surface of the earth is still an active participant—as can be seen in Figure 1.1. Farther along (if the wind is steady) neither any longer "knows" anything about what the ground's doing so the seed hangs below the fluff—the surface of the earth now constitutes a reference frame that's both misleading and unnecessarily complex. People who've traveled in balloons commonly comment on the silence and windlessness they experience and its incongruence with clear visual evidence that the ground is moving beneath.

A more drastic if less commonly observed case is that of a "ballooning" spider. (For a general account of the phenomenon, see Bishop 1990.) A young spider spins a long silk strand that extends downwind from it. Eventually it lets go and drifts along. One might expect that the spider is thereafter pulled behind the silken line until the whole system settles to the earth. What will happen in the absence of gusts, vortices, and gradients is that the spider will fall downward but at a speed that's much reduced by the drag of the line. The line will gradually shift from running horizontally to running vertically, forming a relatively high drag, low weight element that extends *downstream* (here, of course, *upward*) from the falling spider. The line lags behind, slowing descent as a result of its high drag relative to its weight. (Humphrey 1987 estimates that the line has over 75% of the drag of the system while contributing less than a tenth of a percent of the weight.)

Failing to keep in mind a proper reference frame can generate odd misconceptions as well as obscure some real problems. From time to time statements appear in the avian literature about a problem facing (or, per-

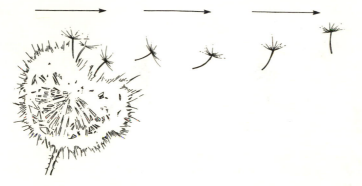

FIGURE 1.1. A dandelion seed carried by a steady wind orients vertically, as if falling through still air. Only during detachment does the fibrous end tilt downwind.

haps we should say chasing) a bird flying downwind—it must somehow keep from getting its tail feathers ruffled. But consider: once launched, a bird simply does not (and, indeed, cannot) fly downwind with respect to the local flow around it. If the wind with respect to the ground goes in the same direction as the bird, then the bird just flies that much faster with respect to the ground. The problem is a really nasty one for slow fliers such as migrating monarch butterflies that can't make progress (once again with respect to the ground) against even modest winds (Gibo and Pallett 1979).

Worse, you might really get stung by confusion about relative motion and frames of reference. Imagine being chased by a swarm of mayhem-minded killer bees, as in Figure 1.2. A decent breeze is blowing, so to get a little more speed you run downwind. Bad move—you're quickly bee-set. Honeybees can fly (despite a lot of lore to the contrary) only about 7.5 m s⁻¹ (Nachtigall and Hanauer-Thieser 1992), but that's equivalent to about a four-minute mile. With a 4.5 m s⁻¹ breeze from behind, they'll go all of 12 m s⁻¹ *with respect to the ground.* You may gain a little from the tail wind, but the bees will automatically get full value. What if you run upwind? You may be slowed down slightly, but the bees will be dramatically retarded—down to 3 m s⁻¹ *with respect to the ground.* Only about a nine-minute mile is needed for you to stay ahead.

The ability to shift reference frames for our convenience can effect more than conceptual simplifications. No special justification is needed for using an experimental system in which the organism is stationary and the fluid environment moves as a substitute for a reality in which the organism swims, flies, or falls through a fluid stationary with respect to the surface of the earth. That's the main reason for the popularity of such devices as wind tunnels and flow tanks for working on flying and swimming. I've used a

FIGURE 1.2. Going downwind, bees are faster than any running human; a head wind slows the human only a little, but the bees are badly hindered—unless they've looked ahead to Chapter 8.

vertical wind tunnel to provide upward breezes just equal to the falling rates of seeds. Shrewd choice of reference frame is an old tradition—assuming (not proving!) a sun-centered system permitted Copernicus to simplify enormously the complex cosmology of Ptolemy.

BALANCING FORCES

When considering solid bodies moving through fluids, Newton's first law has to be put into a somewhat more distant perspective than might have been the practice in one's first physics course. Bodies may continue in steady, rectilinear motion unless imposed upon by external forces; but imposition of external forces is just what fluids do to solid bodies that have the temerity to force passage. Chief among the external forces is, of course, drag.[1] In practice, then, a body traveling steadily with respect to a fluid (ah—frames of reference) must be exerting a force on the fluid exactly sufficient to balance the fluid's force on the body. That force might be provided by the action of wings or tail acting as thrust generators. Or it might be supplied by the action of gravity on the body, as when a body sinks steadily, when the downward force of gravity balances the combined upward forces of buoyancy and drag. Or it might be provided by some other solid structure that transmits forces—a mounting strut supporting an object in a flow tank or wind tunnel or the branches and trunk supporting the leaves of a tree.

In combination, force balance and reference frame can make some subtle but substantial trouble, and the utility of quite a lot of literature is compromised by insufficient attention to them. Say you persuade an insect

[1] Perhaps someone (me, for instance) ought to say a word against the common usage of that needlessly redundant and pretentious term, "drag force." Drag is a force, is always a force, and is nothing but a force.

to beat its wings as hard as it can while attached to a fine wire, and you even manage to attach the fine wire to a device that will measure the force the insect produces. What will this tell you about how fast the insect can fly or about how much force it can generate at top speed? Almost nothing! Even if you (separately) measure how fast the insect can fly, you are on very unsafe ground if you multiply that speed and the force you've measured to get the insect's power output. Similarly, if you want to know the drag of the insect at its maximum speed, you can't just put the thing in a wind tunnel set to that speed and measure the force it exerts on a mount. If the insect is doing exactly what it ought to, then the mount will feel no force at all because the drag of the insect's body will exactly balance the thrust produced by the wings. If it's not beating its wings then you get a force where normally no net force would be exerted. Fortunately, there are ways around such difficulties; what's important at this point is that they be recognized.

I DO REGRET a little the admonitory tone of much of this initial chapter. A lot of it reflects some scar tissue induced by abrasions in the form of written material that I've been expected to evaluate. Even though I quite obviously mean what I've said here, its general character contrasts a bit with my first rule of fluid mechanics—you have to be breezy if you expect to make waves.

CHAPTER 2

What is a Fluid and How Much So?

PERHAPS YOU were long ago told of three common states of matter—gas, liquid, and solid. Perhaps you were also told of a handy rule for distinguishing them—the rule that solids have size and shape; that liquids have size but no shape; and that gases have neither size nor shape. Perhaps you've even heard that while intermediates between solids and liquids are common, intermediates between liquids and gases are certainly not. Where, in all this, do fluids enter? In common parlance, "fluid" is just a synonym for "liquid", a definition the reader must immediately expunge from memory. For present purposes, indeed for all of the science of fluid mechanics, a fluid is either a gas or a liquid but positively not a solid.

Definitions are, of course, in some measure arbitrary, but a bald one such as the present assertion about fluids raises some distinctly nonarbitrary questions. Why do we find it convenient to lump the gaseous and liquid states even when we know of no intermediates? Why do we persist in making a general distinction between liquids and solids when, in practice, we're made up of substances of intermediate character? And, of most immediate concern, how can we reliably distinguish fluids from solids?

A simple experiment can be used (conceptually, anyway) to tell a fluid from a solid, an experiment that makes no qualitative distinction between gases and liquids. Consider an apparatus consisting of two concentric cylinders of ordinary size, such as that shown in Figure 2.1a. Although the outer one is fixed, the inner cylinder may be rotated on its long axis. Between the cylinders is a gap that we can fill with any material of any state. We further assume that the material adheres to or "wets" the walls of the cylinders and that rotation of the inner cylinder is resisted only by forces caused by the presence of the outer wall and transmitted through the material between them.

One naturally expects that the inner cylinder can be turned more or less easily, depending on the material in the gap. Indeed, that's just what happens; not surprisingly, the force required varies over an enormous range. But a less obvious distinction emerges. For some materials, the force necessary to turn the inner cylinder depends on *how far* we've already turned it from "rest." For others, the force is quite independent of how far we've distorted the material and depends only on *how fast* we turn the cylinder. In each case we've imposed a "shearing" load on the material; that is, we've attempted to deform it by sliding one surface relative to another concentric

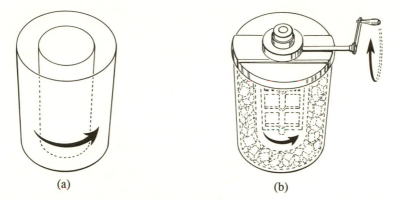

FIGURE 2.1. (a) A pair of concentric cylinders. If the space between them is filled with a fluid, then rotation of the inner one will exert a force tending to rotate the outer one. (b) The practical version, an ice cream freezer. The operator monitors the viscosity increase that denotes progress by feeling the increasing difficulty of turning the dasher relative to the container.

(essentially parallel) one. As it turns out, we've just distinguished solids from fluids.

Solids resist shear deformation—they care how *far* they're deformed. Generally the further one wishes to deform them, the more force is required. To put the matter quantitatively, consider a rectangular block of material as in Figure 2.2 in which the top surface is pushed sideways but the bottom surface is fixed. The pushing force tends to deform the rectangular solid into one in which two opposite sides are parallelograms. If θ (theta) is the angle by which the material is deformed and S the surface area over which the force, F, is applied, then

$$\frac{F}{S} = G\theta. \qquad (2.1)$$

G is called the *shear modulus* (similar to the elastic or Young's modulus) and corresponds to our common notion of shearing distortions such as happen when we twist things. The equation implies that shear stress (F/S) is proportional to shear strain (θ); in short, it's a statement of Hooke's law. If G does indeed stay constant as stress and strain vary, then the material is spoken of as "Hookean."

By no stretch do proper fluids have a shear modulus, high or low— they're infinitely distortable, magnificently oblivious to their shapes, past or present. But deformation still matters, although in quite a different way. What fluids care about is how *rapidly* they are deformed. Thus a much different expression is needed to describe what happens to a fluid when

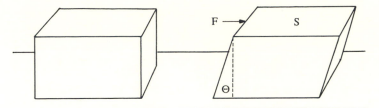

FIGURE 2.2. A rectangular block of material is deformed by a shear force, *F*, or a shear stress, *F/S*, so two faces become parallelograms of tilt θ.

(assuming that by some preposterous prestidigitation nothing leaks out) the same shear stress is applied to the system in Figure 2.2:

$$\frac{F}{S} = \frac{\mu\theta}{t}. \qquad (2.2)$$

θ/*t* is the *rate of shear* and μ (mu) is a property called the *dynamic viscosity*. The latter, to be tediously explicit, indicates how much a fluid resists not shear but rate of shear. If μ remains constant as shear stress (what you do to the fluid) and shear rate (how the fluid responds) vary, the fluid is said to be "Newtonian"—the analog of Hookean behavior in solids. For different fluids, though, the actual value of the dynamic viscosity varies enormously, from low values for gases to exceedingly high values for substances such as tar or glass that, over long periods of time, behave as fluids.[1] Values of viscosity can be usefully determined or assigned to systems ranging from the insides of cells to glaciers or even the mantle of the earth.

This distinction between solids and fluids is precisely the same as that between springs and shock absorbers. That's a very real difference for anyone who installs one or the other on a car—a shock absorber can easily be set to whatever length is needed to slip onto its mounts, but a spring has a forceful preference for a particular length.

THE NO-SLIP CONDITION

The properly skeptical reader may have detected a peculiar assumption in our demonstration of viscosity: the fluid had to stick to the walls of the

[1] Reiner (1964) invented a dimensionless index that provides a guide to whether a system behaves like a solid or a fluid; with biblical allusion, he called it the Deborah number" (*De*). It's defined as the ratio of the time a process takes to the time required for significant plastic deformation of the system. If *De* << 1, you assume you're dealing with a solid; if *De* >> 1, then for practical purposes your system is fluid. *De* rarely appears in the biological literature (but see, for instance, Jenkinson and Wyatt 1992), but I think it deserves to be better known.

inner and outer cylinders in order to shear rather than simply slide along the walls. Now fluid certainly does stick to itself. If one tiny portion of a fluid moves, it tends to carry other bits of fluid with it—the magnitude of that tendency is precisely what viscosity is about. Less obviously, fluids stick to solids quite as well as they stick to themselves. As nearly as we can tell from the very best measurements, the velocity of a fluid *at the interface* with a solid is always just the same as that of the solid. This last statement expresses something called the "no-slip condition"—fluids do not slip with respect to adjacent solids. It is the first of quite a few counterintuitive concepts we'll encounter in this world of fluid mechanics; indeed, the dubious may be comforted to know that the reality and universality of the no-slip condition was heatedly debated through most of the nineteenth century. Goldstein (1938) devotes a special section at the end of his book to the controversy. The only significant exception to the condition seems to occur in very rarefied gases, where molecules encounter one another too rarely for viscosity to mean much.

Yet another peculiarity of this no-slip condition is that the nature of the solid surface makes very little difference. If water is flowing over a solid without an air-water interface to complicate matters, the no-slip condition holds whether the solid is hydrophilic or hydrophobic, rough or smooth, greasy or clean. The nature of the solid surface matters only when we have a liquid-gas interface present as well—in short, where surface tension becomes a factor.

The no-slip condition has a number of important ramifications. In particular, it means that any time a fluid flows across a solid, a velocity gradient is present. That is, velocity varies with distance above the surface; or, put more formally, dU/dz isn't zero. That's what "boundary layers" (Chapter 8) are all about. Again, these velocity gradients are developed entirely within the fluid, not (as with two sliding solids) between one material and the other. If the fluid is an ordinary homogeneous one, its local velocity must smoothly approach that of the surface as the surface is approached; there can be no discontinuity within the fluid. While it may seem a little strange at first, even though the velocity may be zero at the surface, dU/dz, the velocity gradient, isn't. Furthermore the fluid velocity cannot asymptotically approach the speed of the solid, for that would require a variable viscosity—it would have to get lower in fluid closer to the solid surface to account for the increasing steepness of the velocity gradient. In practice, the no-slip condition explains (in part) why dust and grime accumulate on fan blades, why pipes (including blood vessels) encounter trouble from accumulation of deposits rather than from wearing thin, and why a bit of suspended rock is needed in water for the latter to become effectively erosive.

The no-slip condition is as easy to demonstrate as its universality is hard to prove. Simply fill a circular basin with water, stir the water into a smoothly circuitous motion, and inject a small bit of dye on the bottom or side wall. The last layer of dye will remain there despite quite a lot more stirring well above. Eventually, of course, diffusion and dilution will get rid of any adhering colorant, but the time needed is notable. Alternatively, just consider why dishcloths and mops are so much more effective for cleaning surfaces than any mere rinse.

Most often the region near a solid surface in which a velocity gradient is appreciable is a fairly thin one, measured in micrometers or, at most, millimeters. Still, its existence requires the convention that when we speak of velocity we mean velocity far enough from a surface so the combined effect of the no-slip condition and viscosity, this velocity gradient, doesn't confuse matters. Where ambiguity is possible, we'll use the term "free stream velocity" to be properly explicit.

ASSUMPTIONS AND CONVENTIONS

Before proceeding further, I'd like to establish a series of assumptions that will underlie the forthcoming chapters unless one or another of them is explicitly relaxed. They are at least mildly preposterous; but, like panmictic populations and point masses, they are convenient fictions.

1. *Fluids are "Newtonian."* As mentioned, many substances combine properties of both fluids and solids. But we will disallow any trace of solidity as previously defined—the fluids here will have no memory of previous shape nor any elasticity. In effect, we draw a jurisdictional border, putting molasses or treacle and most syrups on the inside and the various jams, yoghurts, and gelatins on the outside. As Shakespeare says in *King Lear*, "Out, vile jelly."

Many biological materials are in the complex, multidimensional continuum of non-Newtonian fluids and viscoelastic solids—whole blood, synovial fluid, mucuses of various consistencies—but we'll largely ignore them and, in doing so, avoid perhaps the messiest and least understood branch of fluid mechanics, rheology. As noted already, a Newtonian fluid is one which shows a direct proportionality, a linear relationship, between the applied shear stress and the resulting rate of deformation. That being the case, a value of viscosity may be found that's independent of the specific test conditions. Air and water, the fluids of main concern here, are virtually perfect Newtonian fluids.

2. *Fluids are continua.* We'll assume that fluids are nonparticulate and infinitely distortable and divisible. We're going to turn the picture of Democritus to the wall and deny molecules. But consider: Is there anything in your everyday world that requires you to make the distinctly odd contrary assumption, that matter is particulate or molecular? Why should cheese, sliced thinly enough, ultimately stop being cheese? Molecules are appar-

ently necessary to explain the physical basis of viscosity; but once viscosity is accepted we have little further need for them. A "fluid particle" will be just a linguistic convenience for specifying an arbitrarily small element of a moving fluid that enjoys our particular scrutiny.

3. *Fluids are incompressible.* This contradicts the experience of anyone who has wielded a bicycle pump. If air were really incompressible, then when you put a thumb over the orifice you wouldn't be able to push the handle down at all. Gases, surely, compress easily, even if liquids are fairly recalcitrant. Put aside these subversive thoughts! While compressing gases with static devices such as piston pumps is easy, getting compression through flow is no small matter. Given the choice, fluids behave as if they would rather flow than squeeze—at least up to speeds within a decent fraction of the speed of sound, about 340 m s^{-1} in air and 1500 m s^{-1} in water.

If you direct a stream of air at a wall, at some point at least the air comes to a halt. Anticipating the discussion of Bernoulli's principle (Chapter 4), we can estimate how much compression is maximally involved in that deceleration. At an ambient pressure of one atmosphere, a flow of 10 m s^{-1} gives a compression of 0.06%; a flow of 20 m s^{-1} gives 0.24%; and a flow of 30 m s^{-1} (67 mph or 108 km hr^{-1}) gives only 0.53%. The latter is about as fast a flow of air as is experienced (or, say, survived) by any ordinary biological system—a report that a deer fly achieved 350 m s^{-1} was carefully demolished by Langmuir (1938), who showed that the fly would either be crushed by its drag or would consume its own weight in fuel each second. So the assumption of incompressibility is quite a safe and conservative one; it affords a huge simplification relative to the world with which airplane designers must contend. Density becomes a constant for most present purposes, with the same value for a fluid in any sort of motion as for the fluid at rest, and with its value varying only with hydro- and aerostatic changes as in deep dives and ascents to high altitudes.

4. *Flows are always "steady."* As we will use the term, so-called steady flow doesn't deny that fluid speeds up or slows down as it flows along. Instead it means that at a given point in space (with respect to some specified frame of reference) the speed of flow doesn't vary with time in any regular manner. The frame of reference may be moving with respect to the earth—one can speak of the steady flow of air above a point on the wing of a gliding bird. And the definition is limited to "regular" variation—the ubiquitous random fluctuations of turbulent flow are for this purpose ignored. Chapter 16 will be given to unsteady, time-dependent flows; otherwise the assumption will be operative.

5. *Fluids make no interfaces with other fluids.* Gases, of course, will not discretely interface with each other, but liquid-liquid and liquid-gas interfaces (especially the latter) are household events. We'll get around to these interfaces late in the book, giving them a whole chapter (17) of their own.

Elsewhere we'll arbitrarily limit our view to situations where they don't occur and thus effect another great simplification at little explanatory cost. Not only won't we have to worry about the chemical characteristics of surfaces (mentioned already), but we can defer phenomena such as surface waves and capillarity. So—bodies of fluid will be considered unbounded except by solids.

One might note that with assumptions (3) and (5) in place the differences between liquids and gases become quantitative rather than qualitative, and no easy and absolute test tells whether a fluid is a liquid or a gas. While that may seem odd or unreal, it proves a useful way to view the world.

In addition to these four assumptions, we'll assume that bodies of fluid are at uniform temperature and otherwise homogeneous, and (usually) that the solids bounding or immersed in them are perfectly rigid.

PROPERTIES OF FLUIDS

Table 2.1 gives representative values of the main properties that will matter for the phenomena we'll consider.

Density

For fluids, using mass is a bit of a mess since they may be unbounded and often move continuously through our frame of reference. In practice, mass is replaced by density, or mass per unit volume. The symbol used is ρ (rho), the fundamental dimensions are ML^{-3}, and the SI unit is the kg m^{-3}. The CGS unit,[2] in this unusual case 10^3 times larger, is the g cm^{-3}. Some contrary sources (even outside the southern hemisphere) use specific volume instead of density; one is the reciprocal of the other.

For both air and water at ordinary temperatures, density drops as temperature increases. Only in air, though, is the effect really substantial; air density is inversely proportional to absolute temperature. But the biological consequences of water's minor density differences are profound and are discussed in almost every book on limnology. In fresh water, density is maximal at about 4° C; it drops slightly when the freezing point is approached, and it drops further with the formation of ice (so ice floats and only a pond's surface freezes). In seawater, −3.5° C is the temperature of maximum density, a temperature at which ice crystals have ordinarily begun to form.

---

[2] While the CGS or "centimeter-gram-second" system is passé, we still have to contend with a considerable body of literature that uses it.

TABLE 2.1 SI VALUES OF SOME PHYSICAL VARIABLES AT ATMOSPHERIC PRESSURE.

		Dynamic Viscosity (Pa s)	Density (kg m^{-3})	Kinematic Viscosity (m^{2}s^{-1})
Air	0°	17.09×10^{-6}	1.293	13.22×10^{-6}
	20°	18.08×10^{-6}	1.205	15.00×10^{-6}
	40°	19.04×10^{-6}	1.128	16.88×10^{-6}
Fresh water	0°	1.787×10^{-3}	1.000×10^3	1.787×10^{-6}
	20°	1.002×10^{-3}	0.998×10^3	1.004×10^{-6}
	40°	0.653×10^{-3}	0.992×10^3	0.658×10^{-6}
Seawater	0°	1.890×10^{-3}	1.028×10^3	1.838×10^{-6}
	20°	1.072×10^{-3}	1.024×10^3	1.047×10^{-6}
	30°	0.870×10^{-3}	1.022×10^3	0.851×10^{-6}
Acetone	20°	0.326×10^{-3}	0.792×10^3	0.412×10^{-6}
Glycerin	20°	1.490	1.261×10^3	1.182×10^{-3}
" 90% aq.	20°	0.219	1.235×10^3	0.177×10^{-3}
Mercury	20°	1.554×10^{-3}	13.546×10^3	0.115×10^{-6}

NOTES: Values for seawater presume a salinity of 35‰ (parts per thousand). Note that for seawater the highest temperature is 30° rather than the 40° for air and fresh water.

Dynamic Viscosity

We met this property when defining a fluid and will now flesh out its definition. One way of viewing it is to envision a large stack of very thin sheets of paper. If shearing is the sliding of each sheet with respect to the one beneath, then dynamic viscosity is the interlamellar stickiness or friction between the sheets. To examine the relevant variables, a slightly fancier version (Figure 2.3) of the last figure is useful. Imagine two negligibly thick parallel flat plates of the same shape and area, separated by a fluid-filled space. The lower plate is fixed, and we push on the upper one with a steady force, F, causing it to move in its plane of orientation at a uniform velocity, U. How is the relationship between F and U affected by the viscosity of the fluid and the geometry of the system? The force required for the upper plate to achieve a given velocity is the product of that velocity, the area of a plate (S), and the property that we're calling the dynamic viscosity (μ), divided by the distance (z) between the plates:

$$F = \frac{\mu US}{z} \text{ or } \mu = \frac{Fz}{US}. \tag{2.3}$$

Dynamic viscosity thus has fundamental dimensions of $ML^{-1}T^{-1}$ or, less fundamentally, (force)(time)/(area). Since SI uses the pascal for force per

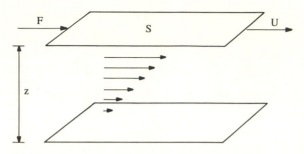

FIGURE 2.3. A pair of thin, flat plates of area S, z units apart, moving at speed U under the impetus of force F. If the lower plate is fixed, a force is needed to keep the upper one moving. The magnitude of that force is proportional to the dynamic viscosity of the fluid between them. The length of each horizontal arrow between the plates is proportional to the local flow speed relative to the bottom plate.

area, the SI unit of viscosity is the "pascal second," Pa s; the unit has no specific name. Much tabulated data are given in "poises," the comparable CGS unit, equivalent to dyne seconds per square centimeter. A poise is ten times smaller than a pascal second.

Minor definitional matters. Very commonly dynamic viscosity is simply called "viscosity" (as already done here), although the adjective in the initial term usefully distinguishes it from "kinematic viscosity," about which more just ahead. Ambiguity about which one is meant can usually be resolved by noting just what units are used. Dynamic viscosity is occasionally referred to as "molecular viscosity" to distinguish it from "eddy viscosity," a larger-scale measure of turbulent intensity used mainly by oceanographers (Sverdrup et al. 1942). Physical chemists traditionally use the symbol η (eta) instead of μ for dynamic viscosity.

Note that the dynamic viscosity of air increases slightly with temperature. In water, by contrast, viscosity drops dramatically as temperature rises. Few physical properties have as extreme a temperature coefficient as does the viscosity of ordinary liquids at ordinary temperatures. Seawater and fresh water behave in a similar manner, with the small differences in viscosity following in close proportion to the concentration of salts.

On the other hand, the dynamic viscosity of either air or water is substantially independent of the value of density—counterintuitive, perhaps, but a great convenience. For water density varies only slightly with pressure: at 20° C the increase in the density of seawater descending from the surface to the bottom of the deepest ocean trench is only about 5%. And the latter is associated with an increase in viscosity an order of magnitude less—only about 0.5% (Stanley and Batten 1968).

Put more formally, dynamic viscosity is the coefficient that relates shear stress (τ or tau) to the local velocity gradient or shear rate (dU/dz); the view given in Figure 2.3 still applies:

$$\tau = \mu \, \frac{dU}{dz} . \tag{2.4}$$

Kinematic Viscosity

Upon first encountering the symbol v (nu) for viscosity you'd naturally ask, "What's v?" As it happens, it's the so-called kinematic viscosity, nothing more than the ratio of dynamic viscosity to density:

$$v = \frac{\mu}{\rho} . \tag{2.5}$$

Kinematic viscosity has fundamental dimensions of $L^2 T^{-1}$; the SI unit is the m² s⁻¹. The CGS unit, the cm² s⁻¹ (10^4 times smaller, which we won't use here) is called the "stokes"[3] or St.

Why bother with such a nearly redundant unit? In many situations the character of a flow happens to depend very much on this particular ratio. It determines the practical "gooiness" of a fluid—how easily it flows, how likely it is to break out into a rash of vortices, how steep its velocity gradients will be. Dynamic viscosity determines the interlamellar stickiness of the fluid, or how much a fluid particle is likely to be affected by any non-synchrony in the movement of adjacent particles. Density determines what might loosely be regarded as the inertia of a fluid particle, or its tendency to continue as it has been going regardless of the activity of its neighbors. Their ratio, the kinematic viscosity, as Batchelor (1967) put it, "measures the ability of molecular transport to eliminate the nonuniformities of fluid velocity."

Note in the table that the kinematic viscosities of air and water are not especially different—only about 15-fold at 20° C. Moreover, air is the *more* kinematically viscous fluid, demonstrating, if nothing else, that this property does not take kindly to raw intuition. In air, kinematic viscosity increases with temperature slightly more than does dynamic viscosity. In water, kinematic viscosity shows the same dramatic decrease with temperature as does dynamic viscosity. So that's what's v.

Additional values of density and of dynamic and kinematic viscosities of gases, liquids, and aqueous solutions (including seawater) are most easily obtained from recent editions of the *Handbook of Chemistry and Physics*. Other such data are given in the appendices of most textbooks of fluid mechanics, hydraulics, aerodynamics, and physical oceanography.

[3] Many otherwise respectable sources call the unit the "stoke." The name honors Sir George G. Stokes (1819–1903), known for the Navier-Stokes equations and Stokes' law. That his name happens to have a final *s* is no excuse for unauthorized truncation.

25

MEASURING VISCOSITY

One way to take an initial look at the consequences of viscosity is to ask how to measure the quantity—really how to make it do something that can be measured. Admittedly, if pure water, seawater, or some simple solution is in use, looking up a value for either dynamic or kinematic viscosity is simpler than making such measurements. Sometimes, though, a solution is used for which tabulated values are unavailable; fortunately, measuring the viscosity of most liquids is not especially difficult. One just needs (who'd ever guess) a viscometer. (The word "viscosimeter" is synonymous but a linguistic barbarism.) Viscosity is, of course, a measure of resistance to rate of shearing; so all viscometers involve some scheme for shearing fluid. The simplest and cheapest commercially available device is an Ostwald viscometer (Figure 2.4). It forces a liquid through a fine tube and counts on the no-slip condition to generate sufficient shear in the flow. Improvised alternatives can be made of capillary tubing, fine pipettes, catheter tubes, and so forth.

For use, the viscometer is mounted in a constant temperature bath, and a known (usually 5 cm^3) quantity of liquid is introduced into the arm without the capillary. The liquid is then sucked up through the capillary tube until the top meniscus is above the upper reference line. The liquid is allowed to fall, and the time is recorded for how long it takes the top meniscus to drop from the upper to the lower reference line. Not surprisingly, the rate of flow is proportional to the density of the liquid—that's just gravity in action. It's also inversely proportional to the dynamic viscosity—that's the situation first described by Hagen and Poiseuille around 1840 (Chapter 13). As a result, *the time one measures is directly proportional to the kinematic viscosity*. The proportionality constant is obtained by timing the descent of a liquid of known kinematic viscosity. If what one wants is dynamic viscosity, one needs nothing more than a separate measurement of density.

If you purchase an Ostwald viscometer from your favorite purveyor of scientific glassware, make sure it's appropriate for the expected range of kinematic viscosities—these aren't all-purpose devices, and aqueous solutions run with dismaying rapidity through meters designed for motor oils. The cheapest viscometers lack a glass connection across the top of the arms of the "U" and as a result are quite fragile. One should provide some appropriate brace before use. Also, these glass devices should be kept scrupulously clean and between uses should be rinsed with acetone or some other solvent that leaves no residue.

Lots of other sorts of viscometers are commercially available, but one should hesitate before getting anything much more complex—unless, say, non-Newtonian fluids are involved. For very viscous fluids such as glycerin or sugar syrups one can get fairly good viscosity values by just timing the

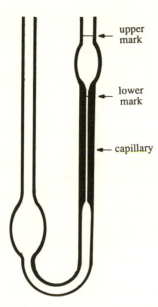

FIGURE 2.4. An Ostwald viscometer, slightly fatter than life. The time needed for a meniscus to drop from upper to lower marks as the liquid passes through the capillary is proportional to the kinematic viscosity of the liquid.

fall of a sphere of known size and weight and applying Stokes' law (Chapter 15); the main precaution is that the walls of the container should be sufficiently far from the falling sphere.

The biologist will rarely have occasion to measure the viscosity of a gas, and the Ostwald device won't work for gases. But an analogous arrangement can easily be contrived; I'll leave the design as an exercise for the reader.

CONSEQUENCES OF THE INVERSE VISCOSITY-TEMPERATURE RELATIONSHIP

At 5° C water is about twice as viscous (dynamically or kinematically) as at 35° C; organisms live at both temperatures and, indeed, at ones still higher and lower. Some experience an extreme range within their lifetimes—seasonally, diurnally, or even in different parts of the body simultaneously. Does the consequent variation in viscosity ever have biological implications? Quite a number of cases are at least possible although very few can be considered well established.

Consider the body temperatures of animals. At elevated temperatures less power ought to be required to keep blood circulating if the viscosity of blood follows the normal behavior of liquids. And, in our case, it does behave in the ordinary way—human blood viscosity (ignoring blood's minor non-Newtonianism) is 50% higher at 20° than at 37° C (Altman and Dittmer 1971). Is this a fringe benefit of having a high body temperature? Probably the saving in power is not especially significant—circulation costs only about 6% of basal metabolic rate. More interesting is the possibility of compensatory adjustments in the bloods and circulatory systems of animals that tolerate a wide range of internal temperatures. The red blood cells of cold-blooded vertebrates, and therefore presumably their capillary diameters, are typically larger than either the nucleated cells of birds or the nonnucleated ones of mammals (Chien et al. 1971). The shear rate of blood is greatest in the capillaries; must these be larger in order to permit circulation at adequate rates without excessive cost in a cold body?

Have some animals arranged for their bloods to behave like "multiviscosity" motor oils, which resist excessive thickening when cold and thinning when hot? That might be helpful if an animal has to be fully functional with the same circulatory machinery at different temperatures. Fletcher and Haedrich (1987) found that the viscosity of rainbow trout blood has an unusually low temperature dependence, but trout seem unlikely animals to encounter highly variable ambient temperatures.

Is the severe temperature dependence of viscosity perhaps a serendipitous advantage on occasion? A marine iguana of the Galapagos basks on warm rocks, heating rapidly, and then jumps into the cold Humboldt current to graze on algae, cooling only slowly. Circulatory adjustments as the animal takes the plunge have been postulated (Bartholomew and Lasiewski 1965), but no one seems to have looked at whether part of the circulatory reduction in cold water is just a passive consequence of an increase in viscosity. A variety of large, rapid, pelagic fish have circulatory arrangements that permit locomotory muscles to get quite hot when they're in use (Carey et al. 1971); blood flow ought to increase automatically at just the appropriate time.

A less speculative case is that of antarctic mammals and birds; Guard and Murrish (1975) found that the apparent viscosity of their bloods changed with temperature even more drastically than in more ordinary animals such as humans and ducks. Antarctic animals must commonly contend with cold appendages, since full insulation of feet and flippers would be quite incompatible with their normal functions. The circulation of such an appendage often includes a heat exchanger at the base of the limb so that, in effect, a cold-blooded appendage and a warm-blooded body can be run on the same circulatory system without huge losses of heat (Scholander 1957). Changes in blood viscosity will reduce flow to appendages when they get cold quite without active adjustments within the circulatory system.

More generally, they have shown that the temperature coefficient of viscosity is certainly a variable that can be to some extent controlled, either in an evolutionary or an immediate sense.

While we're considering the Antarctic, I should say a word about some very peculiar fish that live there, the so-called ice fish (Ruud 1965). They are the most transparent of adult vertebrates, but the transparency appears to have come at the price of loss of red blood cells and a consequent reduction in oxygen-carrying capacity, making them fairly sluggish creatures. Still, there's a compensatory benefit—ice fish blood flows substantially more readily than the bloods of more ordinary fish from the same habitat (Wells et al. 1990).

Might warm, swimming animals release heat through the skin in such a way that kinematic viscosity is locally lowered and drag is reduced? Aleyev (1977) cites Parry (1949) to support a claim that the scheme is well established for cetaceans, but I find no support in Parry's paper beyond some calculations showing that cetaceans produce a whale of a lot of heat, and some evidence that a complex circulation in blubber can actively control the outward passage of heat. The latter is more likely just a scheme to augment heat dissipation during and after high-speed swimming (Palmer and Weddell 1964). Aleyev also cites Walters (1962), who speculated on the possibility that tuna locally release heat behind a structure called the "corselet" more or less amidships to reduce their drag. Walters is dubious, Webb (1975) is dubious, and so am I.

Lower viscosity at higher temperature implies steeper velocity gradients and thus thinner gradient regions on surfaces. In effect, a solid surface is less shielded from free stream flow. That ought to help the exchange of dissolved material between an organism and moving water around it. Does the phenomenon aid gas exchange in fish gills and compensate, in part, for the decreased solubility of oxygen in water at high temperatures?

Similarly, changes in viscosity ought to change the performance of filter-feeding devices. Are the dimensions of such devices adjusted intra- or interspecifically to reflect changes in ambient temperature? In at least one case what changes is the food caught. An antarctic echinoderm larva catches bacteria; a very similar larva from California catches microalgae but not bacteria (Pearse et al. 1991). If California larvae feed in cultures to which a viscosity-increasing polymer (methyl cellulose or polyvinylpyrrolidone) has been added, they then take up bacteriophagy (Pearse, pers. comm.).

Planktonic organisms are quite commonly more dense than the water around them; according to Stokes' law (Chapter 15) they should sink more rapidly at higher temperatures, although, as Hutchinson (1967) pointed out, passive changes in their own densities might offset the effect of altered viscosity. Reportedly (by Sverdrup et al. 1942, for example) plankton from tropical waters are smaller and more angular and ramose than plankton from polar water. That's what one would expect if keeping sinking rates relatively constant were what mattered and if shape and size were the only variables at work. But, as we'll see with a more specific case, this particular world is a little hard to second-guess.

Individuals of many species of the microcrustacean, *Daphnia*, have larger heads, much longer and more curved crests or helmets, or extra spines when in warmer water (Figure 2.5). An old suggestion is that this phenomenon, termed "cyclomorphosis," is an adaptation to viscosity differences between warm and cool water. The difference is evident between generations raised at different temperatures, seasonally within a single species, as seasonal replacement of one group of species by another, and in comparison with the fauna of climatically different ponds and lakes. Similar seasonal polymorphisms are known in dinoflagellates and in rotifers. Certainly a larger head and more flattened body should reduce sinking rates in less viscous water, but that explanation has (as Hutchinson 1967 originally noted) substantial difficulties. Hebert (1978) presented an alternative hydrodynamic explanation involving changes in the muscles used for locomotion to offset sinking (*Daphnia* use their second set of antennae to swim, so muscles in the head are entirely germane). But exposure to chemicals released by certain predators will also induce the development of crests or spines in *Daphnia*. While predation is not as effective on crested or spinose individuals, the latter are less successful in terms of other aspects of their life histories (Grant and Bagley 1981; Havel and Dodson 1987). To complicate things further, the induction of crests and spines involves both genetic as well as environmental factors.

Even the more general observation, mentioned earlier, that cold-water plankton are larger and less surface-rich in shape is hard to tie unequivocally to viscosity differences. The main problem is that sinking rate depends on the difference between the density of the organism and that of the medium, and assuming constant density is clearly unsafe. For instance, embryos of *Euphausia* (another microcrustacean) sink at different rates at different developmental stages (Quetin and Ross 1984); while large specimens of *Daphnia pulex* sink faster than small ones, the differences are much less than considerations of size would predict (Dodson and Ramcharan 1991); and cyanobacteria (which are photosynthetic) change their density with changes in light intensity (Kromkamp and Walsby 1990). To make matters still worse, the density of active planktonic organisms is a bit tricky to measure and isn't commonly done. (Walsby and Xypolyta 1977 describe a technique using a specific gravity bottle and a nonpenetrating tracer to determine the volume not occupied by organisms; alternatively sinking distance might be watched in density gradients of some osmotically inactive material such as "Percoll.")

FIGURE 2.5. The microcrustacean water flea, *Daphnia*: (a) the warm-water form; (b) the cold-water form.

At the very least, one ought not attribute some decrease in activity of an aquatic organism at low temperature to reduction in metabolic rate without checking to see if an increase in viscosity is contributory; the results of Podolsky and Emlet (1993) on swimming speeds of sand dollar larvae certainly make that point. But we're getting ahead of ourselves—less *ex cathedra* treatment of each of these cases requires that additional physical material be developed, which will happen for pipes and blood flow in Chapters 13 and 14, for filter-feeding and sinking plankton in Chapter 15.

Neither Hiding nor Crossing Streamlines

W E ' R E N O W ready to examine what happens when fluids flow, beginning with the most universally applicable notions and then moving
on to increasingly specific phenomena. Especially in this and the next few
chapters, the reader should bear in mind the ultimate artificiality of a
linear narrative. In particular, the biological examples may be nothing but
the truth, but—for want of material to be developed further along—
they're hardly the whole truth.

THE PRINCIPLE OF CONTINUITY

Consider a pipe that's open at both ends. If the pipe has rigid walls, it
must have a constant internal volume. From our assumption that fluids are
incompressible, it follows that, if a given volume of fluid enters one end,
then the same volume has to come out the other. Not only must the volumes
be equal, but entry and exit must take precisely the same time, and so the
volumes-per-time must be equal. This trivial idea, termed the "principle of
continuity," turns out to be surprisingly rich with biological applications
and implications. On occasion, though, it has been less than obvious to
practicing biologists.

To view the matter more formally, consider a pipe whose cross-sectional
area varies from one part to another, as in Figure 3.1a. We'll call the cross
section near the entry S_1 and that near the exit S_2. If a small volume of
fluid, $S_1 dl_1$, enters the pipe in an interval of time, dt, then an equal volume,
$S_2 dl_2$, must leave the pipe in the same period. Thus

$$\frac{S_1 dl_1}{dt} = \frac{S_2 dl_2}{dt} .$$

But any dl/dt is, of course, a velocity, so (using U_1 and U_2 for entry and exit
velocities)

$$S_1 U_1 = S_2 U_2. \tag{3.1}$$

The equation says that the product of the cross-sectional area and the
average velocity normal to the plane of that area is the same in both places
in the pipe. And the rule should hold for any cross section whatsoever—no
matter how the pipe expands, contracts, or changes shape, the product of
cross-sectional area and velocity will remain constant. Put another way, the

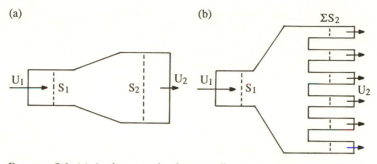

FIGURE 3.1. (a) An increase in the overall cross-sectional area of a pipe must be concomitant with a decrease in the average speed of flow. (b) Dividing the pipe into a parallel array makes no difference. Local velocities are proportional to the lengths of the arrows.

volume flow rate, or volume flux (we'll call it Q), does not change within a conduit.

Now consider a pipe that branches, as in Figure 3.1b. We now have to look not just at entry and exit but at any cross section, adding up the contributions of the parallel conduits. But again fluid inside has no hiding place, so for every volume that enters, an equal volume must leave. So the rule is still a useful one since the sums of the area-velocity products must be the same everywhere—at least if every bit of fluid passes our monitors once and only once:

$$\sum S_1 U_1 = \sum S_2 U_2. \tag{3.2}$$

Note that no assumptions were made about energy or about friction; the argument is purely geometric. That gives it enormously wide applicability. In fact, were incompressibility not assumed, we would merely have had to substitute ρS for S and mass flux for volume flux for the principle still to work. The principle of continuity has the same role and the same generality for fluid mechanics that conservation of mass has for solids; it's really just a specific application of the idea of conservation of mass; it's our principal principle, continuously useful.

One garden-variety device based on continuity is the nozzle attached to a garden hose. By constricting an aperture the water is persuaded to speed up, and its increased momentum carries it a far greater distance than it would go without the nozzle. Volume and thus mass flux is unchanged, except for a little reduction due to the extra friction of the nozzle. But momentum flux, loosely mass flux times velocity, may be increased several-fold. The trick involves no power input by the nozzle; indeed, additional power from a pump upstream might be needed to offset the extra frictional loss. Analogously, with an appropriate input of power one can make

33

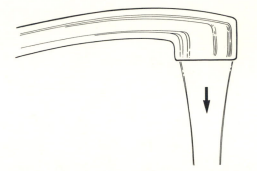

FIGURE 3.2. The column of liquid coming from a spigot contracts as it accelerates downward, as it must according to the principle of continuity.

a stream of fluid constrict without a nozzle. That's what an ordinary axial fan or a propeller does—since flow is more rapid downstream the effective cross section of the moving stream must be less. Which is why, despite the same volume flow rate in both places, you feel a fan's wind more strongly downstream than upstream.

To emphasize the wide applicability of the principle of continuity we might look at a relatively unusual use. If liquid falls freely from a downwardly directed orifice (Figure 3.2), the column of liquid contracts. The effect will be most noticeable with a highly viscous liquid, whose initial speed can be kept low. The column contracts as it accelerates because its cross-sectional area must always be inversely proportional to its velocity. Knowing the acceleration of gravity, it's possible to obtain the rate of discharge of the pipe from nothing more than two measurements of the width of the column and the vertical distance between them. Alternatively, given these latter measurements, a stopwatch, and a container of known volume, one can make a fair estimate of the acceleration of gravity—one just combines equation (3.1) with the ordinary equation for distance covered at constant acceleration beyond an initial speed.

Continuity continually makes large-scale mischief. The old London bridge, the one that lasted from 1209 to 1832, rested on a set of boat-shaped piers so wide that almost half the Thames was blocked. As a result, the flow between the piers should have almost doubled the already rapid tidal currents. In fact, the cutwaters that were necessary to protect the piers from scour made the situation even worse. Only small boats could get through, and these had to take careful aim and considerable risk in doing what was known as "shooting the bridge" (Gies 1963).

Applying Continuity to Bounded Flows

Largish creatures, whether plants or animals, devote considerable anatomy to internal fluid transport systems. The very existence of macroscopic organisms is predicated on such systems, devices to circumvent the nasty difficiency of diffusion for all but very short distance transport. LaBarbera and I (1982) repeated Krogh's (1941) argument on the point rather elaborately, and I've pushed it more recently as well (Vogel 1988a, 1992a). We give these systems that move fluid in bulk various names—circulatory, respiratory, translocational—but they all do much the same job, moving fluid from one site of material exchange to another. Each must contend with several partly conflicting physical imperatives. For decent power economy, long-distance transport is best done in pipes of large cross-sectional area. Large pipes will have greater volume relative to their wall area and thus relatively less surface at which (recall the no-slip condition) viscosity can work its malicious mischief. For effective exchange of material between the fluid and the surrounding tissue, pipes of small cross-sectional area are much better. Exchange is ultimately dependent on diffusion; again, that's an agency effective over only short distances within either fluid or tissue.

Consider an implication of the principle of continuity for such a system. Simply narrowing the large pipes at sites of diffusive exchange would, according to continuity, result in very high speeds. Power losses would be great, and little time would be available for diffusion to occur between tissue and any element of fluid. All internal bulk fluid transport systems, whatever their function, do use both large and small pipes; but in practice fluid always moves fastest in the largest pipes and slowest in the smallest ones. While this may sound like a violation of the principle of continuity, it's not. Rather, organisms just apply equation (3.2) and make the total cross-sectional area of the small pipes very much larger than that of the large ones, as in Figure 3.3 and Table 3.1.

People

We have perfectly ordinary plumbing. The output of each side of the heart of a resting human is about 6 liters per minute (10^{-4} m^3s^{-1}). The ascending aorta and the main pulmonary artery have internal diameters of about 2.5 cm and thus cross sections of about 5 cm^2 (5×10^{-4} m^2). Thus the resting rate of flow in either is around 0.2 m s^{-1}. A single capillary is about 6 μm in diameter or about 30 μm^2 (30×10^{-12} m^2) in cross section. Blood flows through it at roughly 1 mm s^{-1} ($10^{-3}m s^{-1}$). Thus the product of velocity and cross section for a capillary is some 3×10^9 (3 U.S. billion)

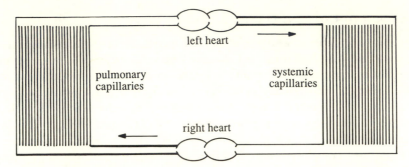

FIGURE 3.3. An extremely diagrammatic view of an avian or a mammalian circulation. Lower speed in the capillaries than in the heart is possible only if the total cross-sectional area of capillaries exceeds that of each half-heart.

TABLE 3.1. THE SIZES AND NUMBERS OF THE PIPES VERSUS THE SPEEDS OF FLOW IN SEVERAL FLUID TRANSPORT SYSTEMS.

	Element Area (mm^2)	Number	Total Area (mm^2)	Flow Speed $(mm\ s^{-1})$
Oak tree, 2.5 cm trunk (Kramer and Koslowski 1960; Lundegårdh 1966)				
Xylem vessels, trunk	7.9×10^{-3}	380	3.0	10
Leaves	5000	350	1.8×10^6	1.7×10^{-5}
Dog circulatory system (Caro et al. 1978)				
Aorta	200	1	200	200
Large arteries	20	20	400	100
Arterioles	2.1×10^{-3}	6×10^6	1.2×10^4	3.2
Capillaries	3.0×10^{-5}	1.9×10^9	5.7×10^4	0.7
Sponge, 2.4 cm³ volume (Reiswig 1975a)				
Ostia (input holes)	3.3×10^{-4}	9.4×10^5	310	0.57
Incurrent canal apertures	0.031	3,400	100	1.7
Flagellated chambers	7.1×10^{-4}	2.9×10^7	2×10^4	0.0087
Excurrent canal apertures	0.11	280	31	0.57
Osculum	3.4	1	3.4	0.07
Idealized human lung (Weibel 1963)				
Trachea	250	1	250	200
5th generation bronchi	10	32	310	160
10th generation bronchi	1.3	1000	1300	38
15th generation bronchi	0.34	3.3×10^4	1.1×10^4	4.4
20th generation bronchi	0.17	1.0×10^6	1.7×10^5	0.3

times less than the equivalent product for the ascending aorta or the main pulmonary artery. So there must be very roughly 3 billion parallel capillaries in the systemic circulation receiving the output of the aorta. These should have an aggregate cross-sectional area of about a tenth of a square meter—approximately a square foot. That total cross section is some 200 times greater than that of the aorta, accounting (by continuity) for the 200-fold drop in the speed of flow of the blood.

Admittedly these figures (from Caro et al. 1978 and Milnor 1990) are very rough ones. At rest not all capillaries are open and operational, and lots of other features of the microcirculation, such as arterio-venous shunts, have been ignored. But however approximate, that estimate of a couple of billion parallel capillaries was bought at very low cost! In our mammalian kind of circulatory system, the volume flow rate through the lungs must be exactly equal to that through the entire systemic set of pipes. Thus if lung capillaries were exactly the same size and conveyed blood at the same speed as capillaries elsewhere, their aggregate cross section would have to be the same tenth of a square meter. As it happens, lung capillaries aren't quite the same; nonetheless, lungs obviously have a lot of capillaries and are exceedingly bloody organs.

Trees

The ascent of sap in the xylem of trees (Figure 3.4a) provides another case where one of the variables in equation (3.2) cannot easily be measured. It's no minor matter to invade a system in which the largest elements are conduits that are only a fraction of a millimeter in diameter, with stiff, strong walls where internal pressures are negative by many atmospheres. To give some idea of what transpires in trees, I've taken some data from Lundegårdh (1966) for the leaf area of an inch-diameter (2.5 cm) oak tree (*Quercus robur*) with 300–400 leaves. I've combined these with sap velocities for the trunk and with transpiration rates for the leaves of *Q. rubra*, from Kramer and Koslowski (1960), to address some questions raised by P. J. Kramer (1959), Zimmermann (1983), and others.

If the tree has two square meters of leaf area (S_1) and transpires water at 1.5×10^{-8} m³s⁻¹ (a liquid volume of about two ounces per hour) for each square meter of area, then the volumetric water loss ($S_1 U_1$) is 3×10^{-8} m³s⁻¹. The 100 μm vessels of the trunk make up about 7% of its total cross section; thus their aggregate cross-sectional area is 1.5×10^{-4} m² (S_2). Dividing $S_1 U_1$ by S_2, we calculate a rate of sap ascent in the vessels of 2×10^{-4} m s⁻¹, a fifth of a millimeter per second. But the rate of ascent of sap is not hard to measure, at least roughly. A pulse of heat is applied to the trunk, and the time is noted for the arrival of heated sap beneath a sensor a few centimeters higher. Such measurements give ascent rates of about a

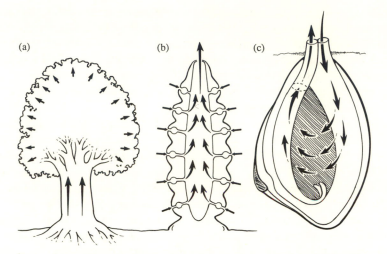

FIGURE 3.4. Changing cross-sectional area and thus speed of flow in (a) a tree, (b) a sponge, and (c) a clam. The numbers in the text and in Table 3.1 should emphasize the impossibility of drawing such hydraulic systems with consistent scales!

centimeter per second, fully fifty times higher than the calculated rate. Apparently the majority of the vessels are filled with air—that is, they have embolized—and are nonfunctional. Air embolisms, mainly resulting from gas release when sap freezes, are a major hazard for systems in which an aqueous liquid is under great negative pressure (see, for instance, Tyree and Sperry 1988). While some embolisms can be repaired, it's not uncommon for embolized vessels simply to be left unused as conduits for sap (Zimmermann 1983).

One doesn't associate rapid flows with plants, but continuity demands that transpiration from the great surface areas of broad-leafed plants be reflected in rapid sap movement in the vessels of stems and trunks. A centimeter per second, while relatively rapid and characteristic only of trees with fairly wide vessels, is far from any record. In vessels of the taproot of wheat plants, Passioura (1972) measured speeds up to 25 cm s^{-1}, and even these rates could be raised 3-fold by experimental manipulations.

Sponges

Sponges are little more than highly elaborate manifolds of pipes, with lots of small pores and one or a few large (commonly apical) openings on their surfaces (Figure 3.4b). Grant (1825) established that flow was unidirectional—into the small openings and out through the larger ones—and he was much impressed by the rate at which a sponge could move water. But what was doing all the pumping? Sponges were known to have flagella, but spongologists persisted in invoking muscles that just as persistently remained undetectable. It just wasn't credible that tiny, slow, flagella could make a sufficient pump. The pumping is certainly impressive—a sponge ordinarily pumps a volume of water equal to its own volume every five seconds (Reiswig 1974), a rate roughly a hundred times that of a human heart relative to body volume. In the fine words of Bidder (1923), a sponge is "a moment of active metabolism between the unknown future and the exhausted past." It was Bowerbank (1864), almost forty years after Grant, who recognized that flagella could do the job. The solution to the problem entailed nothing more than application of the principle of continuity.

Again, let's view the situation quantitatively. (I'm using mainly the morphological data given by Reiswig 1975a.) A small sponge of 100,000 mm^3 will ordinarily have an output opening 100 mm^2 in cross-sectional area and an output velocity of 0.2 m s^{-1}. It's difficult to envision a flagellum only 25 μm long pumping water at more than 50 μm s^{-1}, a velocity 4000 times lower than that of the sponge's outflow. But the total cross-sectional area of the flagellated chambers proves to be nearly 6000 times greater than the area of the output opening. Thus one needn't make any heroic assumptions about the pumping speeds of flagella, much less make muscles mandatory. Interestingly, small sponges put a final constriction on their output apertures. Bidder (1923) recognized these as nozzles that increased the speed and coherence of the output jet and thus minimized the chance that the animal might uselessly refilter its own output.

Similar arrangements are evident in other filter-feeding animals such as bivalve mollusks (Figure 3.4c). In these latter, though, both intake and output apertures have low cross sections and consequently high velocities; input and output conduits are connected by distributing and collecting manifolds to large, slow-speed, ciliary filter pumps. For that matter, one can imagine a circulatory system much like our own but driven by ciliated capillaries. Such a system might achieve the normal speeds in the large vessels in the complete absence of a pumping heart. One drawback of such a scheme is its operating cost—muscles manage much better than cilia or flagella by that criterion. For sponges and bivalves, the functions of pump and filter are combined, so the economic argument must not be so directly applicable (LaBarbera and Vogel 1982). Also, as we'll see in Chapter 14, ciliary and flagellar pumps do better for low-pressure, high-volume applications than for systems with pressures as high as those of mammalian circulations.

Less Completely Bounded Systems

At times the rate at which water is carried by a stream or river (the "discharge") increases substantially. A placid stream may metamorphose into a raging torrent in which the depth and width obviously increase and the velocity seems far greater than normal. How much does the velocity really increase? If it were to increase in proportion to the discharge, then, by the principle of continuity, the river shouldn't rise! Clearly, average flow speed doesn't increase as fast as discharge; in fact, the increase in velocity at a given place that is submerged under both normal and flow conditions turns out to be quite modest. The graphs given by Dury (1969) and by Leopold and Maddock (1953) indicate that mean current speed at a particular station along a river only doubles for every 10-fold increase in discharge. Since peak velocity in a larger cross section will occur farther from bottom and sides of a channel, the actual speeds near the interfaces are likely to increase even less. Being swept downstream by the high water speeds of floods may just not be as worrisome to small attached or bottom-living organisms as we might otherwise think.

As the discussion in Hynes (1970, pp. 224–229) implies, abrasion and alternations of the form of the bottom are more important than velocity increases per se. Thus during a long-term study of an ordinarily well-behaved stream in Denmark, a one-day, 70 mm rainfall caused a brief and rare flood. The effect on the fauna of the gravel and sand on the bottom was slight; the main casualties were due to movement of boulders—some snails (*Ancylus*) and some caddisfly pupae (*Wormaldia*) that lived on those rocks got crushed as they rolled (Thorup 1970).

An analogous situation occurs near an irregular bottom beneath moving water. If a portion of the bottom is elevated above the general terrain, then the speed of flow will be greater across it—it can be viewed as half of a longitudinally sliced nozzle. The constriction effect isn't quite as severe as in a system with a solid boundary all around or even one with an air-water interface just above, but the speeding up of flow and its biological consequences are far from negligible. To give just one example, Genin et al. (1986) looked at seamounts beneath several thousand meters of ocean. Regions of 2-fold increases in flow on the seamounts were associated with 3–5-fold increases in densities of a black coral (*Stichopathes*) and some gorgonian corals. The density differences were attributed to better recruitment and growth of these passive (current-dependent) suspension feeders, since the concentration of edible material on the seamounts was quite clearly lower than on the nearby floor.

STREAMLINES

How might this principle of continuity be applied to situations other than flow through pipes or flow in discrete, homogeneous streams? Can we make full use of the principle for open fields of flow, in particular for cases in which the velocity of flow varies *across* the field of flow as well as in the direction of flow? The principle, if anything, proves even more potent in such situations—in part because we are less likely to use it intuitively, and so the results of its application are less self-evident. The conceptual device that enables us to apply the principle of continuity to open flow fields is something called a "streamline."

A streamline, it should be stressed at the start, has a very special meaning in fluid mechanics, a meaning that bears only an indirect connection to its vernacular use. *A streamline is a line to which the local direction of fluid movement is everywhere tangent.* Loosely then, a streamline traces a path through a field of flow along which some particles of fluid travel. Let's assume we know the direction of flow at all points in a flow field at some instant in time. To create a streamline, we start at some upstream point and draw a very short line in the direction of flow at that point; we then consider the direction of flow at the far end of the line and extend the line a short distance in the new direction. We continue the process until the line wends its way across the entire field, as in Figure 3.5. What have we done by this apparently trivial and obviously awkward procedure? The line follows the stream, and the direction of flow is always along the line. Therefore any component of velocity normal to the line must be zero; in short, *fluid does not cross the line*. Viscosity may move momentum, conductivity may move heat, diffusion may move molecules, but fluid in bulk doesn't ordinarily move across the line.

It's no more difficult to create a second streamline, running in tandem with the first; and with two, we can get to the real point. Streamlines provide a conceptual device for dividing a complex field of flow into an array of pipes with nonmaterial walls. For a simple, two-dimensional flow (no convergence or divergence outside the plane of the paper or the screen of the machine), *the principle of continuity must apply between the pair of stream-lines*. If we want to deal with properly three-dimensional flow, we just need a set of streamlines that surround some fluid, a so-called stream tube, within which continuity is equally applicable. So what all of these lines and tubes lead to is really quite wonderful. Where a pair of streamlines diverge or a stream tube becomes wider, we know immediately that the fluid is traveling more slowly. If streamlines converge or tubes become narrower as one moves downstream, that's a definitive indication that velocity is increasing.

Pathlines and Streaklines

But how can we draw streamlines in the real world? The direction of flow at all points is rarely a matter of public record; indeed, making streamlines is more commonly used to determine flow direction than the other way around. In practice, two fundamentally different schemes can be used. In the first, a visible marker or particle of some sort is released near the

FIGURE 3.5. A streamline (thick line) can be drawn from information about the local direction of flow (arrows) since it's always tangent to the local flow.

upstream end of the flow in question. Ideally, the marker is neutrally buoyant and very small, so it always travels in the local flow direction. A time exposure or repetitive photograph of its travel gives a solid or dotted line recording the history of the marker; the record is called a "pathline" or particle path, as shown in Figure 3.6a. In the second, a continuous stream of particles, dye, tiny bubbles, or smoke is introduced at a fixed point; and some time during the process an instantaneous photograph is taken. It gives the present position of fluid that has, over a period of time, passed by the injection point; it is called a "streakline" or "filament line." Most often, streams are introduced at an array of points normal to the flow direction, so a whole set of simultaneously produced streaklines are recorded in a single image, as in Figure 3.6b.

If the flow is steady, that is, if velocity at all points is constant over time, then pathlines and streaklines coincide, and both mark the streamlines. If flow is unsteady, neither, strictly speaking, marks streamlines, and everything becomes much more complicated. But, as mentioned earlier, we'll deal mainly with steady flows. Photographs showing streamlines both encapsulate a vast amount of information about a field of flow and provide an overall view of complex flows that we visual creatures find intuitively useful and satisfying. Besides that, they are commonly objects of great esthetic appeal—one has only to look at Van Dyke's (1982) collection to be forever convinced of that.

Biologists don't make as much use of this valuable tool as I think they ought to. As qualitative views, streamline patterns are very useful for understanding the complex flows both around organisms of irregular shapes and in common environments. Exploration with a syringe of dye of the flow in any stream with irregularities in its bed leads to a whole new appreciation of the diversity of habitats with respect to flow. Upwellings, downwellings, local upstream flows, places with unsuspected periodic flows, places in which flow either enters or leaves the substratum—all emerge from even a casual and unsystematic survey.

Moreover, quantitative information can be extracted; where organisms resist embellishment with instruments, photographs of streak- or pathlines

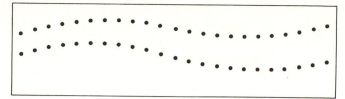

(a) Pathlines

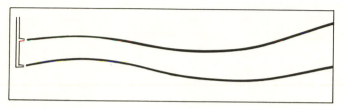

(b) Streaklines

FIGURE 3.6. Two kinds of streamlines. (a) "Pathlines" show the paths taken by particles as they flow through the field of view—the view extends over a period of time, as in a multiple- or long-exposure photograph. (b) "Streaklines" give an instantaneous view of the position of markers (dye streams, for instance) steadily released upstream.

may be the most informative and least abusive way to get information on velocities in their full, three-dimensional vectorial splendor—and from these a picture of drag, thrust, locomotory mechanisms, and so forth. Thus Kokshaysky (1979) induced a bird to fly through a cloud of wood particles while these were photographed with a repetitive stroboscope. The views he and others obtained of the vortex rings behind flying birds have fundamentally changed our view of the aerodynamics of avian flight; we'll have a lot to say about these vortices in Chapter 10. Similarly, a view of wind pollination as a very much more sophisticated business than anyone guessed earlier has come from a series of papers by Niklas (see, especially, 1985 and 1992) on airflow patterns around flowers and conifer cones. Apparently plants contrive shapes that interact with local flows as pollen traps and that may even achieve some degree of specificity for the appropriate kind of pollen.

Some instruments yield data for direction of flow—for instance, tufts of string attached to test objects or probing needles; some instruments give data for speed of flow without specifying direction—hot wire and thermistor anemometers are examples. Streamlines can be used to obtain the complementary data from the other. If one has the pattern of streamlines determined from information on direction together with a single datum for speed, one can use the principle of continuity and the definitional unlawfulness of crossing a streamline to get speed at all other points in the flow field. If one knows a single line of flow across a flow field (such as the location of the substratum), one can use a map of speeds to derive the pattern of streamlines and thus the direction of flow at all other points.

Small size doesn't preclude marking streamlines to get information on speed and direction of flow at specific places. If anything, flow visualization really comes into its true glory on the size scales relevant to individual organisms a millimeter or centimeter long. On the one hand, little else may be available for mapping flows (or measuring forces—see Chapter 4); on the other, both the low speeds and the absence of turbulence greatly simplify matters. Streaklines can be made down nearly to a tenth of a millimeter in diameter by injecting dye from a micrometer-driven syringe through a piece of polyethylene catheter tubing with the tip drawn out. With the injector on a manipulator, the investigator can look at how flow is drawn into and passes through biological filters. Lidgard (1981) used the technique to map colonywide surface currents and functional excurrent chimneys in encrusting bryozoa while watching through a stereomicroscope. And LaBarbera (1981) mapped flow patterns in, through, and out of several kinds of brachiopods only a centimeter or two long, showing how they avoided either internal mixing or any recirculation of previously filtered water. At this scale a stream of fluorescein dye passes right through a lophophore of tentacles and emerges intact as a beautifully laser-straight, very slow (10 mm s^{-1}) green jet.

Pathlines are at least as useful. Figure 3.7 (from Vogel and Feder 1966) gives a view of pathlines around a model of a fruit-fly wing immersed in a moving liquid. This shift from air to water will be described in Chapter 5; for present purposes it brings up the important point that flow visualization, clearly the most direct way to obtain either streaklines and pathlines, is considerably easier in liquids than in gases. Incidentally, the biologist should not overlook organisms as sources of particles for marking pathlines—such things as freshly hatched brine shrimp (*Artemia*) and cultured algal cells are of handy size and density.

Timelines and Isotachs

Two other kinds of flow maps are of considerable use; neither is a form of streamline although streamlines can be derived from either. The first is something usually called a "timeline." At a given instant a marker in the form of a continuous or periodically interrupted line of dye, smoke, or electrolytic bubbles is introduced, usually normal to the direction of flow. That line then moves downstream and is periodically illuminated. A time-exposure photograph thus gives a set of lines, with the distance between

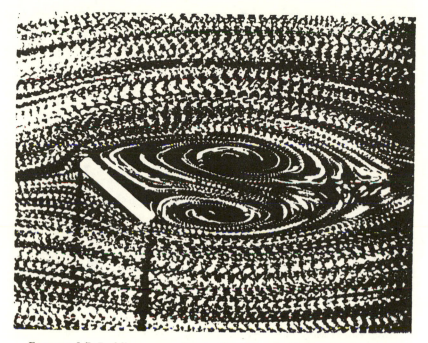

FIGURE 3.7. Pathlines around an inclined flat plate immersed in a rotating bowl of water in which particles have been suspended. The plate was 8.5 mm across, and the rate of flashing of the stroboscope was 12 s^{-1}.

equivalent loci on adjacent ones inversely proportional to the speed of flow. Instead of moving around immersed objects as do streamlines, timelines enwrap them in tangles. They're particularly useful for showing local reversals of the direction of flow, as often occurs behind points of flow separation (Chapter 5).

The second nonstreamline map is one of speeds without consideration of direction, one obtained directly from traverses with a flowmeter or anemometer. Lines are drawn to follow equal values of flow speed, like the isotherms and isobars for temperature and barometric pressure on a weather map. Unfortunately no name for these lines is universally accepted; I like the term "isotach," but I've seen "isovel" and other synonyms at least as commonly.[1] We'll encounter a specific use of isotachometric maps in Chapter 4; Figure 3.8 gives an example of one. Such diagrams may be

---

[1] "Isovel" is a miscegenation of Greek and Latin as offensive as "automobile," which I've been told should properly be "autokineton."

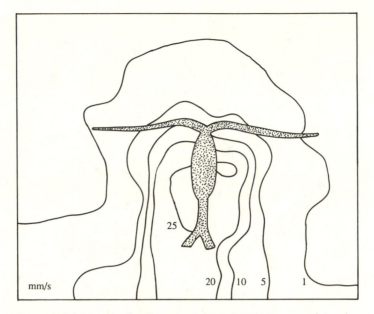

FIGURE 3.8 Isotachs for flow around a swimming copepod (a micro-crustacean), *Pleuromamma xiphias*, made by David Fields and Jeannette Yen. The animal is about a millimeter long and is swimming upward.

uncommon, but they ought to be of special value to biologists dealing with sessile organisms that protrude into spatially irregular flows or with the habitats of organisms that prefer exposure to certain velocity gradients.

LAMINAR AND TURBULENT FLOWS

Turbulence was mentioned as a nuisance for visualizing flow; since it has surfaced, we should no longer postpone the business of distinguishing between laminar and turbulent flow. The existence of these two radically different regimes of flow is another of the strange complications in the behavior of fluids and another phenomenon that gets pretty poor treatment in common practice or parlance. Perhaps the way to begin is to admit that when introducing viscosity in the last chapter, another assumption was tacitly made. We assumed that "layers" of fluid slipped smoothly across one another with all particles of fluid moving in an orderly, unidirectional fashion. And when defining streamlines, I slipped in the word "ordinarily" when forbidding fluid to cross streamlines. These presumptions that all fluid particles move very nearly parallel to each other in smooth paths are strictly valid only for what we term "laminar flow." In it the large- and small-scale movements of the fluid are the same, at least down to the level at which molecular diffusion becomes an appreciable mode of transport.

In "turbulent flow," by contrast, tiny individual fluid particles (a particularly useful polite fiction at this point) move in a highly irregular manner even if the fluid as a whole appears to be traveling smoothly in a single direction. Intense small-scale motion in all directions is superimposed on the main large-scale flow. Turbulence is essentially a statistical phenomenon, and descriptions of overall motion in turbulent flows should not be presumed to describe the paths of individual particles. The easiest analogy is to diffusion, in which Fick's law works admirably for the overall phenomenon but says almost nothing about what any molecule will be doing at any particular instant. The difference between turbulence and diffusion is mainly one of scale—diffusion is a molecular phenomenon while turbulent motions happen on much larger (if still sometimes quite small) scales. By convention—which is to say for convenience—turbulent flows aren't considered automatically unsteady.

In turbulent flow, then, not only is momentum transferred across the flow—which is what viscosity accomplishes on a more limited scale—but actual mass similarly shifts around in directions other than that of the overall flow. The transfer of mass is formally analogous to that of momentum, and something analogous to dynamic or molecular viscosity can be used as a measure of the intensity of turbulence. That's the "eddy viscosity" mentioned in the last chapter; its value is zero for laminar flow. (See Sverdrup et al. 1942, Hutchinson 1957, Massey 1989, or books on meteorology or physical oceanography for further information.) It's the eddy viscosity to which reference is made in some often-quoted doggerel of the meteorologist L. F. Richardson (itself a parody of the more ecological original):

> Big whirls have little whirls
> Which feed on their velocity;
> And little whirls have lesser whirls,
> And so on to viscosity.

The distinction between these two regimes of flow has been recognized for a long time, as has the abrupt character of the transition between them. One can describe it no better than did Osborne Reynolds (1883), who introduced a filament of dye into a tube of flowing water and watched how it behaved as the current was altered (Figure 3.9):

When the velocities were sufficiently low, the streak of colour extended in a beautiful straight line across the tube. If the water in the tank had not quite settled to rest, at sufficiently low velocities, the streak would shift about the tube, but there was no appearance of

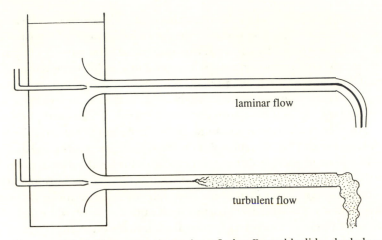

FIGURE 3.9. A diagrammatic version of what Reynolds did—the behavior of a stream of dye in laminar and turbulent pipe flow. The apparatus isn't entirely impractical. Flow in the upper pipe will be slower than that in the lower pipe, and the rates can be adjusted by changing the height of water in the tank so that the upper stream is laminar and the bottom one turbulent.

sinuosity. As the velocity was increased by small stages, at some point in the tube, always at a considerable distance from the trumpet or intake, the colour band would all at once mix up with the surrounding water. Any increase in the velocity caused the point of break-down to approach the trumpet, but with no velocities that were tried did it reach this. On viewing the tube by the light of an electric spark, the mass of colour resolved itself into a mass of more or less distinct curls showing eddies.

What both quotations should emphasize for the reader is that, just as one swallow never makes a summer, one vortex or even a few discrete vortices does not mean the flow is turbulent. Turbulence, again, consists of temporally and spatially irregular motion superimposed on the larger pattern of flow.

While engineers are almost exclusively concerned with turbulent flows, both kinds of flows are of biological interest, as are cases in which part but not all of a flow field is turbulent. In general, small, slow organisms and tiny pipes experience laminar flows, and large, fast organisms and large pipes experience turbulent flows. Reynolds is best known for deducing the basic rule governing the transition point for flow in pipes (see Chapter 5). At this point, I'm mainly concerned that the reader bear in mind that (1) two such regimes exist, (2) the transition is often abrupt, (3) the practical formulas for dealing with the two are quite different, and (4) much of biological consequence occurs near the transition point, where our a priori expectations are least reliable and where, as a result, nature has unusually rich opportunities to surprise us.

Pressure and Momentum

FORCE IS the basic currency of Newtonian mechanics—all three laws of motion are stated in terms of force—and force will play an important role here. For many purposes, though, force proves just a little too abstract. While the force of a blunt knife might be the same as that of a sharp one, the latter may penetrate where the former does not. What most often matters is the force divided by the area over which it's applied. That, though, can be expressed by either of two variables with the same dimensions and the same units, but which are most emphatically not the same. The simpler variable, conceptually, is *stress*, something we met in the form of shear stress when defining viscosity. It's especially useful in solid mechanics—push on an object, and a force per unit area, a stress, is exerted in the direction of the push. In other directions both force and stress are reduced by the usual rules for resolution of forces. Push lengthwise on a rod, and you exert a stress on whatever is in contact with the end of the rod—a stress equal to the force with which you push divided by the area of the rod's end. Any outward or radial force can ordinarily be ignored.

Fluids, of course, have no preferred shape—if you push on a (suitably contained) fluid, it tries to squidge out in every direction with equal urgency. Push on a piston in a cylinder (such as a hypodermic syringe or automotive shock absorber), and you not only exert a stress on the end of the cylinder but on the side walls as well. For such an omnidirectional response we drop the word "stress" and substitute *pressure*. You exert a pressure on the inner walls of the cylinder as well as on the inner walls of anything with which the fluid in it is continuous. Moreover, that pressure is the same everywhere (ignoring gravity)—the force you exert times the area of the face of the piston[1] on which you push, as in Figure 4.1a. Make a tiny hole anywhere in the system, and the fluid will squirt out with equal impetus.

It sounds like something for nothing—force ends up exerted, undiminished, over a very much greater area than that over which it was applied. But no conservation law has been violated since nothing has moved and thus no work has been done; the gain is no different from that brought with a lever or system of pulleys. You just have to get used to this

---

[1] Or, if the face isn't normal to the force, then the normal area of the face of the piston.

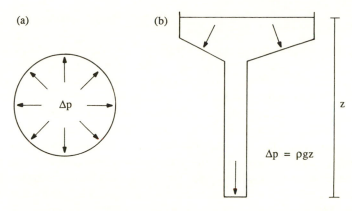

FIGURE 4.1. (a) Stationary fluids exert their pressures equally in all directions, which means that (b) the pressure on the bottom of a column of liquid depends on the height of the column but not on its shape. A wider and thus heavier column has more nonvertical surface somewhere that takes the extra force; pressure, force per area, is unimpressed.

odd omnidirectional character. It's at its queerest when one considers what's sometimes called the "hydrostatic paradox." A column of liquid (Figure 4.1b) of a specific height but of any shape at all will exert a specific pressure on its bottom or on any side near the bottom. That pressure will be determined by three variables only—the height, the density of the liquid, and the acceleration of gravity. Neither the shape of the column nor the total weight of the liquid matters. Odd—but that's the great advantage of using pressure as a variable. Were we to consider the downward force or stress of the column of liquid, the downward component of the force it exerts on all surfaces exposed to it, then we'd have to do some pretty complicated bookkeeping to account for all the orientations of all the bits of surface.

Thus if one dives beneath the surface of a body of water, the pressure rises in proportion to one's depth at a rate of about a tenth of an atmosphere per meter (more specifically, 9800 Pa m^{-1} in fresh water and 10,000 Pa m^{-1} in seawater). As with the enclosed column, only height or depth matters. And the pressure is exerted equally in all directions, even upward; so you aren't awkwardly accelerated in any direction as a consequence of that pressure. Similarly, the weight of the atmosphere above you, approximately equal to that of two elephants, presses on you equally in all directions and thus normally makes no trouble. These pressure are termed "static pressures" in general or "hydrostatic pressures" for the special case of diving.

BERNOULLI'S PRINCIPLE

But what about pressures caused by fluid motion, the sorts of pressures that blow trees over and keep birds aloft? These get a little more complicated; indeed in a sense they're what the rest of this book is about. The most convenient place to start is with a most peculiar notion, the principle (or equation) of Bernoulli, in particular of Daniel Bernoulli (1700–1782), one of several mathematical eminences of a single disputatious family. Recall that the principle of continuity was obtained solely from geometry and is a simple and general concept with no exceptions lurking about to trap the unwary. Bernoulli's principle may be the only bit of fluid mechanics that most of us were taught; it involves such disquieting assumptions that one can reasonably wonder if it is ever trustworthy.

The worst of these assumptions is that of an "ideal fluid," a really oxymoronic notion. An ideal fluid is the name given to a fluid of zero viscosity, and viscosity was the property used to recognize a fluid in the first place. The point is that, lacking viscosity, the fluid doesn't lose momentum to internal friction or to the walls of any container through which it flows.

To get a feeling for the others, it's perhaps best to go through a derivation of the principle.[2] Assume that an element of fluid moves through a length of pipe varying in internal diameter (S) and in height above the earth (z) , a pipe of the form shown in Figure 4.2. The volume of the element of fluid will be Sdl; its mass will be ρSdl. Reverting to differentials, the height difference between the ends of the region of pipe will be dz, and the pressure difference between the ends will be dp.

We apply Newton's second law, the rule that force equals mass times acceleration, to the element of fluid. In this case the overall force is the sum of any pressure force and any gravitational force; since increases in either pressure or height will slow the fluid, both forces must have minus signs:

$$- \text{ pressure force } - \text{ gravitational force } = \text{ mass } \times \text{ acceleration}$$

$$- S\,dp - \rho g S\,dz = \rho Sdl\,\frac{dU}{dt}\,.$$

If the terms are now divided by ρSdl, they then have dimensions of force/mass; dividing and rearranging, we get

$$\frac{dp}{\rho dl} + \frac{g\,dz}{dl} + \frac{dU}{dt} = 0.$$

[2] In the earlier edition I followed the practice of most physics textbooks and worked from conservation of energy. It's easier than what's done here, but I was informed by Stanley Corrsin that it wasn't quite legitimate, it didn't really expose the assumptions, and it jumbled history.

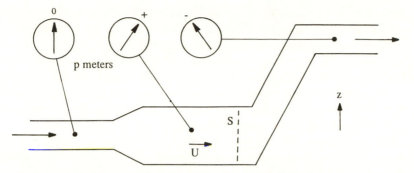

FIGURE 4.2. *Bernoulli's principle*. The decrease in velocity associated with an increase in the cross-sectional area of a pipe causes an increase in pressure (center). An increase in elevation of the pipe (right) causes a decrease in pressure.

This dt is awkward for a continuing process. To get rid of it we assume steady motion so $U = dl/dt$ and that $dt = dl/U$. Consequently,

$$\frac{dp}{\rho dl} + \frac{g\,dz}{dl} + \frac{U\,dU}{dl} = 0.$$

Now we want to integrate with respect to dl; to do that we have to assume that the density, ρ, is constant. We get

$$\frac{p}{\rho} + gz + \frac{U^2}{2} = \text{constant.} \tag{4.1}$$

The result is Bernoulli's equation. It states that the sum of a pressure term, a gravitational (or height) term, and a velocity term remains constant as the fluid flows through the pipe. The dimensions, though, are unhandy, so we multiply by density to get forces over areas, or pressure terms:

$$p + \frac{\rho U^2}{2} + \rho gz = \text{constant.} \tag{4.2}$$

Not only do these terms have dimensions of pressure, but they really do determine the magnitude of the explicit and ordinary pressure in the first term—the gauges in Figure 4.2 can be quite real items in lab or classroom. Decrease the second term by expanding the pipe (using of the principle of continuity) and the first will locally increase. Increase the third by elevating part of the pipe and the first will decrease there.

Bernoulli's equation in pressure terms is used to define some terms (in the verbal sense of "term.") The sum of the "static pressure" (p), the "dynamic pressure" ($\rho U^2/2$), and the "manometric height" (ρgz) remains constant as the fluid flows. *Static pressure* is the ordinary sort that we encountered when talking about hydrostatic pressures, the kind one measures on pressure gauges. *Dynamic pressure* is the pressure invested in movement of the fluid—if the fluid were suddenly brought to a halt without frictional, thermal, or other such shifts in who's-got-the-energy, then that pressure would appear as an increase in static pressure. *Manometric height* is height expressed in terms of pressure—the pressure exerted by a column of liquid of a specified density and height.

The constants in the previous equations are a minor nuisance. One way to circumvent them is to consider the changes in the magnitude of each term as the fluid moves from an upstream point (subscript 1) to a downstream point (subscript 2). We're rarely interested in absolute pressures anyway—how often do you have to use p rather than Δp? Bernoulli's equation then becomes

$$(p_2 - p_1) + \frac{\rho(U_2{}^2 - U_1{}^2)}{2} + \rho g(z_2 - z_1) = 0. \qquad (4.3)$$

And frequently the gravitational term can be omitted; even if it's ultimately relevant, one can often arrange a simplified situation in which the fluid has no net upward or downward motion, so

$$(p_2 - p_1) + \frac{\rho(U_2{}^2 - U_1{}^2)}{2} = 0. \qquad (4.4)$$

Incidentally, the figures that were used in Chapter 2 to argue for the practical incompressibility of air were obtained by using this last equation. One just takes the most extreme case of air with an initial velocity of U_2 brought to a halt so $U_1 = 0$ and calculates a Δp. That pressure difference is then compared to the atmospheric pressure, 101,000 Pa. An initial speed of 20 m s^{-1} would, for example, give a maximum local compression of about half of one percent.

MANOMETRY

To begin exploring how Bernoulli's principle can do useful service in a real world, let's consider some devices based on it, especially machines that can be used to measure speeds of flow. But as a start, assume a fluid at rest; clearly the second term in equation (4.3) will be zero, and

$$\Delta p = \rho g \Delta z. \qquad (4.5)$$

That equation might look familiar—it's the basic formula for column manometry, in which a difference in height (Δz) of ends of a manometric fluid of density ρ is used to measure differences in pressure, as in Figure 4.3a.

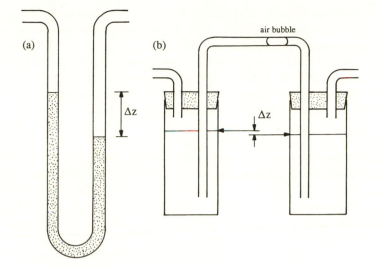

FIGURE 4.3. (a) A U-tube manometer for use with a liquid in air—the difference in height of the columns multiplied by liquid density and gravitational acceleration gives the port-to-port pressure difference. (b) A multiplier manometer, in which a small change in the heights of liquid in the two jars causes (by the principle of continuity) a very large movement of the air bubble in the top pipe.

This is worth remembering on another account as well, since this is the rule you use to get "real" units of pressure, proper forces per areas, from quaint quantities such as earthbound inches of water or mundane millimeters of mercury.

Even a bit of biology emerges here. Systolic pressure at one's heart has to be high enough to supply blood at some minimal pressure to the capillaries in an elevated head and brain, so the Δp at the aorta must be at least this minimal pressure plus the $\rho g \Delta z$ for the height difference between heart and head. (You might imagine that a syphon arrangement could circumvent any difficulty; but it turns out, perhaps because the system can't handle subambient pressures safely, that little if any syphoning takes place—see Hicks and Badeer 1989.) Most mammals have about the same average aortic blood pressure, about 13,000 Pa (100 mm Hg). Large ones, though, have higher pressures, with horses commonly about twice and giraffes almost three times that figure (Warren 1974). Because of the difficulty of supplying an elevated brain, one both expects and finds a constraint on circulatory design—the manometric height of an organism cannot exceed its systolic blood pressure. Manometric height is, of course, a pressure and uses the density of blood, which is about the same as water.

The rule is a somewhat crude one, with the fact that the heart is about halfway up from the ground roughly offsetting the minimum pressure needed at the brain. But it's an interesting constraint, a rule that the dimensionless ratio of manometric height to systolic blood pressure cannot ordinarily exceed unity. One wonders about whether terrestrial dinosaurs could possibly have had typical reptilian blood pressures of around 5,000 Pa. At the same time one understands why arboreal snakes have higher aortic pressures than other snakes and why their hearts are in an unusually anterior position (Lillywhite 1987).

As we saw when calculating the compression they cause in air, the pressure differences involved in low-speed flows are minuscule compared to very ordinary hydrostatic pressures. Both the common mercury sphygmomanometer that warns us of incipient hypertension and the aneroid barometer that warns us of incipient hurricanes prove hopelessly inadequate for the pressure differences resulting from small velocities. Equation (4.5), though, suggests two ways of achieving higher sensitivity. A liquid of low density may be used in the manometer—water, isooctane, or acetone instead of mercury. Or some way of reading very small height differences may be devised. Both schemes have been used in fluid mechanics, and the technology behind inclined-tube manometers, Chattock gauges, and their ilk is discussed in older sources such as Prandtl and Tietjens (1934) and Pankhurst and Holder (1952). The most direct solution is the use of a high-sensitivity electronic manometer, but such internally calibrating, computer-compatible instruments are not for the impecunious.

Still, direct manometry shouldn't be dismissed as anachronistic or unaffordable. For an investment of ten dollars or so, the device shown in Figure 4.3b provides sensitivity to pressure differences as low as a millionth of an atmosphere (0.1 Pa) even under field conditions. Using acetone (to minimize surface tension) in half-pint vacuum bottles (to minimize thermal volume changes of the air inside) and glass tubing of about 3 mm internal diameter, rise or fall of the air-liquid interfaces moves a bubble about 200 times as far. Thus a centimeter of travel of the bubble marks a 50 μm height change in the primary acetone manometer; by equation (4.5) that's a pressure difference of 0.386 Pa. In practice the instrument still runs into a little trouble from surface tension in the bottles and is best used either following specific calibration or in a null-balancing mode in which the apparatus is tilted to restore the bubble's position. With some practice, I got quite useful results from it (Vogel 1985).

For measuring pressures in water, one can take advantage of the fact that water-immiscible liquids can be made up with densities arbitrarily close to that of water. In manometry it isn't really the absolute density that matters (equation 4.8 is a bit oversimplified) but rather the difference between the densities of ambient fluid and manometer fluid. Only when the ambient fluid is air (or any gas) and the manometer is filled with liquid can we safely ignore the density of the former. The main difficulty in making simple liquid manometers for working with pressures in water is that rather wide-bore glass tubes must be used to keep interfacial effects within tolerable limits. Here again sensitivity to tiny pressure change is bought with a requirement for large volume changes—work must be done; there's no free lunch. By the way, these two-liquid manometers may be inverted to form what we might call Ω-tube instruments if it's handier to use a manometer fluid of lower density than that of the ambient water. Oh, yes—be sure to keep air out of such systems or you'll have accidental manometers turning up wherever least convenient.

MEASURING FLOWS WITH BERNOULLI-BASED DEVICES

Just as measuring a change in height can give a measure of change in pressure, measuring pressure can provide a measure of velocity. One just uses a different pair of terms in the Bernoulli equation. And velocity is something we very much need to measure if we're to work with flowing fluids. If the difference between the heights of upstream and downstream points is negligible (or is accounted for elsewhere), then equation (4.4) can be applied:

$$\Delta p = \frac{\rho}{2} (U_1{}^2 - U_2{}^2).$$

Two practical problems become immediately evident. The equation doesn't use velocities per se but the *difference* between velocities. And the pressure change is proportional to the difference in the *squares* of the velocities. The former requires that we worry about two locations, not just one; the latter generates great problems with sensitivity at low speeds, the reason I pressed the issue of high-sensitivity pressure-measuring devices.

Venturi Meters

One very useful instrument makes use of the principle of continuity to circumvent (partly) the differential measurement implied in equation (4.4). It's called a "Venturi meter" and consists of a contraction of known size in a pipe, which locally speeds the flow to U_2 (Figure 4.4). Since S_1 and S_2 are known, equations (3.1) and (4.4) can be combined to eliminate U_2:

$$\Delta p = \frac{\rho U_1{}^2}{2} \left(1 - \frac{S_1{}^2}{S_2{}^2} \right). \tag{4.6}$$

A Venturi meter is worth remembering as a cheap substitute for an expensive in-line volume flow meter. I've made considerable use of one

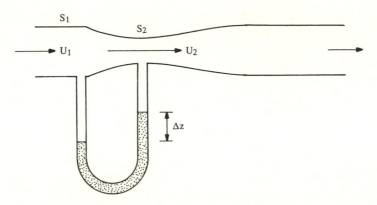

FIGURE 4.4. A Venturi meter, in which a contraction to a predetermined cross-sectional area speeds the flow, lowers the pressure, and thus permits determination of velocity or volume flow in the uncontracted pipe.

built originally just as a demonstration device. At modest speeds even a crudely machined annulus glued into an ordinary pipe seems to give decent results. What's especially important with a crude one is that the manometric port from the uncontracted pipe be upstream from the contraction. Note that no arbitrary constants in equation (4.6) mandate calibration. As we'll see, several organisms use arrangements quite similar to Venturi meters. Laboratory aspirators and automobile carburetors work on the same principle as well; the main difference from a meter in all of these applications is that gas or liquid flows through the bypass that would be occupied by stationary manometer fluid.

Pitot Tubes

Most often the biologist is interested in measuring flow at a single point in an open field of flow rather than across some closed conduit inserted into a system of plumbing; for such point measurements the Venturi meter is nearly useless. How can we contrive an adequate reference velocity in order to apply Bernoulli's equation? The trick is one already mentioned—the dynamic pressure appears as a manometrically measurable pressure if the fluid is suddenly brought to a halt. It then need only be compared with the static pressure that characterizes the local unobstructed flow. And "compare," as with the Venturi meter, implies no more than connection to opposite ends of a manometer.

The device for suddenly bringing a moving fluid to a halt is called a "Pitot tube" or "Pitot-static tube" (Figure 4.5); to it applies a minor variant of the Bernoulli equation,

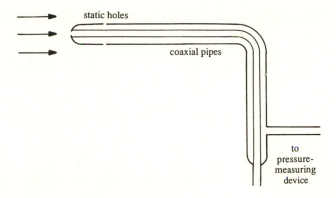

FIGURE 4.5 A Pitot-static tube, in longitudinal (sagittal) section and much fatter than life. One aperture faces upstream, while a ring of small "static holes" are parallel to the flow. If a static hole on a remote surface is used instead of these, the device is simply called a "Pitot tube."

$$\frac{\rho U^2}{2} + p = H,\qquad\qquad\qquad (4.7)$$

where H has the amusing name of "total head" and represents the sum of dynamic and static pressures, the two terms on the left side of the equation. The aperture facing upstream is designed so that it samples fluid locally brought to rest, so it's exposed to both dynamic and static pressures and thus to total head. The "static hole" (or holes), whether adjacent or remote, is exposed only to the local static pressure. The Δp of a manometer located between the upstream-facing and static holes therefore indicates $(H - p)$, or the dynamic pressure. The applicable equation is simplicity itself:

$$\Delta p = \frac{\rho U^2}{2}.\qquad\qquad\qquad (4.8)$$

To use a Pitot tube together with an ordinary liquid-filled manometer, you need only combine equations (4.5) and (4.8). But it's worth remembering that you're dealing with two different densities, that of the ambient fluid (here ρ_a) in equation (4.8) and that of the manometer fluid (here ρ_m) less the ambient fluid in equation (4.5):

$$U^2 = \frac{2g\Delta z(\rho_m - \rho_a)}{\rho_a}.\qquad\qquad\qquad (4.9)$$

The portable version, the Pitot-static tube, is inexpensive and rugged. Much effort has gone into the design of the upstream end to reduce its sensitivity to minor misalignments with respect to the current direction and to determine the best size and location for the static holes. Neverthe-

59

less, for the biologist interested in low-speed flows it has a serious problem of insensitivity; again that *square* of velocity in the Bernoulli equation is trouble. To read a Pitot tube to $0.1 \, m \, s^{-1}$ instead of $1 \, m \, s^{-1}$ requires a 100-fold greater sensitivity of the manometer. Using the acetone-multiplier manometer described earlier, a one-centimeter deflection corresponds to an airflow of about a meter per second, or two miles per hour. For water flow, even using a two-liquid manometer with a density difference of a tenth the density of water and reading it to tenth of a millimeter, the detectable current is still no less than $14 \, mm \, s^{-1}$. In short, the scheme is workable, but only with some manometric audacity. The advantages gained are freedom from any external source of power and from the necessity of a known flow for calibration.

The Pitot Tubes of Life

Quite a few organisms make use of devices that approximate Pitot tubes, although (as with the Venturi-meter analogs) the pressure difference is used to drive some secondary flow rather than to press on a static column of liquid. Wallace and Merritt (1980) note with evident amusement that the larva of *Macronema*, a lotic hydropsychid caddisfly, constructs a Pitot tube in the middle of which it spins a catch net (Figure 4.6a). One opening faces upstream and is exposed to almost full static plus dynamic pressure (a little of the pressure is relieved as a result of flow through the structure or, to put it another way, flow isn't quite halted in front). The other opening is normal to the current and must experience very nearly ambient static pressure. An ascidian, *Styela montereyensis*, has to be a bit fancier, since it lives in shallow seawater where the direction of flow changes periodically. As shown in Figure 4.6b, it attaches itself to the substratum by a flexible stalk and can passively reorient, as does a weathervane, to keep the incurrent aperture facing the flow (Young and Braithwaite 1980). It's likely that plesiosaurs did something similar in aid of olfaction. According to Cruickshank et al. (1991), their palates had scoop-shaped internal nares leading through short ducts to flush-mounted, external nares on the outside of the head. The internal nares strongly resemble nonprotruding air-intake ducts used on aircraft in which a groove deepens in the direction of flow until it ends in a transverse scoop. The most widespread and perhaps dramatic use of such a device is in what's called "ram ventilation" in fishes that swim with their mouths open; we'll return to it shortly.

LIMITATIONS AND PRECAUTIONS

At the start of the chapter, I made some derogatory remarks about Bernoulli's principle; by this point the reader may be wondering if these

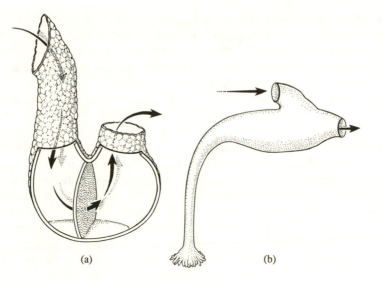

(a)

(b)

FIGURE 4.6. Two natural Pitot tubes: (a) the case and catch-net of the larval caddisfly, *Macronema*; (b) the incurrent and excurrent siphons of the ascidian, *Styela montereyensis*.

were just gratuitous curmudgeonly mutterings. The principle does surprisingly good service on many occasions, but it's still dangerous. Don't forget the assumption of zero viscosity (quite aside from the steady, incompressible flow assumed in our version of Bernoulli). Ultimately energy must be conserved, and in fluids viscosity provides a major route for energy to leave the mechanical domain. Bernoulli's principle gets less and less reliable as the scale of speed and size go down, quite aside from any issue of mensurational sensitivity—for slow, small-scale flows viscosity is particularly significant. It should be applied only along a streamline or within a streamtube, and even there only when pressure taps are close together and shear rates are low. The principle is especially unsafe for points along any traverse normal to the direction of flow, and it's especially perilous in the velocity gradients near walls. After all, these velocity gradients are *caused* by viscosity. Since flow speed drops off near a wall as a consequence of viscosity and the no-slip condition, static pressure rather than total head remains constant there. Speed may drop, but energy is converted to heat in the shear rather than being converted to pressure potential.

I make this point about the nonconstancy of total head with some passion, since I once narrowly escaped misattributing a set of interfacial phenomena entirely to Bernoulli's principle. By the way, the heat from shear is quite real. Some years ago I encountered an annoying problem of drift of a

61

flowmeter being used in a new flow tank. It turned out (after false leads were chased and the investigator was chastened) that the one horsepower being transferred to the water by a pump was warming the water by several degrees per hour. I ended up crudely confirming James Joule's figure for the mechanical equivalent of heat!

Perhaps the worst abuses of Bernoulli occur when people apply it to circulatory systems. In these totally bounded systems pressure drops almost entirely as a result of shear stresses. Flow in capillaries is several orders of magnitude slower than in the aorta, but pressure falls rather than rises as blood flows from latter to former. Textbooks of physiology commonly mention Bernoulli's principle at the start of their sections on circulation, but most of them (fortunately) never apply it to any specific situation. It does have some relevance to the operation of heart valves (see Caro et al. 1978), and it's not negligible for certain pathologies such as local aneurysms (dilations) or coarctations and stenoses (constrictions) of the aorta and larger arteries (Milnor 1990; Engvall et al. 1991). More on this in Chapter 14.

Pressure Coefficients and Pressure Distributions

Recall that dynamic pressure appeared as a measurable pressure when fluid was suddenly brought to a halt. That will occur at some point on the upstream side of any object facing a flow, not just a Pitot tube. This consistent behavior permits us to define a rather handy dimensionless variable. Consider a set of pressure measurements taken at a sequence of locations from upstream to downstream extremities of some solid object in a flow; each is referred to a small static hole in an adjacent flat plate oriented parallel to flow (Figure 4.7). The pressure differences will vary widely, not only with the shape of the object but with such things as the speed of flow and the properties of the fluid. But somewhere up front the measured pressure difference will be exactly the dynamic pressure, and the latter can be calculated from the free stream speed as well as measured. *If all the measured pressure differences are divided by that dynamic pressure, then any graph of pressure versus location will begin upstream at a value of 1.0.*

That's very nice—a measure of pressure that corrects for speed and for at least one of the fluid properties, density. It even puts cases of airflow and water flow on axes with the same scales. One sees, relatively uncontaminated, the effects of shape; and one can always undo the division, multiplying by dynamic pressure to restore pressure dimensions and the original measurements. This dimensionless pressure is called the "pressure coefficient," C_p:

$$C_p = \frac{2\Delta p}{\rho U^2}. \qquad (4.10)$$

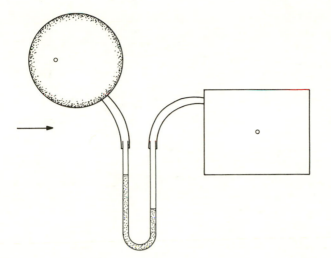

FIGURE 4.7. How to determine the way pressure varies over the surface of a body exposed to a flow. Pressure differences are measured between each of a series of tiny holes (only one is shown) and a static hole well behind the upstream edge of a flat plate oriented parallel to the flow.

Let's use this new variable to look at how pressure varies around some simple shapes. Consider, first, the pattern of streamlines around a low-drag ("streamlined") object (Figure 4.8a). Notice how the streamlines bunch together near the widest part of the body; by the principle of continuity we know that the speed of flow ought to be greatest there. By Bernoulli's principle, we suspect that the pressure ought to be lowest, which in fact is very nearly what occurs (Figure 4.8b). The graph turns out to be nicely general, with only minor differences between the toy water rocket I used (to get something of biologically average size) (Vogel 1988b) and a giant airship (Durand 1936).

The underlying pressures are, of course, forces per unit area; more particularly, they are forces perpendicular to the surface per unit area. Calculating drag from such pressure data (adjusted for orientation of each element of surface) omits what is often a significant component. The kind of drag that results directly from the effect of viscosity, the kind that we used in Chapter 2 as a measure of viscosity, has not been taken into account, and the closer one gets to a flat plate parallel to flow or to perfect streamlining the more important this component gets.

Using ambient pressure as a baseline, as is usually appropriate, a positive pressure coefficient represents a net inward pressure, and a negative pressure coefficient represents a net outward pressure. These inward and out-

(a)

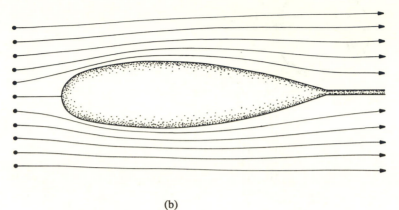

(b)

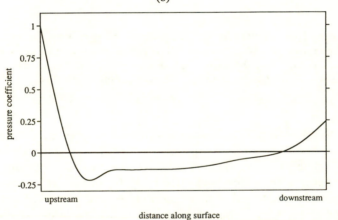

FIGURE 4.8. The distribution of pressure on the surface of an object that has a low drag relative to its size as determined by an apparatus such as that of 4.7. The data were obtained using a toy water rocket with its fins removed, about 30 mm in diameter, in an airflow of 10 m s^{-1}.

ward pressures are perhaps of equal consequence to the drag-related forward and rearward ones, and I'll take a few pages to talk about some of the biology that attaches to them.

PRESSURE DISTRIBUTION AND LIFT

Imagine that the streamlined object in Figure 4.8 is sliced longitudinally in a horizontal plane (a "frontal section" for classically trained zoologists) and the lower half is discarded. Pressure coefficients on top will remain negative, that is, pressure will be outward and upward, while pressure coefficients below will be nearer zero, or ambient. Let's further imagine that we prevent relief of the resulting pressure difference by flow from bottom to top by merely placing the bottom on the substratum. Aha—we've got lift, which is simply to say that pressure above is less than pressure below. And the lift comes from the operation of Bernoulli's principle. Lift-producing airfoils will get the full attention of Chapters 11 and 12. For now let's look just at several situations in which organisms get lift because they form protrusions from a flat substratum over which water flows.

How the Plaice Stays in Place

Plaice (*Pleuronectes platessa*) are bottom-living flatfish similar to flounder. At rest they constitute low, rounded humps on smooth, sandy bottoms. As convex elevations, they experience lift by Bernoulli's principle; and lift isn't exactly the blessing the term usually implies. Despite its fine low-drag shape, a quiescent plaice has some tendency to slip downstream when exposed to flow. This tendency is offset by its friction with the bottom and its submerged weight; it's in fact an especially dense fish. But the weight of a plaice is reduced by its lift, and its lift is ten to twenty times its drag. So slippage is more a matter of lift than drag, since lift reduces purchase on the bottom (Figure 4.9a). In practice it remains quiescent in currents up to a "slip speed" of 0.2 m s^{-1} and beats its posterior median fins in stronger currents—up to a "lift-off speed" of around 0.5 m s^{-1}. Above that speed its net weight is zero or less, and it must either dig in or take off. Rays, which are essentially dorso-ventrally flattened sharks, behave in much the same way (Arnold and Weihs 1978; Webb 1989).

A Slotted Sand Dollar

A sand dollar resting on a sandy bottom presents a hump similar to that of plaice or ray. As a result it also experiences lift and must dig in if faced with currents above a critical value. Several species have slots—"lunules"—of varying numbers that run radially and connect upper (aboral) and lower (oral) surfaces. In a current, water is drawn up through the slots by the reduced pressure on top (Figure 4.9b). Telford (1983) has shown that the presence of slots in *Mellita quinquiesperforata* reduces lift sufficiently to raise the dislodgement speed by about 20%. Again, lift seems to be more of a problem than drag. Small individuals have lower critical speeds than large ones, and young sand dollars select especially dense sand grains and use them as ballast in a kind of weight belt (Telford and Mooi 1987). The slots may also function in feeding, with the upward flow through them helping to draw food-laden water up from the substratum (Alexander and Ghiold 1980).

Feeding by drawing water upward with a mound-shaped body has also

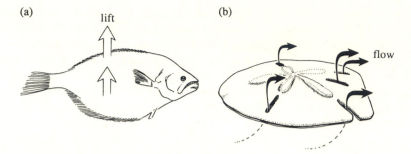

FIGURE 4.9. (a) A plaice or flounder, lying on sand and exposed to flow, develops lift as a result of the operation of Bernoulli's principle.
(b) The same pressure difference draws water up through the holes in a perforate sand dollar, so it suffers somewhat less lift; the water movement may be useful for feeding as well.

been suggested for some other kinds of organisms—for sclerosponges, a relatively poorly known group of Porifera, and for stromatoporoids, a group of uncertain affinities best known from Devonian deposits (Boyajian and LaBarbera 1987). On the upper surfaces of the mounds the sclerosponges have closed canals and the stromatoporoids have grooves running radially from apices to valleys. In a flow, water is drawn from valleys to apices by the lower pressure at the apices; stromatoporoids (assuming the grooves were open in life) could have entrained fluid and directed it upward from anywhere along their surfaces.

More Hoisting and Heisting

The fish and sand dollars just mentioned are motile creatures that live on shifty bottoms. Even for sessile organisms on hard substrata, lift as a consequence of protrusion can present serious problems. Mussels often live in environments subjected to the very high velocities associated with breaking waves, and they're attached with impressive tenacity by their byssus threads. But Denny (1987a) measured pressures immediately above and within a mussel bed in a location whose extremes of wave action and disturbance history he already knew quite a lot about. He showed that the predicted lift maxima were adequate to account for initiating the bare patches that occurred from time to time—by contrast neither drag nor the unsteady forces associated with the waves were seriously disruptive.

A similar problem of lift faces limpets, and Denny (1989) ran into an amusing and instructive case when looking at the hydrodynamic behavior of a bunch of their shells. One shell turned out to have an anomalously low drag, undoubtedly associated with some aspect of its shape or surface sculpture. If one shell could do the trick, why hadn't evolution seized upon the possibility and equipped the other limpets with the facility? Denny argues, and I'm certainly persuaded, that drag simply doesn't matter, that no selective advantage attaches to a low-drag shell, that lift is what dislodges limpets, and since the anomalous shell was in no way unusual in the lift it suffered, the limpet that made it was neither more nor less fit. I gather that the particular shell is carefully preserved—though one hesitates to call it a valuable artifact.

The same problem of lift probably afflicts freshwater organisms as well, especially those that have flat bottoms attached to hard substrata and form protruding mounds—water penny beetles (Coleoptera: Psephenidae), for instance (see Smith and Dartnall 1980). The underlying problem seems to be a geometric one. A shape that avoids drag by hugging a surface and has a large area for attachment will almost automatically develop substantial lift.

Pressure Distributions around Flexible Organisms

Most objects with which our technology deals are fairly rigid. Thus shape is constant no matter what the flow. Organisms are more commonly flexible; as a general rule nature is stiff only for particular ends. The consequence is a vast increase in the complexity of interactions between organisms and flow. Not only do the forces on a living object in a flow depend on its shape, but its shape in turn depends on the forces it experiences, as first pointed out (I believe) by Koehl (1977) for sea anemones. On one hand the added complexity may seem daunting; on the other it gives nature a very powerful additional variable with which to work her adaptive tricks. The point will arise again in chapters to come, particularly when we talk about drag. Here we'll just consider some direct effects of pressure distributions such as that shown in Figure 4.8b.

The Form of Fishes

Consider a fish swimming rapidly through water. The pressure on the head, or at least on its forward part, will be above ambient, that is, inward; the pressure farther back will be subambient, or outward; DuBois et al. (1974) got about the expected results from 0.6 m-long bluefish moving at a little under 2 m s^{-1} (Figure 4.10) through which they'd run catheter tubing to transmit pressure. Quite a lot of fish morphology and some behaviors are at least consistent with this pressure distribution. For instance, each of the fish's eyes is located at a point where the pressure coefficient crosses ambient pressure. At such a location, uniquely, an eye will be neither drawn out from the head by reduced pressure or pressed into the head by increased pressure as the animal swims more rapidly; only there will the

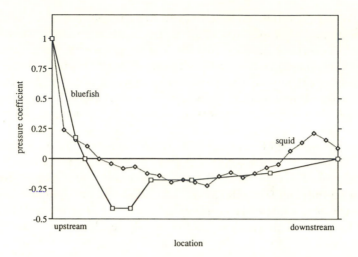

FIGURE 4.10. Pressure distributions (expressed as pressure coefficients) around a bluefish (DuBois et al. 1974) and a model of a squid (Vogel 1987).

pressure not be a function of swimming speed. Where the fish experiences inward pressures, it has a compression-resistant skull, analogous perhaps to the reinforcing battens that are built into the anterior region of every nonrigid airship. Farther back, where net pressure is outward, a flexible but minimally extensible skin is adequate to maintain normal shape. Interestingly, an increase in intracranial pressure in a bluefish, as would happen in rapid swimming, triggers an increase in heartbeat and blood pressure (Fox et al. 1990)—a really reasonable reflex.

Most fish ventilate their gills with water that enters through the mouth up front and leaves through the operculum, located near the point of maximum thickness. While one certainly cannot prove that natural selection has been at work, it's certainly a reasonable arrangement to take in water where pressure is most strongly inward and to eject water where it's most strongly outward. Ram ventilation—augmentation of ventilation by the motion of fish—has been known for many years. For some pelagic fishes such as mackerel it's mandatory—if they stop swimming they suffocate (Randall and Daxboeck 1984). Other fishes such as trout make a sharp behavioral switch at some critical speed from active branchial pumping to ram ventilation, with the shift speed varying inversely with the local oxygen concentration. The phenomenon of switching permits comparison of the cost of the two modes; interestingly, the 10% drop in oxygen consumption attending the switch exceeds the estimated cost of pumping. The suggested explanation, which seems reasonable, is that a fish achieves a slightly lower drag when doing ram ventilation—the pressure-induced flow out the operculum in some way gives hydrodynamically superior flow along the body (Steffensen 1985).

Refilling Squid

A squid jets rapidly by contracting its circumferential mantle muscles and forcing water out a nozzle, as we'll discuss in just a few pages. It refills between jet pulses by drawing in water through a pair of valves on either side of the head. Part of the inward pumping is provided by a layer of short, radial muscle fibers that act as mantle thinners; additional pumping is provided by elastic recoil from the previous contraction. But once under way it seems to use flow-induced pressures to augment the refilling (Figure 4.10). Most of the mantle is located where, according to measurements on models, the pressure is most strongly outward, while the intake valves are located where the the pressure is close to or a little above ambient. So flow will draw the mantle outward and draw water into the mantle cavity. The mechanism may provide over half the refilling pressure for a rapidly moving squid (Vogel 1987). Still, energy saving of the sort we encountered in ram ventilation isn't really available here—outward pressure simply means that more squeeze is needed when the mantle is contracting. The benefit is more likely to be one of quicker refilling and thus shorter pulse cycles and more rapid bursts of swimming.

Engulfing Whales

The biggest users of flow-induced pressures are certainly fin whales, investigated by Orton and Brodie (1987). These rorquals feed as they swim along, first engulfing very large volumes of seawater plus comestibles and then forcing the water through their baleen plates and out near the hinges of their jaws. During engulfing the anterior half of a whale enlarges greatly, mainly by stretching the longitudinally grooved ventral surface of the throat. Whales are well streamlined and ought to have pressure coefficients approximating those in Figure 4.8b. Using those numbers and data from throat-stretching tests, Orton and Brodie calculated that the combination of the dynamic pressure at the mouth's opening and the subambient pressure outside the throat should be just sufficient to expand a whale's buccal cavity if it swims at a little under 3 m s^{-1}, as in Figure 4.11a. Which, perhaps not coincidentally, is about what's been estimated for the speed of feeding fin whales.

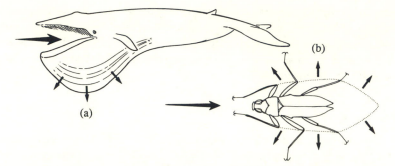

FIGURE 4.11. Putting flow-induced pressures to work to (a) draw water into the mouth of a cruising and feeding fin whale, and (b) maintain an oxygen-extracting bubble around an attached beetle in a shallow, torrential stream.

A Beetle's Bubble

The air-breathing adults of an elmid beetle, *Potamodytes tuberosus*, inhabit rapidly flowing streams in Ghana. These beetles congregate on submerged rock surfaces within a few centimeters of the surface of the water. Most of each beetle is enclosed in a large air bubble supported by the thorax and forelegs and extending a centimeter or two beyond the end of the abdomen (Figure 4.11b). The bubble appears to act as a physical gill, but it's permanent only in moving water saturated with air at atmospheric pressure. In the laboratory, a beetle can maintain a bubble of normal size only when it's near the surface in rapidly flowing, well-aerated water; otherwise the bubble shrinks and the beetle dies (Stride 1955). What happens again reflects the distribution of pressure of Figure 4.8 (with a dash of 5.2). Pressure in the bubble is an average of the pressures from upstream to downstream extremities, so the overall pressure inside is subambient. Thus dissolved gas will diffuse inward, maintaining the bubble—at least if the pressure is subatmospheric as well as subambient. This latter requirement limits this curious adaptation to very shallow and very rapid water: the greater the depth, the greater must be the current needed to counterbalance hydrostatic pressure. Stride showed that at a depth of only a centimeter it took a current of over about 0.8 m s^{-1} to get subatmospheric air pressure inside a bubble. That's pretty shallow and pretty rapid, which explains why the scheme is so rare.

FORCING FURTHER FLOWS

When talking about Venturi tubes, I mentioned that organisms used the same geometric arrangement but commonly used the pressure differences to pump some of the ambient fluid through an open pipe that took the place of the manometer. Basically, flow across the opening of a pipe is used to draw fluid out that opening, with the fluid in the pipe ultimately replaced from another opening less exposed to flow—taking advantage of the drop in pressure coefficient when flow is more rapid, as through the constriction in a Venturi tube (Figure 4.4) or across the wide part of an object (Figure 4.8). Perhaps the easiest way to envision the system is simply to shear off the top half of the main pipe in Figure 4.4, treating it now as an open flow across a substratum that happens to include a bump (Figure 4.12a) as we did with lift-producing protrusions. The direction of flow of the air or water isn't crucial—a difference in speeds is really what matters. In effect, an organism living at a solid-fluid interface can use the energy of the flowing external medium to drive a flow through itself or its domicile (Vogel and Bretz 1972).

This scheme for flow induction helps prairie dogs ventilate their burrows (Vogel et al. 1973) or use flow through the burrows for olfactory assistance; the animals appear deliberately to maintain sufficient asymmetry between openings to develop adequate pressure differences. Open-country African termites use it to help ventilate their mounds, with intake openings around the base and exhaust holes near the apex (Lüscher 1961; Weir 1973). It reduces the cost of filtering water in sponges when, as is usual, they live in currents; as in the termite mounds, the input openings (ostia) are peripheral and the output openings (oscula) are commonly apical or elevated (Vogel 1977). It aids airflow through the thoracic tracheae of some large insects in flight (Stride 1958; Miller 1966), using not only the forward speed of flight but the prop-wash of the beating wings. To make best use of it, suspension-feeding articulate brachiopods orient themselves in currents (LaBarbera 1977, 1981). And it's essentially what's happening in the feeding sand dollars, the sclerosponges, and the stromatoporoids mentioned earlier.

According to Armstrong et al. (1992) the system is used to supply oxygen to the under-marsh rhizomes of the common reed, *Phragmites australis*. Wind across tall, dead, broken-off culms (flower-and-seed stalks) draws gas out of their passageways, which in turn draws gas from the rhizomes (interconnecting roots), which in turn draws air into the low leaves. Oxygen diffusion out of rhizomes is considered desirable as part of good management of wetlands; evidently Mr. Bernoulli is put to use in getting the oxygen down there. (At least one other mechanism of bulk flow is at work in reeds; in general, plants seem far less limited to diffusion for gas transport than has been traditionally assumed.)

If we're invoking a purely physical pumping system and not even demanding moving parts, then we're talking about a scheme so simple that it might crop up even without natural selection. Webb and Theodor (1972) showed that seawater moving across ripples of sand at a depth of about 3 m

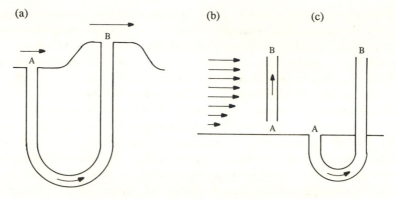

FIGURE 4.12. (a) If a primary flow speeds up when crossing an obstruction, reduction in pressure caused by Bernoulli's principle can drive a secondary flow from A to B. (b) If a small pipe is mounted normal to the free stream in the velocity gradient near a surface, greater viscous entrainment at B will draw fluid from A to B. (c) If the lower opening of the small pipe is looped under the substratum, the arrangement gets a little like that of (a), and the two physical agencies get a bit tangled.

induces water movement in and out of the sand—out at the crests of the ripples and inward in the troughs or the toes of the slopes. This flow induction has effects on sediment chemistry through its presence in relict burrows (Ray and Aller 1985). It's involved in a phenomenon called "breathing" in volcano cones (Woodcock 1987). And it's widely used in the design of houses and storage structures, ancient and modern, in which local airflow is used to provide ventilation (Dick 1950; Bahadori 1978).

In most of these cases, though, Bernoulli's principle isn't the complete extent of the relevant physics—viscosity plays a part as well. As mentioned earlier, pressure rather than total head remains constant in the velocity gradients near surfaces. If a tube is placed transversely in a channel with one end near a wall (Figure 4.12b), then viscous action will move fluid from wall to free stream. Viscosity, you'll recall, amounts to resistance to rapid shear rates. At the end of the tube away from the wall, the velocity and thus the shear rate and thus shear stress will be greater. In response to the shear stress, fluid will move out of the tube and into the channel. At an equal rate, other fluid must move into the tube at the wall end. The phenomenon, rarely of interest to fluid mechanists, is termed "viscous entrainment" (Prandtl and Tietjens 1934). It's one of the reasons why the static holes in Pitot tubes must be small (Shaw 1960).

The difficulty of dissecting the physical mechanism of a given induced flow is evident if the tube in Figure 4.12b is looped around as in 4.12c; the result looks all too similar to 4.12a! In actuality, some mix of the two mechanisms probably always operates, and the nature of the mix is not of overwhelming concern to the biologist (Vogel 1976).

MOMENTUM

Newton quite clearly recognized something basic about a concept of "quantity of motion," and he saw that it was proportional to neither mass nor velocity separately but to their product. To this product of mass and velocity we give the name *momentum*. Newton's second law, of which we've already made use, is commonly stated as an equality between force and the product of mass and acceleration, that is, of mass and the rate of change of velocity. But for present purposes the law is better regarded as equating force and the rate of change of momentum.

Every elementary physics book notes that in all collisions between bodies, momentum is conserved. Energy may be conserved as well, but only in perfectly elastic collisions can we neglect the conversion of mechanical to thermal energy. Conservation of momentum is thus the more potent generalization for purely mechanical problems. To apply energy conservation and stay within our purely mechanical domain we have to presume the tacitly self-contradictory notion of an ideal fluid. By contrast, the use of momentum conservation involves no such risky simplification.

In our macroscopic fluid mechanics, discrete collisions are not usually of much consequence. No matter—momentum is still conserved, and if the fluid is isolated, the total momentum will be constant in both magnitude and direction. But a moving fluid is rarely isolated—it passes walls, obstacles, pumps, and so forth. One can with great confidence state that any change in the momentum of such a nonisolated flow must reflect a forceful relationship with its surroundings. Thus, returning to Newton's second law,

$$F = \frac{d(mU)}{dt} = \frac{mdU}{dt}. \tag{4.11}$$

The last term is the simpler version; but its presumption of constant mass is not always appropriate, for instance for jet propulsion.

The main trouble with the equation is that it's stated in terms of mass and time, neither of them handy quantities when dealing with steady flows of fluids. We can do better by considering what must go on within a stream tube. How fast does mass move through the tube? If we consider uniform flow by taking a small enough tube so transverse velocity differences are trivial, mass flux (m/t) will equal the product of the density of the fluid, the cross-sectional area of the tube, and the velocity of flow—which is the same as the product of density and volume flow rate:

$$\frac{m}{t} = \rho SU = \rho Q.$$

Similarly, momentum flux is obtained by multiplying both sides by velocity:

$$\frac{mU}{t} = \rho SU^2.$$

(This combination of variables, ρSU^2, will occur repeatedly in chapters to come.)

Now we can state an equation, sometimes called the "momentum equation," that puts into a practical form the rule relating force and change in momentum. In terms applicable to a stream tube with entry area dS_1 and exit area dS_2, assuming (as usual) steady flow and assuming that everything of interest goes on in the x-direction,

$$dF = \rho dS_1 U_1{}^2 - \rho dS_2 U_2{}^2. \tag{4.12}$$

To get the force F from dF, one just has to add up the contributions of all the relevant streamtubes that thread their way across the field of flow. The equation (or slightly more complicated versions) can be applied to a variety of problems, including such things as the force exerted on a pipe as fluid passing through it goes around a bend and the force exerted on a surface as a fluid is squirted at it.

We can use this statement of Newton's second law to define some forces. *Drag* becomes the rate of removal of momentum from a flowing fluid. *Thrust* is then the rate of addition of momentum to a stream of fluid. And *lift* is the rate of creation of a component of momentum normal to the flow of the undisturbed stream.[3] These definitions, of course, say absolutely nothing about the physical origin of forces such as drag. Drag, that most awkward subject, will be tackled in the next chapter.

INDIRECT FORCE MEASUREMENTS

But even without really understanding how it happens, we can still talk about this force we're calling drag. Despite the disparaging remarks in connection with displacement of plaice and lifting of limpets, drag is a variable of substantial adaptive significance. Due to the irregular shapes and flexibility of organisms, it can only rarely be calculated or looked up in tables and so most often must be measured. Such measurement, though, may not always be an easy matter under anything approaching normal conditions. Not all creatures can with impunity be mounted with instru-

---

[3] Note that lift is *not* defined as upward with respect to the surface of any planetary body.

ments in flow tank or wind tunnel. But our formal definition of drag, when combined with the notions of continuity and streamlines, can provide an alternative. Working out a procedural formula will provide a little introduction to the use of the momentum equation.

By Newton's third law, if a body exerts a force on another, then the second body exerts a force on the first of equal magnitude and of opposite direction. Thus *if the moving fluid exerts a drag on a body, then the body must remove momentum from the fluid at a rate that just balances its drag*. If this rate of removal of momentum is measurable, then it's possible to get the drag of an organism without detaching or even touching it! The procedure can be a bit laborious and cannot be reasonably expected to give data of better than 5% accuracy, but sometimes nothing else will work.

Consider an attached object subjected to a flow (Figure 4.13), an object that (for simplicity) has its bulk concentrated well away from the substratum, so it's not in a velocity gradient. Consider, as well, two imaginary parallel planes, one upstream (S_1) and one downstream (S_2) from the object, each sufficiently far from the object that flow is normal to its surface. Consider, finally, a stream tube that leads fluid from part of the upstream plane (dS_1) to a corresponding part of the downstream plane (dS_2). Across the upstream plane the velocity (U_1) is uniform, while across the downstream plane the velocity (U_2) varies from a minimal value in the center of the wake of the object. The momentum flux across dS_1 will be, from equation (4.12), $\rho U_1{}^2 dS_1$, and that across dS_2 will be $\rho U_2{}^2 dS_2$. The difference between these fluxes will be the part of the drag acting on that stream tube, so,

$$dD = \rho dS_1 U_1{}^2 - \rho dS_2 U_2{}^2. \hspace{2cm} \text{(as 4.12)}$$

By continuity, though,

$$dS_1 U_1 = dS_2 U_2.$$

Combining these equations to get rid of dS_1 gives

$$dD = \rho U_1 U_2 dS_2 - \rho U_2{}^2 dS_2.$$

All that remains is integrating over the whole plane, S_2. The double integral just means that we have two dimensions to worry about, that $dS_2 = dy_2 dz_2$:

$$D = \rho \int \int U_2 (U_1 - U_2) dS_2. \hspace{2cm} (4.13)$$

While this last equation may look complex, the complexity is more apparent than real. Only rarely can we solve it explicitly; instead, it functions as the procedural formula we've been seeking. It tells us what we have to do to measure drag indirectly. First, we have to measure the velocity, U_2, at a

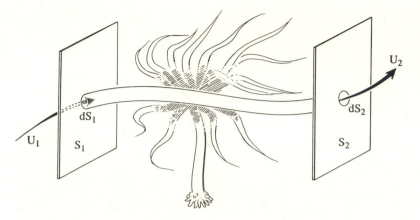

FIGURE 4.13. An attached object removes momentum from a stream tube that passes it. Its drag is the rate at which the object removes momentum from all the nearby stream tubes.

set of points on plane S_2. Naturally it's only necessary to do so in the wake of the object—where U_1 and U_2 are appreciably different—so the limits of the plane of integration present no problems. Second, U_1 must be measured somewhere on the upstream plane; it makes no difference where, since we're assuming that U_1 is constant across S_1. Then all that's left is for each $U_2(U_1 - U_2)$ to be multiplied by its corresponding area, dS_2 (or ΔS_2, to be realistic about it), the products added up, and the result multiplied by the density of the fluid. Voilà, drag!

Several practical matters. First, the location of the downstream plane poses a slight problem. If you're too close to the object, the assumption that the flow direction is uniform and normal to the plane may be violated, so the plane should be distant by at least several times the maximum diameter of the object. If, on the other hand, you're too far away, viscosity will have begun to eliminate the wake altogether, to reaccelerate it to free-stream velocity. A few trial runs are necessary. The requirement for a very large number of measurements (perhaps as many as a hundred) of velocity in the wake may seem daunting. But traverses back and forth with the measuring instrument on a motor-driven manipulator and direct input into a computer make the procedure more reasonable than it might appear.

Thrust can be measured in much the same manner; only the sign of the result is changed. Lift determinations require some way of establishing the deflection of the wake from the free-stream direction. I once measured the lift of a fixed fly wing by first measuring the airspeed at each of a series of points behind it; for each point I then centered the wake of a tiny wire on the airspeed transducer to get the local wind direction. From these data it was a simple matter to calculate the downward momentum flux created by the wing. Further information on indirect force measurements can be found in Prandtl and Tietjens (1934, pp. 123–130), Goldstein (1938, pp. 257–263), and Maull and Bearman (1964).

JET PROPULSION

If you hold a hose with a nozzle out of which water is flowing rapidly, you feel a force in the direction opposite that of flow. A toy water rocket speeds off impressively when pressurized air forces water out through its nozzle. And when a squid reduces the circumference of its outer mantle, the water in the mantle cavity is forced through a nozzle, the siphon, and the squid is elsewhere with remarkable alacrity. Jet or rocket propulsion is clearly effective with water as the working fluid.

No small number of animals use jet propulsion (Figure 4.14), and it seems to have evolved on quite a number of occasions. For that matter, it may have been the earliest truly macroscopic mode of animal locomotion. At least one ctenophore, *Leucothea*, uses it (Matsumoto and Hamner 1988) to get up to un-ctenophoric speeds of about 50 mm s^{-1}. Medusoid cnidarians are, I think, entirely a jet set; see, for instance, Daniel (1983) and DeMont and Gosline (1988). Pelagic tunicates move by jet propulsion; Madin (1990) provides a general view and references. Scallops, although they do it only briefly, can achieve speeds of 0.6 m s^{-1} (Dadswell and Weihs 1990). At least one family of fish, the frogfishes (Antennariidae), jet by ejecting water from a large oral cavity through a pair of small, tubular opercular openings; their speeds (up to 27 mm s^{-1}), though, aren't impressive for fish (Fish 1987). And dragonfly nymphs have an anal jetting system that can get them up to around 0.5 m s^{-1} (Hughes 1958, Mill and Pickard 1975). Of course the best known jetters are the cephalopod mollusks, both the shelled *Nautilus* and the shell-less octopuses, cuttlefish, and squids. In the latter, speeds may reach as much as 8 m s^{-1}, according to what are, admittedly, somewhat anecdotal accounts (Vogel 1987).

It should be easy to make a jet engine by modifying a ventilatory system, and muscular systems that squeeze cavities run innumerable hearts and guts. So perhaps the most interesting question to ask is why jetting isn't still more common, why the really big and fast swimmers mostly use quite a different and less obvious scheme. Considerations of momentum turn out to provide a fairly good rationalization. The underlying problem, and an important consideration in other forms of locomotion in fluids, boils down to a peculiar dissonance between momentum and energy. The force a jet produces reflects its rate of momentum discharge, loosely the product of mass per time and velocity. The cost of producing this jet reflects its energy, proportional to the product of mass per time and *the square of* velocity. Thus the economical way to jet is to process water at a high rate, maximizing mass

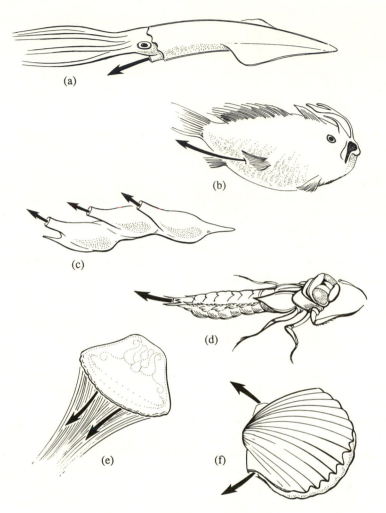

FIGURE 4.14. Jet propelled animals: (a) squid, (b) frogfish, (c) a trio of salps, (d) dragonfly nymph, (e) jellyfish, and (f) scallop.

per time, and to give the water a minimal incremental velocity. A jet, though, tends to do the opposite. Since the water has to be contained in some kind of internal squeeze-bag, its volume is relatively limited. Increasing the girth of the bag means increasing the volume of the animal, increasing its drag, and even reducing (by Laplace's law)[4] the effectiveness of a muscular layer in producing the pressure that will generate thrust.

[4] Laplace's law is the rule that the pressure inside a closed vessel will be proportional to the tension in its wall (what muscles do) divided by the radius of curvature. For more detail, see Vogel (1988a).

Let's discuss the matter with a little more formality. The thrust of a jet is the product of the mass flux and the difference between the jet velocity (U_2) and the free stream velocity (U_1), so

$$T = \frac{m}{t}(U_2 - U_1). \tag{4.14}$$

(We'll stick with the simpler mass per time rather than working through densities, cross sections, and so forth.) Jet velocity is the maximum speed of the jet, usually developed some short distance behind the motor. Power output, then, is the product of the thrust and the speed of travel or free stream velocity:

$$P_{out} = \frac{mU_1}{t}(U_2 - U_1). \tag{4.15}$$

Power input is the familiar kinetic energy per unit time:

$$P_{in} = \frac{m}{2t}(U_2{}^2 - U_1{}^2). \tag{4.16}$$

Dividing output by input gives us, as always, an efficiency:

$$\eta_f = \frac{2U_1}{U_2 + U_1}. \tag{4.17}$$

This particular efficiency is called the "Froude propulsion efficiency" after William Froude, a nineteenth-century British naval engineer, whose name will reappear in quite a different context in Chapter 17. As you can see (and certainly not surprisingly), the jet velocity must be higher than the free stream speed—but it should be only minimally higher for best efficiency. (The ideal, giving an efficiency of 100%, would be equal jet and free stream speeds.) A propeller or a set of external fins or paddles processes a lot of fluid; it can therefore achieve good thrust (equation 4.14) with only a small incremental velocity, and it therefore can operate at high Froude efficiency (equation 4.17). A jet, though, won't do where efficiency is the adaptively significant criterion.

A few more items connected with jet propulsion. Weihs (1977) suggested that a pulsed jet, usually what animals use, can do a little better than the formula for efficiency predicts. He pointed out (and Madin 1990 gives persuasive pictures) that pulses of rapid water entrain additional water as they roll up into toroidal vortices, and he calculated a factor of increase of approximately 30% in efficiency as a result. This entrainment of additional moving mass is not dissimilar to what modern ducted-fan jet engines do to reach higher efficiencies than the earliest jets could manage.

The cost of transport for a squid, jetting along, is about three times that of a trout of similar size or, as a similar comparison, a squid takes twice the power to go only half as fast as the average fish (O'Dor and Webber 1986). And the culprit is clearly what we've just been talking about—Alexander (1977) notes that a trout imparts rearward momentum to about ten times as much water per unit time as does a squid. Still, this odious comparison may be at least a little bit unfair. Smaller or slower jetting animals have substantially lower costs of transport than squid. Nautilus gets its mass around for about a sixth the cost paid by a squid (Wells 1990), and small salps (tunicates) expend only a little more energy than that (Madin 1990). O'Dor and Webber argue that the cost of jetting scales differently than the locomotion of most fish, with jetting looking rather less disadvantageous as size is reduced.

Solids and Fluids: Quick Comparisons

As large organisms whose density far exceeds that of the surrounding medium, most of us regard solids as more palpably real than fluids; our training in physics begins with the solidest of solids, and we forget the fascination of the two-year-old who pours water repeatedly from one container to another. Since we're trying to build a similarly intuitive familiarity with fluids, perhaps it might be useful to list some specific points of comparison, some analogies between the world of solids and that of fluids. While the paired items are certainly not synonymous, the items on the right effectively replace the ones on the left when one deals with fluids instead of solids. Thus . . .

SOLIDS	FLUIDS
mass	density
elasticity	viscosity
interfaces	streamlines
shearing planes	velocity gradients
friction	drag
conservation of mass	continuity
conservation of energy	Bernoulli's principle

Drag, Scale, and the Reynolds Number

S o far fluid mechanics has probably struck the reader as a decent, law-abiding branch of physics. We've touched on viscosity, continuity, momentum; we even dragged in drag and how it might be measured—and all proved to be quite ordinary topics. Only turbulence and streamlining hinted at deferred peculiarities. I now want to pursue the business of drag somewhat further, asking in particular about its actual physical basis. With this most innocent question, Pandora's box springs a leak, and keeping head above water gets harder. So queer is this aspect of the physical world that we'll have to defer most of the relevant biology to the next two chapters in order to spend this one on the physics.

From Whence Drag?

It was easy, in the last chapter, to define drag as the rate of removal of momentum from a moving fluid by an immersed body. Similarly, it's no trick at all to get a dimensionally correct formula for drag from that defini-tion: $\rho S U^2$. Newton suggested just such a formula from just such dimen-sional considerations—but where in it does shape enter? It certainly mat-ters in practice! Alternatively, one might try to apply Bernoulli's principle, adjusting equation (4.8) to work for more than just an upstream point by multiplying both sides by the projected area normal to flow ("face area"). One gets a similar formula: $\rho S U^2$. In physical terms, though, this second approach implies a rather curious situation. Fluid particles collide with the object, in the process losing all their momentum in the x-direction. And that means they disappear, or they're carried with the object like snow ahead of a badly designed plow, or they move precisely sideways without further interactions with the oncoming flow—all distressingly unrealistic scenarios.

Aristotle, by the way, was even further off the mark. He didn't believe in drag at all, figuring that an object needed air to give it thrust, that in a vacuum an arrow would tumble earthward upon leaving the bow. Newton's first law, a thoroughly counterintuitive idea, lay 2000 years in the future.

We might try Bernoulli (with a little more complete accounting) and the ideal fluid theorists of the nineteenth century. Theoretical streamlines can be calculated for steady flow around, say, a circular cylinder with its axis normal to the flow (Figure 5.1). At upstream and downstream extremities

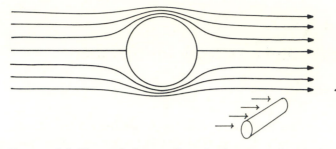

FIGURE 5.1. Theoretical streamlines for flow normal to the long axis of a circular cylinder, as shown in perspective in the inset at the right.

are so-called stagnation points where the fluid is locally stationary with respect to the cylinder. By continuity, the fluid reaches maximal velocity laterally, where the cylinder blocks the greatest part of the path of flow. As a result, at the front and back the pressures on the surface will reach H, the total head. On either side, where streamlines are closest together, the pressure will be less, $H - 2\rho U^2$ to be exact (Massey 1989, pp. 329–331, gives particulars). However, the whole diagram is symmetrical about a cross-flow plane along the axis of the cylinder. Not only do the pressures on each side cancel each other, but those on the front and back do so as well. As a result, no net pressure at all tends to carry the cylinder with the flow—the cylinder feels no drag at all! This amazing result can be generalized to cover bodies of any shape; it's called d'Alembert's paradox, and perhaps is the ultimate pursuit of a will-o'-the-wisp into a cul de sac.[1] Lord Rayleigh (1842–1919) pointed out that according to such theory a ship's propeller wouldn't work but, on the other hand, wouldn't be needed anyhow. Furthermore,

> There is no part of hydrodynamics more perplexing to the student than that which treats of the resistance of fluids. Acording to one school of writers, a body exposed to a stream of perfect fluid would experience no resultant force at all, any augmentation of pressure on its face due to the stream being compensated by equal and opposite pressures on its rear. . . . On the other hand it is well known that in practice an obstacle does experience a force tending to carry it downstream.

As a first step toward reconciling theory and reality, we can return to the pressure coefficients introduced in the last chapter and compare our results for a cylinder with the calculations for an ideal fluid, as in Figure 5.2. For these particular data, the conditions are equivalent to a sapling in a

[1] This phase is borrowed, with affectionate memories, from the late Carroll M. Williams.

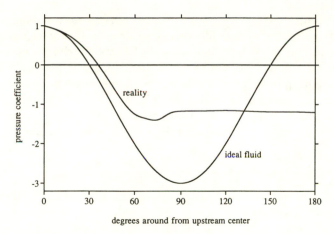

FIGURE 5.2. Two pressure distributions around a long circular cylinder—that for an ideal fluid following the streamlines of 5.1 and for a real fluid at a Reynolds number of 10^5. Theory and reality diverge most at the sides and rear.

light breeze or a human foreleg wading in a very gentle stream. At the upstream center, the pressure is, as expected, H, so the pressure coefficient is unity. Theory and data are in fine agreement, which reaffirms the utility of Pitot tubes, for instance. But proceeding further along the cylinder, the rest of the decrease in pressure does not entirely materialize; and, at about 70°, the pressure actually begins to rise somewhat. From here rearward, reality bears almost no resemblance to ideal fluid theory. About the only point on which theory and data agree is that, for the cylinder as a whole, the net overall force is negative, so the cylinder will tend to expand rather than collapse in the current—what we saw earlier in the air bubble maintained by the beetle *Potamodytes*.

This difference between the dynamic pressure on the front and the *absence* of the predicted counteractive pressure at the rear is what we subjectively feel as drag. We haven't, by this examination and comparison, really solved anything, but we have at least located the problem. The flow of fluid around the cylinder isn't symmetrical, front to back. And it's in the rear where something strange is happening. Not that this pressure difference can't be useful—it causes gases to percolate through cylindrical grain storage buildings (Mulhearn et al. 1976). We'll return to this problem of pressure distribution after introducing some analytical tools; for now just bear in mind that for limiting drag in fluids, discriminating design of the derriere is de rigueur.

Nor is this odd pressure distribution the only peculiarity, even for a case as ostensibly simple as the drag of a cylinder normal to flow. Both the pressure distribution for such an object and the drag itself vary with speed, for instance, in queer and irregular ways. At low speeds drag is proportional to the first power of velocity; at higher speeds it gradually approaches a second-power dependence; it then drops abruptly; and at still higher speeds it resumes its second-power behavior but with a lower constant of proportionality. Clearly our problem isn't simple omission of some simple component of drag, perhaps some direct function of viscosity.

A New Player: The Reynolds Number

Perhaps at this point we should remind ourselves that one objective of science is securing the minimum number of rules that account for the largest variety of phenomena. We don't have and may never have a simple, universal, and practical rule for predicting the drag even of a shape as regular as a circular cylinder. But trying to simplify the situation as much as possible is still worthwhile—anything is better than bludgeoning the problem with massive empirical tables of how drag varies with shape, speed, size, and other relevant variables. For each additional independent variable the number of data will rise by one or two orders of magnitude! And such tables would be of little use for the shapes a biologist is wont to consider. At the very least we ought to ask whether some of the variables have the same effects as others, whether each need be regarded as behaving in a completely unprecedented fashion.

This latter possibility turns out to be both real and useful. Its pursuit leads to the peculiarly powerful *Reynolds number*, the centerpiece of biological (and even nonbiological) fluid mechanics. The utility of the Reynolds number extends far beyond mere problems of drag; it's the nearest thing we have to a completely general guide to what's likely to happen when solid and fluid move with respect to each other. For a biologist, dealing with systems that span an enormous size range, the Reynolds number is the central scaling parameter that makes order of a diverse set of physical phenomena. It plays a role comparable to that of the surface-to-volume ratio in physiology.

This almost magical variable can be most easily introduced in just the way it originated, in the empirical investigation done by Osborne Reynolds (1883) already mentioned in Chapter 3 in connection with laminar and turbulent flow. As you may recall, Reynolds introduced a dye stream into a pipe of flowing liquid. Sometimes the resulting straight streak indicated laminar flow, and sometimes dispersal of the streak signaled that the flow was turbulent. The transition was fairly sudden, both in location in the pipe and as the characteristics of the flow were altered. He found that the flow could be persuaded to shift from laminar to turbulent in several ways: *by increasing speed; by increasing the diameter of the pipe; by increasing the density of the liquid; or by decreasing the liquid's viscosity.* Each change was as effective, quantitatively, as any other; and they worked in combination as well as individually. The rule that emerged was that when a certain combination of these variables exceeded 2000, the flow became turbulent. The particular combination is what we now call the Reynolds number, Re^2, with the length factor, l, taken as the pipe's diameter.

$$Re = \frac{\rho l U}{\mu} = \frac{lU}{\nu}. \tag{5.1}$$

Before going further I simply must point out that, claims in the biological literature notwithstanding, the specific value of 2000 applies only to transition in a long, straight, circularly cylindrical tube with smooth walls, at a decent distance from the tube's entrance. With stringent prevention of any initiating circumstances it may be raised as much as an order of magnitude. With roughened tubing, transition can happen at lower values. So only one figure is at all significant. For external rather than internal flows, the particular datum of 2000 has no relevance at all.

One of the marvelous gifts of nature is that this index proves to be so simple—a combination of four variables, each with an exponent of unity. It has, however, a few features worth some comment. First, the Reynolds number is dimensionless (as you can verify through reference to Table 1.1), so its value is independent of the system of units in which the variables are expressed. Second, in it reappears the kinematic viscosity, ν, (after a three-chapter absence). What matters isn't the dynamic viscosity, μ, and the density, ρ, so much as their ratio, ν: the higher the kinematic viscosity, the lower the Reynolds number. Finally, a bit about l, commonly called the "characteristic length." For a circular pipe, the diameter is used; choosing the diameter rather than the radius is entirely a matter of convention.[3] For a solid immersed in a fluid, l is typically taken as the greatest length of the solid in the direction of flow. But it's a very rudimentary measure of size, which emphasizes the coarse nature of the Reynolds number as a yardstick when objects of different shapes are being compared. As mentioned in Chapter 1, the value of the Reynolds number is rarely worth worrying about to better than one or at most two significant figures. Still, that's not trivial when biologically interesting flows span at least fourteen orders of magnitude.

[2] Take notice that no apostrophe precedes or succeeds the "s" in "Reynolds," that "number" isn't capitalized, and that the abbreviation used in text and equations is Re, not R_e.

[3] Reynolds used the radius, so he figured the laminar-turbulent transition point as 1000, not 2000. His nicely "round" number gives a better sense of the level of precision involved.

TABLE 5.1 A SPECTRUM OF REYNOLDS NUMBERS FOR SELF-PROPELLED ORGANISMS.

	Reynolds Number
A large whale swimming at 10 m s⁻¹	300,000,000
A tuna swimming at the same speed	30,000,000
A duck flying at 20 m s⁻¹	300,000
A large dragonfly going 7 m s⁻¹	30,000
A copepod in a speed burst of 0.2 m s⁻¹	300
Flapping wings of the smallest flying insects	30
An invertebrate larva, 0.3 mm long, at 1 mm s⁻¹	0.3
A sea urchin sperm advancing the species at 0.2 mm s⁻¹	0.03
A bacterium, swimming at 0.01 mm s⁻¹	0.00001

Of greatest importance in the Reynolds number is the product of size and speed, telling us that the two work in concert, not counteractively. For living systems "small" almost always means slow, and "large" almost always implies fast. That's why the range of Reynolds numbers so far exceeds the eight or so orders of magnitude over which the lengths of organisms vary. At this point looking at some real Reynolds numbers might be useful. Since we're interested only in orders of magnitude, Table 5.1 is based on rough approximations for speeds and sizes.

A PHYSICAL VIEW OF THE REYNOLDS NUMBER

This same dimensionless number can be reached by another path. Consider what must have gone wrong with the attempt to explain drag with ideal fluid theory. Ideal fluids have no viscosity, so failure ought to trace to the neglect of forces associated with viscosity. Certainly viscosity involves force—we recognized it in the first place as a retarding force on a moving flat plate that was adjacent to a fixed one. So perhaps two sorts of forces matter when a moving fluid crosses an immersed body. First are *inertial forces*, those derived directly from Newton's second law, which essentially defines inertia, and expressed by an equation that we obtained easily and found practically useless:

$$F_I = \rho S U^2.$$

What we mean by inertial forces are those attributable to the momentum of a bit of moving fluid. Thus, at the urging of its inertia, a particle of fluid keeps on doing its usual thing to the extent that it remains unmolested. Second are the previously ignored *viscous forces*. Fluids don't like to be sheared—if a flow involves shear, then viscous forces will oppose the persistence of the motion just as friction opposes the movement of one solid across another. So viscosity will tend to smooth out any internal irregularities in the flow. Viscous force was defined when we defined viscosity:

$$F_v = \frac{\mu S U}{l}.$$

Put somewhat more metaphorically, inertial forces reflect the *individuality* of bits of fluid while viscous forces reflect their *groupiness*. The former describes the progress of a milling crowd, the latter of a disciplined march. That, in fact, is exactly the distinction between laminar and turbulent flow. So let's make what's now only a small intuitive leap and suggest that what distinguishes different regimes of flow is the relative importance of inertial and viscous forces. The former keeps things going; the latter makes them stop. High inertial forces favor turbulence, with the substantial shear rates inevitably involved. High viscous forces should prevent sustained turbulence and favor laminar flow by damping incipient eddies and other irregularities. To take a rough look at the relative magnitude of the two, we need only divide these equations. What we get is, of all things, the Reynolds number:

$$\frac{F_l}{F_v} = \frac{\rho S U^2}{\mu S U / l} = \frac{\rho l U}{\mu}. \tag{5.1}$$

Here we see its lack of dimensions as the natural outcome of dividing one force by another—the Reynolds number is a ratio of inertial to viscous forces. I say "a" ratio, not "the" ratio as a reminder that we've done little more than a loose dimensional accounting. It is, for instance, distinctly unsafe to assert that a value of 1.0 indicates equality of two entirely specific forces.

Another point should be made emphatically. If, for example, the Reynolds number is low, the situation is highly viscous. The flow will be dominated by viscous forces, vortices will be either nonexistent or nonsustained, and velocity gradients will be very gentle unless large forces are exerted. The value of the Reynolds number is what indicates the character of flow, not the value of dynamic viscosity or even kinematic viscosity per se. A 10-fold reduction in size (length) will increase relative viscous effects with precisely the same efficacy as a tenfold increase in viscosity itself. If, in nature, small means slow and large means fast, then small creatures will live in a world dominated by viscous phenomena and large ones by inertial phenomena—this, even though the bacterium swims in the same water as the whale.

Consider an object of a given shape and orientation immersed alternatively in two flows. *Equality of the Reynolds number for the two situations guarantees that the physical character of the flows will be the same.* A moment's thought should persuade you that this last statement is powerful but still quite reasonable, given the origin of the Reynolds number. As we'll see, equality of the Reynolds number doesn't necessarily mean that forces are unchanged, but it does assure us that the patterns of flow as might be revealed by locating streamlines will be the same. And that holds even if one flow is of a gas and the other is of a liquid.

We come, parenthetically, to a mild curiosity and convenience. It was mentioned back in Chapter 2 that at ordinary temperatures air is about fifteen times *more* kinematically viscous than is water. As a result, for an object of a given size, the same Reynolds number will be achieved at a velocity fifteen times higher in air than in water. I doubt whether it's more than coincidental, but in nature flows of air are very roughly fifteen times as rapid as flows of water—or at least, as a sober assessment, an order of magnitude more rapid. Thirty meters per second (68 mph) is a hurricane of air; two meters per second is a torrent of water. A third of a meter per second in air we find barely detectable; twenty millimeters per second in water appears to be a roughly similar threshold. In short, organisms of any given size operate at approximately the same Reynolds numbers whether they live in air or water, quite a convenience for any biologist who is trying to understand the relationship between flow patterns and functional adaptations. Again, though, I emphasize that equality of Reynolds numbers doesn't imply that the forces of flow are the same—only the patterns of flow. Forces will be dealt with shortly.

This notion that equal Reynolds numbers imply geometric similarity in flow patterns is a potent one. A good cross-sectional shape for a bird's wing ought to be much the same as for the tail of a tuna—the Reynolds numbers are similar. Eel and spermatozoan may look similar, but one shouldn't presume that they swim by the same hydrodynamic mechanism—their Reynolds numbers are too divergent. On the other hand, micturition in humans (Hinman 1968) and the ejection of ink in some ink-jet printers (Levanoni 1977) involve approximately the same Reynolds numbers. So perhaps work on avoiding what are called "satellite drops" by the printer folk may be helpful in understanding the design of a penile orifice that squirts with minimal spraying (Chapter 17). Stokes' law (Chapter 15) is a good rule for sinking rate of pollen and plankton. But if you try to apply it to sinking of trout eggs, you get an erroneous answer of 11 rather than 88 mm s^{-1}—the predictions of Stokes' law increasingly diverge from reality at Reynolds numbers over about 1, and a sinking trout egg operates at about 50 (Crisp 1989).

REYNOLDS NUMBER AND THE DRAG COEFFICIENT

Back to drag. Newton figured that it was proportional to three quantities: first, to the density of the medium; second, to the projected area of the body; and third, to the square of the velocity. And drag does indeed behave in this manner—but only sometimes, and even then only approximately. One can do somewhat better, though, with a dimensional analysis, a mathematical technique mentioned in Chapter 1. (Rouse 1938 has a particularly illuminating discussion of the approach and its application to the present problem.) Dimensional analysis leads to an empirical formula for drag with all the peculiarities heaped upon one composite variable; I'll not go through the formalities but just sketch the logic.

The analysis begins by identifying the variables upon which drag might depend—the *size* of the object, its *speed* relative to the fluid, and the *viscosity* and *density* of the fluid. By a theorem involving the relative number of dimensions and variables, it predicts that the formula for drag will be the product of a term with dimensions of force and a single dimensionless term. The force term proves to be the dynamic pressure ($\rho U^2/2$, remember) times the area of the object; it essentially accounts for the ordinary variations that Newton identified. The dimensionless term accounts in toto for all of the peculiarities in the behavior of drag:

$$D = \left(\frac{1}{2}\, \rho S U^2 \right) \left(\frac{\rho l U}{\mu} \right)^a. \tag{5.2}$$

The dimensional analysis tells us nothing about how this second term behaves, merely that it is the *only* term that should matter. The second term (less its exponent) happens to be the Reynolds number, so the latter has now appeared in a third context!

Thus the variations in the drag of any object of fixed shape and orientation is describable as the product of two variables—the dynamic pressure times the area of the object and some function of the Reynolds number. In practice, that awkward exponent, a, about which we know nothing a priori, is replaced by the so-called drag coefficient, C_d.

$$C_d = (Re)^a = f(Re). \tag{5.3}$$

Of course we know as little about C_d—a multiplicative coefficient is just handier than an exponent. The drag coefficient, then, is a function only of the Reynolds number and handles all the oddities in the behavior of drag. This last statement is a powerful one; put another way, it says that equality of Reynolds number for a given shape and orientation in flow implies equality of the drag coefficient. That's sometimes called the "law of dynamic similitude"; as we'll see, it's the basis of almost all testing of model systems in flow tanks and wind tunnels.

The drag coefficient is most easily envisioned as a dimensionless form of drag, the drag per unit area divided by the dynamic pressure. Or, as usually given,

$$D = \frac{1}{2}\, C_d \rho S U^2. \tag{5.4}$$

That, of course, makes it closely analogous to the pressure coefficients with which the last chapter was replete—they were obtained by dividing measured pressure by dynamic pressure. The only change is the addition of an explicit area.

Equation (5.4) is most definitely *not* the formula for drag, no matter what one sees in the biological literature. It's just a definitional equation that converts drag to drag coefficient and vice versa. The value of a in (5.2) and (5.3) wasn't constant, we have no grounds for declaring C_d a constant, and it doesn't often turn out to be particularly constant. Still, equation (5.4) represents a huge simplification, the one we've been working toward. Instead of having to record how drag varies separately with speed, size, viscosity, and density, we need know only how the coefficient of drag varies with the Reynolds number. The fact that both drag coefficient and Reynolds number are dimensionless may take a little getting used to, but after a while one does acquire a properly contemptuous familiarity with them. As with pressure coefficients, actual drag can be obtained from a value of the coefficient simply by recourse to the defining equation (here, equation 5.4).

The inclusion of a factor, S, for area does present a minor difficulty, one a little trickier than that of picking a length for use in the Reynolds number. Clearly the drag coefficient one calculates from a datum for drag depends very much on the choice. The commonest reference area is the "frontal" or projecting area of an object—its maximum projection onto a plane normal to the direction of flow. Hereafter, we'll use S_f for that area and C_{df} for the corresponding drag coefficient. Frontal area is particularly useful for non-streamlined objects of relatively high drag at high and medium Reynolds numbers, for which drag isn't so very far from dynamic pressure times frontal area.

Three other areas are used, and each has points in its favor. "Wetted area" is the total surface exposed to flow; it's commonly used for streamlined bodies where (as we'll discuss) drag is largely a matter of viscosity and shear, and what matters most is how much skin is showing; for it, we'll use S_w, and for the corresponding coefficient, C_{dw}. For most situations it's more biologically relevant than frontal area, and it's independent of the specific orientation of the organism. The main difficulty in using wetted area is the practical one of measuring it on a real organism. Fish and cetaceans may be fairly smooth sorts, but consider flying insects, swimming crustaceans, nautiloids, trees—they're what we might call "fractomorphic," and any figure for surface area is in part a definitional artifact. One can approximate the shapes of creatures with cylinders, spheroids, and so forth and from these calculate areas; but that may obscure features that have mattered to the selective process.

A third reference, "plan form area" or "profile area" (S_p and C_{dp} here) is commonly used with lift-producing airfoils; it's the maximum projected area of the airfoil, the area that one would see if the airfoil were laid on a table and viewed from above. Like wetted area, it's independent of orientation with respect to flow. That's important if, say, one draws a graph of drag coefficient against values of the small angle between wing and wind—it would be most awkward to have the definition of one variable changing in response to alterations in the other. One should never change the reference area used from one datum to the next, even if something is reorienting in a flow or even if (as with a leaf in a wind) its actual exposed surface area is changing. Also, anticipating a bit, if one is to compare lift and drag using lift and drag coefficients, the same reference area (usually plan form) must be used for both.

Finally, the two-thirds power of volume has been used as reference area, originally for airships, where volume was proportional to lift. We'll designate it as S_v and the drag coefficient with it as C_{dv}. $V^{2/3}$ is probably the most appropriate for organisms, with their fitness a matter of guts and gonads inside; it's also perhaps the easiest to measure, from a simple ratio of mass to density. But it's also the least common. I'll give a strong pitch for its use, and not just because it coincides with my initials.

Picking a reference area is more important than it may at first appear. It can have a major influence on the conclusions one draws, as most cogently illustrated by D. E. Alexander (1990). He compared the drag of specimens of two species of marine isopod crustaceans, *Idotea wosnesenskii* and *I. resecata*. *I. wosnesenskii* is relatively more rotund, 2.9 times as long as wide, while *I. resecata* has an aspect ratio of 4.5:1. Using frontal area as reference he found that the drag coefficient of *I. wosnesenskii* was 77% of that of *I. resecata*. By contrast, using wetted area (and a presumption of oblate spheroidal shapes), the same comparison came out to 142%. Which has the higher drag? It depends on how one does the comparison! So why not just compare drag and ignore the coeffiencients? The problem is that one has then done no normalization for the considerable effect of size (not to mention speed); we're usually not so much interested in drag per se as in comparing the "dragginess" of shapes.

At this point I'll be frankly imperious and declare that no published figure for drag coefficient is of any value unless the reference area is indicated. And, I reemphasize, the results of comparisons among different shapes depend on that choice of area—even for the C_d's of automobiles. For instance, if frontal area is used, then the lower car is at a disadvantage—its lower drag may not result in a lower C_d.

With only two variables for a given shape—coefficient of drag and Reynolds number—graphs are a great convenience. We have, remember, lumped all the peculiarities into the relationship between these variables. If drag behaved as Newton (and much popular literature) believed it should, then all graphs of C_d as functions of Re should be horizontal, lines. In its deviation from such a line, the graph shamelessly dissects out and exposes the queerness. Figures 5.3 and 5.4 give such plots for cylinders and

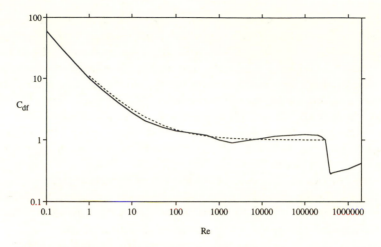

FIGURE 5.3. Drag coefficient (based on frontal area) versus Reynolds number (based on diameter) for flow normal to the long axis of a long circular cylinder. The dashed line comes from a fitted formula, equation (15.7).

spheres, respectively. If your intuition needs a crutch, think of these graphs as representing, for a given object and fluid, drag divided by the square of speed (perhaps "speed-specific drag") on the ordinate and speed on the abscissa.

Biologists may be surprised to find that a simple case such as the drag of a

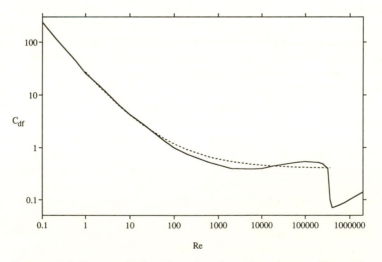

FIGURE 5.4. Drag coefficient (based on frontal area) versus Reynolds number (based on diameter) for flow across a sphere. The dashed line comes from a fitted formula, equation (15.2).

circular cylinder can generate the kind of graph we think of as the special curse of our more complex systems. If a cylinder is messy, what of a barnacle? The abruptness of the transitions and the bumpiness of these graphs of C_d versus Re for simple objects prove to be much more drastic than the equivalent transitions for complexly shaped "biological" entities. The very symmetry and regularity of these simple shapes mean that transitions from one flow regime to another tend to take place synchronously on the whole object rather than first at one location and then at another.

Flow around a Cylinder Revisited

We began the chapter by asking about the flow around a circular cylinder and its drag. We saw that the pressure distribution differs markedly from that expected for an ideal fluid, and we've just seen that the plot of C_d versus Re is, to say the least, irregular. Let's tie up a loose end by looking at the actual patterns of flow that occur as the Reynolds number increases, combining a few paragraphs of verbal description with the illustrations in Figure 5.5. You ought to refer as well to Figure 5.3 as we proceed.

At Reynolds numbers well below unity, smooth and vortex-free flow surrounds the cylinder. Flow looks a little like that expected for an ideal fluid (Figure 5.1), but the resemblance fades on close examination—there's no fore and aft symmetry here, only the same absence of vortices. Notice that the presence of the cylinder lowers the velocities of flow for quite a distance from itself; the cylinder is surrounded by a cloud of retarded fluid that largely obscures any details of its shape. This wide influence can corrupt measurements of drag if a solid wall is anywhere around, and a wall's effect gets worse as Reynolds numbers decrease. Thus for flow across a circular cylinder, White (1946) found that, at $Re = 10^{-4}$, the presence of walls 500 cylinder-diameters away doubles the apparent drag. As the Reynolds number approaches and then exceeds unity, the volume of the disturbed fluid gradually diminishes, and we enter a regime in which the drag curves for different shapes diverge more radically.

At Reynolds numbers between about 10 and 40, the cylinder bears a pair of attached eddies on its rear. With flow from left to right, the upper eddy rotates clockwise and the lower one counterclockwise. Above about 40, the pattern is no longer stable, and the vortices alternately detach, producing a wake of vortices with each rotating in a direction opposite that of its predecessor farther downstream. This pattern of alternating vortices is known as a "Von Kármán trail" and will get more attention in Chapter 16. Periodic shedding of vortices continues up to a Reynolds number of about 100,000, but with an increasingly turbulent wake behind the cylinder.

Somewhere between 100,000 and 250,000 another transition occurs. The wide wake of turbulent eddies narrows rather abruptly, and the drag coefficient concomitantly drops by about two-thirds. The "somewhere" is a

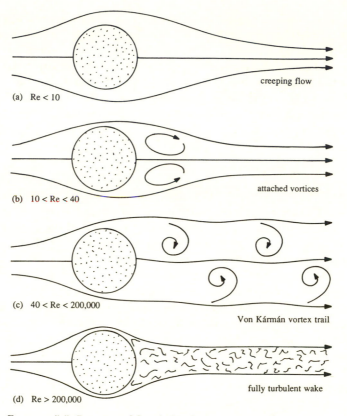

(a) Re < 10

creeping flow

(b) 10 < Re < 40

attached vortices

(c) 40 < Re < 200,000

Von Kármán vortex trail

(d) Re > 200,000

fully turbulent wake

FIGURE 5.5. Patterns of flow behind a circular cylinder. Note the absence of vortices at low Reynolds numbers in (a) and the constriction of the wake between (c) and (d). This last change is concomitant with the drop in drag coefficient—the great "drag crisis"—at Reynolds numbers between about 100,000 and 250,000.

deliberate evasion, for the precise point depends on the circumstances. The less turbulent the basic flow to which the cylinder is exposed, the higher is the Reynolds number for the transition. Not only does the drag coefficient drop, but the drop is usually so sudden that the drag itself briefly undergoes a paradoxical reduction with further increase in the Reynolds number. With still further increase the pattern of flow changes little more—up to numbers so high that compressibility is no longer negligible and so high as to be biologically irrelevant.

The abrupt drop in drag coefficient and the narrowing of the wake is associated with a phenomenon of importance at all Reynolds numbers much above unity—it's called "separation of flow." Recall from Figure 5.2

that the pressure coefficient increases from the widest part of a cylinder toward the rear. Just where and how much it increases depends a bit on the Reynolds number, but there's always some region of increase as one looks farther rearward around the surface. Thus in moving around the cylinder from the forward stagnation point, fluid first goes from high to low pressure, an easy and natural journey. But beyond the point of minimum pressure, fluid must progress "uphill" to make any more progress toward the center of the rear. Such motion against a pressure gradient is possible only at the expense of preexisting momentum, which ordinarily isn't sufficient for the flow to get to the rear stagnation point. The trouble is that fluid has been robbed of its momentum by viscous effects—shear at and near the surface. It's just like a sled that coasts down one hill but, having experienced friction, can't make it up another hill of equal height. What happens instead is that at some point the fluid stops following the surface of the cylinder and instead heads off more or less straight downstream; the place at which this occurs is called the "separation point." It occurs very near the point (see Figure 5.2) at which ideal and real pressure curves diverge most severely.

Downstream from the separation point are eddies, general turbulence, and, further back, the periodically shed vortices of the Von Kármán trail. The noise you hear when facing directly into a moderate wind is the turbulence around your ears, which are just downstream from the separation point for wind blowing around your head (Kristiansen and Petterson 1978). And the coolness felt between the shoulder blades when running shirtless reflects the general stirring about of the postseparation flow around your unstreamlined torso. Immediately behind the separation point, fluid near the surface commonly flows in what's otherwise the upstream direction (Figure 5.6a). Thus the separation point is (going along the length of the cylinder) a line of minimal flow near the surface—flows from front and rear join and move away from the surface. By monitoring the intensity of flow near the surface with something that erodes more swiftly in more rapid flow, you can easily detect this line of reduced fluid motion. Try arranging a piece of ordinary chalk so it protrudes upward from the bed of a steady stream. After a few hours its cylindrical shape will have altered enough to indicate the variations in speed of flow near the surface—in particular, a lengthwise ridge will mark the line of separation.

What happens in the great drag crisis at Reynolds numbers around 100,000 is that the separation point suddenly shifts downstream. At rising but lower Reynolds numbers it had been slowly moving forward as the fluid gradually lost its ability to curve around behind the point of maximum width. Thus at $Re = 300$ the separation point is $121°$ around from the forward stagnation point, while by $Re = 3000$ it has moved ahead to $96°$ (Seeley et al. 1975). But at that transition, turbulence (which has already

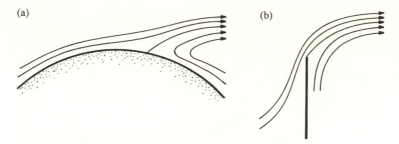

FIGURE 5.6. Separation of flow from a surface. (a) A detailed look at the streamlines near the separation point for flow across a cylinder.
(b) The same for a flat plate broadside to flow. The location of the cylinder's separation point varies with the Reynolds number while the plate's doesn't; concomitantly the drag coefficient of the cylinder varies much more than that of the plate.

been present) invades the region of fluid very near the surface of the cylinder. The result is an increase in the momentum of the fluid near the surface, which permits the flow to follow the surface farther around before separating. And that, in turn, leads to abrupt the abrupt reductions in the width of the wake and in both drag and drag coefficient.

Spheres do much the same things as cylinders, as is evident from Figures 5.3 and 5.4, except that spheres leave no tidy Von Kármán trails in their wakes. The Reynolds number of a sphere's transition point can, in fact, be used as a measure of the turbulence of an airstream and thus as the basis of a figure of merit for a wind tunnel. A lower value for the transition point indicates greater turbulence and, by the usual criteria, a poorer tunnel (Pankhurst and Holder 1952).

SHAPE AND TWO KINDS OF DRAG

The peculiarity of drag, especially its relationship to the Reynolds number, has been blamed on this odd phenomenon of separation. But, as we'll see, separation isn't the whole story—even without it drag remains real. The overall drag of an object, what one measures with a force transducer, can be dissected into two components. Both the "skin friction" and the "pressure drag" ultimately come from viscous effects, but they do so in quite different ways.

Skin friction is the direct consequence of the interlamellar stickiness of fluids, the viscous mechanism by which one plate exerts a force on another (as shown in Figure 2.2). It always exerts its force parallel to the surface and in the local direction of flow. As you might expect, it's more significant in more viscous situations, that is, at low Reynolds numbers. But no Reynolds number is so high that skin friction disappears. As long as the no-slip condition holds, there's a shear region in a fluid near a surface. And the more surface (skin) a body exposes to the flow, the more skin friction it will experience.

Pressure drag is also a result of viscosity, but a less direct one. It occurs mainly because of separation and the consequent fact that the dynamic pressure on the front is not counterbalanced by an equal and opposite pressure on the rear. In effect, energy is being invested in accelerating fluid to get it around the object, but the energy isn't being returned to the object in decelerating fluid near its rear. Instead, the energy is dissipated (to heat, eventually) in the wake. While it's commonly regarded as a high Reynolds number phenomenon, pressure drag is significant at all Reynolds numbers; only its magnitude relative to skin friction is greater when the Reynolds number is high. Of the overall drag of a circular cylinder (yet again), pressure drag constitutes 57% at $Re = 10$; 71% at $Re = 100$; 87% at $Re = 1000$; and 97% at $Re = 10,000$.

Although the occurrence of pressure drag ultimately requires viscosity, it mainly reflects inertial forces. As such it's roughly proportion to surface area (or to the square of a linear dimension) and to the square of velocity. The drag coefficient is inversely proportional to these factors (equation 5.4), so where pressure drag dominates, the drag coefficient will not vary widely. By contrast, skin friction is a direct matter of viscous forces and follows the first powers of linear dimensions and velocity as in equation (2.3). Thus at low Reynolds numbers, where skin friction predominates, the drag coefficient will be inversely proportional to the Reynolds number—as you can see in the descending lines at the left in the graphs of Figures 5.3 and 5.4. In between, at Reynolds numbers of roughly 1 to 1,000,000 are the odd transitions we've just discussed. Airplanes may fly above all these phenomenological bumps, but the range includes an awesome diversity of biologically interesting situations.

An example should sharpen the contrast between pressure drag and skin friction as well as between high and low Reynolds numbers. Consider two objects, the first with negligible pressure drag at any Reynolds number, the second with very high pressure drag at all Reynolds numbers. One object is a long, flat plate of negligible thickness parallel to the flow (imagine a long, thin wing). The other object is the same flat plate now oriented perpendicular to flow. Table 5.2 gives data for the drag coefficients of these objects over a millionfold range of Reynolds numbers, together with the ratio of the perpendicular to the parallel drag coefficients.

The first thing that emerges from Table 5.2 is the rise in drag coefficient at low Reynolds number, what we saw for spheres and cylinders. One then notices the differences from spheres and cylinders—in particular the absence of the sudden drop in the drag coefficient of the perpendicular plate

TABLE 5.2 DRAG COEFFICIENTS (BASED ON WETTED AREA) OF A VERY
LONG, FLAT PLATE EXTENDING ACROSS A FLOW BASED ON WETTED AREA.

| | C_{dw} | | |
Reynolds Number	Parallel to Flow	Perpendicular to Flow	Ratio
1,000,000	0.0013	0.98	740
100,000	0.0042	0.98	230
10,000	0.013	0.98	74
1000	0.042	1.00	24
100	0.13	1.22	9.2
10*	1.1	1.9	1.7
1*	6.2	9.2	1.5

SOURCES: Ellington 1991; Thom and Swart 1940.
*Thin airfoils, actually.

at high *Re*. What causes this omission is the fact that the location of the separation point is fixed at the outer edge of the plate so it can't move upstream or downstream. The constancy of the drag coefficient that results from the fixed separation point can be put to good use. A flat plate or a disk perpendicular to flow can be used as an object of known drag so a force transducer can be used to calibrate the speed of a flow tank or wind tunnel. One might expect a value of 0.5 rather than about 1.0 from all the talk earlier about stagnation points; the increase results from a wake that's actually wider than the plate.[4] What happens, as shown in Figure 5.6b, is that flow moving laterally along the front of the plate is carried by its momentum a short ways beyond the edge of the plate—it doesn't turn the corner sharply as it heads downstream again. For that matter, the width of the wake gives a good first indication of the magnitude of the drag of any object at moderate and high Reynolds numbers.

More interesting yet are the ratios of the drag coefficients (and thus the drag data themselves, since the reference area is unchanged). In general, at low Reynolds numbers, as was noted earlier, skin friction is of great importance. It depends only a little on orientation. At moderate and high Reynolds numbers, pressure drag *can* be overwhelming; but its actual significance depends very much on the orientation of the plate. Thus orientation has a far more potent effect on overall drag at high and moderate *Re*'s.

Which brings us to the question of how drag might be reduced, clearly a vital matter for either nature or the human designer. At moderate and

[4] Flat plates that aren't really long have lower drag coefficients—leakage around an end makes quite a difference. If the plate is only ten times as long as wide, C_{dw} is just 0.68 above $Re = 10^3$; if less than five times longer than wide, C_{dw} is 0.61. A circular disk has a C_{dw} of 0.56 (Hoerner 1965; Ellington 1991).

high Reynolds numbers, any object from which flow separates will experience a relatively high drag; the problem, again, is the energy loss in the rear. A trick, though, has been recognized at least empirically for a very long time. If the object is endowed with a long and tapering tail, fluid gradually decelerates in the rear, little or no separation occurs, and the object is literally pushed forward by the wedgelike closure of the fluid behind it. Instead of being lost in the wake, the energy reemerges as a forward-directed pressure from the rear that nearly counterbalances the dynamic pressure on the front. The trick is called "streamlining" because, loosely speaking, an object such as that of Figure 5.7 is shaped to follow (or, much the same thing, produce) an advantageous set of streamlines. Especially at high Reynolds numbers streamlining can be immensely effective: a well-designed airship may have less than 2 percent of the drag of a sphere of the same frontal area. The sphere, by the way, is an example of a "bluff body," the term antonymous to "streamlined body." What matters is mainly the design of the rear. As Hamlet put it, "There's a divinity that shapes our ends, Rough-hew them how we will."

But as the Reynolds number gets lower, streamlining isn't quite so overwhelmingly beneficial. The reshaping involved exposes more surface relative to either projected area or to volume contained. And more surface inevitably means more skin friction. At the Reynolds numbers encountered by whales, large fish, and flying birds, skin friction is a minor matter. At the moderate Re's of the larger flying and swimming insects, streamlining is still a good thing. At much lower Reynolds numbers the extra surface exposed may outweigh any reduction in pressure drag, which is undoubtedly one of the reasons why very small swimming creatures do not look obviously streamlined.

Perhaps the main disadvantage of streamlining as a scheme for drag reduction, from a biological point of view, is that it assumes a particular orientation to the flow. A minor change in wind direction would more than wipe out the benefits of a streamlined tree trunk. If the direction of flow can be neither predicted nor controlled, then either a bluff body must be tolerated or else some other stratagem brought to bear. More about the latter possibility in the next chapter.

Can anything at all be done about the high drag of bluff bodies such as spheres and cylinders? Curiously enough, the situation can occasionally be improved by roughening the surface. Usually, and the point should be emphasized, roughness is either without effect or it increases drag. At low Reynolds numbers small bumps will be within the slowly moving fluid near the surface and will thus be of little consequence. At high Re's, roughness increases the drag of streamlined objects. But in a certain narrow range roughening can help: a rough surface promotes turbulence close to a surface and can thereby postpone separation as the fluid travels around a

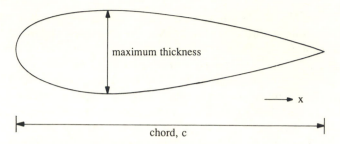

maximum thickness

x

chord, c

FIGURE 5.7. A streamlined shape, with the familiar blunt front and sharp rear. It may be viewed either as a body of revolution, symmetrical about its long axis in the plane of the paper, or as a typical cross section of an elongate body extending far above and below the page. This particular shape has a "fineness ratio," or chord over maximum thickness, of about 3.5. Its maximum thickness occurs 30% of the way from front to rear, that is, at $x/c = 0.3$.

bluff body, just as does turbulence in the flow itself. The transition to a lower drag coefficient that we noted for Reynolds numbers between 100,000 and 250,000 can be pushed down to as low as 25,000 for a rough cylinder or sphere (Figure 5.8). Any resulting increase in drag coefficient below 25,000 won't matter since down there the actual drag won't be especially high anyway. If the object is operating with a top Re of, say, 100,000, then any increase in drag coefficient above perhaps 150,000 won't matter either. Within limits, the rougher the surface, the lower the transition point, but the higher the drag coefficient both before and after transition. Some refining is possible by roughening only the portion just upstream of the maximum diameter, but that requires knowing the direction of flow; however, if the direction can be counted on, then streamlining is a much better bet.

The most familiar example of intentional roughness for drag reduction is a golf ball with its set of dimples. The practice of dimpling originated with the observation that old, rough balls traveled farther than smooth, new balls. At the urging of a driver, a golf ball moves at a Reynolds number of about 50,000 to 150,000, so the scheme is reasonable. One might wonder whether the rough bark of trees is a similar adaptation: slender branches are smooth while trunks and large branches are much rougher. I think it is unlikely—for a trunk or branch 0.2 m in diameter, $Re = 100,000$ will be reached at a speed of about 8 m s^{-1}, less than 20 mph. Speeds high enough to be troublesome would automatically exceed the turbulent transition whatever the degree of roughness, which means that a smooth surface should be better. In any case, leaves (about which more in the next chapter) contribute much more to drag under most circumstances than do trunk

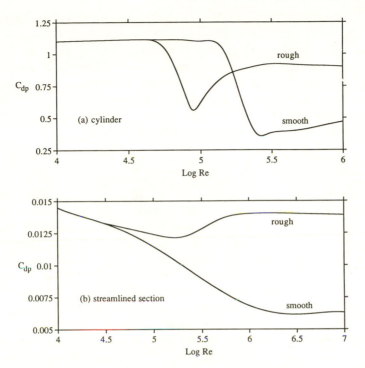

FIGURE 5.8. The effects of surface roughness on the drag coefficient of streamlined and unstreamlined shapes. At low Reynolds numbers roughness is inconsequential in either case. For a cylinder (a), but not a streamlined body (b), surface roughness can lower the drag, but it can do so only over a limited range of Reynolds numbers. Drag coefficients are based on plan form area (for a cylinder the same as frontal area).

and branches. Another possible application of purposeful roughening might be in the design of attached reef or rocky-coastal organisms that are exposed at times to fairly violent water currents—but Denny's (1989) caution that lift is often more important than drag, mentioned in the last chapter, should be kept in mind. At this point I know of no decently documented case in which a bluff organism in nature achieves drag reduction through roughening that might be adaptively significant.

Again, streamlining is a much more potent approach to drag reduction. Quite a number of human cultures have developed streamlined throwing sticks ("boomerangs"), mostly of the nonreturning genre. These are most impressive weapons against small prey and might have really caught on if standardization and mass production had been feasible; not only do they cut a wide swath with lots of rotational momentum, but they travel far greater distances than any hand-thrown, spherical projectiles. I'm told that while it's quite difficult to return a ball from the outfield wall, a boomerang could easily be thrown out of any baseball stadium from home plate. The difficulty with the streamlined stick, once again, is the necessity for knowing or controlling the direction of flow—it loses all advantage if it tumbles.

These matters of drag, Reynolds number, separation, and so forth are particularly well explained by Shapiro (1961), intuitively and anecdotally, and by Goldstein (1938), more formally and thoroughly.

MODELING AND THE REYNOLDS NUMBER

Let's return to two points made earlier. Consider all imaginable situations involving either rigid objects in flows or flows through the interstices of such objects. If any pair of objects are geometrically similar, then equality of the Reynolds numbers of their flows implies (1) geometric similarity of the pattern of flow and (2) equality of pressure and drag coefficients. You might just reread that last sentence. Physical science has rarely given the biologist a better deal or one that's so commonly ignored. What it means is the following. Say you have a situation involving fluid movement that you find experimentally awkward by virtue of inconvenient size, speed, or fluid medium. You are completely free to model that situation with another one of different speed, size, or medium provided only that you maintain the original Reynolds number and preserve geometric similarity. Few restrictions apply to this rule. Mainly, compressibility of the fluid under the circumstances must be negligible, no fluid-fluid interfaces may be involved, and the situation must be isothermal. Beyond these, the rule is both precise and general.

Put in the usual jargon, if the Reynolds numbers for two situations are the same, the situations are "dynamically similar." Drag may be a complex phenomenon, inadequate theory may make recourse to measurements and models mandatory, but dynamic similarity greatly facilitates the business. If both the original situation and the model system involve the same fluid medium, then equality of the Reynolds number is ensured just by maintaining constant the product of length and velocity. If the medium is changed, the variables are three—length, speed, and kinematic viscosity (equation 5.1)—the juggling is only a little more complicated and the options are far wider. Often one can take the same object and test it in a new medium, adjusting the speed to compensate for the change in kinematic viscosity. For instance, flow visualization is far easier in water than in air, especially for small systems, because neutrally buoyant markers are more readily available and because equivalent flows in water are much slower (ordinarily fifteenfold, recall). To work on very small organisms, large models can be made and then tested in a highly viscous medium such as corn syrup, various oils, or glycerin—the shift has been found useful in recent years for several studies on the mechanisms involved in suspension feeding (see, for instance, Braimah 1987). For inconveniently large systems, a smaller model may be tested at a higher velocity.

I've used the scheme on quite a number of occasions. One happened when investigating induced ventilation of the burrows of prairie dogs (Vogel et al. 1973). Such burrows are unwieldy objects, about 0.1 m in internal diameter, about 15 m long, and located more than 2000 km from my laboratory. To look at the effect on internal flow of entrance and exit geometries, we built a model burrow ten times smaller than a real one, a model that fit into the local wind tunnel. To achieve the equivalent of an external wind of 1 m s^{-1} the wind tunnel was simply run at ten times that speed—about 22 mph. Perhaps a lot of stove pipe could have been buried in an open field, but no advantage would have justified the heroics.

As part of a more general look at flow induction, with its mix of Bernoulli effects and viscous entrainment, I wanted to see how the phenomenon scaled with Reynolds number (Vogel 1976). By combining models of a range of sizes with a range of speeds in two flow tanks, I could measure the rate of induced flow over three orders of magnitude. The scheme worked perfectly—the resulting curve had no discontinuities between segments obtained on different models.

Much of the work on flow-induced pressures mentioned in the last chapter was done in wind tunnels despite the fact that all the cases—scallop and algal thallus refilling as well as the squid—were aquatic (Vogel 1988b). It just happened that I had equipment for measuring very low pressures only in air. With scallops I used real shells and so worked at life size, with the normal fifteenfold speed increase. For the squid I made a model 0.3 rather than 0.2 m long, which meant that I could reduce speeds a little. Otherwise I would have exceeded the top speed of the wind tunnel whose use our local mechanical engineers kindly provided, not to mention tickling the tail of the dragon of compressibility. 8 m s^{-1} times 15 is 120 m s^{-1}; reducing that by 0.2/0.3 brought it down to 80 m s^{-1}, a manageable 180 mph.

One additional topic on modeling needs mention. Recall that equality of Reynolds number means equality of drag and pressure coefficients. It doesn't necessarily mean equality of either drag or pressures per se unless the original object and the model are both exposed to the same fluid medium. For a shift from air to water, with both at 20°, drag and pressures will rise, although not enormously—3.5 times if the size of the object remains unchanged. Water may be 800 times more dense than air, but the speed decrease in the shift of medium compensates for a lot of that. (800 divided by 15 squared is about 3.5.) But even that 3.5-fold can spell trouble when dealing with non-rigid objects, not at all an unusual thing for a biologist concerned with organisms. The behavior in flow of a leaf cannot be presumed the same at 1 m s^{-1} in water as at 15 m s^{-1} in air, and sometimes one prefers not to deal with models.

Sometimes a liquid of appropriate viscosity and density can be picked so an object that normally experiences airflow can be shifted to a liquid system without alteration of its actual drag and pressures. Deriving an expression that permits such a liquid to be chosen isn't at all difficult. With subscript 1 indicating the original air and subscript 2 the unknown liquid, constant Reynolds number means that

$$\frac{\rho_1 l_1 U_1}{\mu_1} = \frac{\rho_2 l_2 U_2}{\mu_2} .$$

And constant drag coefficient means that

$$\frac{D_1}{\rho_1 S_1 U_1^2} = \frac{D_2}{\rho_2 S_2 U_2^2} .$$

Using the same object implies that $l_1 = l_2$ and $S_1 = S_2$, so that the S's and l's in equations (5.5) and (5.6) cancel. We can then combine the two equations to eliminate U_1 and U_2, obtaining a simple and useful result:

$$\frac{D_2}{D_1} = \frac{\rho_1 \mu_2^2}{\rho_2 \mu_1^2} . \tag{5.5}$$

In shifting from one medium to the other, drag will be unchanged if D_2/D_1 = 1. Table 5.3 gives D_2/D_1 values for the transition from air to a variety of liquids for an object of a given size. Water at 52° would seem a reasonable choice if the specimen can stand the heat. Alternatively, mixtures of acetone and water or ethanol or of carbon disulfide and methanol might be useful; one might have to get out the Ostwald viscometer (Chapter 2) to get the viscosities of various mixtures.

I found this scheme quite useful when looking at the transmissivity to air of the antennae of the giant silk moth, *Luna*, a few years ago (Vogel 1983).

TABLE 5.3 INCREASE IN FORCE ON AN OBJECT IF SHIFTED FROM AIR TO THE INDICATED LIQUID.

Liquid	Increase Factor	Liquid	Increase Factor
Acetone	0.421	O-xylene	3.052
Carbon disulfide	0.379	Propanol	21.957
Carbon tetrachloride	2.069	Toluene	1.369
Ethanol	6.331	Water	3.467
Methanol	1.556	Water, 52° C	0.994
Octane	1.441	Seawater	4.137

NOTE: Liquids are at 20° C unless otherwise specified.

While I could measure wind speeds just behind an antenna, I couldn't tell exactly where to put my tiny anemometer to pick up air that had passed through rather than gone around the antenna. That called for flow visualization, but the antennae were too flexible to take the requisite force of room temperature water at the right Reynolds number. 52° water with dye markers worked fine as long as I kept immersion times fairly short.

The Drag of Simple Shapes and Sessile Systems

THE SUBJECT of drag has now been introduced, aspersions cast upon its origins, and defamatory remarks leveled at its character. In this chapter and the next we focus attention a bit more specifically on what organisms do about drag—how much drag they encounter and with what adaptations they increase or decrease it. After all the complications of the last chapter, you now appreciate that drag depends on shape in ways so subtle that no single numerical index can relate the shape of an object to its drag. Drag coefficients and Reynolds numbers, one must remember, are mainly devices of practical utility for organizing empirical observations.

A BIT ABOUT THE BIOLOGY OF DRAG

We talk quite a lot about copying nature, but we've really done it both deliberately and successfully surprisingly few times (Vogel 1992b). One of these was the occasion on which Sir George Cayley (1773–1857) devised what seems to have been the first deliberately drag-minimizing shape. Cayley recognized that in "spindle shaped bodies," "the shape of the hinder part of the spindle is of as much importance as that of the front in diminishing resistance." Since the explanation of the phenomenon seemed so mysterious, he quite frankly copied nature, picking a biological system—a trout—in which he guessed that drag was low. He measured the girth of various cross sections of the trout and then divided by three to get rough diameters. From the results he constructed the profile shown in Figure 6.1. Von Kármán (1954) has pointed out that Cayley's profile corresponds almost precisely to that of a modern, low-drag airfoil. The coincidence may seem somewhat surprising, since the body of a trout is a thrust-producing form. But fish do glide or coast, and (as Weihs 1974 has shown) they may improve their energy economy by alternately accelerating and gliding.

The biological relevance of drag needs little belaboring. The relatively high metabolic cost of locomotion in continuous media—swimming and flying—are indisputable; and, except for accelerations, the cost is a direct reflection of the drag that must be overcome. The coincidence in shape between large and fast swimmers and low-drag bits of human technology is profound and demonstrable in sink or bathtub (Kaufmann 1974). It applies to the fuselages of small aircraft, to the bodies of large birds (Pen-

FIGURE 6.1. The profile of a modern, low-drag airfoil superimposed on Cayley's data from girth measurements of a trout, the coincidence pointed out by Von Kármán (1954).

nycuick 1972), and to the hull shapes of ducks and boats (Prange and Schmidt-Nielsen 1970). Only a little less obviously, sessile organisms may pay a substantial price for incurring drag, a cost less easily denominated in units of energy, but evident in elaborate and massive supportive systems. Trees sometimes blow down and corals suffer damage in storms,[1] and trunks and stalks represent as real a metabolic investment as the machinery and operating expenses of locomotory systems. Indeed, much of the difficulty of looking at the biology of drag is that we see the solutions rather than the basic problems, and thus the latter must of necessity be inferred. Measuring the drag of some natural object yields only a datum. Only through controlled alterations of the object and comparisons with models of varying degrees of abstraction can much be said about its design. And any such statements must be based, as well, on information about the characteristics of the flow the living system encounters in nature.

The particular relevance of drag depends very much on the circumstances. For a fish or bird, propelling itself away from a possible contribution to the next tropic level, drag is clearly a Bad Thing. But for the same fish, faced with an urgent need to reverse course, or for the bird, about to land on a fixed branch rather than a runway, drag is undoubtedly most welcome. Achieving especially low (or, as we'll see, high) drag isn't an end in itself—drag isn't power, and so it's even further from the real currency of natural selection, fitness, than is metabolic expenditure. Guts and gonads have to be fitted into the fuselage of a bird—if drag were all that mattered, minimization of size would be the main element of successful design.

Important functions may require structures that incur substantial drag—suspension feeding and photosynthesis are only the most obvious cases. Thus sea fans on reefs are exposed to strong, bidirectional flow. They begin attached life oriented quite randomly but gradually grow to be perpendicular to the currents; apparently this *bad* orientation is important in maximizing the feeding efficacy of their tiny polyps (Wainwright and Dillon 1969). And sometimes survival in the most immediate sense depends on having sufficient drag, with nothing fixing fitness better than

[1] But breakage may be, in part, a reproductive device in corals; so the cost-accounting isn't as obvious as it seems (Tunnicliffe 1981).

parachute or patagium. I once dropped two mice onto concrete from about 15 meters. They certainly seemed no worse from the experience; but as interesting as the lack of injury was their adoption of a nicely parachutelike posture during descent, with all appendages spread wide and with no tumbling at all. Drag maximization clearly looked like a standard feature of their behavioral repertoire.

Several cautions need mention at about this point. Asking whether drag per se is what matters is very important, even in a system in which the forces of flow are obviously substantial. I already mentioned (Chapter 4) a limpet (Denny 1989) for which under some circumstances lift can be a worse problem than drag. But that's only the tip of the iceberg. We'll not talk much about unsteady flows until Chapter 16, but they can't be totally ignored until then. As a wave passes an attached organism, the organism experiences not just the drag attributable to the instantaneous velocity of the water but also a force due to the acceleration of the water crossing it. Perhaps the simplest way to view this "acceleration reaction" is to realize that if an object is accelerated in one direction, an equal volume of fluid or some equivalent of it must be accelerated in the other. So to accelerate an object, one must pay a force tax proportional to its volume and to the density of the medium. Wave surge causes a rapid *increase* in the speed of flow around an attached organism, so the situation is almost the same as that when the organism is accelerated—even if the frame of reference makes the phenomenon oddly nonintuitive. As Denny et al. (1985) point out, drag will increase with the area of an organism and thus with strength-dependent variables such as attachment area. But the acceleration reaction will increase with volume, so it becomes especially significant for large objects—if the flow has sufficient acceleration it may be the greatest force that a large organism must withstand. The moral is that it's as unsafe to ignore unsteady effects as to ignore lift, at least with flows of water. If your flow fluctuates, look at Chapter 16 or read Denny (1988) and try a few ballpark calculations.

We'll be concerned here mainly with minimization of drag, and we'll consider what happens at Reynolds numbers of roughly 100 and higher. Most of the interesting cases of deliberate maximization of drag occur at lower Reynolds numbers and are thus in a rather different physical world; they'll be deferred to Chapter 15. There's human relevance to these questions about the drag of organisms. Jobin and Ippen (1964) emphasized the crucial role of drag on snails in irrigation canals. Only with sufficient water velocity will snails be consistently dislodged and the local inhabitants not be exposed to schistosomiasis. Considerable effort has gone into determining optimal planting patterns for plantations of trees in which large-scale blow-downs ("wind-throw") of evenly aged stands are insufficiently uncommon. And excessive tendency toward "lodging" (the same thing) must be avoided when developing high-yield plants; about the matter, see Grace (1977) and Niklas (1992).

MORE ABOUT SHAPE AND DRAG

The distinction between streamlined and bluff bodies has already been mentioned. For organisms, the functional distinction isn't just a matter of whether drag is to be minimized, ignored, or maximized. It depends, in addition, on whether the direction of flow is constant or at least controllable through passive, behavioral, or morphogenetic reorientation. Streamlining, again, only works for a rather narrow range of relative orientations of object and flow.

And streamlining isn't a simple and definitive "cure" for drag, with one standard shape to replace a cylinder and another to replace a sphere. No—things are more complicated. First, the optimal shapes depend on the Reynolds number. At low Re (below 100 or so) minimization of pressure drag entails a significant rise in skin friction, so one minimizes overall drag with a somewhat stubbier shape than what would give the lowest pressure drag. And second, at all Reynolds numbers alternatives of equivalent drag are available, trading off, as it were, a convexity here for a concavity there; the further from some ideally absolute minimum the system is (for whatever perhaps laudatory reason), the more latitude will exist for such trade-offs. Finally, what is almost always being optimized will be some complex of variables. If, for instance, the object in question is a compression-resisting strut, then it needs a decently high "second moment of area" to resist buckling without excessive weight (for an introduction to such solid mechanics, see Gordon 1978). Consequently a wider cross section may permit a strut that's smaller as a whole, and a higher drag coefficient relative to any of the areas we defined in Chapter 5 may actually be concomitant with a lower overall drag.

Still, the tacit assumption just made that the wider strut has higher drag is reasonably safe at moderate and high Reynolds numbers; it even holds (except around $Re = 100,000$) for bluff bodies. Put in terms of streamlined airfoils (for simplicity assume no lift is being generated), the higher the ratio of thickness to chord (length in the direction of flow), the higher the drag coefficient based on wetted area, as shown in Figure 6.2. Tricks, though, are still possible, such as maintaining laminar flow by sucking fluid in through the porous skin of an object. The latter has been demonstrated in both streamlined and bluff bodies. Hoerner (1965) cites data for bluff bodies in which the drag coefficient (on frontal area) was reduced from 1.2 to 0.15 at $Re = 100,000$, and Riedl (1971a) suggested that the scheme might be used by sponges. Most sponges prefer habitats in which the water moves, and as we saw earlier, they suck in fluid through pores distributed over

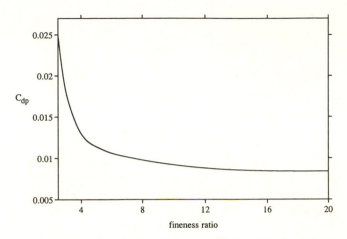

FIGURE 6.2. The variation of drag coefficient, based on plan form area, with fineness ratio (recall 5.7) for an airfoil section at a Reynolds number of 600,000. As an exercise, the reader might try to visualize such a graph with a drag coefficient based on frontal area.

their surfaces. But one attempt to demonstrate drag reduction in a sponge-like model (Susan Conova, pers. comm.) proved unsuccessful. Perhaps both the volume flow rates into sponges (recall Table 3.1) and the Reynolds numbers at which they live are too low for much of an effect. If the value of 100,000 is critical, then one is looking for a sponge 10 centimeters in diameter in a flow of a meter per second, which is pretty big and torrentially fast.

Another trick is nearly the opposite of sucking—ejecting high-velocity fluid in a downstream direction around the normal separation point. Ejection of fluid near the appropriate location happens when fish discharge water through their opercula, either actively or by passive ram ventilation (Chapter 4); and Freadman (1979, 1981) and Steffensen (1985) have shown in bass and bluefish that ram ventilation, at least, produces flow over the body with less turbulence, separation, and drag.

Yet another trick is the use of "splitter plates" (Figure 6.3) behind bluff bodies to reduce the rate at which vortices are shed. Again Hoerner's (1965) compendium supplies data. Benefits are obtained at Reynolds numbers between 10,000 and 100,000, certainly a biologically well-inhabited domain. Chamberlain (1976) suggested that the trailing part of a cephalopod shell acts as a splitter plate, and Dudley et al. (1991) suggested the same for the tail of a gliding tadpole.

At least one difference between the behavior of streamlined and bluff bodies is more apparent than real. The transition to turbulent flow appears

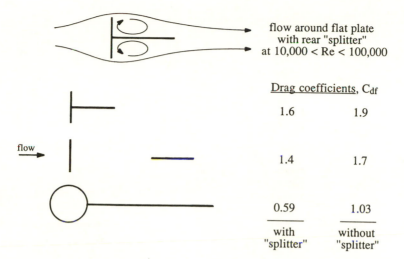

flow around flat plate
with rear "splitter"
at 10,000 < Re < 100,000

Drag coefficients, C_{df}

1.6	1.9
1.4	1.7
0.59	1.03
with "splitter"	without "splitter"

FIGURE 6.3. The influence of "splitter plates" on flow patterns and drag coefficients of long flat plates and cylinders. Reynolds numbers are based on maximum dimensions normal to flow and drag coefficients on frontal areas.

to occur at different Reynolds numbers, between 100,000 and 250,000 for bluff bodies and between 500,000 and 1,000,000 for streamlined ones. But that's mostly an accident of the convention used to pick the characteristic length in the Reynolds number. If, as is usual, maximum length in the direction of flow is used, then the relatively more elongate shape of stream-lined bodies will automatically result in higher Reynolds numbers. The real difference is what happens at transition—for the bluff body the onset of turbulence close to the surface leads to a narrower wake, less pressure drag, and hence less total drag than just prior to transition. For the streamlined body, separation is absent or minor, pressure drag is quite small, and turbulence, if it does anything, almost always increases drag.

Streamlined and bluff bodies are best regarded as the extremities of a continuum of shapes, with degrees of bluffness as well as variations in the effectiveness of streamlining. With a great enough thickness-to-chord ratio, the presence of some posterior pointedness will do little to alter the performance of what is just a minimally disguised cylinder. And the shape of the anterior of a bluff body makes a considerable difference to its drag, although in no version would we refer to it as streamlined. Thus rounding or fairing the upstream edges of a flat-fronted object reduces the width of the wake and the drag coefficient. Figure 6.4 gives some data for bluff bodies at moderate and high Reynolds numbers. For instance, a solid hemisphere with the flat side facing upstream has about the same drag as a

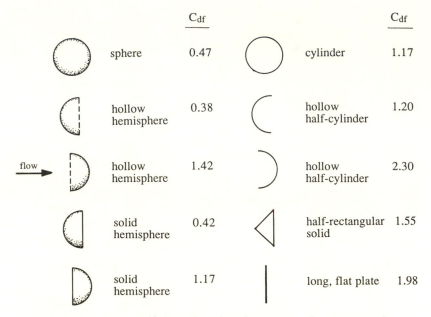

		C_{df}			C_{df}
	sphere	0.47		cylinder	1.17
	hollow hemisphere	0.38		hollow half-cylinder	1.20
flow →	hollow hemisphere	1.42		hollow half-cylinder	2.30
	solid hemisphere	0.42		half-rectangular solid	1.55
	solid hemisphere	1.17		long, flat plate	1.98

FIGURE 6.4. Drag coefficients, based on frontal area, for a variety of three-dimensional bodies and two-dimensional profiles at Reynolds numbers between 10^4 and 10^6. All the data may not be precisely comparable due to variations in the experimental conditions.

circular disk, but it has over twice the drag of a sphere of the same frontal area or, for that matter, the same hemisphere with the flat side facing downstream. A flat front is a bad bluff, whatever the implications in poker.

Still bluffer, as you can see from the figure, is a concave front. A hollow hemisphere facing upstream has about 20% more drag than a solid hemisphere or circular disk, three times the drag of a sphere, and almost four times the drag it would have if facing downstream. This last comparison, by the way, explains why whirling cup anemometers turn. The cup facing the wind has a higher drag, not due to any air it contains, but because it has a much wider wake than that of the cup facing downwind.[2] A corresponding difference occurs with long, hollow half-cylinders as well, which have C_{df}'s of 1.17 and 0.42 when facing upstream and downstream, respectively. This last difference is the basis for the operation of a "Savonius rotor," a pair of half-cylinders on opposite sides of a vertical shaft. It makes an inefficient

[2] Notice, the next time you see one, that whirling cup anemometers have three cups, not two or four. One with two or four cups will work, but it doesn't start as well or run as smoothly. We ran afoul of the problem when designing a current-driven stirrer for small chambers to be located on a bay bottom (Cahoon 1988); it was just another thing obvious only post facto.

windmill, drag based instead of lift based; for a time it enjoyed some countercultural appeal since the open half-cylinders could be nothing more than the halves of an ordinary oil drum that had been cut lengthwise. Structures in which a concave side faces upstream occur in organisms, most commonly in connection with passive suspension feeding. The same pressure drop responsible for their high drag is what drives fluid through the filter—suspension feeders can't have their cake and eat it too. Thus the front of a sea pen (Figure 6.5a) is more or less a hollow half-cylinder (Best 1988), and the cephalic fans of black fly larvae (Figure 6.5b) living atop rocks in rapid streams are cuplike. These fans contribute about half of the total drag of the creatures (Eymann 1988).

Very much the same story applies to axisymmetrical bodies oriented with their axes parallel to flow (Figure 6.4 again). A concave face is worst, a flat face is somewhat better, a convex hemispherical front is better yet, and a parabolic nose has about the lowest drag of all. The same ordering holds for a "rectangular section," a long cross-flow beam with a chord much larger than the thickness. Rounding the upstream edge, much as the exposed edge of a stair tread is rounded, can reduce the drag coefficient by 40%. By rounding the front, the drag coefficient of the original Volkswagen "Microbus" was reduced from 0.73 to 0.44—not the order of mag-

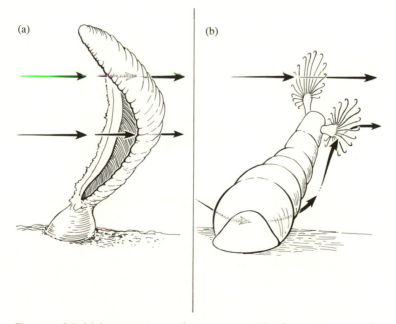

FIGURE 6.5. Living structures whose concave sides face upstream and which thus have high drag coefficients: (a) a sea pen, *Ptilosarcus*; (b) a larval black fly, *Simulium*.

nitude improvement possible with full (and impractical) streamlining, but certainly no inconsequential difference. Air deflectors atop the cabs of articulated trucks function in a similar manner. At least the proper ones do—I've occasionally seen one with a sharp center ridge and the overall shape of a snowplow that might be worse than none at all.

A streamlined body over which flow goes from trailing to leading edge is still a kind of streamlined body, but it's not a particularly bad one in terms of drag. For an ordinary airfoil with a maximum thickness of 12% of the chord, the drag coefficient (based on frontal area) is 0.06 for normal flow and 0.15 for reversed flow. While that sounds bad—over twice as much drag when flow is reversed—it's still only an eighth of that of a cylinder at the same Reynolds number, which in this case is a million. Decent streamlined shapes exist with relatively low but equal drags for opposite flow directions; they have both edges pointed rather than rounded. While irrelevant for aircraft, which are rarely called upon to fly backwards, they matter for underwater protrusions on bidirectional boats (some ferries, for instance) and for the blades of reversing fans. Bidirectional streamlining may be important in the design of fixed and rigid organisms exposed to alternating wave or tidal flows.

Most of the data discussed in the past few pages are strictly applicable only at Reynolds numbers considerably higher than those commonly encountered by ordinary organisms; they're in the range of about 10,000 to 1,000,000. There seems to be a great dearth of comparable data for the range of 100 to 10,000 and almost nothing for 1 to 100. And extrapolation is hazardous—a reasonable way to generate hypotheses but certainly unsafe as hard evidence. Perhaps we biologists can subvert or convert an engineer or two, or perhaps we'll just have to settle down to some exploratory data collection.

FLEXIBILITY—WHERE DRAG DEPENDS ON SHAPE DEPENDS ON DRAG

The structures thrown up by human technology are mostly stiff ones, at least by contrast with most of the bits and pieces of organisms—a point that arose in Chapter 4. While organisms do use stiff materials, there's almost always some special explanation for their occurrence—bones as levers for muscles to pull on; especially cheap material such as the calcium salts of a supersaturated sea used by corals; teeth as cutters, slicers, grinders, and other stress concentrators; some functional intolerance of sag in a flat surface; and so forth. The less stiff, more compliant designs of nature are probably (looked at grandly and globally) more economical of material when measured against strength achieved, which ought to be argument enough for their prevalence.

Compliance, or flexibility, has major implications for what happens to objects in flows. It can be either blessing or curse; what's certain is that it creates complications. And the complications are of at least two sorts. First, shape is no longer a given, an independent variable upon which investigator or analyst can rely. Shape becomes an immediate function of flow speed, and the way drag scales with speed assumes unusual interest. Second, flows themselves behave in odd ways when moving along surfaces that don't stay put, and the forces involved may be either increased or decreased through such interactions.

What must be said with vehemence is that the introduction of flexible solids into the story of objects in flows destroys the tidy distinction, not between solids and fluids as such, but between solid mechanics and fluid dynamics as areas of inquiry. The biologist may be able to ignore compressible flows and make other simplifying assumptions, but this most convenient distinction of traditional engineering is very likely to mislead us. Recognition of the special situation isn't especially recent; Carstens (1968), for instance, noted at least as a possibility the reduction of drag by compliance of an organism. But Koehl (1977) really drew our attention to it—so effectively that at least here at Duke University where she did the work we've taught solid and fluid biomechanics in a single course ever since. Even the title of her paper makes the point: "Effects of Sea Anemones on the Flow Forces They Encounter."

These sea anemones are hydrostatically supported; and, as is commonly the case for hydroskeletal systems, they're not especially rigid when faced with bending loads—their flexural stiffnesses are low. Beyond that, they can deflate and, in extreme cases, retract down against the substratum to form a hemispherical lump. Deflation is only part of a complex behavioral response to drag. A tall anemone, *Metridium senile*, protrudes well into the mainstream, bends downstream at a constriction just beneath the oral disk, and filters with its outstretched tentacles (Figure 6.6). Even just reconfiguration of the tentacles without any alteration in the anemone's trunk causes major changes in drag. For a rigid model equipped with flexible tentacles, the drag coefficient is about 0.9. As the oral disk separates into lobes, the coefficient drops to 0.4; collapse of the oral disk in the manner of an overstressed umbrella reduces the coefficient to 0.3; and retraction of the tentacles reduces it to 0.2. All of these coefficients are based on the original frontal area, so they reflect proportional decreases in drag itself.

We are all too prone to speak of something being *deformed* or *distorted* by the forces of flow, choices of words that carry the bias of a culture with a penchant for stiff things, where alteration in shape is pathological or geriatric and most often irreversible. When dealing with biological structures exposed to flows of varying velocities, I urge the use of the term "reconfiguration," with a neutral to positive connotation to the simpler but

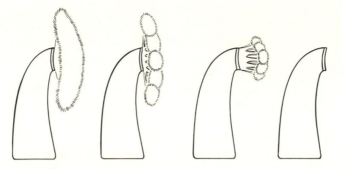

FIGURE 6.6. Successive changes in the appearance of a large sea anemone, *Metridium*, as current increases. Drag coefficients (based on original frontal area) are 0.9, 0.4, 0.3, and 0.2.

prejudice-laden "deformation." We might prefer it otherwise, but choices among words affect attitudes and hypotheses, even in science.

A Measure of Reconfiguration

We now have information about how quite a few flexible organisms alter their shapes and thus their drag coefficients under the influence of the flows they experience. What may be more useful than a summary is to suggest how to compare such systems, to give a generally applicable scheme for putting a number on the degree of reconfiguration. What we need is some variable that indicates how the change in drag with speed of a flexible object is different from that of a rigid object, a variable that draws attention to what's special by correcting for what's ordinary. Drag behaves in such queer ways that "ordinary" may sound like wishful thinking; still, the everyday rule that drag is proportional to the square of velocity does work for some objects on some occasions. Indeed, the rule describes precisely what's going on whenever the drag coefficient doesn't vary with changes in the Reynolds number, just by the way the drag coefficient is defined (equation 5.4). For a sphere or cylinder between $Re = 1000$ and transition and again above transition, or for a flat plate broadside to flow at Reynolds numbers above 1000, the rule works—in short, for bluff bodies at moderate to high Reynolds numbers. For such situations,

$$D \propto U^2 \tag{6.1}$$

and so

$$C_d \propto U^0 \propto Re^0. \tag{6.2}$$

Plotting drag against speed (6.1) not only doesn't adjust for size, but it makes any differences inevitably loom larger at higher speeds simply because at low speeds all curves converge. Plotting drag coefficient against speed or Reynolds number (6.2) avoids this illusion; even better, any deviation from a horizontal line on such a graph is a telltale sign of something other than an ordinary bluff body. We have, in such a graph, a rough-and-ready way of separating the noteworthy from the ordinary.

But the Reynolds number includes a length factor, drag coefficient incorporates an area factor, and biologically interesting objects come in a wide variety of shapes. So looking at changes in drag coefficient all too often gets tangled up in the complications connected with comparisons among objects that differ in both shape and size. But we can get around these problems of shape and size, problems afflicting both the rigid and the flexible, by simply ignoring them. Drag coefficient is by definition proportional to drag divided by the square of velocity, while Reynolds number is proportional to velocity. Plotting drag over velocity squared against velocity gives a graph similar to one of drag coefficient against Reynolds number. For cases where drag coefficient is independent of Reynolds number,

$$\frac{D}{U^2} \propto U^0 \propto Re^0. \tag{6.3}$$

Thus the result will simply be a horizontal line—quite literally, a baseline. If, on the other hand, the line ascends, that means the object has a disproportionate drag at high speeds, at least by comparison with our paradigmatic bluff body. A descending line, by contrast, means that the object has relatively lower drag at high speeds. Even the shape of the line is of interest. Thus for small holly and pine trees I found that at low speeds, this D/U^2, what we might call "speed-specific drag," increases, while at higher speeds it decreases (Figure 6.7). What's happening is that at low speeds the wind mainly disorganizes the originally fairly parallel or coplanar needles or leaves, while at higher speeds they reorganize into increasingly tight clusters (Vogel 1984). Of course at low speeds the real (as opposed to speed-specific) drag is low enough to matter little—one mustn't lose sight of the way we've transformed the data.

A further degree of abstraction permits nongraphical comparisons. If one puts speed-specific drag and speed on a log-log plot, then the slope of a line (or a piece of a line) is the exponent of the relationship between the original (nonlogarithmically transformed) variables. One is, in effect, looking at the value of E in a general version of proportionality (6.3):

$$\frac{D}{U^2} \propto U^E. \tag{6.4}$$

The exponent that we're calling E then serves as the measure of reconfiguration. A value of zero is a baseline, a horizontal line on the graph, a statement that since drag is proportional to the square of velocity nothing worthy of comment is happening. A value above zero says that the system is

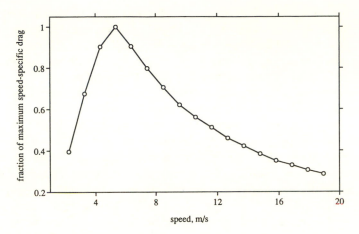

FIGURE 6.7. Speed-specific drag (fraction of maximum) versus speed for a small branch of a loblolly pine (*Pinus taeda*). For speeds above 6 m s^{-1}, $E = -1.13$.

reconfiguring or behaving in a way that makes drag get disproportionately bad at high speeds; thus a value of $+1.0$ indicates that drag is proportional to the cube rather than to the square of velocity. A value below zero signals a reconfiguration that produces less than the expected drag at high speeds—if the value gets down to -2.0, then drag is actually independent of velocity. Table 6.1 gives a a collection of values of E.

Bear in mind that these values of E are not measures of either drag or drag coefficient. What they indicate is how drag changes with speed, quite independent of the specific magnitude of drag or its coefficient—they're a look at reconfiguration, only half the story, but a particularly interesting half in the living world of flexible objects.

In addition, a few cautions about using and interpreting them need to be mentioned. The first is something of a corollary to one made earlier—that drag isn't an end in itself but merely a factor that may or quite possibly may not bear on fitness. A low value of E may be the quantitative signature of a reconfigurational process that lowers drag at high speeds, but that doesn't automatically mean that drag is what matters. Two separate studies of suspension-feeding cnidarians provide a useful exposure to cold water. Harvell and LaBarbera (1985) got nicely negative E-values of around -1.28 for a small, branched, colonial hydroid, *Abietenaria*. At the same time they showed quite convincingly that (1) these creatures would never encounter flow speeds that posed any mechanical hazard to the colony, and (2) the way colonies bent in flow prevented the flows to which individual polyps were exposed from varying nearly as much as the flows facing the colony as a whole. More recently, Sponaugle and LaBarbera (1991) got

TABLE 6.1 VALUES OF E FOR VARIOUS SYSTEMS AND SPEEDS, WHERE $UE \propto D/U^2$.

System	Re or Speed Range	E	Source of Data
Bluff body	<1.0	−1.00	
Bluff body	1000–200,000	0.00	
Flat plate, parallel to flow	10–1000	−0.60	Janour 1951
Flat plate, parallel to flow	1000–500,000	−0.50	
Streamlined body, laminar flow	1000–500,000	−0.50	
Cylinder, axis normal to flow	20–120	−0.29	White 1974
Hedophyllum sessile (alga)	0.5–2.5 m/s	−1.12	Armstrong 1989
Nereocystis luetkeana (alga)	1.3–2.0 m/s	−1.07	Koehl & Alberte 1988
Sargassum filipendula (alga)	0.5–1.5 m/s	−1.47	Pentcheff (pers. comm.)
Laminaria (alga) on mussels	0.12 to 0.62 m/s	−1.40*	Witman & Suchanek 1984
Macroalgae, marine	ca. 2.5 m/s	−0.28 to −0.76	Carrington 1990
Red algae, freshwater	0.2–0.75 m/s	−0.33 to −1.27	Sheath & Hambrook 1988
Pinus sylvestris (pine)	9–38 m/s	−0.72*	Mayhead 1973
Pinus taeda, 1 m high	8–19 m/s	−1.13	Vogel 1984
Pinus taeda, branch	8–19 m/s	−1.16	"
Quercus alba (white oak), leaf	10–20 m/s	+0.97	Vogel 1989
Quercus alba, clustered leaves	10–20 m/s	−0.44	"
Other broad leaves & clusters	10–20 m/s	−0.20 to −1.18	"
Ptilosarcus gurneyi (sea pen)	0.11–0.26 m/s	−1.14	Best 1985
Pseudopterogorgia (gorgonian)	0.13–0.35 m/s	−1.66	Sponaugle & LaBarbera 1991
Abietenaria (hydroid)	0.025–0.40 m/s	−1.28*	Harvell & LaBarbera 1985
Acropora reticulata (hard coral)	1.5–3.0 m/s	+0.26*	Vosburgh 1982
Various limpet shells	0.15–0.45 m/s	0.0 to +1.2	Dudley 1985
Epeorus sylvicole (mayfly larva)	0.4–1.2 m/s	+0.28*	Weissenberger et al. 1991
Simulium vittatum (blackfly larva)	0.1–0.7 m/s	−0.64*	Eymann 1988
Locusta migratoria, antenna	20–120	−0.56	Gewicke & Heinzel 1980

NOTE: Asterisks indicate my calculations from published graphs.

very much the same result on a gorgonian (flexible) coral, *Pseudopterogorgia*. The latter, a single unbranched upright cylinder, is quite different in shape from a hydroid; its E-value, -1.66, is even lower. I don't mean to sound disparaging; by using their flexibility for reconfiguration, these cnidarians are doing something distinctly more subtle and neat than mere drag reduction.

The second caution concerns that baseline of zero—it presumes exposure to moderate to high Reynolds numbers, free-stream flow, and a fairly bluff body; to repeat, it presumes situations in which the drag of a rigid object would be proportional to the square of velocity. At low Reynolds numbers drag gradually shifts toward a direct proportionality, as we saw in Figures 5.3 and 5.4. Thus the baseline ought to drop from zero toward minus one. So the -0.56 of a locust antenna is best judged next to the value for a cylinder at the same Reynolds number, -0.29. Similarly, for a flat plate parallel to flow, the moderate and high Reynolds number E-value is -0.5, not 0.0 (see Table 5.2); the object is essentially a perfectly streamlined object with skin friction only. So streamlining alone, in the absence of flexibility, gives negative E-values. Finally, what about organisms that live in the velocity gradients near surfaces? We'll have a lot more to say about them in Chapters 8 and 9; for now what matters is how the speed of flow at any point deep in that gradient varies with free-stream speed. In these velocity gradients, local speed increases more drastically (even if it's always absolutely lower) than free-stream speed—when the latter increases, not only does dU increase, but dz decreases. Thus the gradient region gets thinner and the gradient steeper, and local speed is proportional to free-stream speed to the power 1.5. So ordinarily drag ought to be proportional to free-stream speed cubed rather than squared. Our baseline therefore shifts up to $+1.0$. By this criterion, Dudley's (1985) rigid limpet shells don't look anomalous, and the mayfly larvae (a comment of Weissenberger et al. 1991 notwithstanding) may be doing something special.

A final caution concerns extrapolation. Bluntly, don't. Look at the graph of speed-specific drag versus speed for a pine branch (Figure 6.7) if you want to see how shaky are the grounds for extrapolating from high to low speeds. And just consider the practical limits of nondestructive reconfiguration if you're tempted to extrapolate to higher speeds. Carrington (1990) makes the latter point, noting that her E-values, obtained at higher speeds, are not as negative as those obtained for macroalgae by most other people.

The Drag of Leaves on Trees

Preeminent among large organisms for whom drag is important are trees. They may exceed 50 m in height, they have enormous surface area, they withstand major windstorms, and they constitute a substantial proportion of the earth's terrestrial biomass. In a sense, trees have been dealt a curiously awkward hand by the way the evolutionary process works. They commonly compete among themselves, both intra- and interspecifically, for their prime resource, sunlight. The tree that gets shaded (at least for the larger players in the game) is the loser, so taller is better—even though, with a sun over 100 million kilometers away, no one gets appreciably closer by all that growth. No anticompetitive treaty seems possible; no way seems open for an agreement whereby each individual foregoes, say, 10 meters of trunk, in order that all be less vulnerable to mechanical failure. Thus the greatest part of their surface area ends up at the greatest distance from the earth. The result, as shown in Figure 6.8, is a huge lever arm converting the drag of the leaves to a turning moment about some axis near the base.[3] As Alexander (1971), among others, has pointed out, the local forces generated in the trunk by winds may far exceed those due to the weight of a tree—a tree may fall under wind stress but is most unlikely to do so under its weight alone.

So what is the drag of a tree? The practical problem is formidable. You can't just measure the drag on a leaf or branch and blithely multiply. If winds were steady, one might pull on a tree with a force-monitoring cable and calibrate its own bending; then the observed bending in a wind of known speed would give the drag, or at least the turning moment. But winds are rarely steady enough, and (as far as I know), the approach hasn't been tried. On one occasion a series of pine trees were tested in a large wind tunnel (Fraser 1962), and drag was measured at speeds between 9 and 26 m s^{-1}. Sometime later the original data from these experiments were further analyzed (Mayhead 1973). As speed increased, the drag in all cases increased, with no discontinuities anywhere in the data—as we'd expect, the parts of such irregular objects experienced no simultaneous sharp transitions in flow regime. But as the wind speed increased, the coefficients of drag (based on original frontal area) decreased since the trees steadily decreased their exposed area. As a result, the increase in drag was more nearly proportional to the first than to the second power of velocity. Put in terms of our measure of reconfiguration, E was -0.72, much closer to -1.0 than to 0.0.

---

[3] The mechanics of what's called "windthrow" in trees is too complex and incompletely understood for easy summary here. There seem to be several rather different schemes for dealing with that evil turning moment. To complicate things further, trees are much more vulnerable to repeated gusts than to steady winds. Blackburn et al. (1988) and Niklas (1992) give a lot more information, but I know of no really satisfactory account of the phenomenon in all its aspects.

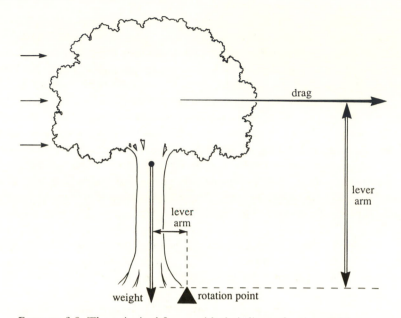

FIGURE 6.8. The principal forces with their lines of action and lever arms caused by a wind blowing on a stiff tree such as a large oak.

It would be very nice to have a tidy comparison between trees and some marine equivalent, in view of the higher overall forces to which a body in an ocean can be subjected (about which more shortly). Most large attached marine organisms—macroalgae and corals, mainly—aren't very treelike in shape and loading regime. But at least one alga, the so-called sea palm, *Postelsia palmaeformis* (see Figure 6.10), has flat photosynthetic structures on top of an upright cylindrical stalk over half a meter high and lives in dense stands on wave-swept shores. Thanks to the work of Paine (1988) and Holbrook et al. (1991) we know quite a lot about both the ecology and biomechanics of *Postelsia*. What we don't know, though, is very much about its drag when subjected to interestingly high flow speeds—and the relevant measurements are only moderately heroic.

Quite clearly the major contributor to the drag of most trees is the drag of the leaves, whether broad or needlelike. A defoliated pine tree barely moves in a breeze that makes a normal tree sway quite dramatically. Around here, in the piedmont of North Carolina, broad-leafed trees blow over much more commonly in summer than in winter—even though in the summer they enjoy the wind-sheltering effect of the leaves of their neighbors. Whether a tree avoids uprooting by having a stiff trunk and broad base, as in the figure, or whether it presses sideways against a deep and sturdy taproot, the real culprits are the leaves. I was surprised to find, a few years ago, that no one had looked at the behavior of individual leaves or clusters of leaves at potentially destructive (to tree or leaves) wind speeds. So I did a few measurements (Vogel 1984, 1989) of drag and areas and took some pictures (Vogel 1993) of the reconfiguration; it proved to be about the easiest bit of science I've ever tried.

What might be regarded as the baseline against which broad leaves should be judged is the drag of a flag rather than the drag of, say, a rigid flat plate. And flags turn out to have very high drag. According to Hoerner (1965), local separation causes flutter, which causes further separation; the result is a substantial pressure drag that wouldn't occur if the surface were rigid, as in a weathervane. For fairly ordinary fabric, a flag as long as it is high, and a Reynolds number of two million, he cites a figure of 10-fold for the increase in drag attributable to its facility for flapping. I got comparable, even slightly higher figures, using a 0.1 m square piece of polyethylene sheet trailing behind a batten—17-fold at 10 m s^{-1} (Re = 65,000) and 12-fold at 20 m s^{-1} (Vogel 1989). (However, the E-value for the polyethylene sheet over this speed range, -0.57, isn't so bad, which reemphasizes the point about how that variable tells only half the story.)

Put most bluntly, leaves turned out not to be flags. Even in very turbulent winds most leaves fluttered little, and they reconfigured into cones and cylinders that became ever more tightly rolled as speed increased up to (in most cases) 20 m s^{-1}. Leaflike models cut from sheets of plastic or stainless steel didn't do quite as well but were still much better than flags—the response clearly involves a lot more than just outline. Shape is obviously a critical part of the game. Those leaves with long petioles that attached to their blades distal to the basal extremities of the blades—tuliptree, maple, sweet gum, sycamore—rolled into cones. It looks as if the free basal lobes catch the oncoming wind and start the rolling process (Figure 6.9a). Pinnately compound leaves—black locust and black walnut—rolled into cylinders, with the leaflets interdigitating somewhat like the scales on a fish (Figure 6.9b). Clusters of leaves reconfigured as well, forming bundles that became tighter as the wind speed increased. Concomitantly, the drag coefficients of these reconfiguring leaves were very much closer to those of rigid plates than to those of flags. In general, cylindrically reconfiguring pinnate leaves had lower drag coefficients than did conically reconfiguring simple leaves, and clusters had lower drag coefficients than did single leaves. A complex and versatile hierarchy of reconfiguration must begin with the individual leaves I looked at and end with the whole trees tested by Fraser (1962).

Almost all the E-values for leaves, clusters, and trees were negative, clearly reflecting the reconfigurational process. The aberrant item, single leaves of white oak, may tell us something as interesting. The latter not only

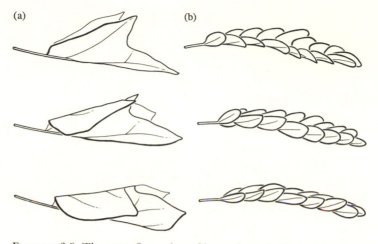

FIGURE 6.9. The reconfiguration of leaves in high winds: (a) tuliptree or yellow poplar (*Liriodendron tulipifera*); (b) black locust (*Robinia pseudoacacia*).

had relatively high drag, but the drag increased with approximately the cube rather than the square of speed—at high speeds it actually exceeded that of the piece of plastic sheet relative to surface area. Individual oak leaves, even though they look sturdy, commonly suffered damage at speeds (15 to 20 m s^{-1}) tolerated by the other leaves. But for one thing, oak clusters weren't quite so bad; for another, oaks may derive a compensatory benefit. In modest winds (either outdoors or on a branch in a wind tunnel) oak leaves maintain their normal, skyward orientations, while others, such as maples, have begun to turn and flutter.[4] In general, some instability at low speeds seems concomitant with good reconfigurational capability when winds increase. The extreme example may be the quaking aspen (*Populus tremuloides*). I found that the locally available congeneric, white poplar (*P. alba*), which with flattened petioles is a proper quaker, does very well at high speeds, with low drag in clusters and a tolerance even by isolated leaves of winds above 30 m s^{-1}. Despite a lot of speculation and measurement on what advantage might derive from quaking, I suspect it's of no particular functional consequence at the speeds at which it occurs. Rather, it may be just the extreme case of low-speed instability associated with good high-speed performance.

Intertidal Macroalgae Undulating in the Waves

Most of us regard algae as tiny organisms, perhaps pelagic unicells or small aquatic encrustations. But especially in rich, cold waters, individual

[4] So—shift from oak to maple and turn over a new leaf.

plants may grow tens of meters long on rocky coasts in water that's far from placid. This adds up to lots of drag and to major mechanical problems. Not surprisingly, they've received a fair bit of attention. In one way at least, the problem facing most macroalgae is simpler than that of trees—the loads imposed by drag are almost entirely tensile and avoid all the complications of more complex bendings and twistings.

For macroalgae, flexibility can be advantageous in a way that escaped notice until fairly recently (Koehl 1984; Koehl et al. 1991); for some particularly long and flexible algae, the problem of drag may be very much less acute than one might imagine. If the length of the alga exceeds the product of average water speed and half the wave period, then the alga may simply never be fully extended in either direction. Beyond a certain length, further length contributes nothing additional to the force pulling on the holdfast—more on this in Chapter 16. Many macroalgae are sufficiently long for the phenomenon to be significant—one might say that they grow to great lengths to avoid drag. Leaves and trees can't play the game since the structures are too small and the speeds and periods far too great.

Can we say anything about macroalgal structure analogous to our comments on leaf shape? One common feature of large laminate algae is a ruffled or "undulate" margin (see, for example, Bold and Wynne 1978 and Figure 6.10). This is the character that results in the folds one sees in the edges of algal fronds pressed flat in museum collections. Ruffling is minor in terrestrial leaves, but it occurs in some submerged leaves of stream vegetation (Sculthorpe 1967). The resulting play on words tempts one to regard an undulate margin as an adaptation to wave action, but enough cases have now been investigated so the hypothesis no longer survives close scrutiny. For instance, Koehl and Alberte (1988) found that the bull kelp, *Nereocystis luetkeana* (Figure 6.10), had flat, straplike fronds in areas of high exposure, with the wider, undulate-margined forms limited to areas with lower currents. They showed that at a speed of 0.5 m s^{-1} plants from exposed sites had only a fourth the drag per unit surface area of plants from more protected sites: the flat, narrow blades stacked tightly in high flows. On the other hand, they suffered more self-shading, with the expected effect on photosynthesis. Interestingly, plants from exposed areas had both lower drag relative to their areas and more negative E-values: -1.07 versus -0.80 at speeds up to 2 m s^{-1}.

Armstrong (1989) got similar results working on a rather cabbagelike alga, *Hedophyllum sessile* (Figure 6.10), quite unlike the elongate *Nereocystis*. Specimens from sheltered sites had broader, bumpier blades and experienced higher drag at any specific speed than those from exposed sites; while both clearly reconfigured, the latter achieved more compact shapes at high currents speeds. Van Tussenbroek (1989) found essentially the same thing in *Macrocystis pyrifera*, with thicker and narrower blades from specimens in exposed sites. Probably the most dramatic intraspecific differ-

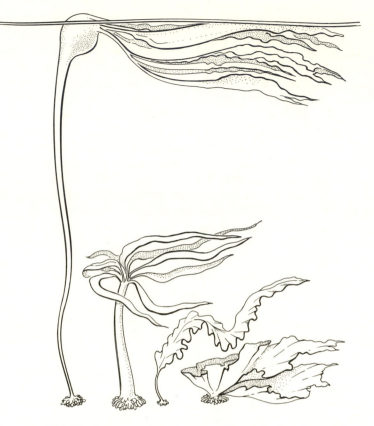

FIGURE 6.10. Marine macroalgae that reconfigure in flow. From left to right: *Nereocystis, Postelsia, Laminaria, Hedophyllum.*

ences are those measured by de Paula and de Oliveira (1982) in *Sargassum cynosum*. In sheltered places plants grew five times as high and had over five times the biomass, yet had holdfasts with nearly 40% less bottom area; plants in exposed areas invested 47% of their dry weight in holdfast while plants from sheltered areas invested about 11%. And a holdfast is just what the name implies—a structure whose sole function is to counteract drag.

A reasonable view, one I think was originally suggested by Gerard and Mann (1979), is that the undulate margin is a device to keep individual lamellae from getting too tightly appressed to each other, with consequent deleterious effects on light interception and (perhaps occasionally) uptake of dissolved substances. But in places when local water movements get really severe, the arrangement is simply intolerable in that the lamellae have to clump tightly during periods of high flow for purely mechanical reasons of drag and strength. What must be emphasized, since algae are not commonly regarded as especially sophisticated organisms, is the complexity of the intraspecific tuning of both overall morphology and the reconfigurational game from habitat to habitat. The magnitudes of the intraspecific differences seem to be substantially greater even than those found (as we'll discuss just below) in terrestrial vascular plants.

GROWTH ALTERATIONS IN RESPONSE TO DRAG

We live in bodies that adjust their structure in response to patterns of use, including mechanical loading. We build muscle through exercise, our bones rearrange their structure after injury or changed loading regime, and people deprived of gravity for much time suffer all sorts of debilitating changes in their mechanical equipment. The phenomenon of readjustment proves to be more widespread and dramatic than one might imagine; it's as good an illustration as I know that organisms are not just some mindless (or feedbackless) constructs of genetic blueprints. We've already seen that macroalgal structure has a wide range of adaptively appropriate intraspecific variation. In these organisms transplant experiments suggest that selection, initial development, and regrowth rather than rearrangement of preexisting structures underlie most of the adjustment. Terrestrial vascular plants appear to make more use of rearrangement—trunks and branches are usually long-lived elements even if leaves often are not. The distinction, though, isn't tidy since much of a large plant is made of essentially dead material external to the active protoplasts. Thus whether differential growth through incremental addition to one place rather than another should be considered rearrangement is in part a definitional matter. Old observations (Jacobs 1954 is usually cited) show that a bit of swaying leads to a tree with a thicker trunk, that guyed trees grow more slender, that exposed, previously sheltered trees very often get blown down after their neighbors have been cut.

Extreme examples of peculiarities in plant growth in windy habitats abound, as do nonmechanical explanations involving such factors as temperature and water availability. Trees at high altitudes where winds are fairly unidirectional are "flagged"; that is, the branches protrude mainly downwind. Whatever the chain of causation, these trees do have lower drag when presented with winds in the direction to which they've been previously exposed (Telewski and Jaffe 1986). In the tropics, a peculiar kind of montane rain forest often occurs on exposed ridges, one characterized by a dense growth of short, gnarled trees with their crowns packed into a low canopy. Lawton (1982) measured tree heights and winds in and above such an elfin forest and did the same farther down on the mountain slopes; either within (despite its density) or above the trees, speeds are much higher in the exposed forest. The suggestion is that energy has to be used to acquire strength as opposed to height to resist the extra wind. These and other cases (such as Palumbi's [1986] observations on sponges) make what I think is an important point. To repeat, drag reduction is not an end in itself. If periods of substantial wind are short and intermittent, then even very extreme reconfiguration through flexibility may be a fine thing—cheaper construction justifies giving up photosynthesis or suspension feeding for short periods, and one should go with the flow. If, on the other hand, rapid flows are chronic, then the cost of going hypofunctional is high, and a steadfast stand may be preferable.

The mechanisms by which plants respond to wind have come in for a lot of attention in recent years; beyond matters of hormones and pathways the main surprise is just how little perturbation it takes to make a substantial morphogenetic difference. A little intermittent stroking or bending is all that's needed. "Thigmomorphogenesis" is the name given to the phenomenon, with "thigmo-" indicating "touch" (Jaffe 1980); the primary response is retardation of lengthwise growth and augmentation of radial growth. Niklas (1992) gives a good account of the mechanical implications of the changes in structure. It happens in the great majority of the plants in which it has been looked for, from herbaceous annual beans to small trees. Thus Rees and Grace (1980a, b) not only found the expected differences between pines grown in a low wind growth chamber and ones grown with an imposed high wind, but they got just the same differences when one set of plants was shaken for 24 minutes per day. And Telewski and Jaffe (1986) found the same parallel between intermittent shaking and wind when looking at Fraser firs in both laboratory and field—increases in Young's modulus and flexural stiffness of trunks, shorter stems and needles, reinforcement of branch bases around the stems, and the usual decrease in height and increase in girth. Six-year-old seedlings treated to a short daily shaking for a year showed precisely the structural changes of specimens that had grown at higher altitudes.

Exposure versus Maximum Flow

For both the marine algae and the terrestrial vascular plants, the prevailing or average flow in their immediate habitat is obviously a matter of substantial relevance. What's often needed is some really cheap-and-dirty device that can be used in large numbers to achieve a nice view of spatial heterogeneity. The high drag of flags was mentioned earlier; that disability has been put to use in what have been called "tatter flags" to get an index of exposure to winds following calibration in a wind tunnel (Grace 1977; Miller et al. 1987).

An equivalent exposure-measuring scheme is available for flows of water—it looks at the rate at which a standard piece of plaster of Paris is eroded by water movement (Muus 1968; Doty 1971). Alternatively, standard hemispheres of carpenter's chalk can be used; I had a flow tank in which all the joints turned blue from a long series of calibration runs someone did on such pieces of chalk. Peppermint candy ("Life Savers") was used to good advantage by Koehl and Alberte (1988). "Exposure" is a little hard to define precisely—is it looking at an average of speed or an average of squares of speed, for instance—but for many inter-habitat comparisons such precise definition is less important than simple standardization.

What may often be more important for absolute survival (as opposed to long-term efficient function) are the maximum flows an organism encounters. In terrestrial situations few macroscopic organisms live where winds are maximal, for reasons of soil and water that probably transcend personal mechanics. On rocky coasts, though, even very exposed headlands support a decent flora and fauna—algae, limpets, and barnacles where even the snails have fled. So just what the maximum flows and flow-induced forces might be in such places has been a matter of considerable curiosity. And the techniques by which maxima have been estimated are of considerable relevance to investigators looking at specific habitats.

The simplest device I know for recording the maximum current is a kind of drogue or sea anchor (Figure 6.11a) used by Jones and Demetropoulos (1968). The device uses a disk normal to the flow trailing behind a spring scale equipped with an arm that remains in whatever extreme position it reaches; the whole device is free to swing downstream from its attachment post. It's inexpensive and can be produced without special equipment from ordinary components in numbers sufficient to survey a decent number of sites over periods of months. It does suffer from a relatively long response time, perhaps a few tenths of a second; it behaves similarly to long, flexible algae. With a device of their design I strongly recommend determining the drag coefficient empirically, using a fast flow tank or towing alongside a boat on a line attached to another spring scale.

The fastest flows reported by Jones and Demetropoulos (1968) were about 14 m s^{-1}; I think that choice of a more realistic drag coefficient would raise them to about 16 m s^{-1}. You should be impressed. 16 m s^{-1} may be only 36 mph, but this is water we're talking about, not air. In water at this speed a broadside flat plate of about the area of an outstretched hand with fingers together would experience a drag of 1700 N, nearly the weight of a 400-pound object. Put another way, if compressibility is ignored, the same drag would require an airspeed of 450 m s^{-1} or about 1000 mph.

Quite a different kind of device was contrived by Denny (1983), who has given this problem of maximum flow-induced forces more attention than

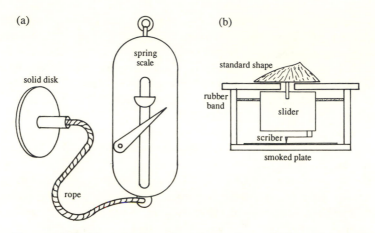

FIGURE 6.11. Meters to record maximum flows. (a) The tethered device of Jones and Demetropoulos (1968) in which a spring scale is modified by adding a lever that remains at whatever extreme position it's pushed. (b) The subsurface device of Denny (1983), consisting of a smoked plate and a scriber that scratches the plate as flow pushes spring-mounted exposed object and slider from their resting position.

anyone else. His design (Figure 6.11b) uses a smoked plate and a scriber in a housing beneath an object such as a limpet shell or a standard drag-suffering shape. It has a much shorter response time, and it records direction as well as magnitude of forces, but it's a bit more complicated to build and calibrate. With an admittedly more complex arrangement, he telemetered continuous records from instruments attached to intertidal rocks in a very exposed location during serious storms (Denny 1985). Accelerations of water were high and had to be taken into account—we talked about them some pages back. On the other hand, the "impact pressures" caused by large volumes of water suddenly changing speed and direction as a wave breaks against a rock prove to be no big deal. Impact pressure doesn't amount to much unless the flow is oriented just normal to the rock and unless the water entrains none of that compressible stuff, air. Neither of these conditions is commonly satisfied. All of the work, along with the relevant background material, is described in his book (Denny 1988); the ecological relevance of the mechanics of both organisms and water is summarized in a short article (Denny 1987b).

But direct measurement of maximum flows may now be unnecessary, which, I suppose, is the ideal outcome of a combination of data, theory, and analysis. Denny (1985, 1987b, 1988) explains how linear wave theory permits prediction of velocities from wave heights. And Denny and Gaines

(1990) have developed rules for predicting the probabilities of different maximal speeds and forces over different time periods—they use a mathematical approach called "statistics of extremes." From a fairly small number of measurements one can make surprisingly robust predictions about maxima.

Shape and Drag: Motile Animals

FOR STREAMLINING to be effective, the organism must continuously face into an oncoming current—which means either weathervaning if it's attached or locomoting if it's free-living. In addition, the organism must have a fairly definite, noncompliant shape. As a result, streamlining is largely the province of animals that propel themselves in continuous fluid media—the swimmers and fliers of the world. Humans have put a lot of effort into designing streamlined shapes; if Cayley's trout is no fluke, nature has done the same. For a sessile organism, resisting drag means resisting only force; deflections can be restored elastically, and little work need be done once the system is constructed. For motile creatures, the price of excessive drag is more immediate and serious; to move, animals must expend energy at a rate that's the product of drag and velocity. If the drag coefficient were halved, then surely power output and most likely power input could be halved. Alternatively, an animal might be able to swim or fly about 40% faster.

STREAMLINING AND THE COMPARISON OF DRAG COEFFICIENTS

For a start, we must bear in mind that drag coefficients and Reynolds numbers, while useful, are imperfect bases for comparing different streamlined shapes. The main difficulty comes from the different conventions used to define the reference area in the drag coefficient together with the fact that just what shape gives the lowest coefficient depends on the choice of reference area. To repeat a bit of Chapter 5, four different areas are in use: (1) frontal or flowwise projecting area, mainly with bluff bodies; (2) plan form, profile, or maximum projecting area, for wings; (3) wetted or total area, for streamlined bodies; and (4) volume to the two-thirds power, for airships. When considering the quality of a streamlined body remember that a change in shape can affect the drag coefficient with no change in drag or the drag with no change in drag coefficient. If drag coefficient is based on frontal area or $V^{2/3}$, then stubbier shapes will look better; if it is based on plan or wetted area, then more elongate shapes will give lower drag coefficients.

The diversity of conventions in the literature makes it useful to be able to convert one sort of drag coefficient into another. Table 7.1 gives some conversion factors; in particular, these may help in using Figures 5.3 and

TABLE 7.1 FACTORS FOR CONVERTING DRAG COEFFICIENTS FROM ONE REFERENCE AREA TO ANOTHER.

	Sphere	Cylinder	Prolate Spheroids 2:1	Prolate Spheroids 3:1	Prolate Spheroids 4:1
Frontal area	1	1	1	1	1
Plan form area	1	1	0.500	0.333	0.250
Wetted area	0.250	0.318	0.146	0.102	0.078
Volume$^{2/3}$	1.208	0.932	0.762	0.581	0.480

NOTES: To convert a drag coefficient, multiply its value by the ratio of the factor for the area in which you want it and the factor for the area in which you already have it.

Flow is perpendicular to the long axis of the infinitely long cylinder; flow is parallel to the long axes of the prolate spheroids, which are described by the relative lengths of their axes.

5.4 for comparisons with data for biological objects. For a sphere or cylinder moving normal to its long axis, frontal and plan form areas are the same. Streamlined objects are a heterogeneous collection, but the factors for prolate spheroids ought to prove to be adequate approximations for rough comparisons. For example, a good low-drag body of revolution (an elongate, radially symmetrical body) at a Reynolds number in the low hundred-thousands will have a drag coefficient, based on frontal area, of about 0.04 (Mises 1945). Using the figures of Mises and of Hoerner (1965), such a body may be approximated by a spheroid whose length is three times its thickness (a 3:1 spheroid), so the drag coefficient based on wetted area will be about 0.004, that based on plan area about 0.013, and on $V^{2/3}$ about 0.02. From the figures given by Pennycuick (1968) and Tucker and Parrott (1970) the 4:1 spheroid should provide a reasonable facsimile of a pigeon and perhaps of other birds. I'm not sure, though, that any of these figures are trustworthy for bilaterally flattened fish.

Beyond the differences in reference areas and the underlying issue of choosing biologically relevant references, other matters cloud the skies and muddy the water. First is the familiar one of lack of physical data. Between Reynolds numbers of 10^4 and 10^7, we have good information on the design of streamlined shapes. This ought to be a good base for work on birds, medium and large fish, and so forth. Between 10^2 and 10^4 we have almost no precise information about the shapes of objects designed for minimal drag. One might guess that just below the transition Reynolds numbers of around 100,000 for a sphere or cylinder, the fairly forward position of separation would mandate a very slender shape in order to prevent that separation. After all, prevention or drastic delay of separation in order to minimize pressure drag is what streamlining is all about. Additionally, one might guess that as Reynolds numbers decrease further and separation moves rearward around spheres and cylinders (recall the data of Seeley et al. 1975 for spheres), the best shapes will be more rotund. In addition, the increasing role of skin friction at low Reynolds numbers will favor shapes that expose less skin, again favoring the fatter. But, again, what's best depends on the objective. Nature is probably more concerned with moving a volume of organism than with frontal or surface area, and this will also favor less elongate shapes.

At this point it might be useful to go through a few simple comparisons of streamlined and nonstreamlined shapes. After all, when considering claims of drag reduction for one scheme or another, one must bear in mind just how monumental the effects of shape change can be—in particular of streamlining. Consider, first, a long cylinder extending across a flow, with the drag coefficients at different Reynolds numbers from Figure 5.3 converted to wetted area with a factor from Table 7.1. Then squeeze it flat, so it's now a flat plate with its width parallel to and its length extending across a flow—the gold standard for streamlining. The drag coefficients at the same speeds (the conventionally defined Reynolds numbers are larger by $\pi/2$ as a result of the squeeze) can be looked up in Table 5.2. At $Re = 100$ (for the original cylinder), drag has been reduced by a factor of 8; at $Re = 1000$, by 19; at $Re = 10,000$, by 65; at $Re = 100,000$, by 215; and at $Re = 1,000,000$ by 180. (The last is a little lower since the cylinder has passed the turbulent transition.)

Our flattened cylinder, of course, is not practical for many applications: what of real struts, as might support a wing or some organism elevated above a substratum? Hoerner (1965) cites data for struts about 6% as thick as they are wide (width is usually called "chord length") that have drags within 20% to 30% of the calculated values for flat plates. For struts whose thickness is 12% of chord length, the drag is about 50% greater than for ideal flat plates. These are still tiny increases compared to the orders of magnitude decreases we just saw for streamlining itself. Pushing practical comparisons farther, though, depends on how the structure we're considering is used. For an elongate body loaded solely in tension, such as a cable, all that matters is cross-sectional area, and a fairly low ratio of thickness to chord will be optimal. For one carrying a lengthwise compressive load, such as an ordinary strut or perhaps a leg, buckling has to be avoided; and that pushes the optimum toward a greater thickness-to-chord ratio—a point made earlier. The specific optima, not surprisingly, depend on Reynolds numbers.

Analogous comparisons can be made between solid bodies of revolution —axisymmetrical or radially symmetrical bodies—and spheres. Using the data in Hoerner (1965), we see that such a body with a length-to-diameter ratio of 8.0 at a Reynolds number of 100,000 has a drag coefficient (referred to wetted area) of 0.006. For a sphere of the same diameter the Reynolds number will be 12,500, and the drag coefficient (Figure 5.4) is about 0.12. So streamlining has reduced the coefficient 20-fold. If we begin with a circular disk instead of a sphere, then the reduction is fully 90-fold. As mentioned in the last chapter, this ratio of length to maximum diameter or thickness is commonly referred to as the "fineness ratio."

These figures, though, are not a sufficiently general guide to the quality of streamlining. How low is the lowest achievable drag coefficient? Recall what streamlining accomplishes—it prevents separation aside and behind an object. Again, the ideal nonseparating flow is that across a flat plate parallel to the flow, so the drag coefficients for such flat plates provide a standard of comparison. Fortunately, convenient and fairly trustworthy formulas are available, and here theory agrees decently with practice. Where flow across the plate is laminar (the plate being smooth, the flow itself nonturbulent, and the Reynolds number not too high),

$$C_{dw} = 1.33 \, Re^{-0.5} \tag{7.1}$$

using wetted area or both surfaces of the plate for reference. For turbulent flow across the plate,

$$C_{dw} = 0.072 \, Re^{-0.2}. \tag{7.2}$$

These formulas are given by Goldstein (1938), Hoerner (1965), and many standard textbooks; Hertel (1966), a widely cited but generally untrustworthy source, misquotes equation (7.2). The transition to turbulent flow occurs somewhere between 5×10^5 and 1×10^7, depending on the circumstances. Equations (7.1) and (7.2) generate Figure 7.1, which illustrates the advantage of postponing the turbulent transition to the highest possible Reynolds number. This is very nearly the same thing as postponing transition to a location as far back as possible from the forward stagnation point or leading edge of a streamlined object, since the Reynolds number, with its length, is proportional to that distance.[1] The equations and graph also point out that drag coefficients typically drop with increasing Reynolds numbers for objects that experience mainly skin friction—something that came up in the last chapter when we talked about the baseline for evaluating reconfiguration.

These equations and their graph have been frequently used as a reference against which the dragginess of organisms is measured, and it's quite a good reference provided its basis is kept in mind. What's being done is to replace an animal with a rectangular flat plate of the same surface area (the C_{dw} part) and the same length (the Re part) for purposes of comparison.

---

[1] We're tickling the edge of something called "local Reynolds number," which will be given proper attention in the next chapter.

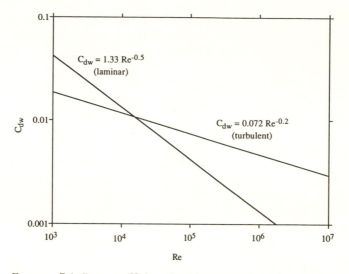

FIGURE 7.1. Drag coefficient (based on wetted area) versus Reynolds number for long, flat plates with long axes extending across the flow and chords oriented parallel to flow. The gentler line presumes turbulent flow across the plate (equation 7.2) and the steeper one laminar flow (equation 7.1). In practice an object in a flow is on the steeper curve at low Reynolds numbers but shifts to the gentler one at some transition value—perhaps 500,000.

For instance, the equivalent of a spherical animal would be a rectangle, parallel to flow, with a width (cross-flow) of $\pi d/2$ and a length (streamwise) of d, where d is the sphere's diameter. (At Reynolds numbers above about 100,000, there's a correction for end effects on the plate; the equations were derived for drag per unit length of a plate stretching infinitely far across the flow, essentially a very long wing or strut. But the adjustment is much less than for a plate broadside to flow and isn't of much consequence on a log-log plot; for more, see Elder 1960 or Webb 1975.) The other matter worth emphasizing about a plot using equations (7.1) and (7.2) is that even the turbulent line for a *plate* represents a very harsh standard of comparison against which to judge any *solid* body—we're letting a porpoise compete in a swimming meet or a kangaroo in the long jump.

DRAG AND THRUST

A motile animal must do more than resist or minimize drag. In particular, it must produce thrust; and under ordinary circumstances drag and thrust must be equal. "Ordinary circumstances" means steady speed, horizontal motion, an otherwise still medium, and only the horizontal component of thrust if it's generating lift as well. That equality of thrust and drag is both opportunity and obstacle for the investigator. On the good side, it means that you can measure either—if you know one, you know the other, which is why figures for drag take on some significance in assessing power output in locomotion. On the other side are all the problems caused by imperfect separation between thrust-generating and drag-sustaining structures.

Sometimes the two kinds of structures are decently demarcated, as with the screw and the hull of a ship. The drag of the hull is nearly (but not quite!) the same whether the ship is being propelled by its screw or being towed by some agency entirely out of the water. The situation is worse but not totally calamitous in such biological systems as flying insects and birds, and in swimming turtles, swans, penguins, and water beetles. In these, the non-thrust-producing structures are subjected, not just to a simple and smooth oncoming flow, but to a complex and temporally varying flow from wings or legs as well. For fish (except for those that keep fairly rigid bodies), for invertebrate larvae with ciliated regions, for most microcrustacea—for all of these the whole body participates in propulsion, and the situation is catastrophic. For them no experimental separation of thrust and drag using models, dead specimens, or other devices can be regarded as a useful dissection of reality.

A number of published studies are irretrievably corrupted by ignoring the problem in organisms where it just can't be swept under the rug—I have a little list—but I won't embarrass anyone with details. Perhaps the way to view a locomoting system is to ask whether the flow pattern around the body when the animal is self-propelled will be reasonably well represented by the flow pattern when the fuselage is isolated and its drag measured. Wu (1977) gives two illustrations of flow around a ciliated microorganism, *Paramecium*, one with the cilia propelling the creature through water and the other of the creature falling under gravity at the same speed. Beyond the facts that in both cases flow is laminar and goes from anterior to posterior, I see no similarities! As one might expect, the problem is worst at low Reynolds numbers; so we'll return to it in Chapter 15.

Put in the usual jargon, what we're talking about is "parasite drag." The term is used in situations where thrust-producing appendages are involved, and it normally refers to the combined drag of all non-thrust-producing structures, in short, to fuselage or body drag. If a lift-based system is producing the thrust, its airfoils suffer in addition from the inevitable "profile drag" of any structure in a flow and an "induced drag" that accompanies production of lift—matters intrinsic to the airfoils. Parasite drag represents the external drag that the thrust producer has to offset. Unfortunately, this usage isn't universal, so one has to check the definitions used by each author. Thus Pennycuick (1989) uses the present convention, while Tucker and Parrott (1970) include profile drag as a component of parasite drag, reserving the term "body drag" for what we're calling parasite drag.

Which brings up yet one more item in this dragged-out accounting: "interference drag." Consider measuring the drag of a body in a flow tank or wind tunnel. The most direct way to do it involves attaching to one side or the rear of the body a mounting strut that leads to a force transducer of some kind. Inevitably some portion of the mount, called (without hymenopterous allusion) by the engineers a "sting," is exposed to the flow. Keeping the unshielded portion of the sting as short as practical, streamlining its profile, and subtracting its drag from that of the body is often sufficient to discount its influence. But the better streamlined the body, the worse the problem. For one thing, the drag of the sting gets relatively higher. For another, the quality of the streamlining is very much a matter of the details of flow around the body, and the presence of a sting alters the flow. Thus the presence of a sting can increase the drag of a streamlined body by more than the drag of the isolated sting. That's the opposite of what one might expect from their mutual shielding, which suggests an underestimate rather than an overestimate of the body's drag when the sting's drag is subtracted. The extra component due to the bad interaction is what's called "interference drag." The item isn't trivial—Tucker (1990a, 1990b) gave the matter a careful analysis and found that the apparent drag of the body of a falcon could be overestimated by more than 20% if no correction for interference drag were applied. While using a wake-traverse system for measuring drag (Chapter 4), I once minimized the problem by considering only the half of the wake opposite the mount. In fact, the phenomenon of interference is a pervasive one that applies even where two bodies don't actually touch; Hoerner (1965) devotes a whole chapter to it.

On a more positive note, one can sometimes get drag figures for swimming animals in a very simple way. A fish or squid doesn't swim continuously; gliding is quite a normal event. If one makes a videotape of an animal that has just stopped active propulsion, one gets a record of its deceleration. As always, force is mass times acceleration—here the force, drag, is rearward, and the acceleration is negative, but Newton's second law still holds sway. Just one extra factor is relevant, and it's often a minor correction. The "virtual mass" of the animal via the "acceleration reaction" now works in the opposite direction from drag, tending to keep the animal going. To deal with it you need to know the volume of the animal (but you need the mass anyway) and the added mass factor, which depends on its shape and can be obtained from Chapter 16. (If you don't correct for added mass, you'll get slight underestimates for drag.) So drag can be determined for an unrestrained animal following its own devices rather than being towed by yours. While you still don't *really* know what the drag is during active swimming, the resulting data should be closer. The scheme has been used for quite a variety of animals, mostly large ones such as penguins and marine mammals. For small creatures such as zooplankton, drag is all too immediate, and deceleration is so dramatic that really high-speed cinematography is necessary. And the latter involves special cameras and lots of costly, nonreusable film rather than virtually free videotape. As Lehman (1977) points out, a 30 millisecond deceleration doesn't partition well at 24 frames per second. But it does work, and Strickler (1977) gives practical advice.

However drag is determined, for results to be applicable to more than just a gliding phase, the drag of the thrust-producing organs must not interfere with the measurement. A flapping wing or paddling leg, like any other structure in a flow, certainly experiences drag. But its drag (together with its lift) appears as thrust—literally antidrag—with respect to the body; I'll withhold the details of this sleight-of-forelimb until Chapter 12. For analysis of a glide, the handiest situation occurs when the animal is accommodating enough to fold or collapse its thrusters. A squid jet is no problem, a penguin's flipper only a little more so. The problem bedevils attempts to measure fuselage drag; the simplest solution, tolerable some-times, is to remove thrust-producing structures prior to measurement of drag.

Looking for Streamlined Struts

Earlier we compared cylinders to streamlined struts, as in Figure 7.2a; the latter in fact prove relatively uncommon among organisms. But they do occur, and they're a good place to start a survey of streamlining.

The so-called torrential fauna, mostly insects, are inhabitants of exposed rocks in rapid streams; they appeared briefly as sufferers from inoppor-tune lift in Chapter 4 and will play a larger role in Chapter 9. While most are flat and cling closely to rocks, at least one sits a bit off its rock and is well-rounded and streamlined—the mayfly nymph, *Baetis*. Long ago (in quite a fine paper) Dodds and Hisaw (1924) compared a series of species of *Baetis* with one another and with relatives from less violent waters. Individuals of species preferring rapid flows are generally smaller and have larger legs; they extend downstream from these legs, swinging to and fro if the direc-tion of flow varies—limited-excursion weathervanes if you wish. From the present point of view what's interesting is the cross-sectional shape of those legs. The femurs and tibias are flattened, thicker at the anterior margins, and thinner posteriorly—in a word, streamlined. While these insects are only about a centimeter long, they're exposed to rapid flows. For the hab-itat of *B. bicaudatus*, Dodds and Hisaw measured speeds up to 3 m s^{-1}; at

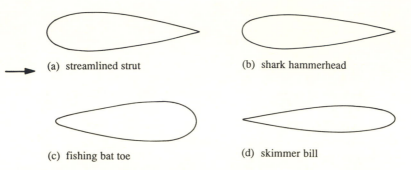

(a) streamlined strut

(b) shark hammerhead

(c) fishing bat toe

(d) skimmer bill

FIGURE 7.2. Low-drag struts: (a) a streamlined section; (b) approximate cross section of the "hammer" that extends outward from each side of the head of a hammerhead shark, *Sphyrna*; (c) the lower bill of a skimmer, *Rynchops nigra*; (d) a hind toe of a fishing bat, *Noctilio*. The last two extend through the air-water interface and prefer to obey the injunctions of Chapter 17 instead of those of the present one.

that speed (and the worst case is probably the relevant one) the Reynolds number for a leg 0.8 mm wide is around 2000, certainly high enough for streamlining to be effective.

Crab legs may sometimes be arranged as streamlined struts. Quite a few crabs either walk around or stand erect in water currents, and they must have only limited purchase on the bottom due to their substantial buoyancy. And flattened legs are quite common in crabs. The problem is that many of them swim at least occasionally, with legs serving as paddles and rudders (see Hartnoll 1971 and Plotnick 1985, for instance). The distinction between paddles, rudders, and streamlined struts is not one I can confidently make. A good animal to look at might be a shore crab, *Pachygrapsus crassipes*, for which data on underwater walking is given by Hui (1992). Another case worth a careful look is the hammer of hammerhead sharks (genus *Sphyrna*). Figure 7.2b is something of a guess, based on my recollection of having fondled a preserved specimen and on published top and front views.

Exposure to spectacularly fast flows of water happens in a pair of strongly convergent cases involving bats and birds. A few species of each fly through the air while dipping gaffing hooks into the water. The bats are fishing bats, genera *Noctilio* and *Pizonyx* (of two different families— another convergence); and their gaffs are hind feet (Figure 7.2c). While these are high-drag hind feet by bat standards, they're also bigger and stronger than the common cave-clinging clamp. The drag coefficients of the hind legs of both fishing and nonfishing bats have been measured by Fish et al. (1991). What's initially curious is that even though those of the fishing bats are lower, about 2-fold at the highest speeds they tried, the coefficients are still high, around 1.0 referred to frontal area. Equally curious are the cross sections, which look like backward airfoils with sharp fronts and blunt backs. We noted already that the drag of reversed airfoils isn't all that bad, but is such complete reversal an inevitable consequence of the morphological revolution that redirected the toenails forward? Fish et al. point out that the hydrodynamic situation is complicated by the fact that the digits extend through the air-water interface and so suffer from some of the special problems of surface ships. The sharp leading edge has positive advantages in reducing drag due to waves and spray, as we'll talk about more specifically in Chapter 17.

The birds are skimmers (genus *Rynchops*), whose gaff is the lower and longer part of the bill (Figure 7.2d). These are major mandibles—a captive bird grew one 17 cm long. In use, they stick down into the water at an angle of roughly 60° below horizontal, at speeds of around 6 to 10 m s^{-1}. Zusi (1962) commented on their lateral compression and knifelike streamlined profiles, clumsy for grasping and holding but great gaffs. Withers and Timko (1977) measured a fineness ratio of about 6.0 for the lower mandible—compared with 1.7 for the upper—and give profiles showing conventional streamlining for the lower part of the bill but sharp leading edges for the part that encounters the surface. Both sources note that these bills are ridged in a curious way that might bear investigation. For both skimmers and fishing bats drag may not represent a big cost in energy— dipping isn't done continuously and the drag of the gaffs is insignificant next to body drag—but it generates an awkward turning moment. In particular, a skimmer with a high drag bill would have to do something special to avoid upending, which in fact it nearly does whenever it catches a fish.

A DIVERSITY OF STREAMLINED BODIES

Two-dimensional streamlined forms—struts—may be rare in nature, but three-dimensional streamlined bodies are the hallmark of virtually every competent swimmer or flyer. The sampling in Figure 7.3 is intended to emphasize the diversity, with representatives of two classes of each of the three great phyla of mollusks, arthropods, and chordates. I'll do a flying survey of drag data, both to illustrate what nature's been up to and to point a finger at gaps and problems of both measurement and interpretation. Table 7.2 is a summary of progress, while Figure 7.4 displays data, converted with any necessary guesswork to wetted area and compared to the standards we established earlier.

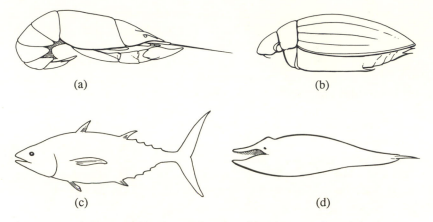

FIGURE 7.3. Streamlined organisms in a flow that goes from left to right: (a) a crayfish going rearward in a rapid escape; (b) a large aquatic beetle; (c) a pelagic fish such as a tuna; and (d) a baleen whale.

Smallish Arthropods

If muscle is muscle is muscle, then animals of all sizes ought to be able to jump to the same maximum height or range—Galileo first made the point—but only if air resistance is negligible. Resistance is a nuisance for baseball or golf ball, reducing the maximum range by 19% and 36%, respectively (Vogel 1988a). For a jumping locust, the reduction is only about 16%—they may be smaller, but they move more slowly. For a generic flea I calculated a range reduction of about 83%, so drag is far beyond being some minor correction and has replaced gravity as the main factor determining range. Bennet-Clark and Alder (1979) shot fleas, a big one (3.7 × 1.7 mm) and a small one (2.05 × 0.7 mm), from a spring gun, calculating drag coefficients from the heights achieved. They obtained coefficients, based on frontal areas, of just about 1.0 at Reynolds numbers between 65 and 205. That's nothing to be proud of, about that of a sphere. Referred to wetted area and assuming 2.5:1 spheroids, that's a coefficient of 0.12, about half that of a sphere. Either wetted area or volume$^{2/3}$ are probably the more appropriate reference; the latter also yields a coefficient about half that of a sphere. So maybe fleas aren't quite as bad as they at first glance appear. Contriving an elaborately specialized jumping apparatus and then neglecting aerodynamics altogether certainly would seem odd. If fleas tumble in local air currents, though, the potential advantages of streamlining might be hard to realize.

The parasite drag of very small insects is of particular interest. As we'll see in Chapter 11, airfoil performance deteriorates as Reynolds numbers

TABLE 7.2 DRAG COEFFICIENTS FOR ANIMALS THAT MOVE THROUGH FLUID MEDIA.

	Re	C_d	*Area*	*Source*
Fleas				
Ctenophthalamus	65–205	0.96	f	Bennet-Clark & Alder 1979
Hystricopsylla	"	1.02	f	
Fruit fly, *Drosophila virilis*	300	1.16	f	Vogel 1966
Locust, *Schistocerca gregaria*	8000	1.47	f	Weis-Fogh 1956
Marine isopods				
Idotea wosnesenskii	2700	0.08	w	Alexander & Chen 1990
Idotea resecata	5500	0.055	w	"
Dytiscid beetles				
Acilius sulcatus	8600	0.28	f	Nachtigall 1977a
Dytiscus marginalis	15,000	0.33	f	"
Tadpole, *Rana catesbiana*	1000–2500	0.36–0.74	f	Dudley et al. 1991
Frogs				
Hymenochirus boettgeri	1500–8000	0.11–0.24	w	Gal & Blake 1987
Rana pipiens	17K–40K	0.05–0.06	w	"
Crabs				
Callinectes sapidus	10,000	0.3	p	Blake 1985
Cancer productus	10,000	0.35	p	"
Ducks, various underwater	420,000	0.028	w	Lovvorn et al. 1991
Cephalopod, *Nautilus*	100,000	0.48	v	Chamberlain 1976
Falcon, *Falco peregrinus*	380,000	0.24	f	Tucker 1990a
Fish				
Trout, *Salmo gairdneri*	50K–200K	0.015	w	Webb 1975
Mackerel, *Scomber*	100,000	0.0043	w	"
"	175,000	0.0052	w	"
Saithe, *Pollachius virens*	500,000	0.005	w	Hess & Videler 1984
Penguin, *Pygoscelis papua*	1,000,000	0.0044	w	Nachtigall & Bilo 1980
Marine mammals				
Sea lion, *Zalophus californ.*	2,000,000	0.0041	w	Feldcamp 1987
Seal, *Phoca vitulina*	1,600,000	0.004	w	Williams & Kooyman 1985
Human, swimming	1,600,000	0.035	w	"

NOTE: The reference areas vary among the sources: f = frontal; w = wetted; p = plan form; v = volume$^{2/3}$.

drop, and small insects are of the order of magnitude of estimates of the lower limit for the practical use of lift. And, as with jumping fleas, drag becomes more important relative to gravity—in part because of the increasing area-to-volume ratio concomitant with small size. Thus parasite drag at ordinary flying speed is 18% of weight in a fruit fly, *Drosophila virilis*, but only 4% in the desert locust, *Schistocerca gregaria*. As part of my thesis project, I measured the parasite drag of this particular fruit fly—the coef-

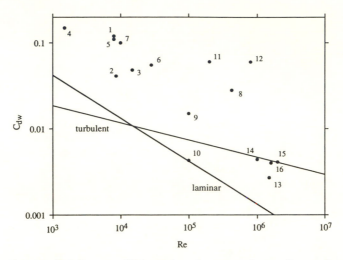

FIGURE 7.4. Drag coefficients, based on wetted area, for motile animals that might be designed for low drag compared with the long flat plates of 7.1 and equations (7.1) and (7.2). 1, desert locust; 2, beetle, *Acylius*; 3, beetle, *Dytiscus*; 4, bullfrog tadpole; 5, frog, *Hymenochirus*; 6, frog, *Rana pipiens*; 7, crab, *Cancer productus*; 8, underwater duck; 9, trout; 10, mackerel; 11, pigeon; 12, vulture; 13, emperor penguin; 14, gentoo penguin; 15, sea lion; 16, harbor seal.

ficient is 1.16 based on frontal area or (assuming a 2:1 spheroid) about 0.17 based on wetted area at a Reynolds number of 300. For comparison I measured the drag of several spheres (bearing balls) under the same conditions. A sphere has a slightly lower drag coefficient based on frontal area, 1.08, but it has a coefficient 60% greater relative to wetted area, 0.27. A flat plate is much better—the calculated coefficient is 0.077, about half that of the fly. So the fruit fly is at least so-so as far as its drag, roughly midway between sphere and plate, and much like the fleas. (Incidentally, reversing a fruit fly in the airstream increases its drag about 10%—that's a crude test of design since the reversed fly has the same frontal and wetted areas but has no reason that I can imagine for low drag.)

The desert locust operates at a Reynolds number of about 8000 and has a C_{df} of 1.47 (Weis-Fogh 1956) and a C_{dw} of about 0.12 (as a 4:1 spheroid)—the former is higher than the value for a sphere, 0.47, while the latter is the same. Locusts are really quite bad; one might wonder about data, but a photograph by Weis-Fogh of a locust body with smoke markers shows quite substantial separation with a large and messy wake. Of course the flow with wings attached and beating might be quite different from that surrounding an isolated body.

The isopod genus, *Idotea*, was mentioned in connection with the choice of reference area for drag coefficient. Alexander and Chen (1990) give the drag coefficients for two species, a slower, smaller, and fatter one and a faster, larger, and more elongate one. Both have about three times the drag of the equivalent flat plate; the smaller has two-thirds and the larger about half the drag of a sphere. For something as poorly streamlined as these isopods, the sphere might provide the better comparison, so we can at least restate the obvious, that elongate is better if wetted area is the reference. A little higher in Reynolds numbers than the low thousands of these isopods are the dytiscid water beetles investigated by Nachtigall and Bilo (1965, 1975; see also Nachtigall 1977a). *Acilius*, operating at Re = 8600, has a typical C_{df} of 0.28; *Dytiscus*, operating at Re = 15,000, has an average C_{df} of 0.33. Assuming 2:1 spheroids, these are C_{dw}'s of about 0.041 and 0.048; around 2.6 times lower than that of a sphere; the comparisons of C_{dv}'s are about the same. Relative to flat plates, they're about 3.6 times worse—again, the animals are about in the middle of the range between spheres and plates. Note, though, that at these higher Reynolds numbers the gap between sphere and flat plate is widening, with increasing wiggle room for nature to play drag reducer. Dytiscid beetles may be as short as 2 mm (by contrast with the 35 mm length of *Dytiscus*), with swimming Reynolds numbers as low as 90. Beetles of smaller species are increasingly round, about what we'd expect as skin friction becomes more important and pressure drag due to separation less so.

At this point I ought to repeat my wish for better information on the optimal design of axisymmetrical bodies—for biologically reasonable criteria such as volume transported or surface exposed, just how good can one do? Is drag minimization simply accorded a low priority in the design of all these arthropods? Or are we looking at their drag in some flawed manner? Or are they really quite near the practical minima? I just don't know of the existence of the basic data that might provide some basis for judgment.

Several Undistinguished Swimmers

Again, let's remind ourselves to keep a little distance from drag, to eschew the prejudice that keeping it down is what life's all about. Perhaps that's best done by looking at some creatures that aren't really noteworthy as swimmers, ones that only intermittently bother doing it, ones that carry other disabling constraints, and ones for whom achieving great speed is probably of little importance.

Consider frogs, the paradigmatic organisms of generations of biology students. The larvae come equipped with tails, but otherwise hatch with the foreshortened form, the low fineness ratio, of adults rather than the elongate shapes of proper swimming amphibia—their slow swimming probably isn't too much of a disability since they're not pursuit predators. And any fish large enough to eat a tadpole will likely be faster on that account alone—maximum speed goes up with size in a fairly regular way in any relatively homogeneous group of swimmers. The adults swim occasionally, but they also walk and jump; they aren't pursuit predators either. Recently Dudley et al. (1991) looked at the drag coefficients of bullfrog (*Rana catesbiana*) tadpoles; these are relatively large as tadpoles go, around 10 cm in length, including tails. C_{df} values ranged from 0.36 to 0.74 at Reynolds numbers (based on body length, excluding tail) between 1000 and 2500. That's equivalent to C_{dw} (presuming a 1.5:1 spheroid) from roughly 0.1 to 0.2, about four times that of a flat plate and about the same as a sphere—not very good at all. And things get somewhat worse when forelimbs and hind limbs erupt.

As part of a study of drag-based locomotion (about which we'll have more to say) Gal and Blake (1987) measured the drag of the adults of several species of frogs, removing feet and dropping them, ballasted, in a column of water. Individuals of the smaller and more aquatic species, *Hymenochirus boettgeri* (1.5 to 2.5 cm snout-to-vent[2]), have C_{dw}'s of 0.24 to 0.11 at Reynolds numbers between 1500 and 8000. Those of the larger species, *Rana pipiens* (7.5 cm), have C_{dw}'s of 0.06 to 0.05 at Re's between 17,000 and 40,000. Given the differences in Reynolds numbers, the drag coefficients aren't notably different. Again the figures are much the same as the drag coefficients of spheres; they're about seven times higher than those of flat plates.

Many crabs are swimmers; like frogs many propel themselves with appendages that do duty as walkers as well, although the particular mix of mechanisms and modes are quite diverse (for surveys, see Lochhead 1976 and Hessler 1985). As noted earlier, most crabs are negatively buoyant—sinkers—not inappropriately for animals that do any bottom-walking. When swimming, lift is needed. Both lift and drag have been measured by Blake (1985) for several species. At Reynolds numbers above about 10,000, *Callinectes sapidus* has a drag coefficient referred to plan form area of 0.3 and a lift coefficient[3] of 1.2—a decent lift-to-drag ratio of 4.0. *Cancer productus* has a higher C_{dp}—0.35 and a lower C_l—0.65, giving it a lift-to-drag ratio of only around 2.0. The C_{dw}'s are around 0.1; at $Re = 10,000$ that's about eight times higher than a flat plate but half that of a sphere. These animals swim sideways; being bilaterally symmetrical they have equal coefficients for either direction. Going forward or backward, drag is approximately doubled.

[2] Called the anus by people who lack appreciation of these finer points of amphibian anatomy.

[3] Lift coefficient is defined the same way as drag coefficient. Plan form area is almost always the reference; whenever lift coefficients are used, the drag coefficients should also use that reference. Details in Chapter 11.

Ducks fly, swim on the surface, and swim under water; again one suspects functional compromises. From the regression equations obtained from underwater towing tests by Lovvorn et al. (1991), I get a C_{dw} of 0.028 at Re = 420,000; that represents a speed of a meter per second. The value is very close to that of a sphere (after the great drop in drag); it's five times higher than that of a flat plate over which flow is turbulent and thirteen times higher than a flat plate in laminar flow. In drag, these ducks are turkeys.

A final group of what must be judged poor swimmers—the shelled nautiloids and ammonoids—are now represented only by the genus *Nautilus*. Chamberlain (1976) towed a *Nautilus* shell and got a drag coefficient, based on volume$^{2/3}$, of 0.48. At Re = 100,000, that's a bit better than a sphere before the big drag drop (0.57), but not much. It's thirteen times worse than a flat plate with turbulent flow and twenty-two times worse than one with laminar flow. One wonders whether these animals are trapped by the inflexible geometry of a conical or spiral shell and about what effect the soft parts, extending downstream, might have on drag. Perhaps moving fast just wasn't necessary to make the paleo scene.

Fish

Webb (1975) collected a large amount of data from his own and other measurements along with a useful discussion of the difficulties in getting reasonable measurements of drag. As even more of a dose of cold water, he describes the theoretical difficulties of using data for body drag to estimate the thrust a fish has to produce. In a typical fish, more than perhaps in any other sort of macroscopic organism, thrust-producing and drag-incurring structures are inseparable. But fish do glide, so what a nonoscillating body does isn't entirely artificial. And quite a few fish do maintain fairly rigid bodies as they swim—some that use pectoral fins for propulsion and a considerable number of large and fast pelagic fish that move only their wide tails. For a 30 cm-long trout (*Salmo gairdneri*), freshly killed and with paired fins amputated, Webb got a drag coefficient of around 0.015 based on wetted area at Reynolds numbers between 50,000 and 200,000; that's about eight times less than that of a sphere before the big drag drop. It's still, though, around twice the drag of a flat plate even with turbulent flow. Again one wonders why nature does no better or whether Cayley just got lucky with his trout.

Webb (1975) quotes a figure obtained by Quentin Bone of 0.0043 for a small mackerel (*Scomber scombrus*) at Re = 100,000. That's essentially perfection, the drag of a flat plate with laminar flow; and mackerel do swim rapidly with nearly rigid bodies. But whether such relatively low drag coefficients can be maintained at higher Reynolds numbers isn't clear. Thus the same mackerel had a C_{dw} of 0.0052 at Re = 175,000, not the ideal

0.0032. Nonetheless they're quite good, and these scombroid fishes (including the tunas) look more like laminar-flow airfoils (with maximum thickness fairly far downstream) than almost any others. Hess and Videler (1984) estimate a C_{dw} for a saithe (*Pollachius virens*, a gadid fish) of 0.005 from a computation of thrust. Like a mackerel, a saithe is a fast, predatory fish that swims with a fairly rigid body, and their result is consistent with the data from towed mackerel. Perhaps it's no coincidence that the lowest coefficients result from testing the kinds of fishes in which drag data from towing or gliding is most relevant to performance in nature. Or perhaps these stiff-bodied fishes are simply better than more flexible, troutlike forms.

The Strange Case of Bird Bodies

Whether hummingbird or eagle, in flight posture a bird certainly look slick and smooth. Some years ago, Pennycuick (1968, 1971) measured the drag on wingless, frozen birds in a wind tunnel; he obtained drag coefficients, based on frontal area, of 0.43, for a pigeon, *Columba livia*, and a vulture, *Gyps ruppelli*. These are an order of magnitude apart in frontal area (36 and 300 cm²); and they were tested over a wide range of speeds, corresponding to Reynolds numbers of 200,000 to 800,000. Now those are pretty draggy bodies—C_{dw}'s of about 0.06 or roughly twelve times the drag of flat plates (at $Re = 500,000$) with turbulent flow—very much in the sphere's sphere. The very constancy of the drag coefficient over a large range of Reynolds number suggests the pressure drag of bluff bodies rather than the skin friction of streamlined shapes.

The question, then, is whether parasite drag, measured in this way, actually reflects the situation in normal flight. A search for some fly in the ointment isn't just a desperate attempt to keep our adaptationist faith—not only do birds look streamlined, but a flock of good candidates might provide experimental culprits. Interference drag has already been mentioned, and it's good for at least 20% (Tucker 1990a,b). All the caveats about measuring drag on thrust-producing machines must apply. And turbulence in the wind tunnels used for testing may also make a difference. Beyond these is that special avian curse, feathers. Thus Pennycuick et al. (1988) found that a generous application of hair spray to a bird body in a wind tunnel reduced the drag by about 15% below C_{df}'s between 0.26 to 0.38 (at Re's around 300,000) obtained on mallard, snow goose, bald eagle, and tundra swan. Tucker (1990a) has now obtained a coefficient of 0.24 for the body of a falcon, *Falco peregrinus*, paying careful attention to both interference drag and feather position. But he comments elsewhere that he was "unable to preen the feathers to lie in the smooth, orderly, overlapping position they assumed on a living Harris' hawk in flight" (Tucker and

Heine 1990). That feathers are major culprits is suggested by his figure for drag coefficient for a smooth-surfaced model of the falcon, $C_{df} = 0.14$. Not that the issue is really settled—the flying bird still has feathers, and we don't know what they might cost in drag. Tucker (pers. comm.) is at this point pursuing the question with additional models. Real birds in flight might possibly do even better than featherless models. Dare I say that while the matter is still up in the air one can't yet see which way the wind is blowing?

Penguins

The cost of moving a body through a fluid is best expressed as "parasite power," the product of parasite drag and forward speed. In flying birds, power output may be divided into parasite power and two other components. These are "induced power," the cost of producing lift, and "profile power," best regarded as the remaining aerodynamic cost of moving the wings; these latter two correspond to induced drag and profile drag, mentioned earlier. At reasonably economical flying speeds, parasite power is much lower than the sum of the other two components. So one can argue that a flying bird might not wish to compromise other functions in order to drive body drag close to some aerodynamic lower limit. What about a bird for whom rapid locomotion in a continuous medium is important but where the medium is 800 times denser? While ducks swim under water, they're also paddlers, walkers, and excellent fliers. A better kind of bird if we're interested in how far nature has pushed the matter is a penguin. Penguins swim with their wings, using an up-and-down motion that produces thrust by a proper lift-based or propellerlike mechanism at maximum speeds of about 2.5 m s^{-1}. A similar bird flying six times as fast (assuming drag is proportional to both density and the square of speed) would encounter less than a twentieth as much parasite drag.

Penguin hydrodynamics has been the subject of quite a few investigations. In the first of these, Clark and Bemis (1979) derived drag from decelerative gliding. For emperor penguins (*Aptenodytes forsteri*) about a meter in body length, they got average C_{dw}'s of 0.0027 at Reynolds numbers around a million and a half. That's about two-thirds of the turbulent-flow drag of a flat plate, although several times higher than the laminar-flow drag—pretty spectacular for such a high Reynolds number. But they didn't correct for virtual mass, so the figure ought to be adjusted slightly upward. And Nachtigall and Bilo (1980), also looking at deceleration, got a figure of 0.0044 for the C_{dw} of a gentoo penguin (*Pygoscelis papua*) at a Reynolds number of a million. In towing tests of wingless carcasses, Hui (1988a) got somewhat higher values, around 0.016 for Humboldt penguins (*Spheniscus humboldti*); rigid resin models were about 20% lower. My guess at this point is that the gliding data are more realistic, and that these penguins do about as well as flat plates in turbulent flow, which is to say very well indeed. Incidentally, Baudinette and Gill (1985) found that the cost of transport (energy expenditure per unit mass and distance) was lower for penguins than for any other aquatic endotherms.

Marine Mammals

Like penguins (and ichthyosaurs, plesiosaurs, and others), these are animals of secondarily aquatic habit, reinvaders of the sea—air breathers that nevertheless do most of their swimming underwater. They're also creatures of romance, myth, and (despite the extreme divergence of habitats) anthropomorphization. So graceful and maneuverable is their swimming that it seems almost effortless. In fact they're highly muscular and have high aerobic capacities (= metabolic scope)—the effortless ease is that of a great ballerina. Over fifty years of looking for "the dolphin's secret" (as one speculative paper put it) has led, as we'll talk about shortly, to the sober suspicion that there really isn't any particular secret—just a lot of good design of a kind that isn't really radical from the point of view of either engineer or biologist (Fish and Hui 1991; Fish 1992). We'd become captive of a kind of wishful thinking that was entirely understandable given an initial and understandable misestimate (Gray 1936), what can only be described as Pentagon and Kremlin paranoia (both!), the enthusiastic optimism that characterizes effective program directors and successful grant applications, and the ever-tempting but rarely realized promise of generating technology by copying nature (Vogel 1992b).

First, the pinnipeds—seals and sea lions. These look quite similar, with casual distinction based on the earlessness (of external pinnae, anyway) of the former. For us the more interesting distinction is that the true seals propel themselves mainly with their tails, like cetaceans, while the sea lions produce thrust solely with their pectoral flippers. We now have a little data on both. Feldcamp (1987) measured the drag of a California sea lion (*Zalophus californianus*), obtaining from gliding deceleration $C_{dw} = 0.0041$ at $Re = 2,000,000$—virtually perfection by the standard of turbulent flow over a flat plate. Williams and Kooyman (1985) measured the drag of harbor seals (*Phoca vitulina*) of several sizes using two techniques, towing and gliding. Interestingly, even though towing involved live, compliant animals biting a soft mouthpiece that they could freely relinquish, the data for gliding yielded substantially lower drag coefficients—consistent with what we saw with penguins. For a gliding adult seal they obtained $C_{dw} = 0.004$ at $Re = 1,600,000$—just the same as the result with the sea lion, which, among other things, gives one real confidence in such data. (Williams and Kooyman also towed a human, who had a drag coefficient of 0.035, about 3.5 times that of a towed seal and almost nine times that of a gliding seal. We are consummately clumsy and ineffective swimmers.)

And then the cetaceans—dolphins and whales. To start with, while dolphins (at least) can achieve high speeds, they can do so only for very brief periods. Lang and Norris (1966) measured a top speed for a bottlenose porpoise (*Tursiops gilli*) of 8.3 m s^{-1} for a duration of 7.5 seconds and a top speed of 6.1 m s^{-1} for 50 seconds. For extended periods this strongly motivated porpoise could manage only 3.1 m s^{-1}. Using the power output per unit weight of human athletes, Lang and Norris calculated that the porpoise had about the drag of turbulent flow over a flat plate.[4] From films of gliding, Lang (1975) measured drag coefficients on two other species of small cetaceans (*Lagenorhyncus obliquidens* and *Stenella attenuata*); again the coefficients were about those of a flat plate with turbulent flow. Quite a few other estimates have been made of drag coefficients based on various models and presumptions. In no case does decent evidence show that a cetacean achieves a lower drag coefficient than that of a flat plate with turbulent flow. Au and Weihs (1980) suggested that drag could be partly evaded and energy saved by repeatedly leaping from the water—"porpoising." But the suggestion has been criticized, and a respiratory function has been proposed for such behavior by these obligate air breathers (Fish and Hui 1991).

TRICKS FOR REDUCING DRAG—HOPE AND REALITY

Extreme flexibility in flow, streamlining, growth to great length in oscillating flow, staying close to a solid surface, expulsion of water through opercula in ram ventilation in fish—these are reasonably well documented ways in which organisms minimize drag. Beyond them are quite a number of schemes, popularly believed to be matters of fact or extreme probability, but for which hard evidence is hardly in evidence. In particular, these schemes describe ways in which organisms, largely large swimming animals, can get body drag down below that of a flat plate with turbulent flow. The potential benefits are large: at a Reynolds number of a million the drag with turbulent flow is *3.4 times* the drag with laminar flow. I'd suggest that the interested reader keep an eye on extant and forthcoming volumes of the *Annual Review of Fluid Mechanics* for developments. Four principal

---

[4] Larger cetaceans can certainly go faster, based on credible reports of 10 m s^{-1} in fin whales (*Balaenoptera physalus*) cited by Bose and Lien (1989). But reports of speeds of swimming organisms include so many egregious overestimates that extreme skepticism is recommended when dealing with the literature. For that matter, quite unrealistic figures for the speeds of running and flying are bandied about as if factual.

arrangements have been espoused; they're discussed rather skeptically by Fish and Hui (1991), a little less skeptically by Bushnell and Moore (1991), and with the full faith of the true believer by Aleyev (1977).

Compliant Surfaces

These were, I think, the first proposed solution to what's been known as "Gray's paradox," alluded to earlier. Specifically, Sir James Gray (1936) suggested that only if flow were laminar could a dolphin's muscles manage to move it at its estimated maximal swimming speed. In fact, the paradox has essentially solved itself. Dolphins don't go quite as fast as Gray thought (a familiar problem!), and striated muscle turns out to be rather better than the figures available to him. But paradoxical or not, drag is something to avoid, so the larger issue is still of interest. The basic idea behind the use of compliant surfaces is the damping[5] of incipient turbulence; it was proposed and promoted by Kramer (1960, 1965). Cetacean surfaces are rather different mechanically from what's typical of mammals, but no one has ever demonstrated that they work in the manner proposed. Moreover, compliant coatings such as he designed have not proven to yield anything like the large drag reductions originally claimed (Riley et al. 1988).

Mucus Secretions

Adding long-chained polymers such as mucopolysaccharides or (a little less certainly, surfactants) to a flow, especially if they're added in the velocity gradients at surfaces, can reduce drag. And many fish are slimy, and dolphin skin produces new cells at a high rate. Fish slimes at concentrations of less than (but approaching) 1% reduce skin friction several-fold, and slimes from faster fishes seem to be more effective (Hoyt 1975). Other natural polymers are effective as well (see, for instance, Ramus et al. 1989). But "effective" most often refers to pipes and rheometers; while not uninteresting, these are not swimming animals. With real fish it turns out to be a distinctly iffy business—sometimes yes (Daniel 1981, for instance) and sometimes no (Parrish and Kroen 1988). It does seem clear that (1) the rate of secretion of polymers adequate to get the ambient concentrations giving good effects is higher than fish could reasonably manage, and (2) the cost of producing such secretions would more than offset any benefit that could be realized, at least over all but the shortest emergencies. The situation is a bit like that of snail locomotion, in which the cost is driven upward enormously by the continuous need to produce nonreusable slime.

---

[5] Not "dampening"; the distinction is lost too often even to let fly a pun—even if a cold shower, which this section is intended to provide, both damps and dampens.

Surface Heating

Recall that the viscosity of water depends drastically on temperature, that skin friction depends directly on viscosity, and that viscosity matters most where shear rates are greatest—immediately adjacent to surfaces. So a hot surface will have less drag (other things being equal) than a cold surface. Cetaceans commonly have surface temperatures somewhat above that of the ambient water, and continuous cutaneous heat loss is necessary to offset production by very active muscles. But calculations of the effect on drag suggest insignificance—flows are too fast and temperature differences too small to transfer enough heat to make a difference (Webb 1975, Fish and Hui 1991).

Surface Morphology

Drag reduction has been claimed for just about every feature of the surface of every large and rapidly swimming animal. The present chief candidate is the ridging characteristic of the dermal scales of sharks. These are claimed to be lined up with the local flow direction (Reif 1985). Experiments with analogous physical systems have been successful enough to result in production of a coating material ("riblets") that has been used on racing yachts. The ridges have apparently evolved separately in several lineages of fast-swimming sharks. It should be emphasized that in both sharks and artificial coating these are tiny ridges, closely spaced—less that 100 μm apart and still less in height—and that what's involved is a reduction of skin friction and not postponement of separation (Bushnell and Moore 1991). Two matters, though, get omitted from popular accounts. First, no one seems to have any direct evidence that the ridges actually reduce the drag of sharks or that they work on sharks by the proposed mechanism. And second, the drag reductions achieved with the artificial coatings are less than 10%, enough to create excitement in the hypercompetitive world of boat racing, enough perhaps to make a difference to fitness in the competitive world of pelagic predation, but nothing approaching the difference in skin friction between laminar and turbulent flows.

Writers of popular material in science are biased toward believing what scientists claim or even suggest. Perhaps they don't appreciate sufficiently the difference between the enthusiasm associated with a novel and exciting hypothesis and the more restrained satisfaction that accompanies decent confirmation and achievement. But we can't escape by shifting blame; I think what's needed at this point is a bio-fluid version of Koch's famous postulates in bacterial epidemiology. A claim of drag reduction should be viewed with skepticism until it (1) has been tied to a plausible physical mechanism, (2) has been shown to work on physical models under biologically relevant conditions, and (3) has been shown to work by some direct test on real organisms under controlled and reproducible conditions. *Much less desirable* alternatives to the third are interspecific comparisons of morphology and correlations of morphological differences with differences in habit and habitat.

Drag-Based Locomotion

At first suggestion, this sounds like a contradiction in terms. But drag in one direction is, of course, thrust in the other. Thus all one needs is some appendage that moves back and forth in the intended direction of progression whose drag when moved one way is less than its drag when moved the other. An appropriate difference in drag demands nothing obscure—recall the data for the drag of flat plates parallel and perpendicular to flow in Table 5.2 and the high drag of some of the forms shown in Figure 6.4. Nor is the motion particularly peculiar. Conversion of a terrestrial walk needs only a recovery stroke close to the body or parallel to flow with a folded or twisted appendage and perhaps some drag-increasing webbing exposed during the power stroke. This kind of secondary swimming is very common in mammals and birds—muskrats, minks, dogs, humans, ducks, and so forth (Williams 1983; Fish 1992). It's common also among smallish creatures, especially swimming arthropods, in which setae are arranged to stick out on one stroke and lie back on the other (Figure 7.5). And it's commonly used by organisms with a lengthwise series of appendages that are moved in either a metachronal rhythm or synchronously.[6] This metameric arrangement may be a really ancient one among metazoa since it's a scheme that can make a walker or burrower as well as a swimmer, and since metamerism is such a widespread morphological arrangement.

Getting high drag for a power stroke is no special trick. Drag coefficients for paddles (on frontal areas) are typically over 1.0, whether the hind legs of frogs (Gal and Blake 1988), water boatmen (corixid hemiptera; Blake 1986), or swimming beetles (Nachtigall 1980, 1981). Getting low drag for a recovery stroke isn't hard either, although information is scanty about just how low the coefficients might be. For the system to work well, the drag coefficient for a good recovery stroke ought to be quite a lot lower than that of the power stroke. Averaged over a cycle, an appendage must move

---

[6] A nonsynchronous, metachronal wave of paddling turns out to be a bit more efficient than synchronous rowing. The nearly universal choice in metazoans is a forward-running metachronism (protozoa are far more variable in their metachronal arrangements), which minimizes both solid and fluid limb interactions (Sleigh and Barlow 1980). But in racing shells, where several humans have to coordinate their efforts, the synchronous scheme is still used.

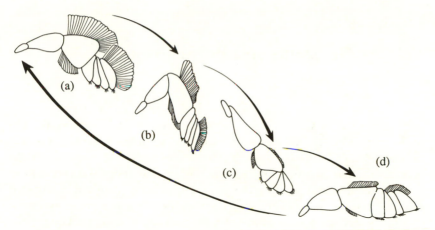

FIGURE 7.5. The rowing stroke of the hind leg of the beetle, *Gyrinus*, viewed from behind. The oarlets fold back to make a low-area recovery stroke in (c) and (d).

forward at the same speed as the body to which it's attached. Thus once the animal is under way, the speed of the appendage with respect to the fluid medium must be greater during the recovery stroke than during the power stroke. Blake (1981) has done a theoretical analysis of drag-based locomotion, and studies comparing lift-based and drag-based schemes are available—Davenport et al. (1984) on marine and freshwater turtles and Baudinette and Gill (1985) on penguins and ducks; in each pair the first uses a lift-based and the second a drag-based system. But we'll defer this most interesting comparison to Chapter 12.

CHAPTER 8

Velocity Gradients and Boundary Layers

$\mathbf{A}$T THE INTERFACE between a stationary solid and a moving fluid, the velocity of the fluid is zero. This, of course, defines the no-slip condition, mentioned back in Chapter 2. The immediate corollary of the no-slip condition is that near every such surface is a gradient in the speed of flow. Entirely within the fluid, speed changes from that of the solid to what we call the "free stream" velocity some distance away. Shearing motion is inescapably associated with a gradient in speed, so in these gradients near surfaces, viscosity, fluids' antipathy to shear, works its mischief, giving rise to skin friction and consequent power consumption. The gradient region is associated with the term "boundary layer," which I've quite deliberately postponed mentioning until it could be defined carefully—which is what this chapter is mainly about. Most biologists who have heard of the boundary layer have the fuzzy notion that it's a distinct region rather than the distinct notion that it's a fuzzy region. The essence of the present task is defining without ambiguity something that's not quite physically distinct.

While these interfacial velocity gradients are inevitable consequences of the no-slip condition, they were undoubtedly recognized much earlier. The boundary layer, by contrast, wasn't so much discovered as it was invented, in the early part of this century, as a great stroke of genius of Ludwig Prandtl. Recognizing the origin of this notion is crucial. In the basic differential equations for moving fluids, the Navier-Stokes equations, some terms result from the inertia of fluids and some from their viscosity. These equations turn out to be difficult to solve explicitly without simplifying assumptions, just as we simplified matters by assuming steady flow in deriving the Bernoulli equation. As we've seen, the Reynolds number gives an indication of the relative importance of inertia and viscosity; and it can be used to determine just what simplifying assumptions in the basic equations might be appropriate to a given situation. At Reynolds numbers below unity, inertia can be ignored and nicely predictive rules nonetheless derived—Stokes' law for the drag of a sphere (Chapter 15) is one. At high Reynolds numbers, one might expect to get away with neglecting viscosity, in which case the Navier-Stokes equations reduce to the Euler equations, which are the three-dimensional analogs of Bernoulli's equation. It may sound neat, but it all too commonly gets us in trouble—results diverge from physical reality, drag vanishes, and d'Alembert has his paradox.

Prandtl reconciled practical and theoretical fluid mechanics at high

Reynolds numbers by recognizing that viscosity could never be totally ignored. What changed with Reynolds number was *where* it had to be taken into account; initially it mattered everywhere, but as the Reynolds number increased well above unity, viscosity made a difference only in the gradient regions near surfaces. These regions might be small, and they might get ever smaller (or, more to the point, thinner) as the Reynolds number increased; but as long as the no-slip condition held, a place had to exist where shear rates were high and viscosity was significant. Prandtl called the place in question the "Grenzschicht" or (the English conveys a little less specificity) the "boundary layer." In general, a higher Reynolds number implies a thinner boundary layer but a higher shear rate in that boundary layer.

Shear entails dissipation of momentum and energy, so another way of viewing the boundary layer is as the region near a surface where the action of viscosity produces an appreciable loss of total pressure head. As mentioned in Chapter 4, Bernoulli's equation cannot be applied to the differences in velocity within a boundary layer because the equation assumes constant total head. Put another way, the velocity variation within a boundary layer comes about in a way that violates the very starting point from which Bernoulli's equation comes.

Of course the main question about boundary layers is just how thick they are. Here's where the real trouble surfaces. The inner limit is no problem—the solid-fluid interface. But there's no outer limit in nature to the gradient region. The speed of flow approaches zero almost linearly at the interface; by contrast, it asymptotically approaches the free-stream velocity with increasing distance from that interface, as shown in Figure 8.1a. So the

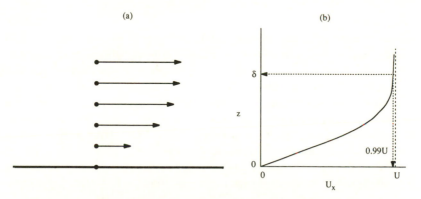

(a)   (b)

FIGURE 8.1. Defining a boundary layer well downstream on a flat plate parallel to free stream flow. (a) The way the speeds of flow vary with distance out from the plate. (b) Defining the boundary layer thickness with a graph of distance from the plate versus local speed—the layer extends z-ward to where the local speed is 99% of the free stream speed.

definition of the outer limit of the boundary layer must be arbitrary since the layer isn't a naturally discrete region bounded by an identifiable physical discontinuity. I emphasized (and will repeat) that the choice of an outer limit depends very much on the circumstances in which one is invoking a boundary layer—don't lose sight of the fact that it is an invented concept. The most commonly used outer limit is where the local velocity, U_x, has risen to 99% of the free-stream velocity, U:

$$U_x = 0.99\ U. \tag{8.1}$$

Figure 8.1b puts the matter graphically. Biologists should immediately recognize that the constant, 0.99, is no more than an artifact of our bipedal habit and pentadactylic anatomy.

THE BOUNDARY LAYER ON A FLAT SURFACE

Just how thick is this boundary layer? Consider a thin, flat plate oriented parallel to a flow for which the Reynolds number based on the distance from upstream to downstream (leading to trailing) edges is under about half a million. (Yes, we're talking about the situation to which equation 7.1 applies, if some sense of déjà vu occurred to you.) We'll call the distance downstream from the leading edge of the plate x, and the thickness of the boundary layer as defined by equation (8.1) δ (delta). An approximate solution to the Navier-Stokes equations gives a formula for the thickness of the boundary layer on one side of the flat plate where flow is laminar and where the boundary layer is defined as in equation (8.1):

$$\delta = 5\ \sqrt{\frac{x\mu}{\rho U}}. \tag{8.2}$$

Notice that the boundary layer thickness increases in proportion to the square root of the distance from the leading edge; its outer limit thus forms a parabola (Figure 8.2). Again let me emphasize the lack of physical discontinuity—nothing out there demarcates the parabola, not even a streamline. Increases in density or free-stream velocity thin the layer, while increases in viscosity thicken it. Authors differ somewhat on the value of the constant in the equation—I've seen values ranging from 4.65 to 5.84, so 5.0 should be a reasonable approximation.

And notice the familiar ratio of viscosity to density—kinematic viscosity—just as in the Reynolds number. One can, in fact, express equation (8.2) in terms of the Reynolds number, only here one uses a mild permutation called the "local Reynolds number," Re_x. The latter is simply a version applicable to a specific place on an object, not to the object as a whole; the only change in figuring it is that the characteristic length is taken as the distance (x) downstream from the leading edge to the specific place.

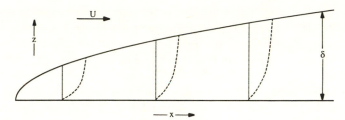

FIGURE 8.2. The thickness of a laminar boundary layer on a flat plate and several velocity profiles within it. The z-direction is much exaggerated; otherwise continuity would require that the flow arrows be tilted upward. Notice how the velocity gradient near the surface gets gentler with increasing distance downstream.

The result of this conversion, equation (8.3), shows that the thickness of the boundary layer relative to the distance downstream (a dimensionless thickness) depends solely on the Reynolds number at the particular point—low Reynolds numbers mean (relatively) thick boundary layers:

$$\frac{\delta}{x} = 5\, Re_x^{-1/2}. \tag{8.3}$$

Thus at a local Reynolds number of 10,000, the thickness of the boundary layer is a twentieth of the distance from the leading edge, while at a local Reynolds number of 100,000, it's about a sixtieth of that distance.

As usual, we have to attach some cautions and conditions to the equations (8.2 and 8.3). First, the derivation assumes that U_z, the velocity component normal to the plate and the free stream, is zero. In effect, the equation presumes that all flow is specified by U_x. Now, if fluid slows down upon entering the boundary layer, by the principle of continuity streamlines will diverge from the flat plate, and fluid not entering the layer will swing out and around it. So the very existence of a boundary layer must create a nonzero U_z. But if the boundary layer thickness is much less than the distance downstream from the leading edge, if $\delta << x$, then U_z will be close to zero. (Note, though, that while U_z may be zero for all practical purposes, the rate of change of velocity normal to the plate, dU_x/dz, will always be significant within a boundary layer—again, that's what boundary layers are all about.)

These equations, (8.2) and (8.3), are usable up to Reynolds numbers around 500,000—to where turbulence invades the boundary layer. To how low a Reynolds number can they be applied? In my experience, they're safe down to where the boundary layer thickness is as much as 20% of the distance from the leading edge, in which case the local Reynolds number is around 600. (Twenty percent may sound pretty generous, but remember that this isn't a stagnant layer, and the outer half has very nearly free-stream velocity.) For cruder estimates of boundary layer thickness, they can be used down to about 100 (Vogel 1962). They work for slightly curved surfaces almost as well as for flat surfaces, but they break down if the flow separates at all.

The most common use of these equations is in the calculation of forces; in fact, equation (7.1) was originally derived from (8.2). The biologist, though, should have a profound interest in specific conditions within the boundary layer. Many organisms live partly or entirely within the boundary layers of either inanimate objects or other organisms. For many creatures flow matters but free-stream velocity is something they never encounter; what's critical in determining where and how they live is the distribution of velocities within a gradient region. As it turns out, the curve describing the speeds of flow within a boundary layer (Figure 8.1b) is not a simply expressed function. Still, our requirements are usually modest; and if we accept a systematic error of up to 5%, the data cited by Rouse (1938) permit adequate approximations. Taking (as before) z as the perpendicular distance from the surface of a flat plate, we'd like to know U_x as a function of z. If z is half or less than half of the thickness of the boundary layer as defined in equations (8.1) and (8.2), then the following linear approximation will serve:

$$U_x = 0.32\, zU\, \sqrt{\frac{\rho U}{x\mu}}. \tag{8.4}$$

If z can be anywhere within the boundary layer, then a slightly more complex parabolic distribution is necessary, with a little less precise fit immediately adjacent to the surface:

$$U_x = 0.39\, zU\, \sqrt{\frac{\rho U}{x\mu}} - 0.038\, \frac{z^2 U^2 \rho}{x\mu}. \tag{8.5}$$

What if our thin, flat plate sports warts? How rough can a surface be without materially affecting the flow over it? Goldstein (1938) gives a pair of formulas that presume the flow is unaffected if the Reynolds number of a projection is 30 or less for pointed projections or 50 or less for a rounded one. The permissible heights of roughness (ϵ) relative to distance downstream from leading edge are then

$$\frac{\epsilon}{x} \leq 9.5\, Re_x^{-3/4} \text{ (pointed)} \tag{8.6}$$

$$\frac{\epsilon}{x} \leq 12.2\, Re_x^{-3/4} \text{ (rounded)}. \tag{8.7}$$

(The Reynolds numbers in the equations are based on distances downstream from leading edges, not on heights of projections.) Organisms are a bumpy, bristly bunch, and totally smooth surfaces are exceptional in nature. These formulas should provide handy tests of the applicability of the equations given earlier.

Clearly, roughness of a given size is more likely to have a disturbing effect if it's near the leading edge. Comparison of the exponents in these last two equations with that of equation (8.3) shows us that the higher the Reynolds number, the deeper within the boundary layer a protuberance must be lest it disturb the flow. As an example, consider a point a centimeter behind the leading edge of a flat plate moving through water at 0.1 m s^{-1}. The local Reynolds number is 1000, and a pointed projection of as much as half a millimeter is tolerable; this is one-third the thickness of the boundary layer. By contrast, if the flow is a meter per second or ten times as fast—a Reynolds number of 100,000—the protrusion can be only about a sixth of a millimeter or a tenth of the thickness of the boundary layer to be without effect.

All these formulas and calculations apply only to laminar boundary layers. Just as we needed a different description of the drag of a flat plate at high Reynolds numbers (equation 7.2), we need another pair of equations for situations above Reynolds numbers of about 500,000, for situations in which the boundary layer becomes turbulent:

$$\delta = 0.376 \, x \, \sqrt[5]{\frac{\mu}{\rho x U}} \qquad (8.8)$$

$$\frac{\delta}{x} = 0.376 \, Re_x^{-1/5}. \qquad (8.9)$$

In practice, a boundary layer may be laminar near the leading edge and then turbulent somewhere downstream, with the particular location of transition dependent (unsurprisingly) on the local Reynolds number, as in Figure 8.3. Within the turbulent boundary layer, right near the surface, a so-called laminar sublayer remains. Incidentally, it's no shear accident that the exponent in equation (8.3) is the same as that in (7.1) and the one in (8.9) is the same as the one in (7.2).

A turbulent boundary layer isn't quite the same kind of beast as a laminar layer, one differing only in the formula for how it happens to thicken in the downstream direction. Fluid in a laminar boundary layer moves consistently downstream as individual bits, not just as a statistical milling crowd. So there's no cross-flow transport of mass or heat as a result of the flow, and it's thus a semistagnant region in which wastes can accumulate, nutrients can be depleted, and mixing is restricted. A laminar boundary layer, in short, is likely to be a substantial barrier to exchange of material or heat. A turbulent boundary layer might provide a similar refuge against the forces of the free stream; but the turbulent motion is automatically associated with cross-flow transport of mass and heat. Thus it provides much less of a

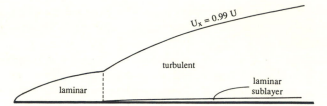

FIGURE 8.3. A boundary layer that's laminar upstream but becomes turbulent farther along, with a laminar sublayer in the latter region. Again the z-distances are greatly exaggerated.

barrier to transport. The biological implications of the distinction are great and will loom large later in this chapter and in the next.

We mustn't forget the arbitrary definition of the thickness of boundary layers stated as equation (8.1) and implicit in equations (8.2), (8.3), (8.8), and (8.9). Since he was interested in calculating the forces of skin friction on a plate, Prandtl took a generous view—he didn't want to exclude any region where viscosity mattered. For other applications, that limit of 99% of free stream may be misleading. In particular, for an organism living in a boundary layer and for a biologist interested in how much of the organism protrudes from the boundary layer, a somewhat thinner definition may be more realistic. Consider a point where the speed of flow is 90% that of the free stream. Something located there is unlikely to be in a region that gets depleted of whatever is supplied by the oncoming flow. Nor does something there get much shelter from drag—the latter will typically be over 80% of what would happen out in the free stream. The matter being ultimately arbitrary, perhaps one should choose 90% of free stream as a functional outer limit for reduced exposure. The difference of 9% may not sound like much, but the shape of the velocity distribution in a boundary layer is a peculiarly asymptotic one (Figure 8.1b again). A boundary layer bound by $U_x = 0.9\,U$ is only 70% as thick as one bound by $U_x = 0.99\,U$. If this smaller boundary layer is used, then the constant of 5 in equations (8.2) and (8.3) should be replaced by 3.5.

The trouble with all of these formulas is that they apply, at least strictly, to situations uncommon in nature. A stone in a stream certainly has a boundary layer, but the stone is rarely flat and as rarely has a sharp and regular leading edge. An organism sticks up into a boundary layer; by doing so it distorts the velocity distribution in that boundary layer; and, of course, it has a boundary layer of its own. A dense array of organisms raises the effective level of the surface; somewhere between a single item protruding z-ward and a solid layer, the location of the surface has to be redefined as we encounter what's often called "skimming flow" (about which more in a few pages). The obvious response to such complications is an empirical one—

to measure rather than to calculate. That's not as difficult as it sounds, since flowmeters and anemometers can be quite small and since boundary layers are thick at low Reynolds numbers.

Even where the geometry seems appropriate, caution and skepticism are needed. Perhaps a few examples—horticultural rather than just hortatory —will emphasize the point; for others, see Grace (1977). Most leaves are fairly flat, and (at least at low speeds) the wind might appear to blow smoothly over them. But . . . (1) in a comparison of soybean leaves and metal models it turned out that the structure of the boundary layer was comparable only at the lowest wind speeds (Perrier et al. 1973); (2) turbulence occurred at much lower wind speeds on a poplar leaf than on a flat plate, and evaporation rates were more than twice those predicted on the presumption of a laminar boundary layer (Grace and Wilson 1976); (3) the turbulence of the natural wind led to a lower-than-expected resistance to heat transfer for a leaf in an apparently smooth and steady airflow (Parlange and Waggoner 1972). So don't perpetuate the practice of equation-grabbing predecessors. Don't blindly adopt the 99% definition; don't use the formulas unless they demonstrably apply; don't be intimidated by the prospect of measuring low flows in small places.

A reasonable compromise between full faith in some revealed truth of physics and the complete agnosticism of empirical measurements is the generation of semiempirical formulas. These can be derived from measurements on models and then tested for their applicability to organisms in nature. The following are some good (and useful) examples, formulas given by Nobel (1974, 1975) for effective, average boundary layer thicknesses on various parts of terrestrial plants to use for predicting water loss and gas and heat exchange.

1. For a flat, leaflike object at Reynolds numbers from 300 to 16,000:

$$\delta = 0.0040 \sqrt{\frac{x}{U}}. \tag{8.10}$$

2. For a cylinder at Reynolds numbers from 1300 to 200,000:

$$\delta = 0.0056 \sqrt{\frac{d}{U}}. \tag{8.11}$$

3. For a sphere at Reynolds numbers from 400 to 40,000:

$$\delta = 0.0033 \sqrt{\frac{d}{U}} + \frac{0.00029}{U}. \tag{8.12}$$

These formulas, incidentally, are not dimensionally homogeneous; as given here they presume SI units.

FORCES AT AND NEAR SURFACES

Avoiding the full wind on a beach by lying against the sand is an experience familiar to most of us. As we've seen, the closer one gets to the surface, the lower is the local wind, with no wind at all right at the surface. Does this mean that the force of the wind, the drag, also approaches zero as an object flattens itself against the substratum? The drag quite clearly diminishes, but it doesn't actually drop to zero. Bear in mind that at the surface U_x may be zero, but dU/dz is not. Thus some skin friction must remain.

This minimal drag of a flat object on a surface can be easily calculated from information already at our disposal. Consider a small spot (not quite a point!) on a flat plate parallel to flow, a spot well back from the leading edge and lying beneath a laminar boundary layer. The velocity gradient above it is just the derivative of equation (8.4) with respect to z:

$$\frac{dU_x}{dz} = 0.32 \, U^{3/2} \, \rho^{1/2} \, x^{-1/2} \, \mu^{-1/2}. \tag{8.13}$$

The drag per unit area is, of course, simply the shear stress:

$$\frac{D}{S} = \mu \left[\frac{dU_x}{dz} \right]_{z=0}. \tag{2.4}$$

We can define a "local drag coefficient," C_{dl}, for the spot at issue just as we did the original drag coefficient:

$$C_{dl} = 2 \left(\frac{D}{S} \right) \rho^{-1} \, U^{-2}. \tag{8.14}$$

We then substitute (8.13) into (2.4) and the combination into (8.14) and get (throwing in the definition of the Reynolds number as well)

$$C_{dl} = 0.64 \, Re_x^{-1/2}. \tag{8.15}$$

This is a formula for the local drag coefficient—the drag of a spot of small but definable area on a large flat plate. (Hoerner 1965 gives the same formula but with a coefficient of 0.664.) Incidentally, if one integrates equation (8.15) from the leading edge of the plate rearward, one gets equation (7.1). An analogous equation works for turbulent flow (Leyton 1975 and other sources):

$$C_{dl} = 0.058 \, Re_x^{-1/5}. \tag{8.16}$$

The skin friction of part of a surface, this thing we're calling local drag, provides a useful baseline with which the measured drag or the force needed to dislodge an attached organism can be compared. If no explicit formula is available for the velocity gradient, it may be measured directly by determining the speed of flow at a few points near the surface. Conversely, one might use the shear on a force platform to infer the velocity gradient of the inner part of the boundary layer or even the boundary layer thickness.

The local drag on a spot on a surface is not a large force. A quick comparison with our previous standard for low drag, a flat plate parallel to free-stream flow, is instructive. Consider a square centimeter of surface, half a meter downstream from the leading edge of a flat plate, with a free-stream water flow of a meter per second. The local Reynolds number is 500,000; by equation (8.15) the local drag coefficient is 0.0009. By equation (8.14), that's a drag of 4.5×10^{-5} newtons. Now take that same square, 1 cm^2 on each side, and mount it in the free stream. The Reynolds number is now 10,000; by equation (7.1) C_{dw} is 0.0133, which corresponds to a drag of 0.0013 newtons.[1] In this case the square centimeter experiences fully *thirty times less* drag when a part of a larger exposed plate than when exposed alone—the same area, the same free-stream speed, the same orientation. And that's by comparison with our gold standard for low drag. By hunkering down against a surface you can't escape drag entirely, but what's left of it isn't exactly monumental.

Naturally, only a rare organism can get sufficiently intimate with the substratum to have a drag as low as that given by equations (8.15) and (8.16). Most creatures protrude to some degree above any surface of attachment, and most creatures are themselves equipped with protrusions—eyes, pinnae, nares, and other such excrescences. Aircraft and ship designers, worried about bolt heads, rivets, and so forth, have given some attention to the drag contribution of minor surface protuberances. Hoerner (1965) gives values of the drag coefficient between 0.74 and 1.20 for Reynolds numbers of 20,000 to 50,000 based on the frontal area exposed to the flow (in C_{df}) and the diameter of the protuberance (in Re); some of these are given in Figure 8.4. The highest value, not surprisingly, is for a plate protruding at right angles to both surface and the local flow. In general, the higher the protuberance, the worse are both the drag and the drag coefficient. Protuberances with rounded tops are better than flat-topped ones. If the flow direction is known and the organism arranged to face it, then improvement of almost an order of magnitude is possible—streamlining works here also. As we'll see in the next chapter, the trick seems to be used by some aquatic insects. In fact, a half-streamlined body is still far from ideal because it suffers quite a lot of interference drag around

---

[1] For that matter, both are pretty low forces. 4.5×10^{-5} newtons is the weight of a 4.5 milligram mass—several fruit flies. 1.3×10^{-3} newtons is a weight of about an eighth of a gram or a two-hundredth of an ounce.

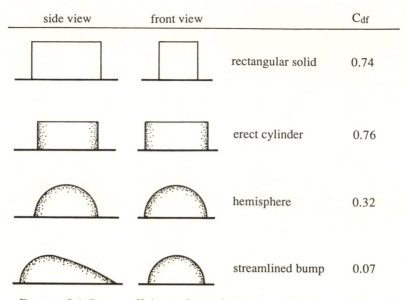

side view	front view		C_{df}
		rectangular solid	0.74
		erect cylinder	0.76
		hemisphere	0.32
		streamlined bump	0.07

FIGURE 8.4. Drag coefficients of protuberances on flat plates, based on frontal area. The values have only relative significance since, in practice, they depend on the heights of the protuberances relative to the local boundary layer thickness, here assumed very thin.

its base, especially in front. It helps to lengthen and flatten the leading edge of the protuberance; by such a maneuver the drag coefficient can be halved again, to a minimum of 0.03 at these Reynolds numbers.

These drag coefficients give some idea of the effects of shape on drag for objects within a boundary layer. But they shouldn't be applied directly to biological situations without checking the local Reynolds number as well as the Reynolds number of the protrusion itself. For the coefficients cited, the local Reynolds number is about two orders of magnitude larger than the latter, implying a turbulent boundary layer. Still, aquatic insects may pupate on large rocks, and epizoans and epiphytes afflict large organisms in rapid flows. In any case, the same general rules most likely apply at more modest Reynolds numbers: (1) rounded tops are better than flat, sharp-edged tops; (2) half-streamlined bodies are better yet, and to avoid separation maximum thickness should be near the upstream rather than the downstream extremity; (3) smooth fairing of the edge of the protuberance into the surface of attachment improves matters; (4) with such fairing, the distinction in shape between upstream and downstream ends of a protuberance should be slight—certainly less than that distinction for well-streamlined bodies in a free stream. Thus good low drag shapes for unidirectional, bidirectional, and even omnidirectional flows may converge. Still, altogether too little hard information is available.

Unbounded Boundary Layers

What happens if our flat plate lacks a leading edge? All the best authorities, antedating even Aristotle, have recognized that the surface of the earth is quite without such an edge. Some sort of gradient region is obviously always present—the no-slip condition isn't easily violated; and one can, as mentioned already, escape most of the wind by lying down even on an unobstructed beach. In fact, only at altitudes of 500 meters or greater is the wind unaffected by friction with the surface and the true geostrophic wind appears (Sutton 1953). Water movement on the bottom of a stream, river, or ocean presents much the same situation and has to be treated in very much the same manner. What seems to be quite a diverse lot of cases reduces to flows across dense arrays of bluff bodies, the latter boiling off turbulent wakes downstream and upward (Rose 1966). Thus superimposed on an overall horizontal, turbulent flow, whatever its speed spectrum, are eddies of various sizes that transport momentum vertically. These eddies prove to be sensitive to the roughness of the surface even when well above it. As a result, the way average speed varies with height depends quite strongly on the character of the surface beneath. All of which sounds dauntingly complex and resistant to simple summarization in predictively useful equations.

But the situation isn't quite as turbulent conceptually as it is physically. Both theory and measurements indicate that the variation of horizontal flow speed with height above the substratum is a logarithmic function, at least if buoyant effects from temperature variations are negligible. And this logarithmic relationship persists up to the winds of destructive gales (Oliver and Mayhead 1974). The following general formula is usually cited:

$$U_x = \frac{U_*}{\kappa} \, ln \left(\frac{z - d}{z_0} \right). \tag{8.17}$$

It will stand a bit of explanation. We're interested, most often, in how the horizontal wind, U_x, varies with height, z, above the ground. Those are variables with which we're already familiar. κ (kappa) is the dimensionless "Von Kármán's constant" with an empirically determined value of 0.40. d is called the "zero-plane displacement"; it accounts for the fact that the logarithmic profile extrapolates to zero velocity somewhere above the ground, especially on thickly vegetated surfaces. The so-called roughness parameter or roughness length, z_0, adjusts the steepness of the logarithmic velocity gradient because it's related to the size of the eddies generated at the surface. Rougher surfaces make bigger eddies, which transfer more momentum in the z-direction and thus reduce the steepness of the velocity gradient. U_* is called the "shear velocity" or the "friction velocity"; it indicates the amount of turbulence, and its value is independent of height for a given surface and free-stream flow.

What's a little odd about equation (8.17) is that the meanings of several variables diverge from their dimensions and names. Thus U_* isn't really a velocity in any normal sense—it's the square root of the shear stress (recall equation 2.4) divided by the density of the fluid:

$$U_* = \sqrt{\frac{\tau}{\rho}}. \tag{8.18}$$

And z_0 is only a nominal length, just as $V^{2/3}$ was used as a nominal area. The equation assumes that the shear stress is constant with height, which turns out to be reasonable for the first 10 to 20 meters above a surface (in air—less in water). This constancy is maintained by a constant downward flux of momentum through turbulent transport of eddies. Figure 8.5 gives a typical profile, plotted linearly in (a) and logarithmically in (b). It assumes that the zero plane displacement is negligible ($d = 0$), which is essentially true for surfaces without a lot of high protrusions—low grass, rocks with barnacles, mud flats with occasional worm tubes. If d matters, the most common procedure (Rosenberg et al. 1983) is to fiddle with its value until the profile comes out properly logarithmic, which is to say until it plots linearly on a graph such as that of Figure 8.5b. An alternative is to use the rules of thumb given by Monteith and Unsworth (1990), applicable to fairly thick layers of vegetation. They suggest that z_0 will be between 8% and 12% of the vegetation height and that d will be 60% to 70% of that same height. Another alternative is to take values from previous studies; Table 8.1 gives some representative values.

But we still don't have a convenient way to determine the velocity profile from the properties of the fluid, some length, and some free-stream speed as we had for the cases with nice leading edges. (Nor, for that matter, are we any longer using that limit of 99% of asymptotic free stream—taken strictly, there's no outer limit with equation 8.17!) U_* presents the most problems; according to Sutton (1953) (concerned with terrestrial systems), it varies from about 3% to 12% of the mean wind speed a meter or two above the surface or zero plane, which is pretty unconstant. On the other hand, the data cited by Ippen (1966) for water flow near the smooth bottom of a channel 10 meters in depth suggest that 3% is a good figure for a decently wide range of average velocities. For rougher bottoms about 5% to 15% of ordinary velocity is reasonable, according to Denny and Shibata (1989). It's just not easy to measure shear stress under appropriate circum-

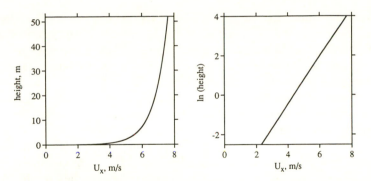

FIGURE 8.5. Two views of the distribution of velocities within a logarithmic boundary layer: (a) a linear plot; (b) a logarithmic plot. Here d, the zero plane displacement, is assumed zero; z_0, the roughness length is 0.005 m; and U_*, the shear velocity, is 0.33 m s^{-1}.

stances. Either one can measure the velocity itself, U, at several heights (z's) and do a logarithmic regression to determine U_* from the slope and z_0 from the intercept, fiddling, as mentioned, with d if the thing doesn't plot out as a straight line (Jumars and Nowell 1984). Or one can use the rules above for estimating z_0 and d or use values from previous studies and a single measurement of velocity at a known height. If one isn't wedded to

TABLE 8.1 VALUES OF THE ROUGHNESS LENGTH, z_0, THE ZERO PLANE DISPLACEMENT, D, AND THE SHEAR VELOCITY, U_*, FOR VARIOUS SUBSTRATA.

Surface	z_0	d	U_* @ U
Very smooth	0.00001	0.007	0.16 @ 5
Lawn grass, to 1 cm h.	0.001		0.26 @ 5
Short grass, 1–3 cm h.	0.005	0.0	0.33 @ 5
Sparse grass, to 10 cm h.	0.007	≤0.07	0.36 @ 5
Thick grass, to 10 cm h.	0.023		0.45 @ 5
Thin grass, to 30 cm h.	0.05		0.55 @ 5
Thick grass, to 50 cm h.	0.09	≤0.66	0.63 @ 5
Long grass, 60–70 cm h.	0.03	0.30	0.50 @ 5
Tall crop, 1 m height	0.2	0.95	0.46 @ 3
Forest, deciduous	1.0–6.0	≤20.0	
Forest, coniferous	1.0–6.0	≤30.0	
Over the sea	0.0003	0.04	
Desert sand	0.0003	0.0	

SOURCES: Sutton 1953; Sellers 1965; Oke 1978.
NOTE: U is taken at 2 meters above the surface or zero plane. In SI units.

commercial equipment one can measure U_*, at least in unidirectional aquatic systems or the shear stress, τ, which is just as good. Statzner and Müller (1989) give the details of a technique for using the threshold of movement of hemispheres of a graded series of densities across standard flat plates as a measure of shear stress.

Why bother with all this semiempirical stuff? For one thing, a graph of how speed, even average speed, varies with height can be a useful thing in an investigation of the forces on some organism that sticks up from a substratum (see, for instance, Denny 1988). The present discussion shows how to get that profile from something less than a full set of measurements of average speed versus height. And such profiles are relevant to phenomena that include fluxes of dissolved materials and settling of planktonic organisms (Nowell and Jumars 1984). For another, U_* itself is an interesting quantity—it is, after all, really a kind of shear stress. Together with the roughness length, it has immediate relevance to similar erosive and depositional processes, whether of snow or of terrestrial or submerged sand (Ippen 1966; Monteith and Unsworth 1990) or of edible particles on a bay bottom (Jumars and Nowell 1984).

For that matter, one can put this peculiar shear velocity, U_*, to good use in yet another version of the Reynolds number, this one called the "boundary roughness Reynolds number," Re_r (see, for instance, Jumars and Nowell 1984). It can give a picture of the character of flow immediately adjacent to the surface (Figure 8.6). Shear velocity simply substitutes for free-stream velocity, and l is taken as the height of the roughness elements —commonly the diameter of particles on the surface. If Re_r is less than about 3.5, then the surface is hydrodynamically smooth with a viscous[2] sublayer of the ordinary sort of thickness $11.6 \, \mu/\rho U_*$. So any exchange of dissolved material has to depend on the phenomenon of molecular diffusion and the value of the diffusion coefficient of each kind of molecule for its final stage. (The effective diffusive sublayer is about 20% of the thickness of the viscous sublayer.) If Re_r is greater than about 100, then the surface is hydrodynamically rough, and no real viscous sublayer exists. For exchange of dissolved material, what then matters is the eddy diffusion coefficient, much larger than molecular diffusion coefficients and independent of the chemical species involved. (For sucrose in water, the molecular diffusion coefficient is 5×10^{-6} cm^2 s^{-1} compared with the eddy diffusion coefficient of about 100, a 20-million-fold difference.) If the boundary roughness Reynolds number is between 3.5 and 100, there's a transitional mess.

The literature on these logarithmic boundary layers is scattered among subjects such as micrometeorology, coastal engineering, limnology, and

---

[2] The "viscous" sublayer mentioned in connection with logarithmic boundary layers corresponds to the "laminar" sublayer of the rest of the fluid mechanical literature.

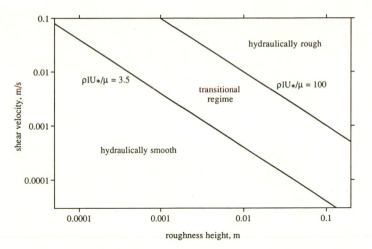

FIGURE 8.6. The character of flow right near a solid surface beneath flowing water, as organized by a plot of shear velocity against the roughness height of the bed.

oceanography. Good sources of general information, besides those referred to specifically, are Lowry (1967), McCave (1976), Arya (1988), Garratt (1992), and Wieringa (1993); a simple and intuitive introduction is given in a book produced by the Open University (1989) entitled *Waves, Tides, and Shallow-Water Processes.*

FLOW RIGHT ON BUMPY BOTTOMS

The top of the canopy of a forest, the top leaves of a uniform field of some crop, the top of a "forest" of tall sea anemones on a subtidal rock face—each of these forms an effective substratum below the logarithmic boundary layer about which we've been talking. Within these stands, matters are complex. Speeds of flow are lower than those above the stand and, as a rough rule, are inversely related to the local density of solid material such as leaves, appendages, tentacles, and so forth, although naturally speeds drop off near the ground (Grace 1977). In open forests with little understory or in a forest of sea anemones, winds beneath the canopy may exceed those within (but not above) the canopy.

In considering surfaces that are either uniform, of low roughness with rare upright elements, or have nearly flow-impervious canopies, we're talking about extremes. Whether the bumps are living or not, one very often encounters intermediate situations, cases where protrusions abound but don't abut. As a rule of thumb, the protrusions have only a minor effect on the speeds of flow near the surface until they occupy about a twelfth of the

171

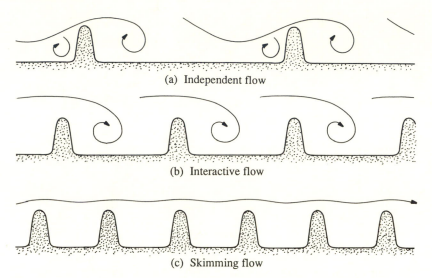

(a) Independent flow

(b) Interactive flow

(c) Skimming flow

FIGURE 8.7. Flow right near a bumpy bottom for increasingly closely spaced bumps. At least three separate regimes can be recognized, loosely (a) independent flow, (b) interactive flow, and (c) skimming flow.

overall (plan or top view) area of the surface. When more than a twelfth of the surface is covered, then eddies don't penetrate well, and one gets what's been called "skimming flow" (Jumars and Nowell 1984). In a sense, two phases of flow of substantially different character then coexist and interact, much as happens in and above a forest.

As will become evident in the next chapter, what happens right down amid the bumps is of enormous biological importance. A finer-grained categorization of phenomena than merely skimming or nonskimming turns out to be a practical necessity; useful approaches are those of Nowell and Jumars (1984), Davis and Barmuta (1989), and Carling (1992). Consider the three rough surfaces of Figure 8.7. If the individual bumps are well spaced, that is, if their heights are much less than the distances between them, then each acts essentially alone, with an anterior downflow that forms a stable vortex and with another vortex behind that also turns (with flow from left to right) clockwise. If the bumps are closer, with their spacing only moderately greater than their heights, then interactions between each rear vortex and the next front vortex create a considerably less stable and regular pattern of flow. If the bumps are still closer, then skimming occurs, with stable vortices in the interstices. All of this, of course, presupposes overall Reynolds numbers high enough so vortices will happen readily and depths (in water) great enough so the bumps don't act like weirs. And it's a bit two-dimensional—with individual rocks instead of cross-flow walls or steep ripples, we encounter horseshoe-shaped vortices wrapped around obstacles and other such phenomena that will resurface in Chapter 10.

Beyond what happens in and between bumps, considerable flow goes into and out of porous substrata, driven by differences in hydrostatic and hydrodynamic pressures on stream beds and terrestrial surfaces; the matter came up in Chapter 4, and we'll have more to say about it when we consider interstitial flows in Chapter 13. I again urge the reader to take some small syringe or baster and a bottle of marker (fluorescein or uranine, food colorant, or even skim milk) to a stream with some bed irregularity and to explore the local patterns of flow. Or, if the opportunity presents itself, toss a bunch of toy parachutes in different directions from the top of a building while the wind is blowing. An otherwise invisible world of environmental heterogeneity will reveal itself.

CHAPTER 9

Life in Velocity Gradients

W E'RE SET TO consider the physical situations and adaptations of a vast multitude of organisms—the world of organisms whose heights above the substratum are less than or of the same order of magnitude as the thickness of the boundary layer. This world, then, is defined not by some velocity but by the ubiquity of a spatial gradient of velocity. It's the world of smallish organisms at the interface between moving air or water and a solid substratum; it's the world, even, of the occasional creature on a solid substratum that's moving through a fluid—the mite on the bird, the barnacle on the whale. If we want to afflict biology with yet another name, we can easily coin one for this assemblage. Expropriating from the Greeks, we might call them *craspedophilic* creatures, from *craspedo-*, an edge or border, and *phile*, fond of—these are the interfacial plants and animals.

In a sense, virtually every organism exposed to a flow lives in a velocity gradient inasmuch as the logarithmic boundary layers of both terrestrial and aquatic habitats are universal and are almost always thicker than an organism is tall. But up to this point we got away with the assumption that organisms were exposed to free-stream velocity—or at least that the spatial variations of velocity impinging on them could be adequately represented by some simple average. Not much heavy weather was made about it, but the assumption enforced a severe selectivity in the examples I used.

Not only are the inhabitants of velocity gradients diverse in size and style, but the functional consequences of these gradients are manifold. A gradient region is both good news and bad news. It's a hiding place from drag, but it's a barrier to exchange of materials and energy. It may be useful as insulation—you can get a burn in a sauna if you're exposed to a local wind. Or it may impede heat loss enough to present serious problems of overheating—that's a difficulty faced by broad leaves when the breeze fails. It may afford some mechanical protection from the impact of material carried by a flow, but on a filter-feeding organism it may cause detrimental deflection of edible particles around the filter. Flow-dispersed seeds, spores, and spiderlings are likely to travel less far if released deep in a gradient region, especially in air, where gravitational settling is a major factor; the same propagules are likely to remain where they settle if velocity gradients are gentle.

Let me repeat for the sake of emphasis that the effects of velocity gradients—and indeed of local currents in general—are diverse. From my point of view, trying to deal with a vast and disparate literature, it seems altogether too easy to do comparative studies with flow as the variable quantity. Comparing different natural habitats, comparing modified and unmodified habitats, comparing artificial streams indoors and out, we now have quite a lot of data that correlate flows with who lives where. Sometimes "who" is intraspecific and sometimes interspecific. Differences in velocity and velocity gradient are sometimes decently documented but more often just roughly categorized. My dissatisfaction is that of the functional biologist looking for mechanisms who is less than satisfied with inference from correlation. Field studies that merely document the distribution of organisms cannot explain distributions in any truly causal sense.

Equally, though, a laboratory study is apt to focus too exclusively on a single facet of multidimensional suites of adaptations. For instance, the strong preferences of an organism for a specific regime of flow could as easily reflect factors such as predation and trophic preference as purely hydrodynamic adaptation. Statzner et al. (1988), for instance, cite literature that documents apparent effects of the current speed of streams on (among others) mate choice, posture, case building, compass orientation, fighting success, net building, phototaxis, rheotaxis, schooling, territoriality, and respiratory movements.

Drag of the Flat and Not-so-flat

Can you stand a little more about drag, mentioned in each chapter since number four? In particular, while we talked about the drag of attached organisms and how many of them contrive to reduce it in Chapter 6, at that point little was said about drag reduction by hunkering down against a surface. If you're located well down within a boundary layer, then flow speed drops steadily as you get closer to the surface; if the layer is a laminar one on a flat plate (so that equation 8.4 applies), then the drop is very nearly linear. Thus drag goes down (for a bluff body, at least) quite dramatically as you approach the surface, reaching the very low value of the local drag of your footprint if you have no height. Mention was made (with something of a promissory note) that in a typical velocity gradient the drag of a bluff body is expected to vary not with the square of flow speed but with an exponent a little greater. We can now see just why. Again consider the inner part of a laminar boundary layer on a flat plate. By equation (8.4), at a given point above the plate (a given x and z), the local velocity, U_x, varies more drastically than the free-stream speed:

$$U_x \propto U^{1.5}. \tag{9.1}$$

And thus for a bluff body at moderate Reynolds numbers (ignoring skin friction on the surface itself),

$$D \propto U^{3.0}. \tag{9.2}$$

What's happening, of course, is that higher free-stream speeds mean thinner boundary layers, so as speed increases, a given location becomes closer to the outer edge (however defined) of the layer. Since drag is proportional to the cube rather than the square of velocity, the baseline for speed-specific drag, D/U^2 (equation 6.4), becomes $+1.0$ rather than 0 for objects in the inner portions of laminar boundary layers.

Before passing lightly over local drag, I might note that I've found few well-documented cases in which the skin friction of something that doesn't protrude at all takes on biological significance. Conversely, practical relevance is obvious, since washing without cloth or brush depends on skin friction to remove any insoluble material. One case that did turn up involved the shear stress needed to get bacteria off the walls of capillary tubes: Powell and Slater (1982) found a shear stress of around 50 Pa necessary to assure removal. Using room-temperature water as the washing medium, that's a shear rate of $50,000$ s^{-1}, the velocity gradient that would occur between two plates a millimeter apart, moving at the enormous speed of 50 m s^{-1} relative to each other. Other things (speed and scale) being equal, using hot water would be counterproductive since its lower viscosity would give a lower shear stress for a given shear rate—recall equation (2.4).

One other matter might be mentioned again: the caution that the lift on creatures at interfaces may be a worse problem than their drag. Even if lift isn't (as it sometimes is) greater than drag, it works in a more awkward direction, outward ("z-ward"), with an awkwardly upstream line of action. Moreover, any upward motion due to lift near the upstream extremity will likely increase lift further (for a while) due to the increased angle of incidence. It will also increase drag and decrease contact area. If that weren't bad enough, drag loads the contact line in shear, of which adhesive layers are typically fairly tolerant, while lift raises the possibility of peel failure, a much trickier business for a glue. (Peeling is the only practical way to get adhesive tape off a roll.) Substantial lift on creatures on surfaces has been documented for limpets (Denny 1989), sand dollars (Telford and Mooi 1987), mayfly larvae (Weissenberger et al. 1991), crabs (Blake 1985), flatfish (Arnold and Weihs 1978), and rays (Webb 1989). And Kirsten Johannesen (pers. comm.) found that periwinkle snails (*Littorina* sp.) oriented to flow in the only position for which they had negative (substrateward) lift—drag was not the primary consideration.

The "Torrential Fauna"

These animals, inhabitants of streams where the free-stream speeds are over about 0.5 m s^{-1}, made a brief appearance (in Chapter 7) when we talked about drag and streamlining of the legs of the mayfly nymph, *Baetis*.

Even the swiftest streams have vegetation on their stones and carry suspended comestibles. Despite currents as high as 3 m s^{-1} (Nielsen 1950), a specialized fauna of both grazers and suspension feeders manages to make a living in such streams without being swept downstream. Most of its macroscopic members are immature insects: mayfly nymphs (Ephemeroptera); caddisfly larvae (Trichoptera); beetle larvae and adults (Coleoptera), especially of the family Elmidae; black fly larvae (Diptera: Simuliidae); netwinged midge larvae and pupae (Diptera: Blepharoceridae); larvae and occasional pupae of some moths (Lepidoptera: Pyralidae); and a few others. Their diversity is far less than that of the animals that live in the same streams but avoid the current by hiding beneath rocks, by burrowing, or by other means. All live in an unusually harsh habitat that demands substantial adaptations to the rapidity of flow and its capacity for impressing forces. Merritt and Cummins (1984) provide a very useful compendium of who lives where, as well as voluminous references; Hynes (1970) and Ward (1992) put the whole subject in a richly ecological context.

Early in this century, Steinmann (1907) suggested that the insects of rapid waters are dorso-ventrally flattened in order to reduce drag and to help them stay attached. The idea sounds reasonable, and it penetrated generations of textbooks, but it doesn't look quite so tidy on close scrutiny and has been questioned in almost every serious contemporary paper on the subject. Part of the problem was that Steinmann viewed the flow as pushing the flattened insect down against the substratum by striking its inclined, upstream-facing surface. The notion isn't intuitively unreasonable, and the nonwhiggish historian bears in mind that modern notions of lift and the concept of a boundary layer were developed in that decade and weren't widely appreciated even among physical scientists (Goldstein 1969). In fact, positive rather than negative lift will ordinarily be developed; to make matters even less obvious, the amount of lift will depend in part on whether flow separates when passing over the organism. So flatness is a two-edged sword—on the one hand it affords a location deeper within the boundary layer of the substratum and thus lower drag plus a greater surface for attachment, but on the other hand it raises the bothersome bugbear of lift. Even separation is a mixed bag—less lift but more drag. To complicate matters further, flatness isn't exactly inappropriate if a creature is to crawl around a stone and get underneath it.

It ought, then, to be no surprise to find that some of the torrential fauna are flattened and some are not and that some nontorrential (lenitic) creatures are nonetheless flattened. The mayfly nymph *Baetis* holds itself off the substratum and has a body that approximates a streamlined body of revolution rather than being appreciably flattened (Figure 9.1a). The species of *Baetis* that inhabit more rapid flows are smaller, but have larger legs, smaller gills, and smaller and fewer caudal cerci (long trailing hairs)—all quite reasonable for organisms found in flows of up to 3 m s^{-1}. And (just so

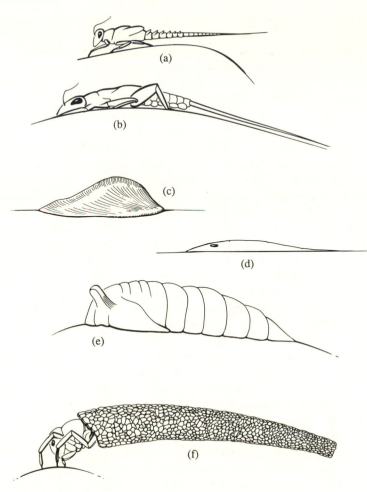

FIGURE 9.1. Creatures of the so-called torrential fauna:
(a) *Baetis*, a mayfly nymph; (b) *Rhithrogena*, another mayfly nymph;
(c) *Ancylus*, a freshwater limpet; (d) *Psephenus*, a water penny beetle;
(e) *Bibiocephala*, a pupal blepharocerid fly; (f) *Neothremma*, a caddisfly
larva.

nothing is too simple) individuals can to some extent flatten themselves
against a substratum (Dodds and Hisaw 1924). Streamlining, as we've seen,
is a potent drag reducer, and *Baetis* should not be troubled by lift. But it's a
relatively uncommon adaptation among aquatic insects, found otherwise
mainly in beetles (Resh and Solem 1984). Other torrential mayfly genera
such as *Iron, Epeorus, Ecdyonurus*, and *Rhithrogena* (Figure 9.1b) are well

flattened dorso-ventrally and live appressed to rocks. Weissenberger et al. (1991) discovered that *Ecdyonurus* sometimes had negative lift, apparently achieved by lowering its head shield and by using its femurs as spoilers.

Mayfly nymphs aren't the only group in moving fresh water to have evolved flattened forms with strong devices for attachment. One thinks first, of course, of turbellarian flatworms, but they're really not an especially good case since flatness is certainly the ancestral condition and since they tend to avoid places with really high currents (Hansen et al. 1991). Too bad—for a planarian the skin friction of local drag might be a real phenomenon. Better examples are several families of freshwater gastropod mollusks, most notably the Ancylidae, which are low-coned limpets (Figure 9.1c). None of the animals we're talking about is especially large; most are less than a centimeter long. The same suite of adaptations, though, happens in even smaller arthropods, the water mites (Hydracarina: Arachnoidea), among whom a millimeter in length marks a fair giant. Species living in streams are smaller than others, are relatively more flattened, have stouter claws, and have fewer protruding hairs (Pennak 1978). Even some tadpoles play the game—several rheophilic groups have independently evolved severely depressed bodies and ventral adherent devices (Lamotte and Lescure 1989).

Perhaps the flattest of the flat are the members of the coleopterous family Psephenidae, the "riffle" or "water penny" beetles. Late-stage larvae run around half a centimeter long but are less than a millimeter thick at the center (Figure 9.1d). They live both under and above logs and stones; being negatively phototactic and positively rheotactic they're exposed mainly at night (and so are sometimes thought to stay burrowed). Smith and Dartnall (1980) found that blocking normal flow through the marginal lamellae on each side led to earlier separation of flow on top of the animals; on that basis they suggested that the animals normally reduce drag by boundary layer control through suction, as described for conventional airfoils. The implications for lift are at this point unknown, although the matter has at least been raised by McShaffrey and McCafferty (1987).

Separation of flow behind the region of maximum thickness seems to be common among these flattened forms. Some early drawings of flows based on observations of moving particles, such as those of Ambühl (1959), don't show the phenomenon. However, more recent work with much less intrusive laser-doppler anemometry (Statzner and Holm 1982) shows significant separation behind the mayfly nymph *Ecdyonurus*, and really substantial separation behind the freshwater limpet *Ancylus*. And, as just noted, Smith and Dartnall (1980) found separation even in unplugged water penny beetles. My guess is that separation reflects a price in drag paid to avoid paying in the harder currency of lift—it ought to reduce lift more efficiently than any other kind of drag-increasing arrangement.

Insect pupae occasionally occur where they're exposed to rapid currents. They don't take on particularly flattened shapes, although some pupae of torrential Pyralidae (Lepidoptera) live in very flat cocoons (Nielsen 1950); probably metamorphosis imposes constraints. But pupae are usually compact and fusiform, and thus, assuming appropriate orientation, they're reasonably well streamlined. Pennak (1978), for example, provides a drawing of a Blepharocerid (Diptera) pupa that looks almost identical to one of Hoerner's (1965) streamlined rivet heads—compare Figures 9.1e and 8.4).

Most of the torrential caddisfly larvae live in cylindrical or conical cases that they construct of tiny stones and bits of vegetation (Figure 9.1f). Dodds and Hisaw (1925) suggested that the caddisflies of rapid waters use stones as ballast; the counterargument is that stones are merely what are available in faster currents (Resh and Solem 1984). In any case, their resistance to being detached and carried away by the current is compounded by active attachment and a passive component, the latter including ballast. Young larvae of *Allogamus auricollis* mainly depend on active attachment, but by the fifth instar active and passive elements are nearly equal and the overall current that can be resisted has doubled—presumably due in part to the larger stones incorporated into the case (Waringer 1989). But of course in the velocity gradients on surfaces the larger organism must suffer greater ambient current. Some species of caddisflies erect elegant nets for catching planktonic prey; the larger the caddisfly, the coarser is the mesh of the net. In addition, nets are coarser in more rapid currents (Wallace et al. 1977). Either larger larvae can eat larger particles and faster currents suspend larger particles (Cummins 1964), or else larger animals, with larger nets, must face something closer to free-stream speed, and the coarser mesh is a way to keep drag from becoming excessive. In short, a larger net may be in part a scheme to prevent destruction of a net even at the cost of missing fine particles. But let's defer further discussion until we get to the subject of filtration, a really low Reynolds number business, in Chapter 15.

And Some Analogs of the Torrential Fauna

A marine equivalent of this torrential fauna is that of the ectoparasites of fast swimmers such as whales. Whales commonly provide a substratum for barnacles of the genus *Cryptolepas*—on a gray whale, *C. rhachianecti* occurs mainly on the head and the anterior part of the back, which I interpret to mean that it prefers quite a steep velocity gradient. In addition, whales harbor so-called cyamids or whale lice, amphipods of *Cyamus* and related genera. These live mainly around the barnacles or are associated with the various slits and pleats on a whale's surface—they presumably prefer less steep gradients (Blokhin 1984). The preference of cyamids for a little shelter may be involved in the genesis of the skin irregularities ("callosities") on the heads of southern right whales (*Eubalaena australis*), the patterns of which prove useful (at least to humans) in distinguishing among individuals. According to Payne et al. (1983), cyamids feed on the surface layer of skin; not only do they excavate pits and depressions, but they avoid projections, which would be exposed to higher velocities. So as a lousy whale makes skin, projections get longer and longer; the degree of relief of the patches of callosities gives an indication of a whale's age.

Drag on a creature deep within a velocity gradient might matter in air as well as in water, and at least one case constitutes a fine parallel to the lice of whales—the feather mites of some large seabirds. Mites are, of course, arachnids, kin to spiders, rather than crustaceans, as are the whale lice (proper lice are wingless insects). According to Choe and Kim (1991), several species of *Alloptes* (Acari: Analgoidea) live on the flight feathers of murres and kittiwakes. The mites prefer to be well back from the leading edge of the wing, they prefer the slightly concave ventral surfaces of feathers, and they prefer the grooves between the individual barbs that make up the vanes of feathers. In each choice, then, they select the aerodynamically more benign location. Nonetheless, they're mighty modified mites—flattened dorso-ventrally, elongated, with reduced dorsal setae, with long lateral setae that seem to help keep them attached, with large suckerlike empodia that also seem important in attachment, and so forth.

Sheets of Water with Really Extreme Gradients

Sometimes flows in nature are essentially all gradient, and not just the gentle gradients of a breeze across the landscape or the current in a river. Consider what happens when a thin layer of water, the swash, runs up a sandy beach and then down again. With the no-slip condition operative, the layer has to have an extremely high value of dU/dz. The same circumstance characterizes sheets of water moving across rock faces as parts of mountain streams. Both situations have called forth rather specific adaptations, although the two are far from parallel: a rock face affords secure opportunities for attachment but permits no escape through burrowing, while the surface of a beach presents exactly the opposite situation.

Ellers (1988) has given the name "swash-riding" to the phenomenon of tidally cyclic migration by using this water movement up and down the face of beaches; he's done a thorough study of how a particular 1 to 2 cm-long clam, the coquina, *Donax variabilis*, does the trick. *Donax* are normally intertidal, but on most marine beaches "intertidal" refers to a situation rather than a fixed location—so the clams must move regularly to maintain that intertidal position. Which they do. Normally they're within the sand at a depth just a little beyond easy reach of beach birds (Richard Coles, pers.

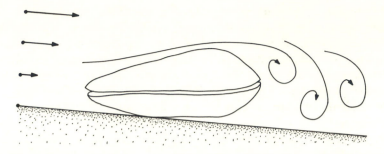

FIGURE 9.2. The tiny clam, *Donax*, oriented in the backwash of a wave on a sandy beach.

comm.). On an incoming tide clams emerge just before the leading edge of the swash reaches them. On a receding tide they emerge directly into the backwash. In the swash they're carried upbeach in the extremely chaotic flow while in the backwash they orient with anterior ends upstream (Figure 9.2) and move down the beach without significant tumbling.

The tiny clams choose exceptionally large waves through acoustic cues, and they maintain their tidal rhythm of sensitivity to wave sounds and their emergence and burrowing behavior in the laboratory for a time, but these neat features of the scheme aren't important in the present context. More germane is the fact that *Donax* are structurally rather unusual clams. For one thing, they're unusually dense, about 13% more so than other bivalves or 45% above the density of seawater. For another, their anterior ends are thinner than their posteriors, and maximum thickness is fairly far back. The combination seems to assure consistent passive orientation and to prevent tumbling in the backwash with the consequent problem of being carried too far seaward without getting a foothold. In their normal orientation in the backwash, separation of flow is fairly severe, although they experience lower drag than in any other position. And lift is downward, according to Ellers' measurements on a large model in the gradient region on the floor of a wind tunnel.

This clam isn't the only swash-rider but merely the one about which we know the most about the mechanics of the behavior. Besides other species of *Donax* and a few other clams, the phenomenon has been described in mole crabs and amphipods. A whelk (a gastropod mollusk), *Bullia digitalis*, uses swash riding to scavenge prey that it has located chemotactically, according to Odendaal et al. (1992).

The inhabitants of sheets of water flowing across rocks have been called "hygropetric" organisms; they include some specialized tadpoles, chironomid (midges) and psychodid fly larvae, caddisfly larvae, and gerrid bugs (water striders). Not all habitats described as hygropetric involve really fast flows—continuous exposure to the spray of a waterfall may result in just a slow seepage—so the designation includes considerable hydro-dynamic heterogeneity whatever its hygrometric[1] consistency. Some of the chironomid flies seem to have secondarily adopted the thin water films of the percolating filters of sewage plants as a habitat, where they're called "sewage filter flies" and can constitute a local nuisance (Cranston 1984; Houston et al. 1989).

SETTLING DOWN IN A VELOCITY GRADIENT

One shouldn't forget that an organism deep within the velocity gradient on a surface knows nothing about the speed of the free-stream flow—or even if a proper free stream exists. What ought to determine whether a particular site is an appropriate abode is the steepness of the gradient—not U but dU/dz. Organisms with both sessile and motile stages inevitably face some kind of choice of velocity gradient, whether behaviorally mediated, as when a moth picks a leaf for an egg or a spiderling picks a site for a web, or mechanically determined, as when a spore lands on a slice of bread whose surface is too rough for it to blow off again. (As incorrigibly motile animals of the almost invariably motile vertebrate lineage, we may easily forget how widespread is an alternation of sessile and motile phases, how attractive organisms have found a distinction between nutritive and dispersal stages.)

Larval Recruitment

In several instances, a proper velocity gradient has been shown to be something chosen by an actively swimming organism looking for an attach-ment site for the next stage of its life history. Barnacles, for example, prefer rocks in areas of high current; but the specific choice depends on local velocity gradients. Crisp (1955) passed cyprids of the barnacle, *Balanus balanoides*, through glass pipes in which the velocity profiles were known. (As we'll see in Chapter 13, the velocity gradient at the inner surface of a circular pipe with laminar flow is an easy thing to predict or manipulate.) He showed quite clearly that free-stream velocity was largely irrelevant—instead it was the current 0.5 mm from the surface that determined whether attachment would occur, a current indicative of the velocity gradi-ent at the surface. A shear rate of 50 s^{-1} was the minimum that stimulated attachment, 100 s^{-1} was optimal, and the cyprids were unable to attach above 400 s^{-1}.

---

[1] In this business it's worth remembering that the prefix "hydro-" or "hydr-" comes from the Greek word for water, while the prefix "hygro-" comes from the word for wet or moist. Thus a hydrometer measures specific gravity (the inverse of buoyancy), while a hygrometer measures humidity.

Lacoursière (1992) found that black fly larvae (*Simulium vittatum*) selected not the greatest speed or even (as with Crisp's barnacles) the steepest gradient at the surface, but the greatest velocity gradient along the lengths of their bodies. As he noted, the choice is one that would give the greatest particle flux through the filters, the cephalic fans, relative to the drag incurred by the rather wide lower (posterior) part of the larval body (Figure 9.3). He looked specifically at settlement on cylinders running across a flow in a laboratory flow tank at Reynolds numbers from 400 to 30,000. (To get the situation back in mind the reader might refer again to Figures 5.3 and 5.5.) Under most circumstances larvae preferred to settle near the separation line. If, for example, separation occurred 81° around from the upstream center, that's where most larvae settled. Fewest settled between 55° to 70°, the region of maximum shear stress and maximum local velocity, and very few settled behind the separation line. Once settled, larvae would shift positions if the cylinder was rotated slightly, maintaining their initial pattern of preference, except that the forward stagnation point was an acceptable location even if only rarely an initial choice.

Larval settlement or "recruitment"[2] has been a subject of considerable interest in recent years, and quite a few studies have implicated hydrodynamic factors—one can get a good impression of both the level of interest and the size of the literature from Butman (1987). Despite our present bias, though, hydrodynamics is not all that matters. Some larvae are simply gregarious; while this sometimes confers a hydrodynamic advantage (as will be described in the next chapter for black fly larvae), it's probably more often a device to assure gametic proximity where the forthcoming adults are sessile. And some larvae prefer surfaces of particular chemical or physical characteristics—the bases for choice are summarized by Woodin (1986). So sometimes local flow patterns are decisive in determining settlement patterns (see, for instance, Havenhand and Svane 1991 on the tunicate *Ciona intestinalis*), and sometimes they're not.

Even the hydrodynamic aspect is turning out to be fairly complicated. For one thing, all the complications peculiar to real surfaces that we noted in the last chapter are operative on the kinds of surfaces on which larvae settle in nature. On thin plates set out just above a sea mount, benthic foraminifera (shelled protozoans) preferred a consistant thickness of boundary layer (or steepness of gradient), but on thick plates a little separation eddy near the leading edge complicated matters in a way analogous to what Lacoursière found for black fly larvae (Mullineaux and Butman 1990). For another, the velocity gradients on real surfaces have their own

[2] "Recruitment," as used by ecologists and population biologists, includes attachment of pelagic larvae, their postattachment metamorphosis, and survival to some later stage at which they're subject to notice by the Authorities.

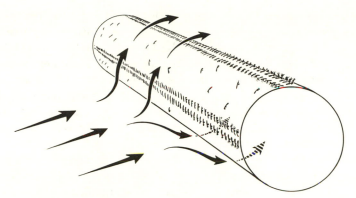

FIGURE 9.3. Black fly larvae, *Simulium*, attached to a cylinder at points of their choosing in a laboratory flow tank.

diversity. Barnacle larvae (in this case *Chthalamus fragilis* and *Semibalanus balanoides*) proved to prefer lower shear when settling on rougher surfaces (Wethey 1986). He suggested that the larvae might prefer high shear above the surface, which would assure rapid growth and adult fecundity, but low shear at the actual microsites of attachment, which would improve their survival prospects early on.

And this microworld is embedded in a larger one, the world of logarithmic boundary layers, shear velocities, and the like. Rocks rather than fine sediment characterize habitats where the ambient velocities are large and the turbulence is substantial; and rocks in substantial currents provide auspicious sites for attachment for organisms with sessile adults. As Denny and Shibata (1989) point out, the speeds of flow at least in surf zones will vastly exceed the swimming speeds typical of pelagic larvae—0.1 to 10 mm s^{-1}—and exceed even more drastically the sinking speeds of the spores of marine algae—0.01 to 0.1 mm s^{-1}. So transport to the surface will happen as a result of turbulent motion. Any laminar sublayer will be of the same order of thickness as the larvae, which is to say insignificantly thick. What matters in surf zones, then, is whether a larva holds on or a spore sticks and whether a larva does any exploration of the surface prior to attachment. Not surprisingly, barnacle cyprids of the genus used by Crisp (1955) turn out to be transported passively up to their initial contact but then explore the surface a bit before settling for good (Mullineaux and Butman 1991). Mere ability to hold on isn't necessarily enough. Just as Crisp found a minimum speed for attachment in his glass tubes, Pawlik et al. (1991) noted that the larvae of a marine tube worm (*Phragmatopoma lapidosa californica*) actively avoided settling when exposed to very low speeds in their flow tank.

A rough surface adds its own complications, and Eckman (1990) has taken a brave stab at a theoretical model for passively transported particles approaching such a surface. In his model, settlement rate should increase sharply with the areal density of roughness features. Settlement should also increase (not unexpectly) with increase in the ratio of the settling velocity of larvae to the shear velocity. Less obviously, settlement should decrease with increasing height of roughness features relative to boundary layer thickness; a high value of this last variable implies less effective mixing immediately adjacent to the surface.

It's just a little frustrating to find that most of the extensive literature on larval recruitment couches things in terms of flow speeds rather than shear velocities and velocity gradients. Admittedly the former are easier to measure, but it really does seem as if the latter are closer to what really matters to the organisms.

Settling from Air

The most relevant physical difference between air and water is in density. The main biological difference is the absence in air of very much that's equivalent to the wide world of pelagic invertebrate larvae. (Denny and Shibata 1989 remind us that fully 70% of benthic marine invertebrates have pelagic larvae.) Settling from air might strike one as trivial—what goes up must come down, as the saying goes. And the terminal velocities of a large number of pollens and spores are matters of public record (see, for example, Gregory 1973 but also some cautionary comments on measuring sinking rates in Chapter 15). Here again, complications rear up before things settle down.

In still air, spores settle randomly on a horizontal, flat surface. But with increasing winds, the efficiency of capture is progressively reduced, with the least deposition on surfaces parallel to flow. The photographs of Hirst and Stedman (1971) strongly suggest that deposition is inversely related to the thickness of the local boundary layer. Several phenomena are probably involved. First, with a thick boundary layer, fewer particles will enter the inner reaches from which they may be captured in a given time. And in a boundary layer of appreciable thickness, as on a leaf at low wind speeds, the downstream thickening produces a divergence of streamlines from the surface, which will tend to deflect dropping particles. In addition, touchdown doesn't guarantee landing, since particles such as pollen and spores can not only blow off but actually bounce off as well (Paw U 1983; Aylor and Ferrandino 1985). Not surprisingly, Chamberlain et al. (1984) found an optimal velocity (and shear velocity) for deposition of *Lycopodium* spores on a nonsticky, ribbed surface.

Tuning Up Nonmotile Systems

That a barnacle cyprid might poke around a little before attaching for the rest of its life doesn't strikes one as startling; more noteworthy is evidence of specific site selection among nonmotile propagules. Still, if both settler and settlement site have a positive stake in the outcome, a biologist shouldn't be shocked. Nevertheless, the idea of fluid mechanical tuning to maximize the chance of contact is novel, even technologically portentous. But that's just what Niklas (1985) found in pine cones—they're aerodynamic pollen traps. Some of anemophilous (wind pollinated) plants even manage some specificity in the trapping mechanism. Niklas and Buchmann (1985, 1987) have evidence for several genera suggesting that individuals of each of two sympatric species preferentially trap conspecific pollen.

In at least one case, this kind of specificity extends to settlement of an epiphyte of no demonstrated value to its host. Pearson and Evans (1990) showed under the controlled conditions of a flow tank that spores of a red (rhodophyte) alga, *Polysiphonia lanosa*, settle at higher than expected rates on the normal host, *Ascophyllum nodosum*, and at lower than expected rates on the sympatric *Fucus vesiculosus*, the latter both large brown (phaeophyte) algae.

MORE DIRECT EFFECTS OF SHEAR

By definition, drag is a force tending to carry an object downstream. In a velocity gradient, though, rather than simply pushing an object, the flow will impose a shearing load that will tend to make the object rotate (Figure 9.4; Silvester and Sleigh 1985). Thus an object that doesn't touch the substrate will rotate as it travels downstream, since what it experiences as a result of the flow is simply a force in one direction (upstream, although it can't easily know that) on the wallward side and a force in the other (downstream) on the streamward side. That's why off-axis blood cells rotate as they flow along in small vessels (Chapter 14; Vogel 1992a). An object protruding from or sitting on a surface will be more effectively rolled than pushed along, quite apart from friction with the substratum. In part, that's why tumbleweeds tumble (Van der Pijl 1972; LaBarbera 1983), and it's an important element in the initiation of erosion (Bagnold 1941).

Sometimes the gradient-induced rotation may be quite a serious stress. Morgan et al. (1976) looked at the effects of severe shear on suspended fish eggs. For most eggs, shear rates (calculated from their data) around 35,000 s^{-1} are lethal with only a few minutes of exposure. Such shear induces very rapid rotation, and the resulting centrifugal disruption may be what causes

187

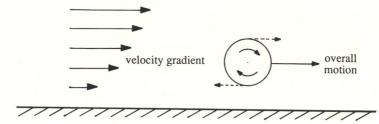

FIGURE 9.4. Rotation of a solid body induced by the shearing motion within a velocity gradient.

the high mortality. Such rates are unlikely to occur in natural or even artificial channels where the turbulence associated with the high shear rates may be important in keeping eggs suspended. But damaging rates are approached near the hulls of fast ships and exceeded in their propellers. They're also of the order of the shear rates occurring in the cooling systems of nuclear power plants, although in these latter I'd guess that simple heating is a far more significant hazard.

The shear stress on the surface itself may be far below what might cause direct mechanical damage, but it proves to be of such enormous importance so close to home that the system must at least be mentioned here; a proper discussion will be deferred to Chapter 14. A circulatory system in which the resistance to flow through the complex manifold of vessels is minimized was shown by Murray (1926) to be one in which the shear rate on the walls of every vessel was the same. Fairly recently it's been shown that the endothelial cells that line blood vessels have a mechanoreceptive function and that they are specifically responsive to shear stress. So all that's needed for self-optimization is some specific threshold, above which a sequence of events leading to vessel enlargement is initiated and below which a sequence leading to vessel narrowing comes into play.

SUSPENSION FEEDING AND LOCAL CURRENTS

Suspension feeding most often takes place in the gradient region adjacent to a surface. (Suspension feeding by some swimming fish, tadpoles, and baleen whales is an exception; see Sanderson and Wassersug 1990.) So it's certainly germane to the present discussion—even though discussion of mechanisms will await Chapter 15, on low Reynolds numbers. Of present relevance are the ways in which interfacial suspension feeders deal with the velocity gradients in which they find themselves.

First, however, several terminological notes. "Suspension feeding" subsumes "filter feeding" as the term was used by older sources. The shift involves a bit more than fashion. "Filter feeding" implies that food is obtained by simple filtration or sieving, but emergence of the details of capture mechanisms has made it clear that the implication is often misleading. In any case, it refers literally to how food is obtained rather than, as with "suspension feeding," to the nature and location of the food—particulate material, typically tiny, typically living, often actively if slowly moving in the ambient fluid, the latter most often water. As we'll see in the next chapter, detritus feeding by using a filter is possible if you employ some device to resuspend particles resting on the bottom; we prefer to exclude these by using an ecological rather than physiological distinction.

Among suspension feeders a distinction is usually made between "passive" and "active" ones, with passive suspension feeders dependent on the motion of the water to move fluid to and through their separation equipment and active suspension feeders working some energy-consuming pump. But these turn out to be either idealizations or extremes (depending on viewpoint or bias), applicable, on the one hand, to a spider or mayfly larva waiting by its net and, on the other, to a clam in a muddy bottom with its siphons barely protruding. In between are a wide range of mixed-media specialists, including whales that swim along with their mouths ajar, barnacles that sweep their cirri through water that's moving in the opposite direction, sponges and brachiopods that are quite competent pumpers but can augment their pumping by using environmental currents (Vogel 1978a). One has to bear in mind just how dilute most natural waters are as sources of particulate food. A marine sponge, which we generally think of as an unsophisticated and slow-growing animal, manages to process its own volume of water every 5 seconds and to extract edible material with almost no loss down to the size of bacteria (Reiswig 1971, 1975b; Vogel 1977). Bivalve mollusks do even better, with scallops and mussels pumping volumes equal to their own volumes every 3 or 4 seconds (Meyhöfer 1985).

Incidentally, the cost of processing a lot of water underlies a rather nicely convergent arrangement occurring in several groups of fishes. The economic advantage of ram ventilation in fishes and its use by rapid swimmers were mentioned in Chapters 4 and 7. Some fishes filter food from water passing through the same mouth-to-operculum stream. Suspension-feeding fishes commonly prove to be obligate ram ventilators—the conjunction occurs in anchovies, menhaden, mackerel, and paddlefish. According to Burggren and Bemis (1992), paddlefish have a minimum metabolic rate about twice that of other fishes their size because they have to keep swimming. Still, the respiratory cost of constant swimming is no additional burden for a filterer; it's really analogous to the cost of swimming in a pursuit predator.

I think suspension feeding is best viewed as a set of compromises among conflicting factors. One was mentioned a page or so ago—low flow right near the surface to facilitate attachment but high flow slightly farther away to provide ample supplies of food. That argues for sites with some surface roughness and sites where boundary layers are thin and gradients steep; and certainly suspension feeders abound on the blades of marine grasses and macroalgae, on wharf pilings, on rocky outcrops, and (under the manipulations of marine ecologists) on small ceramic plates in open racks exposed to tidal currents. The optimal location certainly depends on how an animal deals with fluid. Hughes (1975) measured the currents around erect colonial hydroids (*Nemertesia antennina*) in about one meter of water and found, not unexpectedly, that speeds near their tops were generally above those near the holdfasts. He then made a very thorough catalog of the numerous species that live on the hydroids, noting where each was found. The more passive feeders (some smaller hydroids and others) were more distal or apical; the more active feeders were more proximal or basal. Really active pumpers—some bryozoans, bivalves, ascidians, and sponges—were usually found only on the holdfast. Furthermore, the suspension feeders on the holdfast treated food differently from the higher and more passive sorts. The former were much less selective in initial capture but had well-developed mechanisms for subsequent rejection of unwanted stuff. Clearly there's a secondary disadvantage of being on the bottom. Suspended plankton is more likely to be nutritive than particles that have dropped out of the water column and are tumbling along near the bottom. A large and scattered literature addresses this matter of choice of habitat by suspension-feeding animals.

Another compromise must lie between maximizing rates of fluid processing and ensuring that processed, food-depleted fluid is prevented from reentering the separator. The problem will be most severe where the suspension feeders live deep within the velocity gradients—where, as at substantial depths and in sheltered locations, ambient flows are low, and where flows are both nonturbulent and bidirectional. Both the problem and one solution seem first to have been recognized by Bidder (1923), working on sponges. He identified the constrictions on the excurrent openings (oscula) as nozzles that increase flow speed, and he noted the role of the apical locations and upward orientations of these openings. The siphonal openings of many ascidians and bivalve mollusks certainly work as jets in the same way and for the same purpose, as mentioned in connection with the principle of continuity back in Chapter 3.

Making a really small nozzle or jet that carries fluid any decent distance is hard because the area of "contact" between the jet's sides and the main fluid will be great relative to the jet's volume (and thus its momentum flux). Some groups of small suspension feeders get around these pernicious problems of scale and viscosity through cooperative flow management. Some of the ascidians, the Botryllidae, have a colonial discharge system to get a single,

(a)   (b)

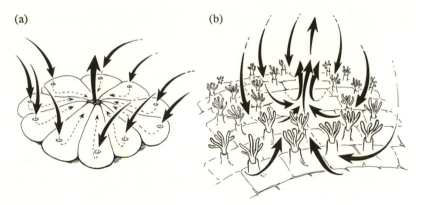

FIGURE 9.5. Colonial discharge systems—common jets: (a) an ascidian, *Botryllus*; (b) an encrusting bryozoan, *Membranipora*.

large jet instead of a half-dozen or a dozen individual ones (Figure 9.5a). More impressive still are the cooperative arrangements of some species of bryozoans that form thin colonial encrustations on the flat surfaces of marine macroalgae. Since the individuals are only a fraction of a millimeter tall and since they form a continuous and fairly smooth layer, these animals certainly live well within local boundary layers. A neat device has been described by Cook (1977) and investigated in more detail by Lidgard (1981). While the individual zooids have no gaps between them, a colony of *Membranipora villosa* has regions of nonfeeding, degenerate individuals every few millimeters (Figure 9.5b). These regions act as excurrent chimneys for colonywide currents. Water from near the surface of the colony is pumped downward by the cilia on the tentacles of the individuals. The water then moves laterally along the substratum between individuals toward the chimneys. It emerges from the latter as jets of substantial speed (25 mm s^{-1}) that broaden and themselves coalesce well above the zooids, above and as well separated from incurrent water as if a collecting manifold of physical chimneys were present. So, except perhaps in the stillest of still waters, no zooid reprocesses water previously used by other zooids.

An alternative to a jet is elevation of the filtration apparatus itself. Wherever external nets or tentacles occur, the device is almost inevitably concomitant. One can point to the various erect colonial coelenterates and bryozoa, to sea anemones such as *Metridium*, whose drag was mentioned in Chapter 6, to several families of polychaetes living in tubes and extending feeding appendages upward (see Fauchald and Jumars 1979), to sea lilies among the echinoderms, to goose barnacles among the crustacea, to the stalked solitary or colonial protozoa such as *Vorticella* and *Carchesium*, and to the black fly larvae that keep their cephalic fans uppermost. In some members of this diverse assemblage the stalk or other elevation device is retractile, in others it isn't.

Still another kind of compromise comes to play in an animal's choice of flow speeds at which to feed. At low (and concomitantly less turbulent) ambient flows, decreased supply, local depletion, and increased reprocessing of food should have a detrimental effect on the cost-benefit ratio for suspension feeding. At high ambient flows, several adverse factors become more significant. A larger fraction of high density, nonedible material may be entrained into the medium (a problem for clams at least, according to Turner and Miller 1991); drag and the consequent difficulty of keeping the food-trapping device properly exposed will be greater; and the greater momentum of passing particles may make them harder to capture. And even where organisms don't seem to take advantage of local currents to induce internal flow, the rates of suspension feeding often depend quite strongly on ambient flow. An increase in the rate of active pumping with increases in local flow was found by Walne (1972) in five species of clams, mussels, and oysters—perhaps the more rapid the flow, the likelihood of a richer food resource and the desirability of a greater investment of energy in pumping. But Walne used flow speeds that seem to have been below 10 mm s^{-1}; at substantially higher speeds, pumping may decrease, as found for mussels by Wildish and Miyares (1990). A large literature documents great influence of current speeds on suspension feeders (see, for instance, Shimeta and Jumars 1991), although a search for mechanisms and general patterns is hampered by inconsistency in choice of variables to measure—rates of fluid processing, rates of food acquisition, growth rates, and so forth.

DISPERSAL AND VELOCITY GRADIENTS AS BARRIERS

Passive dispersal of propagules through the use of winds and water currents is practiced by at least one member of every major phylum of animals and every division of plants. Aerosol particles, polluting and otherwise, likewise enjoy the benefits of the free transit system in the sky. It looks as if all a propagule has to manage is a sufficiently low sinking rate through high drag, lift production, upcurrent detection, or density reduction. In fact such factors address only part of the problem. Somehow a propagule has to get liberated into the current in the first place, and it needs to alight on a surface at the end of the ride. In both, velocity gradients must be crossed, and we turn now to the extraordinary collection of devices nature has contrived for accomplishing that first step in aerial transport.

The basic problem may be easiest to see if we consider a situation in which spores simply sit on a surface in a wind. Grace and Collins (1976) let

Lycopodium (a club moss) spores settle on leaf models of paper, and they then tried to blow them off again in a wind tunnel. These relatively large (30-μm diameter) spores protruded an average of 15 μm above the surface. A spore 40 mm from a leading edge needed a free-stream wind of 2.7 m s^{-1} to get the necessary 18 mm s^{-1}-threshold wind speed on its center— fully 150 times higher. To get 50% of the spores off a piece of paper of roughness comparable to spore diameter took a wind of 6.8 m s^{-1}, no mere zephyr. In general, increasing the turbulence level in the wind tunnel improved matters, even with these tiny particles well down in the boundary layer. Nonetheless, in nature things might not be quite so bad—Grace (1978) has shown that boundary layers on leaves will be thinner than as-sumed from work in wind tunnels. In a nearly smooth free stream, turbu-lent flow across a fluttering white poplar (*Populus*) leaf will start at 8.6 m s^{-1}, while in a flow with a turbulent intensity of 20% to 25%, flow over the leaf will get turbulent at only 3.8 m s^{-1}. The lesson in this is to be cautious in using the carefully crafted wind tunnels of aerodynamicists, who want minimum turbulence in order to model systems that ordinarily propel themselves through otherwise still air. For present purposes their worst wind tunnel may be our best! When looking at the drag of leaves in high winds (Vogel 1989) I deliberately worked in the messy flow downstream of the propeller for just this reason.

To experience flow, getting even a little way up through a boundary layer helps a lot. Consider a location 40 mm downstream on a flat surface parallel to flow in a wind of a meter per second. The local speed is 4 mm s^{-1} at 10 μm above the surface, 40 mm s^{-1} at 100 μm above the surface, and 400 mm s^{-1} at 1 mm above the surface (equation 8.4). And elevation is the most common tactic. Spores, pollen, seeds, and so forth are commonly pre-sented to the wind on some sort of elevation, whether we consider the spores of a *Penicillium* fungus held less than 100 μm high or the elaborate arrangements to expose pollen to wind in many anemophilous higher plants. The newly hatched first instars of a scale insect or coccid, *Pul-vinariella mesembryanthemi* (Homoptera), stand on their hind legs facing downwind with their bodies well elevated when they decide to enter the air column for dispersal. Standing erect, a coccid is only 0.34 mm tall and still in the region of the local boundary layer in which velocity increases linearly. Washburn and Washburn (1984) tested them in a wind tunnel and found that free-stream winds of over 2 m s^{-1} were needed to produce enough drag to tear their hind tarsi from the substratum.

Perhaps these coccids hold on strongly enough to guarantee that they won't get wafted off by too gentle a breeze to assure effective dispersal— their sinking rate is fairly high, 0.26 m s^{-1}, even with extended legs and antennae. The case seems similar to that of a fungus, *Helminthosporium maydis*, that afflicts corn. Aylor (1975) found that it took a force of 10^{-7}

newtons to detach conidia from corn leaves, which may sound trivial but which requires a free-stream wind of 10 m s^{-1}, both by calculation (using Stokes' law) and by direct measurement. I think a useful scheme lurks here. An animal might begin by holding tightly and gradually relax its grip, thereby picking a high wind if one is available but not being overchoosy if it can't do as well. Similarly, a wind-dispersed seed might gradually reduce the strength of its attachment, eventually just coming loose if no decent wind ever occurred. Alternatively, a propagule of constant tenacity might be borne on a stalk that gradually grows away from its substratum into higher flow speeds.

Nor does elevation as a benefaction top out right above the boundary layer of a leaf or flower—the logarithmic boundary layer is orders of magnitude thicker. Remember that the tassels are on top of a corn stalk and the male cones are highest in a pine tree. When I mow my scruffy lawn, the highest items are the structures bearing the wind-dispersed seeds of grass, dandelions, and plantains. And both lycosid spider young and gypsy moth caterpillars are reported to climb as high as they can before spinning their silk threads and letting themselves go in the breeze (Tolbert 1977).

A logical alternative to getting some height in a velocity gradient is to pick circumstances in which the gradient is especially severe. According to Matlack (1989), the seeds of black or sweet birch (*Betula lenta*) are released in the winter on days that are below freezing and have substantial wind and low humidity. In a deciduous forest, winter dispersal gets around much of the problem of low wind beneath the canopy; and, after falling, these seeds can slide over the smooth and nonadhesive snow to increase the area of dispersion more than 3-fold—an especially useful device for a gap-colonizing species.

Yet another way to defeat the velocity gradient for wind-dispersal is brief use of an engine for propulsion through it. One-shot engines are fairly common among plants; most often a gun of some sort (rather than a jet or rocket) gives an initial push to the seeds or spores. Just among seed-producing plants twenty-three different ballistic mechanisms have been described in as many different families (Stamp and Lucas 1983). Among this array of phyto-artillery we can loosely distinguish two sorts. In some cases the distance over which the propagules are aerially dispersed corresponds closely to the range of the gun, and ambient fluid motion is of little consequence. At the other extreme the only real function of the gun is to get away from Mama and her semistagnant miasma.

The first arrangement is clearly the most common; but, while quite a lot is known about the diverse propulsive schemes, little attention has been given to any aerodynamic specialization of the projectiles. One might expect that projectiles could deviate from simple sphericity in ways that enhance stability in flight and, by limiting tumbling, permit streamlining.

Certainly with Reynolds numbers in the thousands, streamlining might usefully lengthen their range. But whether any such drag-reducing scheme is used is unclear. *Arceuthobium*, a dwarf mistletoe, manages a muzzle (discharge) speed of around 14 m s; it has an ellipsoidal seed, 2.9×1.1 mm and thus an initial Reynolds number of around 3000 (Hinds et al. 1963). But the seeds tumble during flight, going at most about 15 meters rather than the 20 or so that I calculate would be achieved in the absence of drag.

The other arrangement, using a gun simply to get out of the parental velocity gradient, is less common for, I think, a very basic fluid mechanical reason. A good projectile needs relatively low drag and high momentum. The simplest way to get that combination is to be large, to be much denser than the medium, to have a high initial velocity, and to operate in air—in short, to operate at fairly high Reynolds numbers. A particle suitable for effective dispersal by ambient fluid motion needs almost the opposite characteristics. It must have high drag so as to settle out very slowly, which argues for small size, low density, and a preference for water over air. So the requirements for shooting and settling are substantially antithetical. One suspected case is that of the ascomycete fungus, *Sordaria*—at least the performance of the gun is well established, and one can think of few other reasons to shoot so hard for such a short range. *Sordaria* seems to hedge its bets by shooting between one and eight spores per projectile. An eight-spore ball is about 40 μm in diameter and goes about 60 mm (Ingold and Hadland 1959), from which I calculate a muzzle speed of 30 m s^{-1} (a program for doing such computations is given in Vogel 1988a). In street terms that's a range of two and a half inches consequent to an initial speed of nearly 70 mph, a catastrophically draggy shot. Concomitantly, the initial angle for the maximum horizontal range is about 6°, not the 45° one calculates for projectiles that experience no drag. That, incidentally, is a general rule—the worse the force of drag relative to the force of gravity, the lower is the angle that gives the greatest horizontal range. Put in intuitive terms, where drag causes severe deceleration, a projectile must achieve most of its horizontal distance while it still has decent speed. Some calculated trajectories for *Sordaria* spore clusters are shown in Figure 9.6—parabolic they are not.

A partial evasion of the problem of reconciling projectile and particle sizes can be had by shooting a mass that then breaks up. A puffball (*Lycoperdon*) works this way. Indenting its thin and flexible wall forces out a jet of air and spores; in nature raindrops are probably the immediate stimuli (Gregory 1973) in this bellowslike scheme.

The conflicting requirements, bad enough in air, appear to be truly disabling in water. Relative particle densities are, of course, much less, and drag is much greater. Going with the flow is so easy that settling out rather

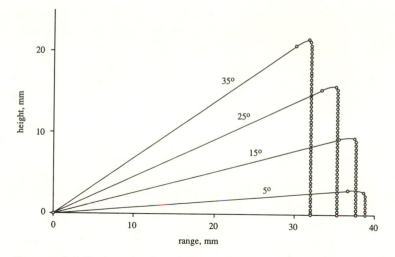

FIGURE 9.6. Trajectories for microprojectiles in air, calculated assuming the drag of a sphere and the characteristics of an eight-spore cluster of the fungus, *Sordaria*. Successive circles on each line mark positions at approximately equal intervals of time.

than staying aloft is what takes explanation. Conversely, guns work very badly. The nematocysts of the coelenterates aren't really guns in the usual sense but rather devices that stab the victim after it touches their triggers. And even our macroscopic and bellicose species uses self-propelled torpedoes rather than cannon to convey explosives under water.

DIFFUSION ACROSS THE VELOCITY GRADIENTS AT SURFACES

Never is a velocity gradient adjacent to a surface of greater biological consequence than in the presence of diffusive exchange between the surface and the flowing medium. And such exchange is enormously common. Diffusion is the basic transport scheme at the cellular level (cyclosis and a few other phenomena notwithstanding), but diffusion is pitifully slow over all but the shortest distances. The almost inevitable concomitant of elaborate macroscopic and multicellular organization is some scheme to augment diffusion with convection—in other words with bulk flow, as pointed out by Krogh (1941). Whether we call the systems circulatory, respiratory, translocational, or other, they have in common this circumvention of the speed limitations of diffusion over macroscopic distances (LaBarbera and Vogel 1982). But such systems only augment diffusion; and each inescapably retains a final pair of diffusive links. One, of course, is diffusion through whatever cellular or membranous barrier divides transport medium from utilization site. The other occurs across the innermost portion of any laminarly flowing transport medium.

Since the steepness of a gradient determines transport rate, whether as a factor in Fourier's law for heat transfer or Fick's law for diffusion, the steepness of the velocity gradient takes on special significance. Bird et al. (1960) do an especially good job on the interrelationships of the transport of mass across a concentration gradient, of heat across a temperature gradient, and of momentum across a velocity gradient. Not only are these transport phenomena interrelated, but often one can be used as a quantitative model for another—for instance, using heat conduction for oxygen diffusion in egg masses of snails (Hunter and Vogel 1986).

A simple consequence of diffusion through a velocity gradient, in this case the boundary layer on a flat plate, can be seen when a photographic print is developed. As the print is gently agitated in a tray of developing solution, the image first appears near the edges (Figure 9.7) because the necessary chemical reactions proceed more rapidly where the steeper gradient supplies reactants more rapidly. The phenomenon can be made to serve as a way to measure concentration or velocity gradients at surfaces (Dasgupta et al. 1993).

Will Diffusion across a Velocity Gradient Limit a Process?

The two diffusive links just mentioned are arranged in series, and more often than not other diffusive elements extend the chain. The implications of a sequential array are perhaps easiest to appreciate for transpirational water loss from leaves. Air within a leaf is water saturated, and water vapor diffuses through tiny holes of adjustable aperture, the stomata. Thus stomatal resistance and so-called boundary layer resistance are additive in the manner of serial electrical resistors.[3] Does wind increase water loss? Sometimes yes and sometimes no, depending as much on the resistance of the stomata as on the magnitude of the wind, as Grace (1977) very nicely explains. Where stomata are closed and their resistance high, wind has little effect—if resistors of a thousand ohms and of ten ohms are in series, decreasing the latter even by an order of magnitude makes little overall difference. Similarly, opening stomata beyond a certain width usually has little effect on water loss since boundary layer resistance has become the dominant element.

I've seen some oddly worded explanations of how the stomata, most often making up less than 1% of total leaf area (Meidner and Mansfield

---

[3] I'm simplifying by judicious omission. "Cuticular resistance" is an element in parallel to stomatal resistance that's enough higher than the former unless the stomata are closed to be of little importance. Then there's "mesophyll resistance" and other series elements.

FIGURE 9.7. A photographic image appearing, edges first, in a tray of gently agitated developing solution.

1968), are equivalent to a surface entirely open to water loss on account of interactions of the vapor caps outside each stomate—to me what they're saying is simply that boundary layer resistance is now dominant. In addition, at high levels of illumination another factor sometimes makes the situation appear anomalous. Vaporization of water increases with temperature, and leaf temperatures can get as much as 20° above ambient when the wind is very low. Thus an increase in wind speed can lead to a reduction in transpiration by decreasing leaf temperature more than enough to offset the lower boundary layer resistance. For a leaf, still air and full sunlight are in any case mutually exclusive conditions because of solar heating and the convection the consequent temperature difference induces. Diffusion of water vapor through boundary layers has been of special concern to plant physiologists, who periodically produce reviews of the state of the art. For further information, I suggest consulting Meidner and Mansfield 1968, Leyton 1975, Grace 1977, Rand 1983, or Monteith and Unsworth 1990.

Sometimes the resistance imposed by velocity gradients may be unimportant as rate-limiting elements. For instance, Tracy and Sotherland (1979) showed that the boundary layer of bird eggs is of little consequence in water loss; under no circumstances is its resistance more than 10% of the total.

The Minute Currents That Matter in Water

The diffusive resistance associated with interfacial velocity gradients may be important in air, but it's overwhelming in water. The diffusion coefficients of small molecules through water are about 10,000 times lower than through air. An alveolus in a lung is of the order of a hundred times the diameter of a pulmonary capillary (the square root of 10,000, in accordance with Fick's law); that's the difference it takes to ensure that transport in air and in blood are reasonably balanced. Concomitantly, flow makes more difference in water, and much lower currents have noticeable effects on exchange between an organism and its environment.[4]

Very low flow rates may make such a difference that still water is no simple thing to achieve as a control. Schumacher and Whitford (1965) measured both respiration rates (in the dark) and photosynthetic rates (with light) for four classes of attached algae; in all cases slight currents greatly increased these rates. In *Spirogyra*, a filamentous green alga, a current of 10 mm s^{-1} increased the photosynthetic rate by 18% compared to the rate in still water. Even lower speeds were found effective by Westlake (1967), who could detect an increase in photosynthetic rate in *Ranunculus pseudofluitans* (an aquatic dicot) when the current was raised from 0.2 to 0.3 mm s^{-1}. The latter is about a meter per hour, well below what we'd normally notice as moving. Perhaps we should regard truly still water as an idealization comparable to a completely rigid body.

Conversely, if high flow rates increase productivity over rates found with more moderate flows (as has sometimes been found for large marine algae), the explanation cannot be assumed to be augmentation of diffusive transport without some more specific demonstration or systematic rejection of alternatives. Changes in exposure to light, reduction in fouling with both living and nonliving material and other less tidy explanations must first be given decent burial.

We saw that morphological adaptations were clearly central in avoiding excessive drag; only a little less certain is that another suite of structural features may alleviate the problems of diffusion through velocity gradients. Small leaves, which would on the average have thinner boundary layers, are variously reported to have higher photosynthetic rates than large leaves. Aquatic plants often have smaller, thinner, or more highly dissected leaves than their terrestrial relatives (see Sculthorpe 1967 or Haslam 1978). The possibility of reduced damage from flow forces comes immediately to mind, but many of these plants never experience rapid currents. So their shapes and sizes may well be primarily adaptations to the

---

[4] Nonetheless, diffusion is slow even in air—the classroom demonstration using the spread of perfume is a fraud, with convective transport in the ubiquitous air currents conveying the aroma over all except the last small fraction of a millimeter at the nasal epithelium. Indeed, a decent demonstration of diffusion takes great effort to achieve even in a liquid system, much less in air—convection can't be turned off without stringent control of temperature and without a long wait to let any history of motion become adequately ancient. I've usually resorted to a little agar or gelatin in the water to keep the medium from stirring itself up.

much greater difficulty of diffusive exchange in an aqueous medium. But by contrast with drag, here current helps matters. As mentioned earlier, at least some of the insects of torrential currents have gills of reduced area (Dodds and Hisaw 1924); when rapid flow is normal, then gills of ordinary size would just be a drag.

Closer to home phylogenetically than photosynthesis in aquatic plants is cutaneous respiration in amphibians. Quite a few of these animals stay submerged for long periods (overwintering submerged in some cases), and most deliver blood to the skin from a branch of the aorta that supplies the lungs as well—skin is a major element in oxygenating blood. For bullfrogs (*Rana catesbeiana*) the external velocity gradient constitutes a highly significant 35% of the total resistance to oxygen uptake in an ambient flow of 50 mm s^{-1}. In even slower water the situation is yet worse—at 1 mm s^{-1} it makes up fully 90% of the resistance. Pinder and Feder (1990), whose figures I quote, point out that putting an immobilized frog into still water is equivalent to putting it in fully anoxic water. In still water bullfrogs do a certain amount of spontaneous and seemingly undirected movement, which turns out to make quite a lot of difference to oxygen availability. Another frog, *Xenopus laevis*, not only moves about more in still water but makes voluntary dives of less than half as long as in moving water. Lungless, gill-less salamanders (*Desmognathus quadramaculatus* in particular) have an especially drastic way of coping with the problem. When they get into stream microhabitats with low flow rates they cope with the chronic hypoxia by substantially reducing their metabolic rates (Booth and Feder 1991).

Diffusion through velocity gradients is important in the world of cell biologists as well. I can't resist citing one iconoclastic experiment (Stoker 1973). On a culture of fibroblasts a strip was removed, leaving a denuded patch. Cells at the border of this denuded zone increased their growth and invaded it. Conventional wisdom would assume the growth to be caused by reduction of contact or diffusional inhibition between cells. But Stoker showed that one needn't invoke so biological an explanation—nothing more is responsible than higher concentration gradients in the stirred medium at the edges of the patch due to the steeper velocity gradients.

The "Unstirred Layer"

Dropping down in organizational level from our usual concern with whole organisms we encounter another important aspect of velocity gradients. Textbooks of physiology often discuss something called the "unstirred layer," something that (in some cases at least) sounds a little like the distinct region beneath a surface of discontinuity that I fussed about at the start of the last chapter. But the unstirred layer turns out to be yet another useful fiction ("still water," again). What it really amounts to is the region adjacent to a surface through which transport is diffusive rather than convective, the region where there is a significant concentration gradient of whatever material is under consideration. Which is to say that it's what happens in a laminar velocity gradient near a surface.

As explained by Barry and Diamond (1984), a gradient in the relative significance of bulk flow or stirring can be modeled for analytic purposes as a completely flowless region outwardly bounded by an abrupt transition plane. In the present context especially, it's important to appreciate that this unstirred region is defined in terms of concentrations and concentration gradients rather than velocities and velocity gradients. Obviously the two are related, as we'll get to just ahead, but definitionally at least the relationship is incidental. The specificity of the notion is perhaps best put with a graph (Figure 9.8) and a formula. Using δ_u for the thickness of this idealized unstirred layer adjacent to a surface, c_b and c_m for the concentrations of the solute in the bulk solution and at the interface (the "membrane"), respectively, and $(dc/dz)_m$ for the concentration gradient at the interface, the thickness of the unstirred layer is defined as

$$\delta_u = \frac{c_b - c_m}{(dc/dz)_m}. \tag{9.3}$$

As the graph illustrates, what's being done is to extend the surface concentration gradient linearly until the concentration reaches that of the bulk solution. That coincidence point is then taken as the location of the transition plane between unstirred layer and bulk solution.

One could, of course, define a velocity boundary layer in an analogous way, extending the velocity gradient at the surface out to where it reaches the speed of the free stream. The relative thickness of such a laminar boundary layer would still vary inversely with the square root of the Reynolds number as in equation (8.3). But it would be considerably thinner than one whose outer limit is reached at a speed of 99% of free-stream speed. From equation (8.4) it's a simple matter to calculate the thickness of this velocity boundary layer—it comes out to about 60% of the thickness of the conventionally defined layer. Coincidentally, this is about the thickness of a layer whose outer limit is set at 90% of free-stream speed and whose use was recommended in the last chapter. Thus the "diffusive boundary layer," as the term has been defined in the studies of cutaneous respiration in amphibians (see also Feder and Pinder 1988), corresponds to the "unstirred layer" as defined by Barry and Diamond (1984) and to the 90% velocity boundary layer as defined here.

Besides referring to a model with a discontinuity rather than a reality without one, the term "unstirred layer" can be a little misleading in another way. One is likely to presume that substantial mixing takes place outside the layer, that all else is "stirred" with some form of cross-flow bulk transport.

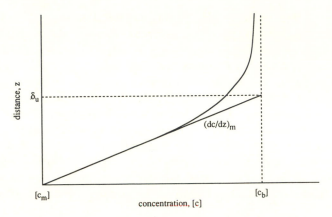

FIGURE 9.8. Defining the thickess of the "unstirred layer" for diffusion adjacent to a membrane across which fluid is moving—the concentration gradient at the membrane is extrapolated to the concentration of the un-affected bathing solution.

If the unstirred layer corresponded to the laminar sublayer in turbulent flow, the image might be fairly realistic, but the concept is ordinarily used for physiological situations in which Reynolds numbers are sufficiently low that flows are assuredly laminar. The region of "mixing"—one really should put quotes around it—doesn't necessarily have cross-flow transport of mass and momentum by turbulent eddies. It's just a region with a free-stream concentration of solute and no gradient. With exchange across a membrane happening in a particular place rather than along an infinitely long pipe or surface, free-stream concentration can be maintained wher-ever flow from offstage is sufficient to offset diffusive depletion. Beware of explanations that presume or specify turbulent mixing outside the un-stirred layer—smile and calculate the Reynolds number for yourself.

A useful index for deciding whether exchange in a given situation is dominated by diffusive or by convective transport, the Péclet number (sometimes known as the Sherwood number), will be introduced in Chap-ter 14.

INTERACTIONS AMONG ORGANISMS AND THE GRADIENTS NEAR SURFACES

By now the notion that the presence of an organism almost inevitably alters the local flow field needs no belaboring—the idea is implicit in all that's been said about topics as diverse as interference drag, colonywide currents, the location of ectoparasites, and variations in shear velocity and zero-plane displacement. From the collection of references from which

I'm working, I find I can assemble a very large pile devoted to flow in and among . . . well, everything from forests to eel-grass communities in marshes. Each of these points out that the alteration of flow resulting from the presence of some kind of large organism has a major influence on community structure. I have another group of references that document flow-mediated interactions among organisms—cases in which competition for food, light, attachment space, and so forth takes the form of competition (both intra- and interspecific) for adequate flow; schemes to use someone else's flow (with or without detriment to the host); arrangements in which the forces on some organism depend very much (positively or negatively) on epiphytes or neighbors; cost-benefit analyses of relative location with respect to flow; and systems in which various forms of chemical (and even physical) communication are flow mediated.

Perhaps the appropriate point to emphasize at the end of what has turned out to be quite a lengthy chapter is one that emerges from the wild diversity of these ecological and behavioral phenomena. We've come a long way from some idealized diagram of the distribution of flow speeds on a flat plate as far as physical complications, consequences, and implications go. What ought to be transparently clear without further examples or belaboring is the still greater complexity of the biology that's built on these complications, consequences, and implications. The central message I'd push is that without some appreciation of the world of flow one is unlikely to recognize what organisms are up to in this world of velocity gradients.

CHAPTER 10

Making and Using Vortices

A T ONE LEVEL we all know about vortices. Water doesn't just drain downward from the bathtub, but instead goes around and around before disappearing, like a dog circling a time or two before lying down.[1] Much larger in scale are various violent storms—hurricanes, typhoons, tornadoes—called cyclonic on account of their vortical motions. Clearly organisms are exposed to vortices—quite beyond Edgar Allen Poe's horror story about a boat sucked into a maelstrom. And some creatures even *contain* vortices, for instance in the larger and more rapid parts of internal fluid transport systems of large animals. In addition, they deliberately either make vortices or use naturally occurring ones for quite a number of useful purposes. In particular, flying and the more macroscopically popular forms of swimming are inseparably linked to the existence and behavior of vortices.

Like just about every other aspect of fluid mechanics, this business of vortices is a distinctly peculiar one, although I suspect that the oddness comes mainly from our usual lack of familiarity with the details of the motion of transparent media. A vortex is quite unlike a rotating solid. Neither fluid nor solid minds going in circles; indeed it can be argued that truly noncircular motion must be a special case in any bounded system. But the solid stolidly rotates as a solid, while the rotion of a vortex gets more vortiginous as one approaches the axis. Let's take a few pages to build a picture of the events and concepts surrounding vortices. Incidentally, an entertaining and eminently penetrable book (Lugt 1983a) describes how vortices lace together just about everything in the larger subject of moving fluids.

TWO WAYS TO GO IN CIRCLES

If a solid body rotates about an axis that passes through it, all bits of the body have the same angular velocity. Consequently the tangential velocity of any bit is proportional to its distance from the axis of rotation. You can

---

[1] The bathtub vortex, commonly ascribed to the Coriolis pseudoforce that results, off-equator, from the rotation of a spherical earth, is supposed to rotate counterclockwise in the northern hemisphere and clockwise in the southern. In fact, the expectation is realized only with the kind of painstaking circumvention of confounding factors that precludes making relevant observations in an ordinary (Archimedean) bathtub. With no

hit a golf ball or a baseball hardest with the longest club or bat, assuming the same angular velocity of swing. That's because angular momentum (the product of mass, angular velocity, and radius squared) varies in proportion to the square of the distance from the axis of rotation. Since angular momentum is ordinarily conserved, a figure skater rotates more rapidly when hands and feet are brought closer to the axis of rotation; with mass remaining constant, any decrease in radius must be accompanied by a great increase in angular velocity.

A body of fluid may certainly rotate about an axis in just this manner. Spin a bowl of water and, after a little time for viscosity to get the water up to the angular velocity of the bowl, the system rotates without any particular pattern of internal motion. A tiny boat on the surface of the water will go around with it, facing first east, then south, then west, then north (or, of course, vice versa). The surface, as in Figure 10.1, is higher around the periphery, with a shape that looks spherical but is really a paraboloid. Its parabolic section is a result of the pressure differences caused by that square of the radius in the formula for angular momentum. We say that the rotating water constitutes a vortex.

But a body of fluid may move about an axis in another way, one rather hard to persuade a solid to do. Fix the bowl of water, insert into it a cylinder with a vertical axis, and make the cylinder rotate at a steady rate, as in Figure 10.2. After a time, viscosity and the rotation of the axial cylinder induce motion throughout the bowl (assumed, for simplicity, to be very large or without outer walls). A very different situation has been set up. As would be expected from all that was said in the last two chapters, a velocity gradient extends outward from the inner cylinder's surface. So the angular velocity of a bit of fluid isn't constant but now decreases with distance from the inner cylinder. Likewise, tangential velocity now drops with distance from the center. It does so asymptotically, specifically in a hyperbolic fashion. As with fluid rotating without internal shear, the height of the water, indicating relative pressure, is lowest nearest the middle and highest peripherally. Here, though, the shape of the surface is hyperbolic, not parabolic; and this kind of vortex looks a lot more like a whirlpool or a bathtub drain.

What's perhaps most startling about this latter vortex is the motion it gives a tiny boat on its surface. No longer does the boat face each of the compass points in turn. Instead it keeps facing in the same direction, as it moves around a short distance outward from the inner cylinder. What

---

small effort, Shapiro (1962) in Boston and Trefethen et al. (1965) in Sydney obtained the proper opposite rotations, but both used shallow tubs almost 2 meters across and had to wait about 24 hours before residual motion subsided sufficiently for a consistent direction to be obtained.

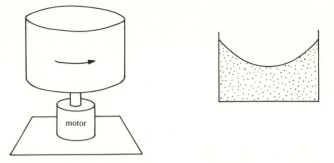

FIGURE 10.1. Making water go around in a cylindrical bowl by rotating the bowl. At right is the water's surface profile across the bowl.

we've made is something called an "irrotational vortex." To realize that motion around and around might be appropriately called "irrotational" is a little jarring, but the designation is no more than proper recognition of the difference between translational and rotational motion. In translation, an object changes its *location*—it moves from place to place. Nothing in that definition restricts the object from periodically revisiting an earlier position. In rotation, by contrast, an object changes *orientation*, whether or not it ever comes back to a place it previously occupied. That the objects here are just arbitrarily small elements of a fluid makes no difference to this definitional distinction.

Translation in a circle without rotation isn't something that only a fluid can do. While the wheels and cranks of a bicycle rotate, the pedals just translate. The flagella of bacteria truly rotate; those of protozoa merely translate in circles. For that matter, translation around some repetitious orbit is an almost inevitable feature of appendage motion in legged locomotion.

That hyperbolic variation of tangential velocity is the critical item in rendering a vortex irrotational. Consider an object some distance from the axis of such a vortex (such as our boat). For it to maintain its orientation, the water nearer the axis must be moving faster than the water farther from the axis. More specifically, it can be shown that *for the irrotational condition to occur, the product of tangential velocity and radius must be constant*—hence the hyperbola.

But a constant product of velocity and radius implies that velocity approaches infinity as the radius approaches zero—not an agreeable prospect, although one that victims of hurricanes and tornadoes might find credible. In practice, however, any real vortex in which the fluid extends inward to the axis of rotation has a rotational core. Viscosity sees to that,

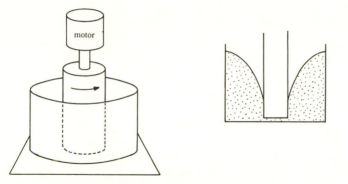

FIGURE 10.2. Making water go around in a cylindrical bowl by rotating a smaller cylinder sticking coaxially down into it. At right is the water's surface profile across the bowl.

ince the shear rate gets greater and greater as the axis is approached. In a ense, our original driving cylinder in the bowl has some physical reality; ne can usefully imagine the cylinder as hollow, with a rotational vortex nside and the irrotational vortex surrounding it—essentially the distribu-ion of speeds shown in Figure 10.3. The hurricane does, after all, have an :ye, whose low winds give a respite before it moves farther and hits you with a great gust from the opposite direction.

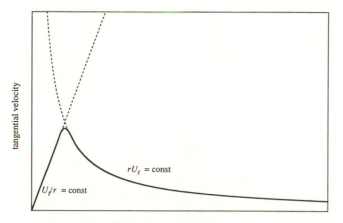

distance from center of vortex

FIGURE 10.3. The variation of tangential velocity with distance from the center of an ordinary vortex—irrotational but with a rotational core.

MORE ABOUT IRROTATIONAL VORTICES

In moving fluids these irrotational vortices are the really interesting ones. Not that rotational ones or intermediates don't occur, but these irrotational vortices crop up under an amazingly wide range of circumstances and are rather curious creatures. For one thing, they can be made in several ways. A rotating cylinder will naturally surround itself with an irrotational vortex in any fluid with finite viscosity; in fact, any other rotating object such as a flat plate will do likewise, the only difference being the greater complexity of the vortex it makes. Velocity gradients can make them. And, as we'll see in the next chapter, even some nonrotating objects such as inclined flat plates and airfoils can induce irrotational vortices.

From the skater's scheme to increase the rate of spin comes yet another way to make such a vortex. Angular momentum, remember, is the product of mass, angular velocity, and the square of the radius. If a bit of fluid with some small angular momentum, that is, a little spin on it, moves inward in a rotational vortex, it will increase in angular velocity—it will spin faster. If lots of bits of fluid move inward, a gentle rotational vortex can be converted into a substantial irrotational one. Remember that at any point in the motion of a bit of fluid, its tangential velocity will be the product of its angular velocity and the radius. Thus constancy of angular momentum generates the requisite condition for the irrotational vortex, a constant product of radius and tangential velocity.[2]

Since fluids, in our domain, are incompressible, simultaneous inward motion of bits of fluid from all directions sounds pretty unlikely. But that's only if one lives in Flatland (Abbott 1885)—in a properly three-dimensional world, fluid can move axially through the middle of the vortex. It may then either leave the system entirely, as through the drain beneath the bathtub vortex. Or it may move peripherally again somewhere else, perhaps to be recycled in toward the axis again, as when cold milk is gently added to the middle of a stirred cup of hot coffee. The axial and upward component of flow is perhaps the worst feature of the most violent of all meteorological phenomena, tornadoes.

What happens to such a vortex if it's left to its own devices? In a frictionless, inviscid world any vortex, irrotational or rotational, ought to persist indefinitely. Galaxies, though perhaps irrelevant here, are not short-lived phenomena. But friction is the great enemy of vortices, and it's at its worst for irrotational ones. A patch of fluid spinning as if solid need only suffer

[2] For instance, if inward movement halves the radius, constancy of angular momentum means that the angular velocity will increase 4-fold. With radius down 2-fold and angular velocity up 4-fold, tangential velocity will by increased 2-fold. The latter will just offset the 2-fold decrease in radius to maintain the constant product. We are, of course, making our usual presumption of constant density.

interaction with its surroundings; by contrast, in an irrotational vortex shear is ubiquitous since angular velocity varies with radius. So it takes energy to keep an irrotational vortex going—to offset the pernicious effects of viscosity quite beyond the cost of any interaction with its surroundings. Deprived of an adequate energy source, an irrotational vortex deteriorates toward a rotational one as the original radial variation of angular velocity decreases. Thus the distinction between rotational core and irrotational periphery is gradually lost—in effect, the core grows at the expense of the periphery as overall angular momentum drops.

This requirement for energy to counteract the shear stresses of viscosity has another consequence. Where viscosity is greatest, relative to momentum, sustaining a vortex is hardest. Therefore vortices at low Reynolds numbers have a proportionally larger rotational core and a greater hunger for energy—cut off their supply and they quickly grind to a halt. And at really low Reynolds numbers proper vortices simply don't happen. I think the minimum Reynolds number at which vortices have been produced is about 10^{-2}. Organisms manage to crowd the lower limit. *Vorticella*, a stalked protozoan (Figure 10.4) beats its cilia in a way that puts it in the center of a large, food trapping vortex (Sleigh and Barlow 1976); the Reynolds number is of the order of 10^{-1}.

One can view fluid motion as having three domains. In one, the world of turbulent flow, vortices form so readily that there's little else around. In a second, vortices occur but need substantial provocation and are relatively orderly and free of smaller vortices within themselves. In the third, the phenomenon of a vortex is simply out of the range of practicality except inasmuch as something might be immersed in a larger one. Note that the latter two are both characterized by laminar flow—under conditions of laminar flow vortices are often but not always possible. In a sense they're an example of the kind of quasi-periodic state that characterizes the transition region between order and chaos in many physical domains (Van Atta and Gharib 1987).

So, high Reynolds numbers are propitious for vortices. What would happen at a Reynolds number of infinity, in that impossible but occasionally revealing idealization called an inviscid or ideal fluid? In a world of unbounded inviscid fluids vortices would have a curious sort of quasi-reality. We used viscosity to get ours started, and it turns out that nothing else will do it. On the other hand, once somehow started, without viscosity a vortex would last forever since there's also no way to stop it. Since no inviscid fluids occur anyway, the fact that such a fluid couldn't generate a vortex might be said to make the situation no less real; and quite a bit of analytical work has been done on such vortices in inviscid fluids. In fact, many of the conclusions prove quite useful in understanding what goes on in the world of real, viscous fluids, so we're not just snickering at games

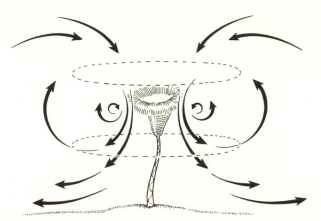

FIGURE 10.4. The unusually small toroidal vortex with which the stalked protozoan, *Vorticella*, draws edible particles within reach.

played by people whose fascination with mathematics distracts them from science.

In an unbounded, inviscid fluid a vortex not only has no temporal end but it has no spatial end either. The only way to stop the thing this side of infinity is to loop it back on itself in the form of a closed loop—whether a tidy torus, some pair of elongate whirls, or a complex snarl is immaterial (Figure 10.5a). That's very much what happens in the real world, where vortex rings (smoke rings and so forth) are much more the rule than the exception. A droplet of colored water released just above the surface of a larger body of water will usually form a vortex ring as it falls. A pulse of fluid forced out of a sharp-edged orifice into a larger body of fluid will almost always make a vortex ring. If you imagine that some system contains a vortex, you ought immediately to ask about its ends. If you observe a whirl in some two-dimensional view, then you ought to be alert to the location of a twin that makes ends meet. And, as reference to the figure or a moment's consideration should reveal, the members of any such pair of vortices should have opposite directions of circulation (whoops, I almost said rotation!)—a pair is really just a slice of a vortex ring.

Since vortices (at least when viewed in section) commonly occur in pairs, one is naturally curious about interactions between the members of such a pair. As it happens, a pair with opposite circulations will repel each other. On the sides facing each other, the flows will be in the same direction, so the relative speed of flow will be minimal; by Bernoulli's principle, pressure will therefore be high. The closer they are, the greater the repulsion, so a vortex ring will tend to form a proper circular torus.

A toroidal vortex, besides rounding up nicely, likes to move through the

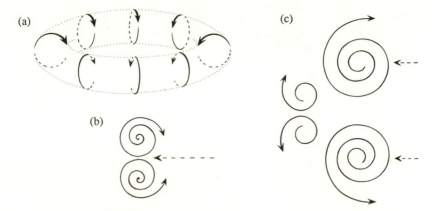

FIGURE 10.5. Vortex rings: (a) a single torus; (b) a travelling torus;
(c) a pair of vortices, one of which has just passed through the other.

fluid in which it lives. What drives the motion is shear in the extensive region of contact between the periphery of the vortex and the rest of the fluid—shear is diminished if the vortex moves axially in a direction such that the fluid on its outside is going at a speed closer to that of the rest of the fluid, as in Figure 10.5b. Thus fluid on the inside of the torus moves forward relative to the movement of the torus as a whole, while fluid on the outside moves backward relative to the overall movement. So this inner fluid moves in the same direction as the torus to which it belongs, but it moves at a higher speed. To put the matter another way (and one that will take on relevance when we talk about soaring), in the inner part one could fly or swim against the current indefinitely without getting out of the toroidal vortex of which the current is part. This kind of vortex ring does slow down over time if not maintained by some force such as might be provided by buoyancy—it gets bigger by entrainment of surrounding fluid and slower by loss of its original impulse, as described by Shariff and Leonard (1992).

Despite their predilection for rings and loops and hence, in effect, pairs of opposite circulation, a vortex can sometimes encounter another that has the same direction of circulation. Here, instead of being repulsive and essentially stabilizing, the interactions are attractive, even salacious. Flows are opposite on the sides that adjoin, and the biologist can easily recognize a basis for attraction. But the biological analogy doesn't completely apply— two small vortices enwrap each other but the normal result is merger into a single larger one. A particularly odd interaction is what's been called "leap-frogging" (Shariff and Leonard 1992). If two vortex rings have the same rotational sense they will travel in the same direction. If one follows another of the same size, the rearward one will catch up and attempt to pass through the forward one. If it succeeds (without merging), then the new follower will in turn catch up and try to squeeze through the middle of the new leader. (See Figure 10.5c.)

MAKING AND USING VORTICES NEAR INTERFACES

Biologists needn't worry much about whether or how vortices might form in an unbounded fluid—for most of us what matters is the immediate vicinity of some solid substratum or air-water interface. In such places, provided Reynolds numbers are decently above unity, vortices are extraordinarily easy to make. Indeed, they're the rule rather than the exception, and what takes careful design is avoidance rather than instigation. Streamlining, for instance, is in one sense a matter of preventing the formation of vortices associated with and just downstream from a separation point—bluff bodies form and shed more and bigger vortices than streamlined ones. As a general rule, anywhere there's a z-ward gradient of velocity one ought to be alert to vortex formation. The matter came up in the last chapter for spores rolling across leaves, where the side of a particle farther from the surface met a higher flow speed, and the whole particle was therefore set into rotational motion. The situation applies to bits of fluid in quite the same way, and it applies whether or not either chunk or bit actually touches the surface. Very simply, when the lines of action of resistance to motion (resulting from the no-slip condition) and the main motion-inducing push (from the free stream) don't coincide, a turning couple is produced, as we saw in Figure 9.4. Let's look at some of the possible vortex-inducing circumstances at interfaces and the forms of the vortices to which they give rise—all have potential biological relevance even where documented cases are at this point lacking.

Elongate Grooves

Consider flow across a long furrow or trench in the substrate (Figure 10.6a). The resistance of the fluid to shear between what's inside (and initially stationary) and what's outside (and moving) will cause the fluid inside to rotate in a vortex that runs the length of the furrow. The situation came up briefly in Chapter 4 (specifically, in Figure 4.12) in connection with viscous entrainment and the movement of fluid out of a hole in the substratum, and again in Chapter 8 in connection with the issue of skimming flows. Here attention is directed to the existence of persistent vortices, with all the attendant possibilities of particle resuspension, enhanced material exchange, induced flow in and out of porous substrata, and so forth. Such vortices should be expected whenever air or water flows across corrugated surfaces, whether plowed fields or rippled sand on dunes or the

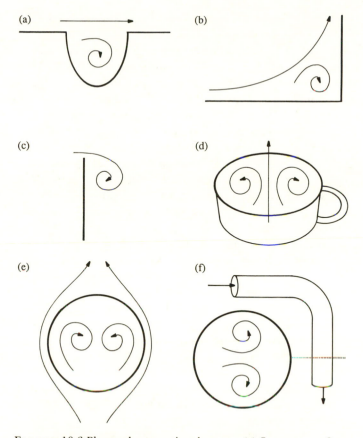

FIGURE 10.6 Places where vortices happen: (a) flow across a furrow
(b) flow making a turn upstream from a sharp corner; (c) flow just
behind a sharp cross-flow edge; (d) flow across and within a pit or cup;
(e) internal circulation in a fluid drop falling through another fluid;
(f) secondary flow just downstream from a bend in a cylindrical pipe.

bottoms of streams or oceans If the furrow or trench is narrow but deep,
the vortex can even generate another beneath it, one circulating in the
opposite direction. The phenomenon isn't limited to long trenches in otherwise smooth surfaces across which fluid flows; the same vortex induction
can take place in a sharp corner, where a vortex (or occasionally a set of
vortices of diminishing diameters) is generated by fluid making a gentle
turn (Figure 10.6b). And it can take place behind a sharp corner as a form
of separation of flow (Figure 10.6c); this last case is relevant to flow around
the rear edges of airfoils tipped substantially with respect to the free
stream.

Cups and Funnels

A trench makes a nicely two-dimensional case; with the inescapable three-dimensionality of a hole in the substratum (as in Figure 10.6d) things get a little more complicated. Blow gently across a full cup of coffee with a little unstirred milk in it to mark the flow. Coffee will move downwind across the part of the surface that's widest in the direction you're blowing. But it may move upwind on each side of that widest portion, indicating that something other than a vortex with a horizontal axis has been set up. What happens is that the area of contact between fluids is most extensive in the middle, so there interfluid shear is especially effective in making the coffee move. Conversely, the portions lateral to it are influenced by shear with the walls of the cup as well as by the motion of breath and midstream coffee. As a consequence the circulatory axis bends upward on either side, forming a torus that's sliced off at the coffee-exhalation interface.

This pattern in which flow is downstream across the (free-streamwise) widest portion of a cup and upstream to either side is particularly easy to form in a conical cup. According to Brodie and Gregory (1953), it's put to use by a lichen, *Cladonia podetia*, which bears its propagative elements, soredia, on the insides of upright cups. The cups don't just keep soredia from falling to the ground—the wind-induced vortices in the cups are important in getting the soredia out. They found that a free-stream wind of 2 m s^{-1} removed soredia from cups but was inadequate to blow them off a flat surface. Using model cups and lycopodium powder (especially appropriate for an investigation of lower plant dispersal) they found that the system cared little about whether the wind was smooth or turbulent and that a cone with walls diverging about 60° was most effective. The same system seems to work in at least one slime mold, where wind at no more than 0.5 m s^{-1} proved adequate to blow spores out of cupulate sporangia.

These shear-driven vortices in cups and funnels are sites of enhanced deposition of low-density suspended material—they're self-full-filling feeding troughs; the process has been described by Yager et al. (1993) and will reappear in a few pages in connection with a mayfly larva.

The Insides of Droplets

A complete and regular torus occurs in a fluid droplet if another, immiscible fluid blows across it. That happens, among other instances, when a raindrop falls through air at Reynolds numbers above about 10 (Figure 10.6e). The passage of air upward around the periphery of the drop induces upward flow of water at the sides of the drop; that in turn forces downward flow in the center. The overall effect, aside from the mixing, is a little like a violation of the no-slip condition at the droplet's surface—the surface itself goes with the flow. And that very slightly decreases the drag relative to what a solid body would experience. The effect is more substantial for a droplet of gas ascending in a liquid; as we'll see in Chapter 15, the internal toroidal circulation can reduce the drag by up to a third from what Stokes' law predicts. In a large bubble of gas this internal circulation induced by passage through the water may be sufficient to deform it into a proper torus.

The Wakes of Jets

As mentioned earlier, another way to form a toroidal vortex is by expelling a pulse of fluid in a jet into an otherwise stationary body of the same fluid—one can make something that initially looks quite a lot like a jellyfish with an eyedropper of milk and a glass of water. The fact that the structure appears a little ways from the dropper is exploited, it appears, by cuttlefish and other ink-squirting cephalopods. A persuasively writhing blob is a particularly nice target to offer a predator if you're decently distant from it and if that distance is progressively increasing. And increase it will, as a result both of the opposite momenta of blob and creature and of the normal progression of a vortex ring.

Bent Pipes

Quite a different situation in which interfacial velocity gradients generate vortices involves flow through pipes that go around curves. In a pipe, flow is (as will be elaborated in Chapter 13) fastest along the axis and nonexistent at the walls. That means that momentum is greatest on the axis. So in a bend the axial fluid turns less readily than the peripheral fluid, and it moves toward the wall on the outside of the curve. That forces peripheral fluid centripetally, and thus fluid leaves a bend with a pair of vortices distorting the normally parabolic distribution of speeds across a pipe's radius (Figure 10.6f). Pipes with curves are certainly not uncommon items in organisms—you need look no further for an example than your own aorta. A good introduction to vortices in and downstream from bends is given by Berger et al. (1983).

The Rears of Cylinders

Back in Chapter 5 we considered flow around circular cylinders in connection with the cylinders' oddly bumpy curves of drag coefficient versus Reynolds number. Vortices are generated on either side of a cylinder whose axis extends across a flow as part of the separation process (Figure 5.5)—at least above a Reynolds number of about 10. The pair of vortices are of equal strength (we'll get to the matter of quantifying the intensity of a vortex shortly) and of opposite signs. But under ordinary conditions this nicely symmetrical system is unstable above Reynolds numbers of around 40, and the vortices are alternately shed into the wake to form the Von Kármán trail mentioned earlier. (Goldstein 1938 has a good discussion of how the instability comes about.) So this is yet another way of generating vortices, one that might be important on account of the prevalence of cylindrical elements in organisms, but one that, at least for a stable pair, seems limited to a narrow domain.

In fact, stable, paired vortices behind cylinders may occur at higher Reynolds numbers than general books on fluid mechanics state. Goldstein (1938) cites experiments where testing in narrower channels postponed the start of shedding—for instance, where the channel width was only ten times the diameter of the cylinder, shedding didn't occur below $Re = 62$. More or less equivalent to a narrower channel would be other cylinders with parallel axes across the flow. In addition, shedding is discouraged when a cylinder extends upward from a solid surface through a velocity gradient (Lugt 1983a). Either way (or both) the proximity of surfaces has a stabilizing effect. In casual observations in a small flow tank, I've found that tipping a protruding cylinder back with respect to the flow has a further stabilizing effect. So one catches the odor of biological relevance—a wider range near interfaces, Reynolds numbers appropriate to the situations of quite a few organisms, an erect cylindrical structure that we commonly find, a posture consistent with what unstiff structures would normally do in a flow, potential for communal advantage, and so forth.

A cylinder sticking up from a substratum but still largely located in the boundary layer of that substratum has another feature of interest both fluid mechanically and biologically. Flow in its vortices has a strong upward component in addition to its circulating motion—fluid enters near the substratum and leaves near the top. Vortex shedding may be limited to the top, where it can be anything from a conventional Von Kármán trail to a pair of vortices bending rearward and detaching in tandem, as in Figure 10.7a.

And creatures demonstrably use these ascending, paired vortices. We talked about suspension feeding quite a lot in the last chapter, but we've not yet put in a word for another way in which an animal can acquire minimally motile, particulate food. That's detritus feeding, taking advantage of the fact that many things worth eating are a little denser than water and tend to settle out on any available substratum. The main problem facing the detritus feeder, the rate of deposition of its resource, is at least partly circumvented in a flow by transport across the substratum. Such transport in and by the slow flow just above the surface may have the additional benefit of sorting lower-density edibles, higher up in the flow and thus faster, from higher-density inorganic particles, lower and slower. Several cases are now known in which erect, tubular organisms resuspend low-density particles

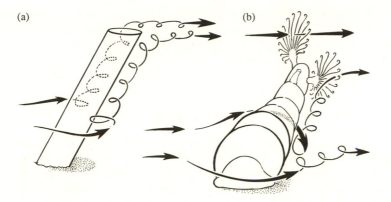

FIGURE 10.7. (a) Ascending vortices behind a cylinder that protrudes from a substratum. (b) How a larval black fly uses an ascending vortex to feed on detritus with one cephalic fan while suspension feeding with the other.

in the vortices, which bring the food up to structures that one would ordinarily regard as suspension-feeding devices. One (worked out by Carey 1983) involves a marine terrebellid worm, *Lanice conchilega*, that makes a tube a few centimeters high and about half a centimeter wide emerging from a sandy substratum. The head of the worm, with an array of tentacles of which Medusa would have been proud, sticks out the upper end of the tube; and the worm feeds on suspended material delivered by the flow. At the same time, it takes food from the ascending vortices—detritus feeding and suspension feeding with the same equipment. A phoronid worm (not an annelid), *Phoronopsis viridis*, operates similarly (Johnson 1988), again using suspension-feeding equipment to feed on detritus. These creatures live in dense populations, just short of densities that would produce skimming flows; and, according to Johnson (1990), the communal arrangement enhances turbulence and upward water flow and thus the feeding of all the individuals.

An analogous but fancier case (also including communal augmentation) is that of the black fly larvae (*Simulium vittatum*) of rapid streams that have appeared earlier in several related contexts. These attach to rocks at their posteriors and stick up in the form of tapering cylinders that terminate with two cephalic fans. Chance and Craig (1986) showed that they twist lengthwise and bend so that the anatomically paired cephalic fans end up not on either side in the flow but instead above and below each other (Figure 10.7b). The top fan is a normal suspension feeder, but the bottom one (still near the top of the erect larva) feeds from a vortex that has ascended from the substratum and will subsequently detach into the wake.

Matters are complicated a little by the fact that larvae are bent a bit across the flow (as well as downstream) so the paired vortices are above and below as well as side by side—only the upper one passes through the lower fan. Besides their role in larval feeding, the vortices are put to respiratory use by the pupae, which stick their gill filaments into them (Eymann 1991).

THE VORTICITY OF VELOCITY GRADIENTS

At the start of this chapter we distinguished between rotation and translation, noting that any element of a vortex not including the axis (outside the core) usually translated in circles without rotation even though the vortex as a whole was obviously rotating. That was possible because fluids permit internal shearing motion, in this case slip between concentric tori. Under another circumstance as well the occurrence of internal shear implies that vorticity is present in a fluid. Well over a century ago, Sir George Stokes explained the matter as follows. One suddenly solidifies a tiny sphere of fluid and looks to see whether it's rotating; if so, the fluid is said to have vorticity. Consider an ordinary velocity gradient of the kind that occurs adjacent to planar solid surfaces. As mentioned a few pages back, because of the shearing motion in the gradient, such an element in such a gradient will be subjected to a turning couple (Figure 9.4). *Thus vorticity is normally associated with a velocity gradient.*

The presence of a vortex implies that a fluid has vorticity (even if the vorticity is concentrated as a vortex line running along the axial core of the vortex); on the other hand, the presence of vorticity isn't a perfect indicator of vortices. While vortices don't happen without vorticity, with vorticity they are likely but not inevitable. Do boundary layers generate vortices? Quite often they do, with all sorts of effects on bed erosion, cross-flow mixing, particle transport, and so forth. Much of the puffiness of gentle breezes at the surface of the earth represents the passage of what might be called "rollers"—vortices where the boundary layer has, as it were, rolled up. The resulting periodicity is certainly conspicuous in the motion of a field of tall grain or grass. One might naively expect that in an eastbound breeze (Zephyrus by name) the flow near the actual surface might head westward, but that's just a misapprehension caused by a stuck-in-the-mud frame of reference. Remember that the rolling vortex was formed as a result of variation in the intensity of the eastbound flow, so even the lower part will normally head that way, albeit more slowly. Light-wind lake sailors know that as a rolling vortex approaches, an especially calm moment typically precedes the puff.

We noted a bit earlier that vortices didn't easily form ends within a fluid. Together with the present business about vorticity near a surface, this creates some very odd phenomena (or sorts out some very ordinary phenomena that are otherwise hard to rationalize, depending on your point of view). Imagine vortices rolling (or vorticity flowing) along a surface from which something protrudes. To get around the protrusion requires an interruption of the vortex. But vortices don't like to make even temporary ends any more than a skier (as in a famous Charles Addams cartoon) can pass a tree with one track on either side. So they get tangled around protrusions, with the part just upstream of the protrusion staying in place. As a result snow counterintuitively fails to accumulate on the upwind side of a tree—the trapped vortex has a locally downward direction at the tree's surface and scoops out a hollow. Similarly, if you stand facing the ocean in the swash on a beach, after passage of a few episodes of backwash you notice your heels sinking in. The receding water, with a severe velocity gradient, has scooped out sand from the upstream side of your personal protrusions. Somewhat less obviously under most circumstances, the vortex that's doing the digging bends around tree or leg to take on the shape of an elongated horseshoe, as in Figure 10.8, with regions of scour (upstream and laterally rearward) and regions of deposition (just downstream and farther laterally). Such horseshoe vortices, with erosion and deposition, happen around scallops in scallop beds, with implications for passive transport and other processes (Grant et al. 1993).

One consequence of such vortices is that an erodible bed beneath moving air or water may not necessarily be stabilized by worm tubes or emergent vegetation. Only if the protrusions are sufficiently dense will skimming flow, mentioned earlier, have a stabilizing effect. In practice, "sufficiently dense" means a cross-sectional area parallel to the substratum more than about a tenth or twelfth of that of the substratum (Eckman et al. 1981; Nowell and Jumars 1984). At lower densities, protrusions usually have a

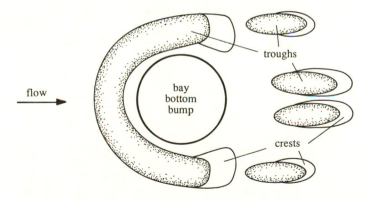

FIGURE 10.8. The action on an erodable bed of the "horseshoe vortex" that's associated with a protruding object in a shear flow, here taken from data of Grant et al. (1993) showing what happens beneath a sedentary scallop.

destabilizing effect. And the destabilization generates exceedingly heterogeneous conditions on the surface, with—as Eckman (1983) has shown—major effects on both nonbiological processes such as sedimentation and on biological ones such as larval recruitment.

Nor are such general considerations the extent of the biology associated with these horseshoe vortices. Soluk and Craig (1988) have described the behavior of a mayfly larva (*Ametropus neavei*) that lives in the surface layer of unstable, sandy sediment on river bottoms. The insect seems to require a flow in order to feed; given a flow, it faces upstream and excavates a shallow pit in front of its head. A vortex forms in the pit, its formation aided by the postures of forelegs, head, and antennae; and the larva suspension feeds from the vortex (Figure 10.9a). The vortex seems to assist feeding in several ways—it deflects fluid downward, enabling the animal to remain largely within the substratum and sheltered from hydrodynamic forces that might otherwise preclude life on a soft bottom. It also increases the residence time of suspended particles within reach of the forelegs and their filtering setae. Also under some circumstances the combination of pit and vortex may act as a depositional trap and resuspension device so

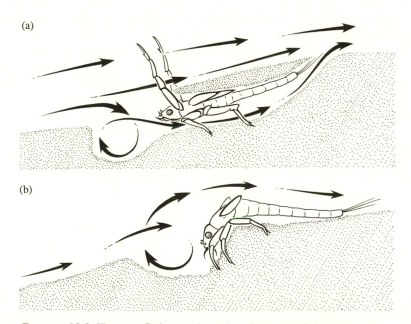

FIGURE 10.9. Two mayfly larvae that take advantage of the upstream (center) part of a horseshoe vortex. *Ametropus* (a) makes a pit for its vortex, which then traps edible suspended particles. *Pseudiron* (b) arranges itself so the vortex digs the pit while it eats any excavated prey. It moves slowly backward (downstream) so excavation proceeds continuously.

suspension-feeding equipment can effectively detritus-feed as well, as we saw earlier with terrebellid and phoronid worms and the black fly larva.

Another mayfly larva (*Pseudiron centralis*) of a similar habitat plays a different variation on the horseshoe vortex theme. It feeds on larval chironomid flies that live in the sands, but it hasn't any decent digging appendages. According to Soluk and Craig (1990), the insects are dependent on erosion to expose prey, but they make sure that erosion happens when and where they want it. With adequate current, larvae arch their bodies and lower their heads and thus create the conditions for vortices to dig out upstream pits. From these they seize any prey that get exposed (Figure 10.9b). Excavation is a continuous process, so as a *Pseudiron* larva moves slowly backward it leaves a shallow groove as a trace. By contrast with *Ametropus*, *Pseudiron* lets the current do all the work.

THERMAL VORTICES

We talked a few pages ago about toroidal vortex rings, but mainly about small ones. They exist on much larger scales as well; indeed, the mushroom cloud of a nuclear explosion is a partially formed vortex resulting from the interaction of rapidly rising hot gas and the surrounding atmosphere. More benignly, ascending vortex rings commonly form when winds are light and the ground heats the lowest part of the atmosphere. Such hot air beneath cooler air is unstable. Here and there a bubble of hot air will detach itself and rise, forming, in a way now familiar, a vortex ring that ascends because of its buoyancy and the normal axial motion of such a structure. "Here and there" may not be entirely random. A tree in otherwise low vegetation can generate a convective updraft as a result of the heating of its leaves (Gates 1962 describes the process); that updraft can initiate these thermal vortices (usually simply called "thermals"). A plowed field, more absorptive of solar radiation than surrounding vegetation, provides a common site for initiation of thermals. Highways through vegetated areas are similar initiation sites. The events involved in starting a thermal vortex are described in most books on microclimatology (see the references on planetary boundary layers in Chapter 8) or where these structures take on biological relevance, as in Pedgley's (1982) *Windborne Pests and Diseases*.

The most conspicuous users of thermals are large terrestrial birds that soar. The process (Figure 10.10) is nicely described by Cone (1962) and Pennycuick (1989). Since air in the inner part of the circulating and ascending torus is moving upward faster than the overall system, a bird can be descending with respect to the local air while still ascending with respect to the ground. What it must do is to keep turning with a sufficiently narrow radius to stay in that inner, locally ascending, part of the torus; large raptorial birds such as hawks, vultures, and eagles are quite good at such

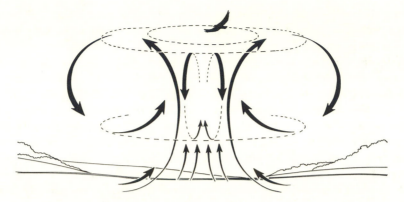

FIGURE 10.10. Thermal soaring. A bird (here a bit larger than life) can, by circling within a rising vortex ring, descend with respect to the local air but remain within the ring and ascend with it.

maneuvers. Human-carrying gliders may have better overall lift-to-drag ratios (more about these in the next chapter), but they travel faster and can't turn as sharply, so they either require larger thermals or suffer a shorter residence time in each. Where I live, thermal soaring is common, especially among vultures; and the sight of one or more often marks the location of a freshly cleared or plowed field beneath. Major highways, cutting swaths through our local forests, may provide a bonanza for such birds—a bird flying along such a road on an appropriately sunny day of low wind will sooner or later encounter a thermal. It could then take the elevator up and glide slowly downward until running into the next thermal—all the while watching for the road-kills thoughtfully provided by the same highway.

These thermal vortices also seem to be used for transport by quite different kinds of creatures. Some spiders and moths (most notoriously the gypsy moth, *Lymantria dispar*) exude long silk threads out into a breeze when in their first instars (the immediate posthatching stage). Under particular meteorological conditions they then let go of the substratum and drift off—the phenomenon is called, a bit inappropriately, "ballooning." They can go remarkable distances—Darwin noted a landing of spiders on the *Beagle* 60 miles from the coast of South America. Now no amount of thread can produce true buoyancy—what these tiny spiders and caterpillars are gaining is drag. Which they get in abundance: adding a thread of less than 0.1% of body weight quadruples the drag of the system (Humphrey 1987; Suter 1991). Release into rising air might just be a chance occurrence, but quite clearly it's not. After climbing as high as possible (Tolbert 1977) they wait for the right sort of atmospheric instability. One preferred set of conditions is low wind and a rapid rise in temperature on a sunny day after a cold night. These are conditions, as Vugts and Wingerden (1976) pointed out, that favor the production of thermals. Neither spiders nor caterpillars select especially high winds, as they would if simple wind-drift were the objective: ballooning ceases when winds exceed about 3 meters per second.

To bring the story around to some topics that came up earlier, ballooning spiders find still another condition auspicious for flight—a steep vertical wind *gradient*, that is, a high value of dU/dz (Greenstone 1990). That implies turbulence near the ground and, more particularly, rolling vortices with horizontal axes. Greenstone suggests that the loose silk strand ought to permit a spider contemplating takeoff to determine the local wind direction and thus pick the side of the vortex (the rear) in which air is ascending. But a somewhat more startling suggestion was made to me by Lloyd Trefethen. He figures that a silk strand won't follow streamlines in a vortex since it resists extension. He also points out that thermal vortices have some horizontal motion around their axes as well as the toroidal circulation. In combination, spiderlings ought to be drawn in toward the center, where air is ascending faster than the overall vortex, and they ought to be able to do the same trick as the soaring vulture—to descend locally while ascending globally. The stunt should work, I think, in Greenstone's rolling vortices as well, with the spider acting as the gravitational analog of a sea anchor beneath a thread that gets wound into and pulled along with the vortex, whose axis is always just ahead.

Something like ballooning in aerial arthropods may happen in marine mollusks. A young postlarval mussel (*Mytilus edulis*) produces a byssus thread of the kind that will later be used for anchorage (the "beards" one cleans off mussels before cooking), but a single long one instead of a whole pelage of tie-downs. A mussel half a millimeter long may produce 70 millimeters of thread with a diameter of about a micrometer. These threads are clearly used to increase drag during passive dispersal (Sigurdsson et al. 1976), and mussels may be capable of rapid deployment of thread and quick detachment into upward currents (Lane et al. 1985). A gastropod (*Lacuna* spp.) may manage something similar using a stretchy mucous thread and active foot-raising to release itself into oscillating water currents (Martel and Chia 1991).

On a very much larger scale, locust swarms usually move downwind. Given the normal vortical motions of the atmosphere, that will move them toward low-pressure areas, whose updrafts generate a rain and new vegetation (Rainey 1963; Pedgley 1982). Such movement, though, seems to require ground cues; and just how locusts (either the desert locust, *Schistocerca gregaria,* or the migratory locust, *Locusta migratoria*) accomplish the requisite navigation (Baker et al. 1984) is still unclear.

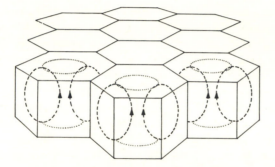

FIGURE 10.11. Bénard cells, a packed array of density-gradient vortices. You can make them yourself by mixing a little pearlescent liquid soap with water in a dark frying pan and, after allowing flow to stop, heating the pan very gently and uniformly from beneath.

On a very much smaller scale, density-gradient vortices occur in small, shallow vessels of liquid. If a pan of water is very gently and uniformly heated from below, the same kind of instability occurs that made thermals in the atmosphere. A pattern of point upwellings and peripheral sinkings such as that in Figure 10.11 often develops—they're called "Bénard cells" (or "Rayleigh-Bénard cells"). In addition to thermal gradients, cultures of motile microorganisms can generate these cells. What's required is for the organisms to be negatively buoyant so they passively sink, but at the same time to be either negatively geotactic or positively phototactic so they actively swim upward (Pedley and Kessler 1992). Laboratory cultures thus develop local areas of concentration, rather like what happens in the Langmuir circulations near the surfaces of lakes and oceans (Chapter 17). Sometimes these "bioconvective" structures can be seen in dense algal blooms in small natural puddles.

CIRCULATION AND VORTICITY

If something is constant, it gets named. That's no more than a reflection of the way we look for elements of order and rationality in a superficially messy and irrational universe. The constant product of speed and distance in an irrotational vortex is given the name "circulation," and its value provides a measure of the intensity of such a vortex. Put properly, circulation (capital gamma, Γ) is the product of circumference and tangential velocity; thus for any streamline surrounding the core of an irrotational vortex,

$$\Gamma = 2\pi r U_t.$$

(10.1)

Circulation has dimensions of distance squared per unit time.[3] It turns out that the value of the circulation doesn't depend on going around the core on a single streamline. So one can give a more formal and more general definition of circulation as the line integral of the component of velocity on and tangential to a closed curve lying entirely within the fluid:

$$\Gamma = \int_1 U_t dl. \tag{10.2}$$

The value of the circulation defined in this way comes out the same for any closed loop that encloses the core of an irrotational vortex. And it comes out to zero for any closed loop that doesn't encircle the core. Very tidy, except that U_t isn't the most convenient thing to measure inasmuch as it requires determination of both speed and direction without a handy shortcut to either. I can't recall any instance in which a biological investigation used it as a primary datum. Still, the notion of circulation as a proper quantitative variable is important, and the concept plays on a major role in explaining the origin and magnitude of lift.

Another odd notion, vorticity, came up a little earlier. It, too, is defined as a properly quantitative variable, one even more abstract and general than circulation. In explaining how a boundary layer, with shear but with all flow going in the same direction, could have vorticity, we invoked Stokes' device of "freezing" a tiny bit of fluid and looking at whether it rotated. That's the basis of one definition of vorticity—the angular velocity of matter at a point in space. In slightly different terms, vorticity is the circulation around an infinitesimal circuit divided by the area of that circuit[4] (and hence it has dimensions of time^{-1}). Vorticity is an enormously important concept in theoretical fluid mechanics, but it's inseparable from a level of mathematical formality quite beyond both the level of this book and the sophistication of its author; in short, we won't make specific use of it, and the interested reader is referred to the standard textbooks of fluid mechanics.

THE ORIGIN OF LIFT

Let's return to our rotating cylinder sticking down into a body of fluid (Figure 10.2). In an otherwise stationary fluid it will be surrounded by an irrotational vortex whose speed distribution was given in Figure 10.3 and the strength of whose circulation was defined by equation (10.2). Now

---

[3] As do both kinematic viscosity and the diffusion coefficient. No particular relationship between the three is implied, and the coincidence is best regarded as coincidental.

[4] This latter, the formal definition, makes vorticity equal to twice the mean angular velocity of particles at a point.

imagine superimposed upon this circulation an additional motion, a translation in a line, either by moving the rotating cylinder sideways through the fluid or by moving the fluid past the cylinder. No surprise—the cylinder now incurs a certain amount of drag. But the force on it has an important difference from most of the cases of drag that we've previously considered. Apparently as a consequence of the rotation—for what else could be responsible?—the direction of the force is no longer exactly in a direction opposite that of the motion. For convenience (that's all, really) we can resolve the force into two components. One is the familiar drag, the force component opposite the motion that pushes the cylinder downstream; the other is a component normal to the flow, what we define as lift, pushing it across the flow. (It's important to keep in mind this definition of lift as a force normal to flow. In the vernacular, "lift" implies an upward, contragravitational direction that is often at variance with the present, technical use. We really need two terms, and we'll use "upward force" for the everyday version of lift, reserving "lift" for use as just defined.)

A look at the resulting streamlines (Figure 10.12) clarifies what's happening in this superposition of rotation and translation of a cylinder. On one side of the cylinder the two motions in the fluid oppose one another, so the velocities are lower and the streamlines are farther apart. On the other side, the motions are additive, velocities are increased, and the streamlines are closer together. By Bernoulli's principle pressure will be elevated on the side where flow speeds are lower and will be reduced on the side where the speeds are higher. Thus a net pressure or force will act in a direction normal to the free-speed flow—in short, lift. Note, though, that while circulation is fundamental in generating lift, no actual fluid particle need travel all the way around the cylinder. A limited amount very close to the surface may do so as a result of viscosity, but not enough to merit more than mention.

Upon what does the magnitude of this lift depend? Within certain practical limits, it's directly proportional to the rate of rotation of the cylinder—to the circulation, which is really why the concept was introduced. It's also directly proportional, again within limits, to the speed of the cylinder's translational motion. A formal statement, the Kutta-Joukowski theorem, puts it in tidier form: "If an irrotational air stream surrounds a closed curve with circulation, a force is set up perpendicular to that air stream, and the force (per unit length) is the product of the fluid's density, the free-stream velocity, and that circulation":

$$\frac{F}{l} = \rho U \Gamma. \tag{10.3}$$

This phenomenon, the lift of a rotating cylinder moving through a fluid, is called the "Magnus effect," after H. G. Magnus (1802–1870).

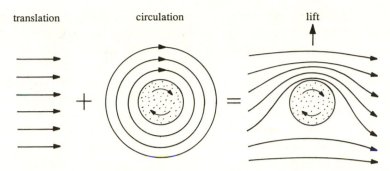

translation   circulation   lift

FIGURE 10.12. If a solid body such as a cylinder rotates as it translates (here right to left) through a fluid, the resulting asymmetry of flow generates a force normal to the free-stream flow. We call the force *lift*.

The Magnus effect (at a little lower intensity) works for spheres as well as for cylinders. It's a really big deal in sports in which spheres are thrown, hit, or otherwise put into motion since (except for a golf slice) a confusingly nonstraight course is distinctly meritorious. Two pleasant books on such contemporary compulsions are *Sport Science*, by P. J. Brancazio (1984), with good references, and *The Physics of Baseball*, by R. K. Adair (1990), with more on the Magnus effect specifically. Incidentally, according to the latter, outfielders find balls with some spin more predictable than those without—baseballs have stitched seams that protrude a little, and so the paths of nonspinning ones can be even worse.

FLETTNER ROTORS

The Magnus effect can be put to more serious use. Aircraft have been designed with rotating cylinders sticking out of their fuselages in place of conventional wings. Rotation of the cylinders so that the upper side moves rearward will generate—engines willing and plane moving forward—an upward force to sustain the craft in the air—if conventional airfoils didn't yield so much lift with so little drag we might use such aircraft. Actual application of the Magnus effect was made in several ships that had one or more large, rotating, vertical cylinders in place of masts, built in the early 1920s by Anton Flettner. The last of these, the *Buckau*, displaced 550 metric tons and had fore and aft rotors 3 meters in diameter and 16 meters high that rotated 100 times per minute (specifications and pictures are given by Herzog 1925 and in Flettner's fascinating autobiography, 1926). A sailboat that required an engine to sail sounds like the worst of both worlds, but the idea would have been quite viable twenty or thirty years before. Hybrid sail and screw freighters were at that earlier time quite common, since coal was bulky and not cheap at out-of-the-way ports. A Flettner ship used a lot less coal than a propeller-driven steamer; at the same time the sailing apparatus required far fewer crew (with attendant space and facilities) to operate. Flettner claimed that in a storm the rotors caused a ship to heel less than did bare, sailless rigging. What made Flettner's design unattractive were the easy availability of liquid fuel and a shipping glut in the '20s; it has since been consigned to textbooks to provide a little levity in the long series of abstractions leading to lift and airfoils. There is, by the way, nothing special about the use of a cylinder; Hoerner (1965) says that a rod of X-shaped cross section works as well.

And when, as McCutchen (1977a) has pointed out, Nature anticipated Flettner, she hasn't used cylinders. Flat plates, sometimes with lengthwise twist, take their place; and the necessity of a rotating joint is avoided by having the entire craft rotate about its long axis. These rotorcraft are certain of winged seeds (or samaras), gliding down and away from the parent tree by autorotating. In the tree-of-heaven (*Ailanthus altissima*) the samaras are blades about 3 to 4 cm long, twisted lengthwise, with the seeds in thickenings in the center (Figure 10.13a). When released they rotate about their long axes and either travel away from the parent or descend in a wide helix slowly enough to achieve some dispersal by wind. The trick requires little in the way of specialized structure—if you drop a small index card it will almost always begin rotating about its longer center line and will thereafter move sideways as well as downward. Cutting the card lengthwise to increase the length-to-width ratio gives earlier rotating and a bit better glide angle; cutting the card a bit diagonally makes it descend along a helical path. According to Lugt (1983b), who gives a good review of autorotating flat plates, an aspect (length-to-width) ratio greater than 5.0 is best, and a plate needs certain minimum moment of inertia to autorotate.

Two other autorotating samaras common where I live are those of ash (*Fraxinus*) and tuliptree or yellow poplar (*Liriodendron*). Each has its seed not in the middle but at one end, and descent is inevitably in a tight helix that looks very much like what's done by a maple (*Acer*) samara (Figure 10.13b). The maples, though, get their lift from conventional airfoils with identifiable front and rear (leading and trailing) edges, while these autorotating samaras are symmetrical front to rear. Both sorts "autogyrate" downward, essentially gliding along a helical path as if on a twisted sliding board. In addition, though, the symmetrical ones autorotate, getting lift as Flettner rotors by Magnus effect. In comparisons of performance in either nonhelical gliding (McCutchen 1977a) or helical descents (Green 1980), the symmetrical autorotators seem a little inferior to the asymmetrical nonautorotators. It's been both argued (Green 1980) and disputed (Greene and Johnson 1990b) that the symmetrical ones are, on the other

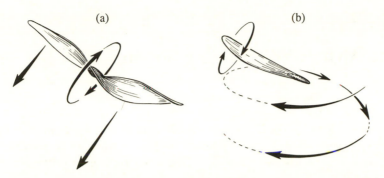

FIGURE 10.13. Two samaras that work as Flettner rotors: (a) *Ailanthus* rotates about its long axis and thus moves with some horizontal component instead of descending directly. (b) *Fraxinus* autorotates similarly but, with the seed at one end, its centers of lift, drag, and weight no longer coincide; so it descends in a tight helix, gaining longer exposure to ambient winds rather than direct dispersal distance.

hand, more stable and do better when turbulence or obstacles complicate descent.

Many long, thin leaves act as Flettner rotors when they're shed. Around my place the willow oaks (*Quercus phellos*) are particularly conspicuous autorotators, and their leaves are carried well beyond the tree even with no noticeable breeze. It's not clear whether the behavior has any adaptive significance—perhaps excessive deposition of leaves above the roots is in some way detrimental—or is merely an incidental consequence of a leaf shaped by selection for other functional characteristics.

Lift, Airfoils, Gliding, and Soaring

I N OUR DISCUSSION of vortices and circulation, we explained lift in a way far removed from wing of bird, bat, insect, or even pterosaur. To explain how any of these (and airplanes, too) produce circulation-based lift, we still have several sticky problems left to resolve:

1. Things other than rotating devices can be observed producing lift. I refer, of course, to airfoils[1] or wings that stay properly fixed— wings that may translate and thus transport but that quite obviously don't rotate while doing so. Do these also work by the superposition of circulation and translation, or do they depend on a different physical mechanism?

2. For a flat, inclined surface to deflect an airstream and produce lift seems intuitively reasonable. But a curved surface with its convex side upward commonly produces an upward force even if its leading (upwind) and trailing (downwind) edges are at the same horizontal level, with no inclination at all.

3. For rotating cylinders lift is proportional to the first power of translational velocity (equation 10.3), while for airfoils lift is more nearly proportional to the second power of the translational velocity. Is the Kutta-Joukowski theorem underlying the equation applicable to nonrotating, lift-producing devices?

CIRCULATION AND AIRFOILS

The route around these awkward points requires that we explore the behavior of this thing we're calling an airfoil, the relevant terminology for which is given in Figure 11.1. According to F. W. Lanchester (1868–1946), an airfoil is a device that can produce circulation in its vicinity without itself actually rotating. Just how can this happen? The fact that the lift for an airfoil is proportional to the square rather than the first power of the free-stream speed suggests that the circulation itself must be proportional to the speed of the oncoming airstream. If so, then somehow the shape and orientation of the airfoil must interact with the oncoming air to produce

---

[1] All of the material to follow applies to lift production in water as well as air. Using "airfoil," "airstream," "wind," and so forth should be regarded as just a linguistic convenience or ancient prejudice of the author.

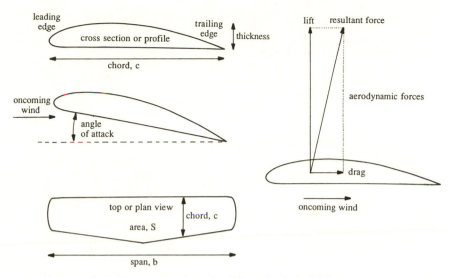

F IGURE 11.1. The terminology for lift-producing airfoils.

the circulation—the faster the air, the more intense the circulation. A rotating cylinder, of course, generates its circulation quite independently of any translating airstream.

What determines the value of the circulation and hence the lift was first realized by Joukowski (or Zhukovskii) early in the present century. He pointed out that for an airfoil with a sharp trailing edge, only one pattern of flow permitted the air to slip off the rear of the airfoil without any discontinuity or without having to turn the sharp corner at the trailing edge. This pattern had to determine the amount of circulation and lift. Pressure might be lower on top than bottom because of the sharp trailing edge and the rapid rearward flow on top—but fluid, with momentum, just doesn't find it feasible to sneak around and relieve the difference. So instead everything else shifts, creating a net circulation, as shown in Figure 11.2. It's only a net circulation, of course—fluid doesn't actually travel around the airfoil—but it's still a flow pattern equivalent to that which would be generated by the superposition of translation and circulation. In short, a wing can be viewed as surrounded by a fictitious vortex (fictitious because it's only a *net* vortex), and the strength of that vortex (its circulation) is proportional to the airfoil's translational velocity. Hence the dependence of lift on U^2—one U is that of translation itself, the U in equation (10.3); the other U comes from the circulation, since $\Gamma \propto U$.

In the last chapter, vortices were made with the aid of viscosity, whether by rotating a cylinder in a liquid and capitalizing on the no-slip condition at

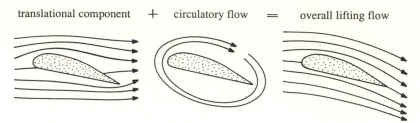

translational component + circulatory flow = overall lifting flow

FIGURE 11.2. The overall flow over an airfoil that's producing lift consists of a translational component across it and a circulation around it.

its surface, or by using the vorticity of the velocity gradients just above the ground, themselves a result of viscosity. In an ideal fluid, circulation can be neither created nor destroyed. While real fluids are more accommodating, the apparent paradox does have some reality and consequences. Prandtl and Lanchester first recognized the existence of a "starting vortex," equal in strength and opposite in direction to the "bound vortex" of a wing (Figure 11.3). Thus the system as a whole need have no net circulation, and the phenomenon of lift (unlike drag) needn't depend on a distinction between real and ideal fluids. In the 1920s Prandtl even made movies showing starting vortices; the movies also showed that when an airfoil was stopped, the bound vortex slipped out from around the airfoil and became a free and obvious "stopping vortex." The reality of these starting and stopping vortices gives added credibility to the notion of a bound vortex, a reality sometimes uncomfortably evident: they have the unfortunate habit of hanging around airports on still days until eaten up by viscosity, buffeting small craft that are landing or taking off.

For reasons that will become clear shortly, a wing of finite length sheds "tip vortices" at its ends that amount to continuations of the bound vortex or circulation about the wing. These tip vortices extend back to the starting vortex, completing a vortex ring like those already considered. Naturally, the vortex ring is almost always much longer than wide, about 100,000 times so for a large plane completing a transcontinental flight. (Of course, in such a situation, as a result of viscosity the starting vortex gets dissipated long before the stopping vortex is shed.) But we needn't postulate any discontinuities in the flow, and ideal fluid theory (of which Bernoulli's principle as the basis for lift is a part) isn't seriously abused.

No great mystery then attaches to the business of getting lift from curved plates whose leading and trailing edges are at the same level with respect to the free-stream motion. If a plane is convex on the side facing upward, and if the flow goes smoothly from leading to trailing edges both above and below it, then the streamlines will be squeezed together above and drawn

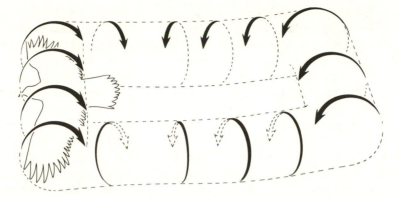

FIGURE 11.3. The bound vortex around the wings, the tip vortices, and the starting vortex for a gliding aircraft. In all, the vortices form a complete vortex ring.

apart below. That means, by Bernoulli's principle, lower pressure above and higher pressure below and therefore an upwardly directed net force acting normal to the free stream—in short, lift. Higher velocity above and lower below—that's what we recognized as equivalent to a superposition of translational and circulatory components of flow. Thus we've used the notion of circulation to get a consistent view of lift, one in which airfoils and Flettner rotors in free-stream flow develop their forces by the same mechanism, the mechanism that we noted back in Chapter 4 for flow across a local elevation of a substratum. Autorotating *Ailanthus* seed, gliding bird wing, and bottom-resting flounder all develop their normal forces by the same process.

Incidentally, this use of an upwardly convex surface to develop "upward lift" (not, remember, a tautologous term) explains why the contour of the upper surface of a lift-producing airfoil is generally of more consequence than that of the lower surface. Flat lower surfaces work aerodynamically about as well as concave ones; being flat below mainly gives an enclosed volume useful for fuel tanks, bracing, muscles, control equipment, and so forth. A sharp trailing edge, as already noted, is crucial for the functioning of an airfoil as a lift generator as well as for drag minimization. And a rounded leading edge discourages separation, quite as important for producing lift as for limiting drag. The asymmetrical streamlined cross section of airfoils useful at moderate and high Reynolds numbers and subsonic flows is thus rationalized.

The distribution of lift on the surface of a wing is of considerable interest, whether viewed as part of the circulation argument or as a practical matter of the location of the overall force vector that passes through the wing. The detailed distribution of lift can be determined in the same way as can the pressure distribution around a cylinder (Chapter 4)—a series of holes is drilled in the surface and led by internal tubes to a manometer. The only difference is that one must now look specifically at the component of force normal to the free stream. One finds (1) that more of the lift comes from the reduced pressure on the top than from the excess pressure on the bottom, something quite consistent with our notion of the role of the convex upper surface; (2) the center of lift is relatively near the leading edge of the airfoil, usually at or in front of the point of maximum thickness; and (3) as the angle at which air meets wing (the angle of attack) increases, the center of lift shifts forward, with consequent alteration of pitching moments and effects on stability.

Lift Coefficients and Polar Diagrams

The lift of a wing is approximately proportional to the square of its velocity through the air. It's also nearly proportional to the area of the wing and to the density of the air. All of this sounds very much like the way drag behaves at moderate and high Reynolds numbers. So we can conveniently de-dimensionalize lift in just the same way we did drag (Chapter 5), dividing lift per unit area by dynamic pressure to obtain a dimensionless coefficient, this one called the "lift coefficient" and analogous to the drag coefficient of equation (5.4). Similarly, one shouldn't be misled by regarding this as a "formula for lift" since it does nothing more than standardize the interconversion of lift and lift coefficient. Thus

$$L = \frac{1}{2} C_l \rho S U^2. \tag{11.1}$$

The convention for designating an area, S, for the equation is fairly specific—plan form or profile area, the area one would see if viewing from above an airfoil lying on a horizontal surface. Like the drag coefficient, the lift coefficient is a function only of shape, orientation, and Reynolds number. The two coefficients, though, depend on these three factors in rather different ways, and the consequent interplay of lift and drag underlies much of the subtlety of airfoil design.

For a given airfoil at a given Reynolds number, orientation determines both lift and drag coefficients. Orientation, in practice, hinges mainly on the angle between a line from leading to trailing edges (the "chord") and the direction of the oncoming wind—what was defined as the "angle of attack" in Figure 11.1. Lift (and its coefficient) increases from zero at an angle of attack near 0° (how far from zero depends on the top-to-bottom asymmetry of the airfoil) to some maximum at an angle of attack of around

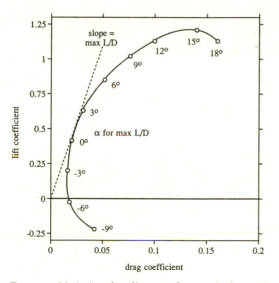

FIGURE 11.4. A polar diagram for an airplane wing. The lift coefficient is plotted against the drag coefficient, both referred to plan form area and usually with different scales, and the angles of attack are noted on the curve. Any line that passes through the origin is a line of constant lift-to-drag ratio. The one tangent to the curve gives the maximum obtainable ratio, and it touches the curve at the angle of attack that gives that maximum ratio.

20°. It then drops off again, either sharply or gradually, depending on airfoil and circumstances. Drag is of course never zero, even at 0°, and it increases continuously up to angles of attack so high that lift has long since become (relative to drag) negligible.

One might look at these variables by drawing graphs of lift and drag coefficients against angle of attack. But such a pair involves an unhelpful and unnecessary redundancy since the choice of angle of attack determines both lift and drag. Gustav Eiffel (the famous tower builder) had a better suggestion. Lift coefficient (as ordinate) can be plotted against drag coefficient (as abscissa), with the angle of attack treated parametrically and merely noted on the curve. Figure 11.4 gives an example of such a "polar diagram"[2]—a fast, easy view of several of the characteristics of an airfoil. Besides the obvious horizontal and vertical tangent lines that define maximum lift and minimum drag, two most interesting data practically jump out. First, a line through the origin that's tangent to the curve is a line whose

---

[2] Polar diagrams, a little confusingly, do not use polar coordinates.

slope gives the maximum ratio of lift to drag possible with that airfoil. As we'll see, maximizing that ratio is often a primary objective in airfoil design and operation. And second, the point at which that tangent line touches the curve gives the angle of attack at which the maximum L/D occurs.

To achieve its promised directness and utility, a polar diagram must be constructed from lift and drag coefficients based on the same reference area. So the convention is consistent (and thus rarely mentioned)—plan form or profile area is used for both, with the same maximum projected area however the airfoil might be tilted.

WHAT DETERMINES AIRFOIL PERFORMANCE?

An intimidating number and diversity of variables influence the performance of a lift-producing airfoil, determining (among other things) the form of its polar plot. But none really flies in the face of an intuitively reasonable explanation, so we'll work our way through them on a verbal and semiquantitative level with an intuitive "feel" rather than real analytic rigor as objective. At a next-higher level, the reader might look at Mises (1945) for a purely physical and technological view and at McMasters (1986), Pennycuick (1989), and Norberg (1990) for integration with the relevant biology.

Aspect Ratio

Figure 11.1 includes a definition of the "aspect ratio" of a wing—in simplest form it's the ratio of tip-to-tip length ("span") to average width ("chord"). Thus long, skinny wings have high aspect ratios. Since wings, especially those of organisms, vary widely in shape, it's usually handiest to multiply span over chord by span over span and thus slightly redefine aspect ratio as span squared over area. Even so, the definition has a certain looseness, since for aerodynamic as well as mechanical reasons wings taper in various ways from center to tip; but it proves adequate for practical purposes. Aspect ratio is far more important than one might guess, and its consequences need a little explanation.

Measuring the lift and drag of something approximating an infinitely long wing can be done: one just puts big end plates on a model or extends the model from one side wall to the other in a wind tunnel. If real wings, with aspect ratios much lower than infinity, are compared to an infinite (or "two-dimensional") airfoil, the real ones are inevitably the poorer aerodynamically—they produce less lift and suffer more drag. What a finite wing has that an infinite wing lacks are tip vortices. And, indeed, they turn out to be serious culprits—intuitively quite reasonable since the stubbier the wing, the more it ought to be influenced by what happens at its tips.

The Cost of Lift

But the more basic issue underlying the inferiority of finite airfoils is one that came up in Chapter 4 in connection with jet propulsion. Recall that power output was the product of thrust and jet velocity, while power input was kinetic energy per time as delivered by an engine. The ratio of output to input is the Froude propulsion efficiency,

$$\eta_f = \frac{2U_1}{(U_2 + U_1)}, \tag{4.17}$$

where U_1 is the free stream velocity and U_2 is the jet velocity.

Now consider what the fixed wing of an airplane does. It produces an upward force, lift, by creating downward momentum at an adequate rate—by making a downward jet. But since no downward component is a preexisting part of the free stream ($U_1 = 0$), the Froude propulsion efficiency for staying aloft must be zero. That may strike you as odd, but after all no power is absolutely required to produce a force. A properly designed machine shouldn't use any energy to stay aloft; certainly the chain holding up the chandelier consumes no power and requires no fuel. On the other hand, helicopters and hovering hummingbirds are profligate consumers of energy.

The sad circumstance that determines why power is required is that the spans of the bird's wings and of the helicopter's blades are not infinite. The reasoning is essentially the same as that used to derive the propulsion efficiency. To produce downward momentum at a given rate one might use either a large mass of air per unit time and give it a small downward velocity or one might use a small mass per time and give it a large downward velocity—all that matters is the final mU/t. But the power input to make that momentum flux is energy per unit time, proportional to mU^2/t. So keeping the downward velocity as small as possible is best, which can be done by dealing with a large amount of air. Ideally, with m/t infinite, no power at all would be needed to stay aloft. Thus, since the long, narrow wing intercepts more air per time than does the short broad wing, the former consumes less power. The limit, again, is the infinite wing, which takes an infinite mass per time and imparts to it a negligible downward velocity.

Induced Drag

How does a finite wing signal its requirement for power? The only way it can do so is by having more drag than an infinite wing and thus demanding that the engine (assuming for ease of explanation a separate propeller) impart more momentum per time to the passing (horizontal) airstream. The extra drag is termed the "induced drag"; it is *the drag incurred as a*

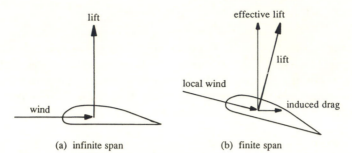

(a) infinite span   (b) finite span

FIGURE 11.5. The origin of induced drag in a wing of finite span. Since the wing gives the air a significant downward push in (b), the net local wind direction crossing the wing is no longer the same as that in (a) or of the free stream. Thus the lift vector is tipped back and resolves into a slightly lower vertical force ("effective lift") and the induced drag.

consequence of producing lift with a less-than-infinite wing. A useful way of looking at the induced drag is to recognize that in giving air a significant downward velocity, the airstream crossing a wing is deflected from the horizontal, as in Figure 11.5. The lift vector, normal to the local airstream, is no longer normal to the overall free stream or (in level flight) to the horizon. It can be resolved into two components—an upward (literal lift) vector normal to the free stream and the induced drag, a horizontal drag vector parallel to the free stream. Referring to a polar diagram, reduction of the aspect ratio pushes a wing's curve toward the right and toward the abscissa, with less lift, more drag, and a lower maximum lift-to-drag ratio. Put another way, induced drag times free stream velocity is the price in power paid for staying aloft with a wing of less than infinite span or aspect ratio—the measure of the energetic inferiority of short, broad wings.

The induced drag and the resulting induced power vary in an odd but not unreasonable way with the speed of the airfoil through the fluid. As speed increases, the airfoil comes in contact with more air per unit time. Since its requirement for lift is nearly constant (equal to the weight of the craft), the airfoil needs to give the air passing it less of a downward deflection at higher speeds. If lift is kept constant as speed increases, the local airstream gets more nearly horizontal, the lift vector gets more nearly vertical, and the induced drag is less. The induced drag, then, is greatest when speed is least; it's almost inversely proportional to the square of speed of the airfoil relative to the undisturbed fluid. This dramatic increase in induced drag at low speeds is one of the main reasons why flying slowly is so expensive.

For a wing of a given aspect ratio protruding directly outward from a fuselage, the induced drag is minimized if the wing has a particular elliptical plan form. It ought to have its maximum chord at the base and taper outward toward the tip, with most of the taper very near the tip. And the wings of many low-speed aircraft have shapes that approximate this ideal ellipse. Natural airfoils are more complex structures, especially those called on to produce thrust as well as lift (Chapter 12), and no elliptical plan form is ordinarily obvious. Recently, though, Burkett (1989) has shown that a tapered wing with its tips swept backward can experience an induced drag under reasonable operating conditions about 4% less than that of an elliptical wing. By contrast with elliptical wings, tapered, aft-swept wings are conspicuous in nature; and they occur in creatures for which high speed seems a primary consideration—in the lunate tails of the fastest marine fish, in cetaceans, in swifts, and in many seabirds (Figure 11.6).

Body Lift

How low can the aspect ratio be and still develop some useful lift? The question may seem impractical if the design of wings is at issue, but it's of relevance in a slightly different context. Rather than being symmetrically streamlined, the bodies of many flying animals are somewhat more convex on their upper surfaces. Hocking (1953) seems to have been the first to suggest that an appreciable fraction of the lift of a fly could come from airfoil action of the body; more recent workers have viewed with some skepticism his claim that up to a third of total lift could be fuselage generated. After all, the aspect ratios for most flying animal bodies are well less than unity, which is pretty horrid. And Jensen (1956) showed that a desert locust, an elongate creature with a great wing area, gets no more than 5% of its lift from the body and has a lift-to-drag ratio for the body of 0.8. Similarly, Dudley and Ellington (1990b) found that at its top speed of 4.5 m s^{-1}, a bumblebee worker's body has a lift of just 4.7% of its weight (see Table 11.1). Nachtigall and Hanauer-Thieser (1992) found a small but significant body lift in honeybees; they give a useful summary of other such data.

But these low figures shouldn't be regarded as definitive since they were obtained on isolated fuselages, and thus they ignore any effect on body lift of the complex airflow generated by attached wings. An extreme case of body lift occurs in swimming rays where, without interaction between beating wings and a pitching and rebounding body, insufficient lift would be generated to produce enough thrust to account for the speed of swimming (Heine 1992).

Body lift may play a substantial role in the "swooping" flight of small birds, in which the wings are periodically folded and held close to the body. In zebra finches, according to Csicáky (1977), as much lift as drag is produced when the body is pitched head up at about 20°. Body lift should be

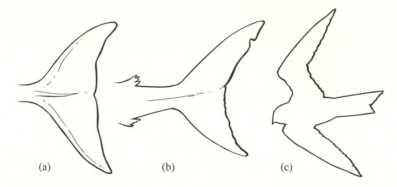

FIGURE 11.6. Natural airfoils whose tips are swept back: (a) the flukes of a whale (from the top); (b) the tail of a shark (from the side); and (c) the wings of a swift.

mainly significant in descending body-gliding. If a bird keeps the initial heading that gave it a negligible body angle at the start of a body glide, its increasingly downward trajectory would increase the body angle and lift as the glide or swoop progresses. Rayner (1977, 1985) has analyzed the phenomenon from a theoretical viewpoint. Another case in which body lift proves useful is in extending the flight of ski jumpers. Leaning forward over the skis and inclining them at an angle of attack of about 25°, a jumper can achieve a lift-to-drag ratio of about 0.3 (Ward-Smith and Clements 1982).

Profile Drag

Besides induced drag, a queer sort of drag that's produced only when a finite airfoil is developing lift, there remain the two familiar sorts of drag— pressure drag and skin friction—that were introduced in Chapter 5. Together they're usually called "profile drag," as mentioned in Chapter 7. Like induced drag, profile drag increases sharply with increases in angle of attack. Like parasite drag (but unlike induced drag) it increases nearly in proportion to the square of speed. As a result, if lift is kept constant by appropriate adjustment of the angle of attack, the overall drag of an airfoil (profile plus induced) passes through a minimum value as speed increases. The power required to offset that drag passes through a minimum value at a speed just a little lower (due to the extra velocity factor in power) than that giving minimum drag, as in Figure 11.7. The energy expended per unit distance traveled (the cost of transport) will be minimized at a slightly higher speed than that giving the minimum power or energy expended per

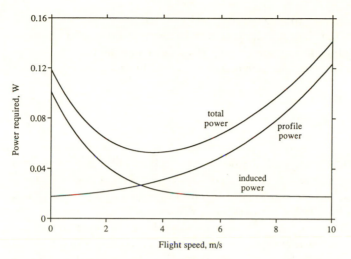

FIGURE 11.7. Profile, induced, and total power as functions of speed for the wing of a bat that's producing a constant amount of lift, from the data of Norberg (1990). (Power is the product of speed and the thrust needed to overcome drag.)

unit time. In short, while total drag goes through a specific minimum, just what constitutes the optimal speed of flight depends a bit on the criterion used (Pennycuick 1989).

One other factor complicates consideration of the relative importance of profile drag and induced drag, and that's (yet again) the Reynolds number. As a consequence of increasing skin friction, drag coefficients begin to rise markedly when the Reynolds number drops below about 1000. Since skin friction is part of profile drag, the latter increases at low Reynolds numbers and may become substantially larger than induced drag. As Ennos (1989a) perceptively pointed out, the long, skinny wings that by minimizing induced drag give best performance in conventional (large scale) fliers will not be optimal for smaller craft such as gliding insects and seeds, statements in Vogel (1981) to the contrary notwithstanding. The problem is that long, skinny wings develop more skin friction since more of their area is near the leading edge where velocity gradients are steepest. So wings of somewhat lower aspect ratio in small natural gliders are entirely reasonable.

Perhaps at this point we ought to categorize all the kinds of drag that have now been considered for lift-producing systems:

1. Parasite (body) drag, made up of skin friction and pressure drag

2. Airfoil drag (not a common designation)
   a. Profile drag, made up of skin friction and pressure drag
   b. Induced drag, the consequence of producing lift
3. Interference drag, the consequence of hooking airfoils on bodies

Stall

Assume you want your aircraft to take off at a fairly low speed. Lift at least a little greater than weight is required, and the obvious way to compensate for the low lift associated with low speed (as in equation 11.1) is with an increase in angle of attack. This might be accomplished by rotating the wings lengthwise with respect to fuselage, by changing the effective pitch of wings with trailing ailerons, or by raising the nose and thus increasing the pitch of the craft as a whole. Lift is nearly proportional to angle of attack, and a poor lift-to-drag ratio matters little in the short run if sufficient power is available for a short takeoff run. So operation at an angle of attack well above that giving the greatest L/D is a fine idea for takeoffs. But lift increases with angle of attack only up to a point. Figure 11.4 shows that at angles above some critical value, lift drops; it may crash abruptly. In physical terms, the flow has begun to separate on the upper surface of the airfoil, the circulation is much reduced, and the airstream is not so effectively deflected downward. So the stall point, the angle of attack giving maximum lift, has considerable practical relevance, and aircraft are commonly equipped with stall-warning consciousness-raisers.

The stall point, though, is not inviolate, and a wide variety of devices for postponing stall to higher angles of attack have been proposed, tested, and even used; Mises (1945) has a particularly useful discussion of stall-deterring and other high-lift devices on low-speed aircraft. Not surprisingly, candidates for antistall devices have been proposed for living fliers. The most frequent of these is the "alula" of bird wings, a group of feathers attached to the anatomical thumb on the leading edge of the wing. Nachtigall and Kempf (1971) found that when the angle of attack is high (30° to 50°), as just prior to landing, the alulae are commonly erected and give lift increases of up to 25%. Their photographs of flow show the expected mechanism—air is persuaded not to separate so readily as it flows across the top of a wing at high angles of attack. The alulae seem to have little effect on lift or drag at lower angles of attack. Blick et al. (1975) reported that leading-edge barbs copied from those of great horned owl wings (and previously presumed associated with the especially quiet flight of owls) could eliminate the sharp drop in lift at an angle of attack of 12°, replacing it with a gentle leveling of lift (at a somewhat submaximal value) out to at least 25°.

Wing Loading

How might lift be varied either in addition to changing angle of attack or without leading to submaximal lift-to-drag ratios? Adjustment of wing area is obviously attractive, so much so that even our stiff technology uses the device. Modern jet transports use some remarkable area-increasing excrescences during landings and takeoffs. But even these are minor compared to the dramatic alterations of wing area in birds as their speeds change—watch one try to keep lift equal to body weight as it lands on a zero-length runway. Pennycuick (1968) has carefully documented the phenomenon in gliding pigeons. For that matter, wing area is altered continuously during the flapping cycle of virtually all birds and bats.

Quite beyond any use as a controlling variable, wing area is an important factor in determining the performance of an aircraft. Slow craft have large wings and fast ones have small wings in order to produce the requisite lift without operation at uneconomical angles of attack. A common measure of the relative area of a wing is the so-called wing loading—the weight of the craft divided by the area of the wings. Successful people-pedaled planes operating at 5 or 6 m s^{-1} have the lowest values among human-built aircraft, about 20 N m^{-2}. A 747 jet transport operating at about 240 m s^{-1} has a wing loading of about 6000 N m^{-2}. Wing loading, though, is a peculiar kind of variable. It does give a direct look at how much lift a given area of wing must produce since lift has to equal weight in steady-state flight, and it bears strongly on the mechanical stresses and structural requirements of a wing. But it can't be considered a direct indication of wing quality—only at constant speed might a higher wing loading be taken to reflect a better lift-producing design.

Also, wing loadings for conventional aircraft aren't directly comparable to those of flying animals since the wings of flapping fliers produce thrust as well as lift and since small flappers beat their wings much more frequently than do large ones. But, for the record, an Andean condor has a wing loading of about 100 N m^{-2} (McGahan 1973), a wren of 25 N m^{-2}, a bumblebee of 50 N m^{-2} (Greenewalt 1962), and a fruit fly of 3.5 N m^{-2}. McMasters (1986) gives lots of other examples.

Consideration of wing loading does, though, draw attention to a basic problem of scaling. The lift needed by a flying animal ought to be proportional to its body weight and thus to the cube of a typical linear dimension. But an airfoil ought to produce lift in proportion to its area (equation 11.1, again) and thus to the square of a linear dimension. Constant wing loading therefore requires that shape change with size and implies that the larger flier will need disproportionately large wings. Among animals, larger ones do in fact have relatively larger wings than do smaller ones, but the differences aren't sufficiently great to keep wing loading constant, as indicated by the figures just cited (Alexander 1971). The main compensation is in flying speed—larger creatures typically fly a bit faster, and since lift varies with the square of speed even a small speed increase goes a long way. An incidental consequence is the greater difficulty of takeoff in larger fliers. This scaling problem implies that angels must be either of unusually low mass, supersonic, or buoyed up by contemplation of their divine mission.

The Effects of Reynolds Number

Conventional aircraft airfoils are designed or have been selected (by a process resembling natural selection) for operation at Reynolds numbers, based on chord length, above about a million. Birds, bats, and insects operate at values below about half a million—about 500,000 for large birds, 50,000 for smaller birds, 5000 for large insects, 500 for medium-sized insects, and 50 for tiny insects. We've already noted that because of increasing skin friction, profile drag gets worse at lower Reynolds numbers. In fact, just about everything about a lift-producing airfoil gets worse as the Reynolds number goes down through this biological range. The ultimate culprit is viscosity through its effects on velocity gradients. With gentler gradients, the carefully contrived shapes of airfoils are increasingly obscured by a cloud of low-speed fluid. With increasing shear stress for any given gradient, the rotational cores of vortices are larger and circulation is more difficult to maintain. So the maximum obtainable lift coefficient decreases. With greater skin friction, profile drag and thus total drag gets greater. Consequently the maximum lift-to-drag ratios achievable get lower. For an airfoil whose best L/D (at infinite aspect ratio) was 80 at $Re = 6,500,000$, the best L/D dropped to 47 at $Re = 310,000$ (Goldstein 1938). Even the stall angle gets lower. An airfoil for which the stall angle was 18° at $Re = 3,300,000$ had a stall angle of 12° at 330,000 and 9° at 43,000.

To a large extent this deterioration is real and probably unavoidable, but some of the data of the kind just cited must simply be the result of using specific airfoils under conditions for which they're inappropriate. Some effort has gone into the design of airfoils for model airplanes (Schmitz 1960), which might serve better as comparisons for the wings of birds. At $Re = 42,000$, a moderately cambered flat plate whose thickness was 5% of its chord proves far better than a conventional airfoil—the cambered plate has a best lift-to-drag ratio of about 11.0 while the conventional airfoil reaches only 4.5. Only in minimum drag coefficient are they similar. Conversely, at $Re = 168,000$, the conventional airfoil is better in all respects. Further comparisons and discussion of the matter are provided by Lissaman (1983)—but the reader should bear in mind that when it comes to Reynolds numbers, "low" may mean quite different things to engineers and biologists.

At high Reynolds numbers, even minor surface irregularities may have a great effect on the performance of an airfoil, and competitive glider pilots take fastidious care of the tops of the wings of their craft. But airfoils such as insect wings, selected for operation at low Reynolds numbers, are nothing if not irregular. While exceedingly thin (a fruit fly wing has a mass of about a gram per square meter), insect wings are usually somewhat corrugated. Not only do veins protrude, but the veins are not coplanar, giving greater stiffness to these light structures that move at several meters per second and change directions several hundred times each second. In large insects eddies may form in the pleats, as suggested by Newman et al. (1977) for dragonfly wings. Still, considering the magnitude of the irregularities, their effects seem usually to be fairly subtle. Nachtigall (1965) measured only small differences in the polar plots of butterfly wings with and without scales. And Rees (1975) compared the performance of a model of cross sections of a hover fly (syrphid) wing to a smooth model representing an envelope around the corrugations. Again little difference appeared, and his polar plots are nicely consistent with my data (Vogel 1967b) on fruit fly wings and the data of Dudley and Ellington (1990b) on bumblebees. Apparently at his Reynolds numbers of 450 and 900, the flow treats the folds as if filled.

The Limits of Circulation

At some point the deterioration of airfoil performance as Reynolds numbers drop will preclude flight using this circulation-based scheme for producing lift. As we'll see shortly (and is obvious by a glance forward to Table 11.1), the maximum achievable lift-to-drag ratios are substantially lower in smaller creatures. To how low a Reynolds number can flight by the present mechanism be pushed? The only data I know of for airfoils at really low Reynolds numbers are those of Thom and Swart (1940) on cambered airfoils at $Re = 10$ and $Re = 1$. At $Re = 10$, their best L/D was only 0.43—the drag was over twice the lift—and it occurred at an angle of attack of 45°. At $Re = 1$, the best L/D was 0.18, again at about 45°. Despite the very high angles of attack, the deflection of the wake of the airfoil was never more than about 11°.[3] In this world, fluid is highly resistant to being set into circulation by any fixed (non-Flettner) airfoil. And, as far as I know, the use of the Flettner type of rotating airfoil has never been investigated at Reynolds numbers this low. Nor can I think of a biological case. Flettner devices seem to exist only among plants, and really small wind-dispersed

---

[3] Still, one ought to bear in mind that the airfoils used in this investigation were chosen fairly arbitrarily, certainly with none of the fastidious testing of natural selection working on a major functional feature.

seeds apparently prefer drag maximization to lift production as a means of slowing their descents.

Other mechanisms for generating lift are available at Reynolds numbers around unity, but most of the possibilities are unpromising. We'll return to the issue in Chapter 15.

MORE ON BIOLOGICAL AIRFOILS

The discussion so far has looked at lift-producing airfoils in general; let's now turn more specifically at the performance of those of organisms. They're quite a diverse group. Their Reynolds numbers extend over four or five orders of magnitude; the equivalent for all human-carrying subsonic aircraft is less than three orders. And they serve a wide range of functions, with the production of lift through circulation about the only common feature—in at least one case we'll see that even the generation of force is irrelevant. Finally, they've evolved from a wide variety of precursors in a large number of animal and plant groups. I mean to emphasize these latter points to suggest that many biological lift-producing airfoils have escaped identification as such because of insufficient appreciation of how they might look and what they might do.

A Lift-producing Sand Dollar

First, a brief look at what must be the least obvious use of an airfoil to have emerged so far. A certain sand dollar, *Dendraster excentricus*, living subtidally on the Pacific coast of North America, routinely adopts a most un-sand-dollar-like posture when exposed to currents. It buries its anterior edge in the sand and erects its posterior in the flowing water above— instead of lying, half-buried, like a bump on the bottom. And it likes company: these sand dollars cluster at densities of 200 to 1000 per square meter (Figure 11.8b).

While this peculiar habit was well known, its function remained enigmatic until O'Neill (1978) showed that each animal formed a lifting body. A sand dollar, flat or slightly concave on its lower (oral) surface and convex on its upper (aboral) surface, develops a substantial circulation (Figure 11.8a). Since one end is in the sand, it has only one tip vortex and an effective aspect ratio about double that of a dollar in a free stream. The lift, of course, is directed horizontally, and as a force proves to be of little interest. Nor is drag of direct consequence. What mainly matters is that the circulation developed by the cambered profile brings food particles closer to the tiny feeding appendages on its oral surface. Aggregation of individuals turns out to be valuable through mutual improvement of feeding currents; usually the oral surface of one individual faces the aboral surface of its

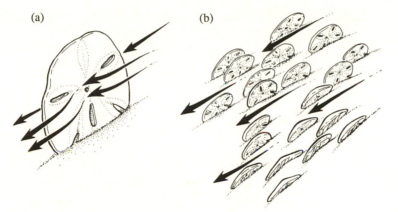

FIGURE 11.8. Flight isn't the only use for lift. (a) At least one sand dollar stands on its anterior edge. (b) Dense groups of these dollars orient with respect to the flow and each other to operate as a colonial, multiwinged craft and bring feeding currents closer to their oral surfaces.

nearest neighbor, much like the wings of a multiwinged aircraft. The sand dollars seem to regulate their density to maintain optimal gaps between individuals, adjusting their spacing from 10 to 100 mm in proportion to the square of the average surge velocity. As hydrofoils with identical leading and trailing edges that get good circulation at near-zero angles of attack, they're well suited to take advantage of bidirectional wave surge through their mutually formed channels. Their tuning to flow is impressive—populations from sheltered locations are slightly more cambered and have slightly higher lift coefficients than those from more exposed localities.

Wings

Figure 11.9 gives the polar plots and Table 11.1 the salient aerodynamic characteristics of a variety of animal airfoils, mainly obtained from measurements on isolated but real specimens in wind tunnels. Several of the points made earlier can be seen in these data. For the wings,[4] the greater minimum (profile) drag at lower Reynolds numbers is clearly evident, as is the increase in the angle of attack that gives the best ratio of lift to drag. But

---

[4] Insects have two pairs of wings. In locusts, most of the lift and thrust are produced by the hind wings; in bees, each functional wing consists of fore- and hind wings hooked together; in beetles, the elytra, or forewings, commonly function as fixed wings ahead of the propellerlike hind wings; in flies, the hind wings have been modified into tiny clublike organs that have no direct aerodynamic function or consequences.

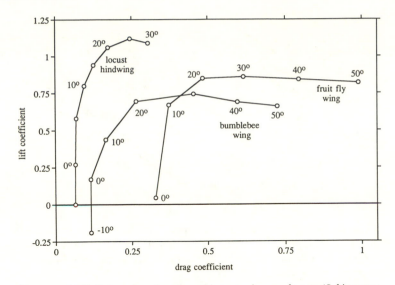

FIGURE 11.9. Polar plots for several insect wings: a locust (*Schistocerca gregaria*) hind wing (Jensen 1956), the linked fore- and hind wings of a bumblebee (*Bombus terrestris*) (Dudley and Ellington 1990b), and a fruit fly (*Drosophila virilis*) wing (Vogel 1967b).

the data have to be viewed with a certain cautious skepticism. Bird wings in wind tunnels usually do rather badly by any reasonable aerodynamic standards—the maximum lift-to-drag ratios of 3.8 for hawk and duck and those of 2.7 obtained on another duck and a sparrow by Nachtigall and Kempf (1971) try the adaptationist faith of this biologist. The problem almost certainly comes from the subtlety of feather configuration and its better feather management by bird than biologist—something that came up earlier (Chapter 7) in connection with the parasite drag of bird bodies. As we'll see in a few pages, such data can be estimated from tracks of live, gliding birds in the field or wind tunnel; and these versions look quite a lot better. Maximum steady-state lift coefficients around 1.5 to 1.6 have been measured on gliding birds and bats (Pennycuick 1989; Norberg 1990). Maximum lift-to-drag ratios of 8.5, 10.4, 13.7, 15.5, and 18 have been reported for petrels (Pennycuick 1960), falcons (Tucker and Parrott 1970), condors (McGahan 1973), vultures (Pennycuick 1971), and albatrosses (Pennycuick 1982), respectively. Moreover, these latter figures are for entire animals and so are deflated by inclusion of parasite drag in their denominators.

By contrast, the data for insect wings are as tidy a set of numbers as one might wish, especially since they represent the results of as many investiga-

TABLE 11.1 CHARACTERISTICS OF BIOLOGICAL AIRFOILS.

	Re	AR	$C_d\,min$	$C_l max$—@α		$L/Dmax$—@α		Source
Buteo lineatus	10,000 to							
(hawk) wing	50,000	6.0	0.074	1.0	25°	3.8	6°	1
Aix sponsa								
(wood duck) wing	"	6.2	0.096	0.90	20°	3.8	8°	1
Chaetura pelagica								
(swift) wing	"	7.8	0.030	0.80	8°	17.0	5°	1
Schistocerca gregaria								
(locust) hindwing	4000	5.6	0.06	1.13	25°	8.2	7°	2
Tipula oleracea								
(crane fly) wing	1500	6.9	0.10	0.82	35°	3.7	13°	3
Bombus terrestris								
(bumblebee) wing	1240	6.7	0.13	0.78	30°	2.48	15°	4
Melalontha vulgaris								
(beetle) elytron	1100	4.2	0.17	0.83	25°	1.9	15°	5
Drosophila virilis								
(fruit fly) wing	200	5.5	0.33	0.87	30°	1.8	15°	6
Petarus breviceps								
(flying "squirrel")	100,000	1.0	0.10	0.6	34°	2.0	34°	7
Amusium japonicum								
(scallop) shell	79,000	1.3	0.036	0.85	25°	5.47	12°	8
Bombus terrestris								
(bumblebee) body	6200	0.3	0.34	0.22	40°	0.33	32°	4
Velella velella (sailor-by-								
the-wind) sail	1000	1.4	0.2	0.95	35°	0.23	10°	9
Alsomitra macrocarpa								
seed-leaf	4000	3.5	0.04	0.87	14°	4.6	—	10
Acer diabolicum								
(maple) samara	1400	4.3	0.1	1.6	19°	3.3	12°	11

SOURCES: (1) Withers 1981; (2) Jensen 1956; (3) Nachtigall 1977b; (4) Dudley and Ellington 1990a,b; (5) Nachtigall 1964; (6) Vogel 1967b; Zanker and Gotz 1990 on *D. melanogaster* is similar; (7) Nachtigall et al. 1974, Nachtigall 1979b; (8) Hayami 1991; Millward and Whyte 1992 on *A. pleuronectes* is similar; (9) Francis 1991; (10) Azuma and Okuno 1987; (11) Azuma and Yasuda 1989.

tors as species and they span some thirty-five years. Still, even these data are likely to represent underestimates of actual flight performance. For wings that normally flap, steady-state measurements lack several crucial elements of reality. Even beyond nonsteady effects (to get attention in the next chapter), flapping causes the wind direction to vary spanwise, something quite hard to simulate in static measurements. I'd make a strong pitch for a presumption of asymmetrical bias in such experimental results. For a machine designed by the evolutionary process, the only appropriate operating conditions are its natural conditions, and data obtained under other circumstances are more likely to show substandard than superior performances.

In comparing the five insect wings, we see the pernicious effect of low Reynolds number on performance discussed earlier. Drag gets much worse relative to lift as the Reynolds number decreases and the drag becomes almost all profile drag (as indicated by $C_d$min) rather than being about half induced drag. Thus making longer and more slender wings is of little help, although it might still give benefit by reducing the interaction of the wind of the beating wings ("prop-wash") and the fuselage. Concomitantly, the best lift-to-drag ratio drops. More interestingly, this shift of polar curves to the right pushes the best operating point up to higher angles of attack, essentially to the stall point for an insect the size of a fruit fly. The benefits of stall-resisting arrangements ought to be greater; and, indeed, some evidence points to such devices in small insects. A fruit fly wing doesn't stall in the usual sense; its lift just levels off above an angle of attack of about 20°. What seems to be happening is that with so much drag, very little momentum flux is left behind the wing to be deflected downward. But a flat or cambered plate at the same Reynolds number does stall (although with less dramatic effect than for wings of larger craft), and the lift coefficient drops at angles above about 25°. Just what's different about a real wing isn't clear, although some indirect evidence implicates the tiny hairs (microtrichia) on the surface of the wings (Vogel 1967b). One way or another, such a wing clearly doesn't quite work like an amorphous paddle in a cloud of attached air.

Lifting Bodies

Table 11.1 also gives the characteristics of a gliding phalanger, *Petaurus breviceps*, an engaging marsupial of extraordinarily close resemblance to the North American flying squirrel, *Glaucomys volans*; its polar plot is quite similar to that for the fruit fly wing in Figure 11.9. These data were obtained on whole, performing animals and are probably quite reliable. The interesting point about this airfoil is its very low aspect ratio (Figure 11.10a), a rather bad state of affairs if these animals are viewed as purely aerodynamic devices. The low ratio limits maximum lift and increases induced drag—what pushes its polar curve to the right is increased induced drag rather than profile drag as in small insects. The result, though, is the same—the best operating point is essentially the stall point. And anything that either postpones stall to higher angles of attack or reduces the suddenness of separation and stall should be of substantial benefit. Nachtigall (1979a,b) has some evidence that the phalanger has special fur with such a function.

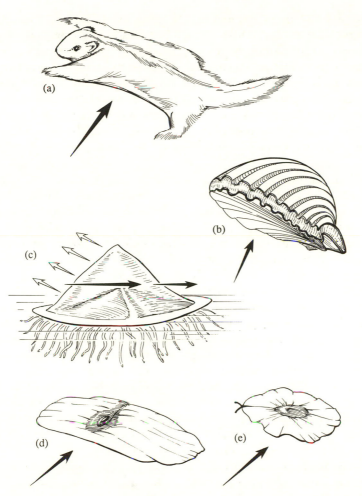

FIGURE 11.10. Lifting bodies in nature: (a) a gliding phalanger, *Petaurus breviceps*; (b) a scallop, *Amusium japonicum*; (c) the by-the-wind-sailor, *Velella velella*; (d) the seed-leaf of the Javanese cucumber, *Alsomitra macrocarpa*; (e) the fruit of an elm, *Ulmus*.

The table gives, as well, the performance of a scallop (Figure 11.10b). Scallops can swim briefly but surprisingly rapidly (well above half a meter per second for some) and are denser than seawater. Swimming is by jet propulsion through apertures near the hinge; they make some use of vectored thrust at the gape to get the upstream end initially elevated. Some scallops have markedly asymmetrical shells, with the functional upper one rounded and the functional lower one flat (anatomically these are left and

251

right valves), and these asymmetrical scallops are in general the better swimmers (Millward and Whyte 1992). The best seem to be several species of *Amusium*, although not by any overwhelming margin. *Amusium* both looks and behaves like quite a competent airfoil; the main limitation on its performance is the low aspect ratio, a variable that none of the scallops seem to go to any great adaptive lengths to improve relative either to each other or to other bivalves. According to Hayami's (1991) data and calculations, *Amusium japonicum* could manage level swimming without any downward jet thrust at a speed of only 0.45 m s^{-1} and an angle of attack of just 5.3°.

And then there's the bumblebee fuselage as a lifting body. We talked about body lift a few pages back; it seems to be significant without being a really big deal. What ought to be mentioned is that a bumblebee can't just pick a body angle (pitch) that gives the best performance as a lifting body. In many insects, including bumblebees, the pitch of the body is adjusted to set the pitch of the plane of beating of the wings and thus the ratio of lift to thrust. At high speeds, where body lift might be greatest, the body is pitched down to a nearly horizontal posture (Dudley and Ellington 1990b).

There is also *Velella*, perhaps the most off-beat of animal airfoils. *Velella* (Figure 11.10c) is a pelagic, colonial coelenterate, a bit like the better-known Portuguese man-of-war, *Physalia*. But instead of an emergent balloon, it has a thin, leaflike sail obliquely mounted above a less conspicuous float and skirt, beneath which dangle its tentacles. *Velella* reportedly can sail as much as 63° off the direction of the wind as a result of the lift of the sail and its asymmetry above the float—it's about the closest thing to a sailboat that nature has contrived. But, as Francis (1991) has shown, it's not a really great sailboat; on the other hand, what tack can one take to reach for adaptive advantage for good cross-wind sailing?

Two botanical airfoils are included in Table 11.1, one the seed-leaf of the Javanese cucumber, the other a samara (the fruit) of a maple. The former is a simple glider and will be discussed shortly; the latter is an autogyrator and will reappear in the next chapter.

GLIDING AND SOARING

An unpowered airfoil can't travel horizontally through still air at a steady speed—that's why engines and muscles are so useful. Before putting power sources to work, though, let's consider the possibilities of getting a free ride, first from gravity and then from atmospheric motion. We'll follow the common practice of calling the first scheme "gliding" and the second "soaring." Gliding, to be explicit, is defined as a situation in which an airfoil moves through still air, while losing altitude just rapidly enough to maintain a steady course by producing a vertical force the same as its weight. What's important is that no forces are unbalanced, so the aerodynamic resultant force, the combination we ordinarily separate into lift and drag, precisely balances the weight of the craft.

How to Glide

Knowing that weight and aerodynamic resultant are equal and opposite, the angle at which the craft descends can be easily derived, as shown in Figure 11.11. Both resultant force (R) and weight (W) are obviously vertical. The oncoming wind (U), equal and opposite the path of the craft, and the drag (D) are both obviously perpendicular to the lift (L). By elementary geometry the angle between resultant and lift vectors must therefore be the same as the angle between horizon and wind. The former is the angle whose cotangent (the reciprocal of the tangent) is the ratio of lift to drag, while the latter is the angle of descent, called the "glide angle." In short, the glide angle, θ, is set by the lift-to-drag ratio:

$$\cot \theta = \frac{L}{D} = \frac{C_l}{C_d}. \tag{11.2}$$

Isn't it nice that something as simple and familiar as the lift-to-drag ratio determines something as critical as the glide angle—to minimize the glide angle, just maximize the lift-to-drag ratio. Incidentally, we now know how to determine lift and drag coefficients on freely gliding birds. One needs only glide angle, flying speed, wing area, and weight; and by equations (11.1) and (11.2) the coefficients fall out.

If the air is otherwise still, then minimizing the glide angle maximizes the distance a simple glider will go before reaching the ground. An albatross, with a lift-to-drag ratio of 18, released 1 km above ground, could glide a horizontal distance of 18 km. A phalanger (*Petaurus*) with a best lift-to-drag ratio of 2.0 would go at most 2 km. Long, skinny wings and the consequent low-induced drag are clearly advantageous (at least where Reynolds numbers are high enough so profile drag isn't a big concern). Whether bird or person-carrying craft, good gliders typically have wings of high aspect ratio. One high-performance sailplane with an aspect ratio of 20 has a lift-to-drag ratio of 39, and so can go 39 km horizontally for each kilometer of descent (Tucker and Parrott 1970).

This business of distance traveled versus descent is counterintuitive in at least one important respect. For moderate and high Reynolds numbers, speed affects both lift and drag in much the same manner, so their ratio isn't especially speed dependent. As a result, the glide angle is also largely independent of speed, and a very heavy glider descends along nearly the same path as a light one! Weight, though, must be balanced by lift, and the

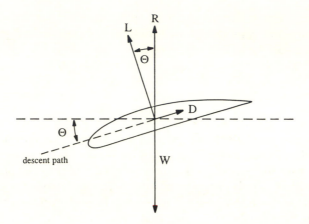

FIGURE 11.11. The relationship among the glide angle (θ), the lift, the drag, and the weight supported by an airfoil in a steady glide.

latter is roughly proportional to the square (or a little less) of speed. While the heavier glider may fly the same path, it must of necessity travel faster, so it covers its horizontal distance in less time. A rapid descent is usually not desirable, since gliders are typically looking for opportunities to soar. Therefore most gliders, living or not, avoid excessive weight. Sometimes, though, staying aloft poses hazards. Military gliders were usually heavily loaded affairs, balancing the pitfalls of high-speed impact against the inefficiency of carrying only light loads and the insecurity of a long, slow flight over hostile territory.

The relationship between glide angle and lift-to-drag ratio explains, in part, why gliding animals are fairly large. Small size means relatively more profile drag and, as we've seen, a lower maximum L/D. So small size almost invariably implies a higher (worse) glide angle. An albatross may descend at about 3°; a falcon can do no better than 5.5°, a pigeon about 9.5°. Small insects are far worse—even if the fuselage contributed no drag at all, a bumblebee would descend at 22° and a fruit fly at nearly 30°. Simple gliding, based on circulation and lift, just isn't very good at low Reynolds numbers. In general, the larger members of most animal groups are the more recently evolved—large size constitutes a specialization, and likely ancestors tend to be unheroic in stature. Exceptional are those groups that engage in active flight—birds, bats, and insects—for a reason that the present discussion makes persuasive. If in each case active, powered fliers evolved through ancestral forms that were exclusively or mainly gliders, then those ancestors ought to have been fairly large.

The Glide Polar

A convenient way of viewing the performance of a glider is with a so-called glide polar, a curve on a graph that has flying speed (airspeed, not necessarily ground speed) on the abscissa and sinking speed (increasing downward) on the ordinate, as in Figure 11.12. On such a graph, any straight line extending from the origin is a line of constant lift-to-drag ratio. The one tangent to a craft's curve (the line of least slope) gives, at the tangent point, *the flying speed and sinking speed at the minimum glide angle*. Craft are characterized by curved rather than straight lines mainly as a result of the way drag varies with speed—again at low speeds induced drag dominates, while at high speeds profile drag dominates, with an overall minimum (and hence maximum C_l/C_d) in between. Increasing the weight of the craft would shift its curve rightward and downward without much change in the slope of the tangent line.

The glide polar permits easy comparison of several other factors as well. Time aloft may be more important than maximum still-air glide distance. A horizontal line tangent to the curve gives, at its tangent point, *the flight speed at which sinking speed is minimized and thus time aloft maximized*; at its intersection with the abscissa it gives the specific sinking speed. Notice that the greatest time aloft is achieved at a lower speed than the least glide angle, so a slow glider can do as well as a faster one that has a better lift-to-drag ratio. If time aloft is the important factor (as we'll see that it is in soaring), then birds don't look quite so bad when compared with sailplanes. By this criterion even large insects aren't disastrous gliders. A desert locust, despite a lift-to-drag ratio of only 8.2 for a hindwing (and less for the entire insect) descends at 0.6 m s^{-1}—in the same range as a vulture or sailplane (Jensen 1956). A monarch butterfly has an overall lift-to-drag ratio of 3.6 and therefore a steep glide angle of 15.5°, but it travels at only 2.6 m s^{-1} and consequently descends at only 0.68 m s^{-1}—again in the same range (Gibo and Pallett 1979).

Another variable that emerges from the glide polar is the *minimum glide speed*. It's the left end of the curve, the point at which combination of the maximum wing area (if variable) and the maximum lift coefficient provides just enough lift to balance the weight. One can encapsulate the conditions for minimum glide speed in a formula recognizing that lift must equal weight and using the definitions of wing loading (W/S) and lift coefficient:

$$U_{\min} = \left(\frac{2W/S}{\rho C_{l\max}} \right)^{1/2}. \tag{11.3}$$

Just where an animal might choose to operate on its glide polar depends in part on what the wind is doing. If the animal wishes to maximize distance traveled and is assisted by a tail wind, a lower flight speed than that which

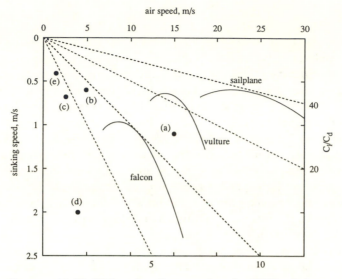

FIGURE 11.12. Glide polars for several craft—SHK sailplane, vulture, and falcon (Tucker and Parrott 1970). Also shown are single data points for some other gliders: (a) Andean condor; (b) desert locust; (c) monarch butterfly; (d) flying phalanger; and (e) Javanese cucumber seed-leaf.

would maximize L/D is the best tactic. If it faces a head wind less than its flying speed, it's best off picking a higher flight speed than that maximizing L/D. If it faces a head wind greater than its flying speed, the best course is to land as quickly as possible (Tucker and Parrott 1970).

The same considerations, of course, hold for plants. The most famous simple glider is that of the seed-leaf of the Javanese cucumber, *Alsomitra* (= *Zanonia*) *macrocarpa* (Figure 11.10d), for which data are given in Table 11.1. The gliding seed formed the model for a successful series of gliders around the turn of the century (Bishop 1961)—the death of Otto Lilienthal in a birdlike hang-glider focused attention on the need for stability and the clear instability of birds.[5] The *Alsomitra* seed-leaf, with a span of about 140 mm and a weight of 210 mg , glides at 1.5 m s^{-1} as it descends at 0.41 m s^{-1}—at a lift-to-drag ratio of 3.7 and an angle of 15°, according to Azuma and Okuno (1987). They found that a seed-leaf could reach L/D = 4.6 if forced to elevate its nose a bit, but it then suffered substantially reduced

[5] The instability of a bird is almost certainly concomitant with its maneuverability and must have evolved in parallel with highly sophisticated controls, according to the aerodynamicist-turned-evolutionary biologist, John Maynard Smith (1952). A seed leaf, by contrast, must be intrinsically stable. But the model didn't lend itself to the addition of an engine, and the design was returned to historians and botanists.

stability. Of particular interest in the present context is that normal operation, at $C_l = 0.34$, is between the points on polar plot and glide polar that give minimum glide angle ($C_l = 0.27$) and minimum rate of descent ($C_l = 0.53$).

But simple gliding is a lot less common among wind-dispersed, lift-producing seeds than are autogyrating, autorotating, or their combination. McCutchen (1977a) argues (and I see no reason to disagree) that simple gliders are inferior in the presence of any turbulence. So they should be useful mainly in the still air of the interior of a rain forest. In fact, smaller and less elegant plain, winged seeds with some central concentration of mass are not all that uncommon (see, for instance, Augspurger 1986, 1988); examples are the fruits of the hoptree, *Ptelea trifoliata*, and of the elms (*Ulmus*) (Figure 11.10e). These seem to be simple gliders, if smaller and less stable than *Alsomitra*.

Reduction in sinking speed by lift production is, in any case, a great deal less common among plants than is reduction by drag production. The two schemes seem to represent a major fluid-mechanical dichotomy since lift is most useful when drag is least; that is, a high L/D makes the present scheme work best. Matlack (1987), looking at the situation from a comparative viewpoint, pointed out that plumed drag maximizers are more common at weights under 45 mg, while samaras and other lift maximizers more commonly weigh more than 45 mg. In addition, drag maximizers are more common among herbs and lift maximizers among trees, perhaps reflecting differences in optimum investment of mass per dispersal unit. That larger size makes gliding work better is, of course, the same argument made earlier to rationalize the evolution of small active fliers from large ancestral gliders.

Gliding and Parachuting

Gliding of some sort is practiced by many animals other than birds, bats, and insects—a few are shown in Figure 11.13. Flying squirrels and phalangers have already been mentioned; detailed information about the flight-related morphometrics of the former has been collected by Thorington and Heaney (1981). In addition there are flying lemurs (*Cynocephalus*), several separate lineages of flying frogs (*Rhacophorus*, *Phrynohyas*, etc.), a colubrid snake (*Chrysopelea*), and several geckoes and lizards, among which the best is certainly the aptly named *Draco volans* (Norberg 1990). Gliding ability is quite variable, and Oliver (1951) suggested that a distinction be made between gliding and parachuting, based on whether the still-air descent angle is greater or less than 45°. The point of division isn't entirely arbitrary—where data exist, it appears that animals with descent angles over 45°, such as the arboreal lizard, *Anolis carolinensis* (68°),

FIGURE 11.13. Gliding animals from groups that don't go in for active, flapping-wing flight. A frog, *Rhacophorus*; a lizard, *Draco*; and a fish, one of the Exocoetidae.

and the tree frog, *Hyla venulosa* (57°), don't have particularly obvious aero-dynamic structural adaptations (Oliver 1951).

Draco, the flying lizard of the Philippines, has a broad patagium supported by rib extensions. Hairston (1957) estimated its glide angle at 11°, certainly well within the range of flying squirrels. A Malayan flying frog, *Rhacophorus nigropalmatus*, with huge, webbed hands and feet, glides at 25° to 37°, moving about 5 m s^{-1} (Emerson and Koehl 1990). Curiously, gliding vertebrates are especially diverse and common in Indo-Malaysia; Dudley and DeVries (1990) argue that gliding is especially useful in the unusually high dipterocarp forests found there.

Probably the best gliders among animals that don't also actively fly are flying fish. Some (the Exocoetidae) use only their pectoral fins as wings. Larger ones (such as *Cypselurus*) use pelvic fins as well to form a staggered-wing biplane; according to Davenport (1992), the main contribution of the pelvic fins is to assure pitching (fore and aft) stability since the pectoral fins are in front of a fish's center of mass. *Cypselurus*, among others, initially leaves the hypocaudal lobe of the tail in the water, beating fifty to seventy times per second. During this "taxiing" phase, it goes from emersion speed of about 10 m s^{-1} (in a large specimen) to takeoff speed of 15 to 20. Single flights may be up to 50 meters long, and repeated flights in a sequence can extend up to 400 meters. The aspect ratios of flying fish are usually high (3 to 20); estimates of lift-to-drag ratios also run high but are almost certainly unreliable. One can't do a quick calculation from glide angle since these creatures don't do anything close to simple gliding. They lose little altitude during a glide, since they don't have much to start with, and probably lose speed instead. They glide close to the ocean's surface, so ground effect probably augments performance considerably. In addition, they may practice slope soaring, taking advantage of local updrafts upwind of waves. Much of this information comes from Fish (1990), the other modern source on the subject.

Soaring

Atmospheric motion, of course, includes much more than just horizontal wind. And for organisms, the speeds of air movement are of the same order as or higher than their own flight speeds. The combination must, on one hand, enforce respect for local air movement but, on the other, permit all kinds of schemes for soaring in the moving medium. Their general utility is obvious—air in motion with respect to the ground provides a source of energy; and flight, per unit time, is the most energy-intensive form of animal locomotion. In simple gliding, an organism heads earthward, the only real question being its time, speed, and place of arrival. In soaring, as commonly defined, altitude is maintained or even gained by using the energy of the wind. The problems involved in soaring can perhaps best be appreciated by reading an instruction manual for sailplane pilots such as Conway (1969).

Two general forms of soaring can be usefully distinguished. In the simpler, "static soaring," a craft or creature takes advantage of a region in which air is moving upward. Somewhat trickier is "dynamic soaring," which involves no upward air movement but only a temporal or spatial gradient in wind velocity from which an animal extracts the power necessary to stay aloft. Good accounts of soaring are given by Cone (1962), Pennycuick (1972, 1975), and Norberg (1990); I'll just briefly describe the possibilities in order to emphasize their diversity.

The simplest sort of static soaring is "slope soaring," or flying in air that's moving upward along the side of a hill. The updraft may extend above the crest of the hill and may have a fairly complex structure, but the successful craft finds a place where it can sink continuously with respect to the local air without sinking with respect to the earth—that is, where the magnitude of the vertical component of the wind is equal to or greater than the sinking speed of the craft (Figure 11.14). Slope soaring is the usual kind practiced by hang-gliders. Petrels and albatrosses do it along ocean waves, as do (probably) the flying fish mentioned earlier (Pennycuick 1982). Sometimes

FIGURE 11.14. Slope soaring—if air moves up a hill, then a bird can descend with respect to the local air without descending with respect to the earth. Thus it can maintain position without active propulsion.

a version of slope soaring is possible in waves and vortices in the lee of a range of mountains, and it may be possible in the lee of ships and elsewhere. Birds have been reported to move long distances along ridge lines, almost certainly by slope soaring. The most careful analysis of free-flying birds is that of Videler and Groenewold (1991), who measured wind profiles as well as bird performance for kestrels hanging in fixed positions over a Dutch sea dike; they calculated that by hanging rather than actively flying, a kestrel can reduce its energy expenditure by fully two-thirds.

A second kind of static soaring is "thermal soaring"; it was described in the last chapter when we considered vortices. Yet another is what might be called "sea anchor soaring," suggested for petrels by Withers (1979). Petrels get their name (from "Peter") from their apparent ability to walk on water. What they're more likely doing is getting enough drag from their feet in the water to be pushed only slowly downwind as they face into the wind with outspread wings. A bird, then, can gain enough lift (with a little ground effect helping) to get the body (except the feet) entirely out of the water—it operates much like a kite (the toy, not the bird), which stays aloft only as long as a tether is provided to allow it to maintain airspeed. The trick may also be used by another kind of seabird, the prions (Klages and Cooper 1992), which suspension feed on copepods—they get their bodies nearly out of the water with the lift of spread wings while keeping heads submerged for feeding.

Of the possible types of dynamic soaring, the best understood is that done in the altitudinal wind gradient near the surface of the ocean. Again, note that a vertical *gradient* in wind speed is used, not a vertical component

FIGURE 11.15. Dynamic soaring in a wind whose horizontal speed increases with altitude. The bird alternately ascends and descends, extracting energy from the gradient. The bird reverses its heading at the marked points.

of velocity. This "gradient soaring," as done by an albatross, has two alternating phases—a downwind, downward glide and an upwind, upward glide (Figure 11.15). A bird well above the ground or ocean glides downwind, gaining airspeed as it loses altitude and encounters slower, lower air. Nearly at the lowest point it turns into the wind, and dives a bit further. It then reinvests its high airspeed in the upwind ascent, which brings it into air that's moving increasingly rapidly with height. The bird then turns downwind again, and repeats the process. The overall motion is a series of vertical loops that progress downwind.

Other kinds of dynamic soaring are also possible—a bird ought to be able to use a temporally unsteady wind to soar by flying through gusts. Similarly, it ought to be able to fly in and out of breeze fronts and other such irregularities. A large amount of energy is dissipated in the lowest part of the atmosphere, and extracting some of it is possible with appropriate machinery and behavior.

The Thrust of Flying and Swimming

T HE ORIGIN OF drag came in for attention quite a few chapters back; the origin of lift occupied much of the last two chapters. Our tour de force has one other major variable yet to explain. Real airfoils can make lift only by paying some price in drag. So what we now have to do is to make some antidrag to counteract its nefarious (except for parachutes) influence. The polite name for antidrag is *thrust*.

THRUST FROM FLAPPING

The notion of an oncoming wind that wasn't horizontal proved central in explaining how simple gliding works. In gliding, though, that oncoming flow was never far from either the horizontal or the line of progression of the craft. Generating thrust requires that an airfoil encounter a flow decently distant from the line of progression of the craft. As will become clear shortly, a craft can't produce thrust without some cross-course wind. Such a situation can be made to happen if the airfoil moves with with respect to the craft—it must rotate, flap, wiggle, undulate, or otherwise move about. But if it does move in an appropriate manner, then, mirabile dictu, thrust is exerted on the craft by a structure that, in its own frame of reference, always suffers drag, usually produces lift, and never feels thrust.

The Origin of Thrust

Before turning to reciprocating airfoils such as flapping wings and beating tails, let's first look at a simpler case. Consider an element of a rotating propeller, the latter turning about an axis that happens to coincide with the overall movement of the craft, as in Figure 12.1. What we mean by an element is a cross section normal to the long axis of the propeller, usually drawn as if the craft were behind the page, the propeller sticking through it, and the craft progressing from right to left—so the free-stream flow is, as usual, left to right. The oncoming wind seen by this blade element, U_w, is the resultant of two component winds, U_f, the wind due to the craft's forward motion (the free-stream flow) and U_r, the wind due to the rotational (here downward) motion of the propeller blade.

If the airfoil, this propeller blade element, is set at an appropriate angle to the line of progression of the craft (usually called either the "pitch" or

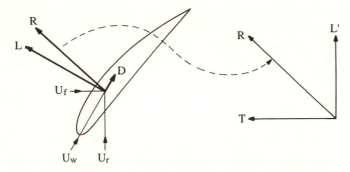

FIGURE 12.1. The origin of thrust. If the overall oncoming wind strikes a blade element from a sufficiently nonhorizontal angle, then the aerodynamic resultant may be tipped forward (upstream) of the vertical. If so, it has a component opposite the free stream flow—negative drag, or thrust. The thrust component appears explicitly when the resultant is reanalyzed, as at the right.

the "angle of incidence"), it can have an angle of attack with respect to the resultant oncoming wind that gives it good lift and reasonably low drag. But, it must be emphasized, such lift and drag are defined relative to the oncoming wind—drag parallel and lift normal—and the oncoming wind is anything but horizontal. So we immediately combine the lift and drag, concentrating on what's real: the resultant force on the airfoil as a result of its shape, pitch, and motion.

The resultant force may (as in the illustration) be tilted forward, which means that it has a component opposite the progression of the craft. By any reasonable definition such a component is thrust. Thus, even though the airfoil knows only lift and drag, the craft as a whole gets pushed forward by thrust. Again, only when the oncoming wind relative to the blade element has a component coming from above or below the direction in which the craft is moving can thrust be produced—just as a conventional sailboat can't make headway directly into the wind.

What we've done is to re-resolve the resultant of lift and drag on the airfoil section into a pair of forces on the craft—a thrust and a residual component normal to the thrust. This latter component can be viewed in several ways. If the blade element is moving downward, then it's a kind of lift. But in a rotating propeller, downward movement of one blade element is always balanced by upward movement of some other blade element (or elements). So in an aerodynamic sense, this other force isn't any useful kind of lift. What it does is to determine the cost of spinning the propeller, since it directly opposes that spin. Thus the component (L' in Figure 12.1) multiplied by its distance from the axis of rotation of the propeller (behind the

263

page) gives the torque that the engine must supply to keep that element of the blade moving. If the analysis is extended to all elements of all blades, the sums give (1) the overall thrust that (at steady speed) must just balance the drag of wings, fuselage, and all else except the propeller, and (2) the torque required of the engine.

A practical propeller is twisted along its length from hub to tip. The reason is simple enough: nearer the hub, the spinning propeller encounters a wind mainly due to the craft's progression, while nearer the tip it meets a wind whose direction reflects more of its own rotation. So the farther from the hub, the more the wind a propeller encounters will deviate from the direction of motion of the craft as a whole. Thus, to maintain an optimal angle of attack along its length, the propeller ought to be twisted with its flattened faces more nearly aligned with the craft's motion near the base and with the plane of the propeller's motion near the tips. Assuming a constant rate of rotation, little twist will be needed at low flying speed but more will be necessary as speed increases. Aircraft of the nonliving sort do a crude approximation of such variable twist by lengthwise rotation of their rigid propellers—they vary the pitch. Flying organisms with nonrigid beating wings—insects and hummingbirds most clearly—manage to change the *degree of lengthwise twist*, even (as necessary of course) while reversing the direction of twist between downstrokes and upstrokes (see, for instance, Ennos 1989b). Ennos (1988) showed very neatly that one needn't invoke special machinery in the insect thorax—the wings are designed to be just sufficiently flexible in torsion so the inertial and aerodynamic forces of the wingbeat are sufficient to cause the twisting. But not all fliers do the twist. It seems to be nearly or entirely absent in some wings that operate at low Reynolds numbers—for example, fruit fly wings (Vogel 1967a) and maple samaras (Norberg 1973). At low Reynolds numbers, the lift-to-drag ratio just isn't so sensitive to small changes in angle of attack, so twist matters less, as one can see from Figure 11.9.

The Plane of Flapping

In birds, bats, and insects, flapping wings combine the functions that airplanes divide between fixed wings and propellers—in a sense, they're closer to helicopters than to airplanes, and it's all too easy to be misled by our habit of calling the propulsive appendages "wings" rather than "propeller blades." But they aren't quite like ordinary propellers either, since flapping wings produce both thrust and lift directly, rather than producing thrust directly and getting lift by diverting some of the thrust to pay for the drag of fixed, lift-producing wings. That composite function, as well as their reciprocating rather than rotational motion, means that the motion of flapping wings is inevitably complex. In general, the stroke takes the form of an inclined ellipse or figure eight if projected onto the surface of a sphere, or a saw-tooth pattern if projected (to account for forward motion) onto the surface of a cylinder, as in Figure 12.2. The downstroke moves a wing forward as well as downward and produces mainly upward force but usually some rearward force as well. The upstroke goes backward as well as upward, producing mainly rearward force but often some upward force.

For a flying animal we need several variables to deal with items that are invariant in an airplane. For one thing, the airplane's propeller turns on a shaft that closely parallels the direction of flight, so the plane in which the propeller turns is always one that's normal to the craft's progression. In a flying animal, what's called the "stroke plane" is almost inevitably tilted—as just described, its top is farther aft than its bottom. That tilt, though, is quite variable, ranging from almost horizontal to nearly vertical. In fact, the tilt of the stroke plane is often a function of flying speed, with a near-horizontal plane (small angle) characterizing hovering and with larger angles typical at faster speeds of flight.[1] Early work on tethered insects suggested that the stroke plane angle was the principal variable used to set the ratio of lift to thrust in hummingbirds and in insects such as fruit flies but not in others such as hover flies and locusts. Now that more data are available from freely flying animals, things look less simple—stroke plane angle is only one of the variables that can adjust that crucial ratio of lift to thrust (Ellington 1984a; Ennos 1989b); the main alternative is really subtle alteration of the phasing of lengthwise wing rotation during the stroke. Ennos, in fact, found that a hover fly, *Eristalis tenax*, can hover with either a horizontal or a near-vertical stroke plane. Horizontal-stroke hovering is energetically cheaper, but he figured that vertical-stroke hovering permits quicker transition to fast forward flight—like a sprinter on starting blocks instead of a distance runner standing upright. Of course in horizontal-stroke hovering wings don't really go up or down on upstroke or downstroke; a classical anatomist would call the half-strokes dorsad and ventrad and never get confused.

Hovering is in almost every instance the most expensive kind of flight an animal can do, and work on hovering (especially the monumental treatise of Ellington 1984a) has been of the greatest importance in understanding the special features of animal flight. The high cost is concomitant with that high induced drag at low speeds discussed in the last chapter. With no component due to forward speed, wind over the wings is slower, and lift coefficients have to be higher to get enough lift to offset body weight. Also (as you can see in Figure 12.3), in hovering a wing has to rotate lengthwise

---

[1] Confusion can arise all too easily between this *stroke plane angle*, the angle between the plane in which the wings beat and the horizontal, and the *stroke angle*, the amplitude of wingbeat, about which more shortly.

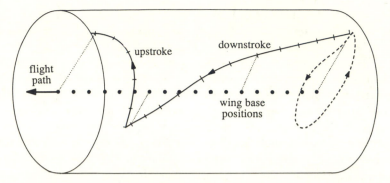

FIGURE 12.2. The wing motion of a insect flying forward, as the path of the wingtip might trace a path on the surface of a cylinder. (We're neglecting movement off that surface resulting from the forward and backward motion of a wing of fixed length.) The dashed curve uses the insect's body as a frame of reference, while the solid line refers to the earth as the insect flies along. Notice that the curve gets nearer the top than the bottom of the cylinder—opposite wings nearly meet at the top of the stroke.

more at the top and bottom of each stroke to achieve decent angles of attack in the two half-strokes—U_w is almost entirely a matter of the U_r component, and the latter reverses between half-strokes. So a larger fraction of the stroke must be spent in such shifts. With a limit of about 250 W kg^{-1} (Weis-Fogh and Alexander 1977) for the power output of muscle, hovering for more than very brief periods isn't possible for creatures much larger than hummingbirds (Weis-Fogh 1977)—wing loading is higher, so a larger downward velocity has to be imparted to the air, so the Froude propulsion efficiency is lower, so the cost of sustaining body weight is higher. Still, quite a few larger birds and bats hover at least briefly (incurring oxygen debts) in connection with feeding, landing, and other intermittent activities (Norberg 1990).

Advance Ratio

Obviously no thrust can be generated if the relative wind striking a rotating propeller is too close to the axis of progression—the aerodynamic resultant (Figure 12.1, again) will tip backward, not forward. That's almost certainly why many insects (mosquitoes and crane flies, for instance) have wings that are narrow near their bases and broad farther out, and why maple samaras put most of their area well away from their axes of rotation. More importantly, the necessity of having a resultant wind well off the direction of progression sets an upper limit on the speed of a craft. What's

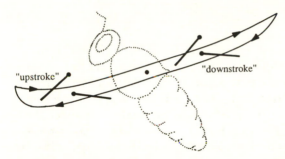

"upstroke"   "downstroke"

FIGURE 12.3. The wing motion of an insect hovering with a horizontal stroke, a little idealized to fit on a flat page as in a Mercator projection of a global map.

critical is the ratio of the forward speed of the craft to the speed of the blades of its propellers. The usual index of this quantity is the "advance ratio," J:

$$J = \frac{U_f}{nd}. \tag{12.1}$$

U_f is the craft's forward speed, d is the diameter of disk swept by the propeller blades, and n is the rate of revolution of the propeller. For ordinary airplanes, advance ratios are generally under 4.0—if you want to go faster, you should increase either the length or the rotation rate of the propeller. For a propeller without provision for pitch adjustment, the angles of attack (and thus the L/Ds) of its blade elements will vary with the advance ratio, and so its operating efficiency will depend very strongly on flying speed—it's common to plot propeller efficiency as a function of advance ratio. Mises (1945) gives a very good introduction to the blade element view of propellers.

Flying animals use reciprocating rather than fully rotating wings, so equation (12.1) proves inadequate. Every rotating propeller blade makes 360° per revolution, but a reciprocating blade or wing can be operated over a range of stroke amplitudes. So both amplitude and frequency must be specified. (Typically an insect wing, for instance, goes up and down through an arc of around 100°, or 200° per full stroke cycle.) Using R for the length of a wing, ϕ for the amplitude (usually referred to as "stroke angle"), and n for wingbeat frequency, Ellington (1984a) gives the following formula for the advance ratio:[2]

[2] This formula, now in general use, differs by a factor of π from that given in the earlier edition. Thus the figures I gave earlier should be divided by π. I suggest reconverting the present figures (multiplying by π) if one wants to compare them to data for airplanes.

$$J = \frac{U_f}{2\phi nR} \, . \tag{12.2}$$

Using this formula, a bumblebee (*Bombus terrestris*) reaches an advance ratio in free flight of 0.66 (Dudley and Ellington 1990a), a black fly (*Simulium* sp.) 0.50, and a fruit fly (*Drosophila melanogaster*) 0.33 (Ennos 1989b). All are figures for insects in free flight; tethered ones typically give somewhat lower values. Below about bumblebee size, the smaller the insect, the lower is its best advance ratio—with higher profile drag, it takes more wing beating to put a decent forward component on the aerodynamic resultant force.

The wingbeat frequencies of small insects are, in fact, very high. Fruit flies beat their wings around 200 or so times per second, mosquitoes around 300 to 600 (Greenewalt 1962), and midges up to about 1000. The all-time record for rate of reciprocation of any animal appendage is held by a midge that had its wings almost completely amputated to reduce their moments of inertia—it achieved 2218 wingbeats per second (Sotavalta 1953). On the other hand, the higher frequencies are almost entirely offset by shorter wings—tip velocity, the denominator in (12.2), doesn't change all that much with size.

The advance ratio provides a handy way to make crude estimates of the top speeds of insects. One just assumes a value of J appropriate for the size of the insect together with a moderate amplitude (say 2 radians). Wing length is easily measured, and data on frequency have been collected by Greenewalt (1962), so U_f can be calculated from (12.2). The results may be crude, but they're a lot better than some of the grotesque guesses buzzing about in the entomological literature!

Four Kinds of Moving Airfoils

The present explanation of how beating wings work began with airplane propellers, noted the addition of variable amplitude and variable stroke plane, and then drew on helicopters to deal with hovering. For a better perspective, let's look at the entire realm of moving airfoils that deal with thrust and power. Besides propeller and helicopter blades, there are windmill and autogyro blades—four types in all, of which three have clear biological examples. Figure 12.4 considers a blade element of each of these devices, first generating an aerodynamic resultant as the consequence of the relative wind striking it, and then re-resolving that resultant into components parallel and normal to the axis of rotation. We have, then,

1. A propeller with (ideally) a horizontal shaft, which inserts power into a horizontal airstream;
2. A windmill, also with a horizontal shaft, but which extracts power from the horizontal airstream;

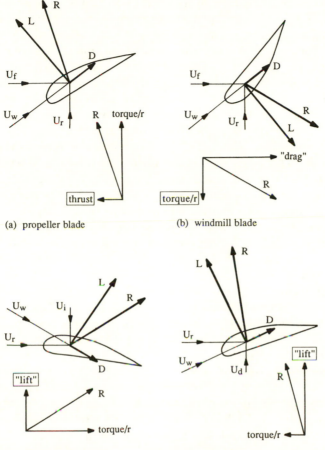

(a) propeller blade

(b) windmill blade

(c) helicopter blade

(d) autogyro blade

FIGURE 12.4. Four kinds of moving airfoils, analyzed as in 12.1. U_i is the wind component induced by the action of a helicopter blade as it pushes air downward; U_d is the wind component due to the descent of the autogyro. Other symbols are the same as in 12.1. Variables in quotes refer to the earth rather than the airfoil, and the desirable components of the resultants are boxed.

3. A helicopter rotor (here hovering) with a vertical shaft, which inserts power into the airstream, taking air from above and thrusting it out below the plane of the rotor; and

4. An autogyro rotor (here descending) with a vertical shaft, which extracts power from the airstream, taking air from below and retarding its passage upward through the plane of the rotor.

We've already mentioned the propeller (Figure 12.4a); its operation is essentially the same whether it rotates as an airplane propeller, oscillates as a beating wing, or reciprocates as the dorso-ventrally elongate (lunate) tail of a mackerel or tuna. The flow comes from ahead, and the aerodynamic resultant generates a force that engine or muscles must counteract. Notice that to keep a positive angle of attack, an airfoil that's not symmetrical top to bottom must have its convex side facing more upstream and its concave or flat side facing more downstream. Only when one considers the direction of the oncoming wind does the arrangement appear logical. The rule, by the way, is general and quite useful for all axial fans and blowers in home or lab: *blades should be concave downstream*. You can't reverse the direction in which a fan blows by reversing the fan hub on its shaft; all the operation does is reduce the fan's effectiveness because an inverted airfoil has a lower L/D_{max}![3] To reverse a fan either the motor must be reversed as well as the hub or the lengthwise twist and chordwise curvature (camber) of each blade must be reversed. Reversing fans usually blow these latter problems under the rug by using airfoils that are symmetrical top to bottom and have no lengthwise twist.

A windmill (Figure 12.4b), on the other hand, must have the "lower" flat or concave side of its airfoils facing the wind (which is why it's lucky that the sails of a sailboat are naturally concave upwind). Thus a propeller makes a poor windmill unless the hub is reversed on the shaft or the flow direction is reversed. If the hub is reversed but the flow direction kept the same, the propeller-now-turned-windmill will, of course, revolve in the opposite direction. In windmills, the wind produces a resultant force that can be separated into a torque (per unit radius), now the useful component that drives whatever the windmill is connected to, and a drag parallel to the overall wind. The latter is irrelevant to power generation and matters only in judging the necessary strength of blades, bearings, and tower. A windmill, incidentally, works as much because its feet are on the ground as because its rotor is in the air—a velocity difference is what it's taking advantage of.

I know of no close biological analog of a windmill. Wheel-and-axle arrangements are unknown in organisms except among bacteria, so we ought to cast about for reciprocating analogs of beating wings. Leaves such as those of aspens and poplars rotate a little as well as translate in circles in a wind, restrained by petioles that can't rotate freely, but nothing suggests that useful power is being extracted in the process. Aquatic organisms are probably better candidates than terrestrial ones, and probably the power

[3] I once took an ordinary ventilating fan and reversed the hub on the shaft, whereupon the speed dropped from 0.8 to 0.2 m s^{-1}. The direction of flow, naturally, was unchanged.

extracted would have to be invested in something such as induction of fluid movement that requires only a minimum of transducing machinery. Biological windmills certainly do exist in the general sense of devices that extract energy from velocity gradients in fluid media, as mentioned in Chapter 4. But they don't seem to include any quixotic devices that do so by direct application of lift-producing airfoils. Birds performing dynamic soaring extract energy from velocity gradients, but again the mechanism is a long way from a windmill.

Helicopter rotors (Figure 12.4c) are the closest technological analogs of hovering fliers. By contrast with propellers and windmills, the wind comes from above, so these airfoils must be concave downward and downwind, with their leading edge elevated. During pure hovering, rotation produces an upwardly tilted resultant; the latter resolves, in turn, into a vertical force (lift with respect to the earth) and a torque (per unit radius) that, as with a propeller, the engine must counteract. Since the only downward component of the wind is that produced by the rotor, helicopter rotors should be as long as mechanically practical in order to intercept the maximum volume of air per unit time and thus minimize their induced-power appetite. This preference for long blades is the main reason for the impracticality of aircraft in which propellers used for fast forward flight are made to face upward for hovering or takeoff. That's of course the same argument (ultimately based on Froude propulsion efficiency) we used to explain why a flying animal capable of reasonably fast forward flight finds it so expensive to hover.

Autogyros (Figure 12.4d) may be less familiar to the reader than the previous three devices, but they're completely competent flying machines. Superficially they look like helicopters, but the overhead rotor isn't powered, another propeller is necessary for level flight, and they can't hover or ascend vertically. An autogyro, moving forward, has the plane of its passive rotor tilted back a little so that the approaching air encounters the underside of that plane of rotation (a helicopter uses the opposite tilt). Air passes upward through the rotor and, in being retarded, it generates lift and drag with respect to the craft. The (separate) propeller counteracts that drag and so keeps the craft moving. Unpowered, an autogyro descends slowly with the rotor passively turning in a horizontal plane; in effect it's a set of airfoils (the blades) gliding earthward in a tightly helical path. As in any of these devices, air must strike the concave side of the blades, so the blades have to be concave downward (as in helicopters). But in contrast with helicopter blades, the leading edges of the blades of the autogyro must be depressed rather than elevated.

If a helicopter loses power, it can revert to autogyration by rapidly shifting the pitch of the blades to get their leading edges downward. Otherwise the rotor will turn backwards, and its profile will be wrong for action as a good airfoil. If pitch is adjusted, though, both camber and lengthwise twist are proper, unlike the shift from propeller to windmill.

THE FLIGHT OF SAMARAS

An unpowered autogyro, again, is a kind of glider that happens to descend along a helical path. Like any glider, it can have no unbalanced forces under steady-state conditions. Initially, the aerodynamic resultant is tipped forward, pointing in the direction of rotation, and produces a torque that accelerates the blade or blades, as shown in Figure 12.4d. As the blades speed up, the oncoming wind becomes more nearly horizontal, and the resultant tips back until it's vertical and just equal to the weight of the craft.

Autogyros of this kind are quite common in nature as seed dispersal devices ("samaras") in plants; as mentioned in Chapter 10, some are Flettner-rotating autogyros, but more are simple lift-producing airfoils. These latter range in blade length from about 10 to 180 mm, have from one to three blades, and span several orders of magnitude in mass. The best-known samaras are those of the maples (Aceraceae) (Figure 12.5), but they also occur in some tropical Juglandaceae and Leguminoseae as well as, among conifers, in many Pinaceae and Cupressaceae. The definitive study of the flight of samaras is that of Norberg (1973); for more on the aerodynamic characteristics of their wings, see Azuma and Okuno (1987) and Azuma and Yasuda (1989), and on their flight see Seter and Rosen (1992). Quite a few people have recently looked at performance variation and its ecological implications—for instance Green (1980), Guries and Nordheim (1984), Augspurger (1986, 1988), and Greene and Johnson (1989, 1990b), most of whom have been mentioned earlier.

For an autogyro, a low sinking speed is obviously desirable, which implies that a low rather than a high advance ratio indicates good performance. A fairly large, one-bladed samara of the Norway maple, *Acer platanoides*, with a functional blade length of 37 mm, falls at 0.9 m s^{-1}, and revolves thirteen times per second; its advance ratio (equation 12.1) is 0.9. The smaller samara of the Norway spruce, *Picea abies*, has a blade length of 11 mm, falls at 0.64 m s, and revolves twenty times per second; thus it has an advance ratio of 1.5 (data from Norberg 1973). I measured an advance ratio of 1.3 for red maple (*Acer rubrum*) samaras, whose functional blade length is about 17 mm. That's as it ought to be—the advance ratio should increase as the Reynolds number decreases because the poorer lift-to-drag ratios at low Reynolds numbers mean relatively less force is available to spin a samara and more drag opposes the spin. These advance ratios don't vary greatly, so just as one can estimate insect flight speeds from wingbeat

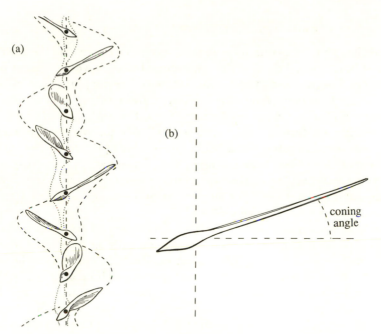

FIGURE 12.5. (a) The autogyrating descent of a maple samara.
(b) The low cone it describes as it rotates about the seed as an apex.

frequencies, one can make good guesses of spinning rates (and Reynolds numbers and so forth) from measurements of falling speeds.

If the advance ratio stays fairly constant, then sinking speed and rotation rate will remain in the same proportion. That's really a version of the statement made in the last chapter about the insensitivity of the glide angle in simple gliding to the weight of the craft. For a constant velocity descent, the upward force will equal weight. Lift, of course, is nearly proportional to the square of airfoil speed—to both sinking speed and rotational speed—and overall upward force will nearly follow airfoil lift. Thus sinking speed should be proportional to the square root of weight. And that's about what happens. For a bunch of red maple samaras of different degrees of dryness, with masses from 11.2 to 57.2 mg (5-fold), I measured sinking rates ranging only from 0.62 to 1.15 m s^{-1} (less than 2-fold). In fact (as you can tell from these numbers), the sinking-rate variation is even a little less than just predicted since faster sinking means higher Reynolds number and an improved (lower) advance ratio. This relative insensitivity of sinking rate to weight is a rather nice feature of the autogyrating scheme!

MOMENTUM FLUX AND ACTUATOR DISKS

In talking earlier about drag, we took several viewpoints on the interaction of fluid and solid, ultimately equivalent but each with different perspectives, insights, and applications. We looked at the particulars of skin friction and pressure drag after looking at the more global business of momentum flux. One can do the same for a thrust-producing or power-extracting rotor through which fluid passes. Up to this point we've taken a close-up view, a largely qualitative glimpse of what's properly called "blade element analysis." For some problems that's unnecessarily particular, and returning to a consideration of momentum gives more insight—specifically, looking at the rate at which the momentum of the fluid changes as it passes through the thrust-producing device.

To look at momentum flux, it's customary to set up an ideal "actuator disk," a plane extending across the airstream with an area equal to that swept by blades or beating wings, as in Figure 12.6. Matters can be simplified further by assuming that the disk accelerates fluid only axially, that is, normal to its plane; that fluid velocity is uniform across the disk; and that a complete, frictionless discontinuity exists between fluid that passes through the disk and the rest of the fluid. In passing through the disk, a stream of fluid is accelerated; and therefore, by the principle of continuity, the stream contracts. According to the momentum theory of propeller action, the velocity of the fluid passing through the actuator disk is just halfway between that well upstream and well downstream. Consequently half the cross-sectional contraction occurs upstream and half downstream from the disk.

The main utility of analysis using momentum flux through actuator disks is in estimating such important items as induced velocity and power. To hover, for instance, an animal must balance its weight by making downward thrust, and the latter is a matter of the rate at which it imparts downward momentum to the airstream. Thus if you know wing length, you know disk area; from the latter you can figure the velocity increment (the induced velocity) that the wings must impart to the air—at least given the simplifications implicit in the approach. A useful variable in this regard is what's called "disk loading," weight over disk area, the equivalent for a thrust-producing propeller or beating wing of wing loading (Chapter 11) for a lift-producing airfoil. Induced velocity, U_i, turns out to be proportional to the square root of the disk loading,[4] W/S_d; more specifically,

---

[4] One can combine this statement with the earlier one about the sinking speed of a samara being proportional to the square root of its weight, declaring (as did Norberg 1973) that the sinking speed of a samara is proportional to the square root of its disk loading.

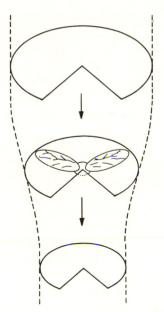

FIGURE 12.6. Replacing the beating wings of a hovering animal with an "actuator disk" that inserts momentum at a certain rate into a downward airstream.

$$U_i = \left(\frac{W}{2\rho S_d} \right)^{1/2}. \tag{12.3}$$

Since induced power is equal to weight times induced velocity, then

$$P_i = \left(\frac{W^3}{2\rho S_d} \right)^{1/2}. \tag{12.4}$$

And this can be read as yet another way to say that low weight and long blades are best for hovering, whether in hummingbird or helicopter. But I'll declare further details beyond our present scope, suggesting reference to Mises (1945), Ellington (1984a), and Norberg (1990).

Ellington, incidentally, recommends against using a fully circular disk for animal flight, replacing it with two pie-shaped sectors to account for the less-than-full-circle stroke angle and the tilted stroke plane.

A WAKE OF VORTICES

Another view of flapping flight focuses on the vortex structures left behind in the wake. We've already talked (Chapter 11) about bound vortices, tip vortices, and starting vortices, and the way in a fixed-wing aircraft the three are linked in an elongate vortex ring. We've noted (Chapter 10) that a vortex ring will progress through the fluid on account of its own circulating motion—a ring with its outside going downward and its inside upward, such as a thermal vortex, will move upward as a whole, even if it's not buoyant. Similarly, a ring with its inside moving downward will move downward as a whole—and that's the kind made by an airfoil producing lift. And lift, as just noted, must be associated with downward movement of fluid. So there's a nice congruence among these matters.

Where the vortices get really interesting is in flapping flight. Consider, again, a hovering flier with a near-horizontal stroke plane. Reciprocating wings have a problem quite unknown among rotating propellers. Reversing the direction of wing motion, as happens twice during each stroke, without at the same time reversing the direction of thrust requires that the circulation about the wing must also reverse. What was on the downstroke the airfoil's upper surface, with net circulatory flow from leading to trailing edge, becomes on the upstroke its lower surface, with circulation from trailing to leading edge. Reversing the bound vortex requires shedding the old vortex as a stopping vortex after each half stroke; since successive half-strokes have opposite circulations, successive vortices shed into the wake will turn oppositely. The wake of a hovering flapper, as pointed out by Ellington (1978) and Rayner (1979), must contain a downward-moving stack of vortex rings, each one generated by a half wingbeat, as in Figure 12.7a. Looking at hovering in terms of actuator disks and momentum flux ignores all these periodic effects and thus gives what amounts to a minimum estimate of induced velocity, thrust, and power.

A bird flying rapidly beats its wings mainly up and down, getting lift on both strokes but more on the downstroke than the upstroke. It leaves a wake that's closer to that of a fixed-wing craft, a single vortex ring with tip, starting, and bound vortices (Figure 12.7c). The variation in lift leads to an in-and-out wobble of the tip vortices, closer together on the upstroke and farther apart on the downstroke. A bird flying more slowly leaves something in between in its wake—a zigzag ladder of vortices, with horizontal cross-pieces between tip vortices corresponding to each reversal of wing direction and the concomitant reversal of circulation (Figure 12.7b). In addition, the direction of rotation of the tip vortices reverses with each half-stroke.

The reality of these vortices is well established. Kokshaysky (1979) was the first to persuade birds to fly through clouds of particles and then make photographs of the vortices. Spedding (1986, 1987) then did a very much more elaborate analysis of such moving particles for a variety of birds under a variety of conditions. Ellington (1980b) showed that cabbage white butterflies (*Pieris brassicae*) flung vortex rings earthward as they beat their broad wings up and down in slow flight. More recently, Brodsky (1991)

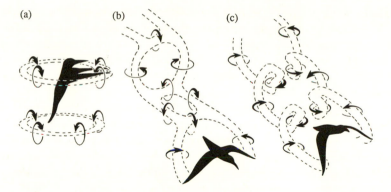

FIGURE 12.7. The vortices behind flying birds. (a) A hovering bird leaves a vertical stack of separate vortex rings beneath itself. (b) A bird flying slowly forward leaves a zigzag ladder of vortices, with tip vortices reversing direction between half-strokes and with adjacent rungs of the ladder having opposite directions of circulation. The stronger vortex rings correspond to the lift-producing downstroke. (c) A bird flying rapidly forward leaves a pair of tip vortices that wobble inward when shed during upstroke and outward on downstroke—since downstroke produces more lift.

found quite a complicated set of coupled and uncoupled rings, depending on flying speed, in another butterfly, *Inachis io*. These vortices constitute a record of what airfoils have just done, and we're learning to make them visible and to read their messages. This wake-vortex view of flapping flight has proven most fruitful in the dissection of the roles of various nonsteady phenomena involved in such flight—to which we'll now turn. It seems to hold considerable promise, as well, for analyses of swimming in fishes.

NONSTEADY EFFECTS IN FLAPPING FLIGHT

The difference between a propeller and a pair of beating wings, again, is that the wings reciprocate, changing direction and pitch twice in each stroke. And therein lies a set of problems for analyses of animal flight even more vexatious than anything so far considered.

On one hand, if an airfoil starts suddenly from rest, full circulation about it doesn't develop for the first two or three chord lengths of travel. In this "Wagner effect," the starting and bound vortices initially interact destructively, so the lift doesn't develop right away (Weis-Fogh 1975). Since wings may start and stop a hundred times each second and since they travel no more than a few chord lengths before reversing, the lift actually available in flapping flight may be considerably overestimated by steady-state measurements on isolated wings in wind tunnels. On the other hand, if the angle of attack of a wing is suddenly increased above an angle that would produce stall under steady-state conditions, the wing can travel several chord lengths before separation and stall occur; and lift can reach values around 50% greater than the steady-state maximum ("delayed stall"; Ellington 1984b).

Of what significance are such nonsteady effects? They have long been invoked as a kind of deus ex machina to cover oversimplified views, inadequate data, and ex cathedra statements about bumblebee lawlessness. The first really relevant measurements were made by Jensen (1956), working with Weis-Fogh. Jensen painstakingly showed that forward flight in desert locusts could be fully accounted for by appropriate summing of the force coefficients derived from steady-state measurements on isolated wings—what's called a "quasi-steady analysis." But at least two troublesome matters kept this from settling the issue, both emphasized by (among others) Ellington (1984a, b). First, Jensen's work failed to find only that nonsteady phenomena needed to be invoked and didn't disprove their involvement; and second, his work didn't even do that for the more demanding case of hovering flight.

Estimating the average lift coefficients needed to support a hovering creature's weight isn't especially difficult, although the exact figures depend a bit on the particular simplifications used—one needs to know only a few things such as body weight; wing length and area; and wingbeat frequency, stroke angle, and stroke plane angle. The results in all too many cases are steady-state lift coefficients well in excess of those such as we saw in Table 11.1—maximally only 1.0 or a little more for insects and perhaps up to 2.0 in birds. And bear in mind that an animal need hover only very briefly to make trouble, since we're talking about short-term lift, not aerobic metabolism. Ellington (1984a) and Ennos (1989b) got values ranging between 1.2 and 4.4 for quite a wide range of hovering insects. Norberg (1975, 1976) estimated a coefficient of about 5.0 for the pied flycatcher, *Ficedula hypoleuca*, and over 3.0 for a bat, *Plecotus auritus*. The most extreme cases are animals hovering with inclined rather than horizontal stroke planes since they produce most or all of their lift during downstroke.

Delayed stall, the most conventional tactic, apparently doesn't play a major role in achieving high lift coefficients. Of more significance is a scheme that not just avoids the Wagner effect of delayed onset of circulation but that gets especially strong circulation going right at the start of the downstroke, where high lift coefficients matter most. Weis-Fogh (1973) first proposed this "clap-and-fling" mechanism for very small insects such as the tiny wasp, *Encarsia*, and fruit flies. (The original evidence included some pictures I had published without having noticed anything of significance!) The scheme is really quite simple, once one is accustomed to think-

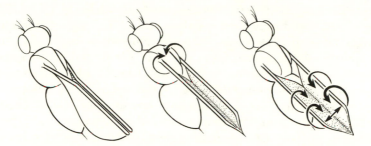

FIGURE 12.8. Weis-Fogh's "clap and fling" mechanism for the initiation of circulation through interaction of a pair of wings. The wings are shown cut chordwise about halfway along their spans. At the start of downstroke the two wings part first in the front, and air passes around the leading edge to fill the gap. That air goes from what will be the undersides to what will be the tops of the wings, so the direction is appropriate for lift-producing circulation.

ing in terms of circulation and vortices. Recall that as it first develops lift, a wing becomes surrounded by a bound vortex and leaves behind a starting vortex of the same magnitude but opposite spin—vorticity (at least in the short run) is conserved. What a beating wing can do is to use its opposite number on the other side to account for the opposite vortex, rather than having to leave one at the starting gate—each wing aids development of the bound vortex of the other. What happens is that at the end of the upstroke the wings "clap" together above the thorax, either touching or coming very close. They then rotate ("fling"), peeling apart from their leading edges, pivoting about the trailing edges, as in Figure 12.8. The space between them must be filled with air, and that air can most easily come around the leading edges—so it constitutes a circulation about each wing, well begun at the very start of the stroke. This circulation is created quite independently of the classical scheme described in the last chapter. What determines its magnitude is the angular velocity of the lengthwise rotation of the wings, not their chordwise, translational speed—the Flettner scheme momentarily reenters the picture. So here at least the low wind speeds of hovering flight impose no handicap.

The clap-and-fling mechanism has been found in all small insects thus far examined, as well as some (but not all) larger creatures, including some moths and butterflies and the hind wings of locusts in climbing flight (Ellington 1984b). Birds that can hover only briefly, such as the pied flycatcher (see above) and doves and pigeons, commonly bring their wings together with an audible clap—Weis-Fogh (1975) noted Virgil's allusion to the phenomenon in the *Aeneid*—and are almost certainly using the mechanism.

A second rotational lift mechanism occurs farther into the stroke. The leading edge moves downward faster than the trailing edge—as the wing accelerates downward, the relative wind gets more nearly vertical, so keeping an appropriate angle of attack requires lengthwise rotation as well as the basic downward motion of the beat. This wing rotation causes additional circulation of air around the wing, and thus generates additional lift (Ellington 1984b). At this point there's no reason to believe that all unconventional mechanisms of lift production have been uncovered; indeed, people working in the area are upbeat about the chance of recognizing others.

SWIMMING

Swimming, the other mode of locomotion in a continuous medium, is far more common among animals than is flying. It's certainly easier, whether in terms of power expenditure relative to body mass or in terms of minimum necessary modification of, say, a walking creature. A swimmer can at any point stop swimming (with at most a minor problem of buoyancy or respiration), while in flight the consequences of sudden termination of powered motion are more immediate and dramatic. Conveniently, for most of the ways that animals swim, the principles developed for flying ought to be applicable.

Swimming may be easier to do, but it has not proven easier for analysis. Part of the difficulty is the greater morphological diversity in equipment used for swimming, but the greater obstacle is an all-too-common characteristic of the way macroscopic swimmers operate. The problem was alluded to back when we talked about the drag of motile animals—thrust-producing and drag-inducing structures cannot easily be separated. Perhaps these swimming systems come especially close to an ideal, what's been called a "zero momentum wake" (Elliott 1984), in which drag and thrust cancel locally as well as globally, the Froude efficiency is unity, and nothing is left behind to indicate that an animal has swum by. A fish such as a trout bends its body in producing thrust to overcome the drag of its body. As pointed out in Chapter 7, one can tow a dead fish and measure the drag, but the figures come out unreasonably high, and the fish in any case wobbles unrealistically. Stiffening the fish leads to lower figures for drag (Webb 1975), but what assurance have we that these latter are relevant to active swimming and not just to decelerative gliding? Considerations of momentum transfer suggest that they underestimate the force that thrust must offset, and this argument is consistent with the common alternation of active swimming and stiff-body gliding (Webb 1988).

This "burst-and-coast" behavior, in fact, can be used to get a crude estimate of swimming drag. Videler and Weihs (1982) showed that saithe

(*Pollachius virens*) and cod (*Gadus callarias*) choose initial and final velocities for bursts of swimming close to the values expected if swimming drag is three times greater than coasting drag. For a saithe about a third of a meter long swimming at five body-lengths per second, they found that burst-and-coast swimming is about 2.5 times cheaper than continuous swimming. A really small fish, though, should gain no advantage—stiff-body drag is too high relative to swimming drag. As he predicted, Weihs (1980) found a gradual transition from continuous to burst-and-coast swimming as larval anchovies (*Engraulis mordax*) grew larger.

For fish alone—much less other aquatic creatures—swimming modes are diverse. Various classifications have been used, mostly derivative of the scheme of Breder (1926); summaries are given (with less than total consistency) in quite a number of places, including Lighthill (1969) and Webb (1975). A mode is most often named after an exemplar of that mode, which implies a certain amount of pure memorization for the nonichthyologist—presumably at the start no one else was expected to have any interest in fish swimming! Thus passing bending waves down a body one or more wavelengths long is referred to as "anguilliform," after the eel, *Anguilla* (Figure 12.9a). Waves move backward both with respect to the fish's body and with respect to the free stream. The mode is also used by some largely or entirely limbless amphibians, by aquatic snakes, and by the smooth-bodied polychaete worms that lack conspicuous parapodia (Clark and Hermans 1976). The cost of transport (power per unit distance covered) appears to be higher for this mode than for the ones that follow (Webb 1988). Eel elvers may make long migrations; but, according to McCleave (1980), they have to take advantage of water currents.

What we usually regard as the ordinary mode of fish swimming is termed "carangiform," after *Caranx*, the jack (Figure 12.9b)—and is used by the perches, trout, cichlids, mullets, and so forth. It's the mode regarded as the best generalist arrangement, combining versatility in possible movements and control, good starting acceleration (using the "C-start" and other devices about which more shortly), good maximum speed, and good Froude propulsive efficiencies of between 50% and 80%. A wave of bending is passed backwards, but these fishes are shorter relative to the wavelength than the eel-like ones. (Sometimes "subcarangiform" is used for cases where the body is more than half a wavelength long.) The wave increases in amplitude as it passes, from negligible at the head to maximal at the tail. Both necking just ahead of the tail and bilateral flattening of the body minimize recoil and thus reduce body oscillation relative to tail oscillation. In both anguilliform and carangiform locomotion, vortices are passed down the body, one on each side per full wave, and shed with their associated momentum into the wake—thus propelling the fish forward.

A useful distinction separates the ordinary carangiform mode and

281

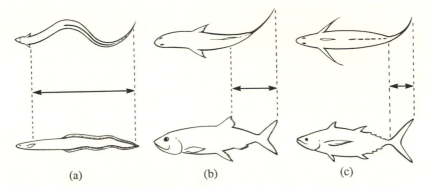

(a) (b) (c)

FIGURE 12.9. Modes of swimming in fishes: (a) the "anguilliform" motion of an eel; (b) the "carangiform" motion of a trout; (c) the "thunniform" motion of a tuna.

what's most often called "carangiform with lunate tail" (Figure 12.9c) or "thunniform" after the tuna (*Thunnus*). Practitioners of the latter are distinctive in shape, with a tail that extends across the flow a considerable distance beyond the body. In addition, they have somewhat stiffer and rounder bodies and a more substantial narrowing (the caudal peduncle) between body and tail. Trunk muscle seems arranged primarily to power the tail beat rather than to bend the body. Thus for all practical purposes the tail functions like an oscillating propeller or a pair of beating wings; the main difference is the fluid-mechanically minor one of a pair of hydrofoils moving laterally and together rather than swinging above and below their separate articulations. Aspect ratio is once again a relevant parameter, and these fishes have fairly high values of 4.0 and above (Chopra 1975). This latter group includes many large pelagic fish that go especially fast both with respect to top (anaerobic burst) speed or sustainable (aerobic) speed—the mode appears to be the most force- and power-efficient system for high-speed swimming. Besides these tunas, mackerels, marlin, and the like, the arrangement is used by some sharks and by whales and dolphins—these latter with dorso-ventral rather than lateral tail oscillation; and it's clearly indicated by the shapes of fossil ichthyosaurs. A good analysis of how the tail of a fin whale (*Balaenoptera physalus*) works as a propeller was done by Bose and Lien (1989). The flukes of their 14.5-meter whale had an aspect ratio of 6.1; it got its best propeller efficiency at an impressively high advance ratio of 5.0 to 6.0.

These three modes are far from exhaustive of the arrangements that have been described and named. What they (and most of the others) certainly have in common is the business of generating and shedding vortices. Almost certainly they produce the zigzag vortices observed for birds flying forward (Figure 12.7). Since they're making almost pure thrust and doing it with (except cetaceans) side-to-side oscillations, any ladders ought to have vertical rather than horizontal cross-flow rungs. Visualizing these vortices has proven even more difficult than for birds; the pictures published by McCutchen (1977b) are crude but very nearly all we have. Recently Charles Pell (pers. demonstration) has begun producing models out of flexible plastic that, when forced to oscillate, produce lots of both thrust and vortices and swim in a persuasively fishy fashion; if these prove good models of what fishes do they could really teach us a lot by permitting far easier wake visualization.

We'll return to fish swimming in Chapter 16, which will deal with unsteady flows and the phenomenon of "added mass" or "acceleration reaction," alluded to earlier in Chapter 7. For now, just bear in mind that when locomoting in a medium of density comparable to that of the craft, the resistance of the medium to acceleration can be of the same order of importance as its momentum once it is in motion.

Drag-based versus Lift-based Thrust

Consider two ways of using paired, flattened, elongate appendages to produce thrust. They might be moved back and forth along the axis of progression, oriented broadside to flow on a backward power stroke and parallel to flow during a forward recovery stroke. The backward stroke develops much more drag, and rearward drag is equivalent to forward thrust—a proper analysis of the scheme is provided by Blake (1981). You can easily row a boat this way, feathering the oars on the recovery stroke without taking them from the water. It's also very close to what a paddling duck does with its webbed feet. Alternatively the appendages might be moved up and down, that is, in a plane normal to the axis of progression. As we've seen in this chapter, by rotating the appendages periodically along their length an appropriate angle of attack can be maintained to generate enough lift via circulation so there's a net forward thrust. That's what a swimming penguin does with its short wings.

Clearly nature makes no absolute choice between these mechanisms, so superiority is likely to be circumstance-dependent. Thus a comparison of their respective performances might be enlightening—it might even have evolutionary implications. One can, in the manner of classical comparative biologists, make guesses about function based on the different life-styles of what we'll call paddlers and flappers; but models and mechanical analyses seem a surer basis for understanding.

As a start, let's therefore consider a very crude pair of models. Imagine a pair of rectangular plates, each 10 mm by 100 mm, arranged to extend outward from a fuselage in water. They thus have an effective aspect ratio

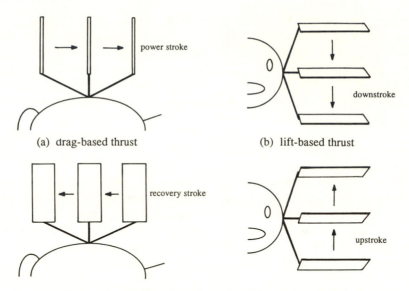

FIGURE 12.10. The simplified motions of paddles and hydrofoils assumed in our comparison of (a) drag-based (top view) and (b) lift-based (front view) thrust making.

of 20 and each has a plan form area of 0.001 m². As in Figure 12.10, they might move back and forth or up and down. (We'll ignore any swinging about a point of articulation.)

Case (a): Drag-based Thrust

The plates go back and forth relative to the free stream, oriented perpendicular to flow going backward and then parallel to flow going foreward. We assume a speed of movement, relative to the craft, of 2 m s⁻¹ and a duty cycle of 50%—that is, the recovery stroke occupies half the time. We'll also assume that the drag coefficient, C_{df}, during the power stroke is 2.0 (Ellington 1991) and that drag during the recovery stroke (negative thrust) is negligible. At Reynolds numbers in the present range of 10,000 to 20,000, that's not unreasonable: the drag of an elongate plate parallel to flow is only a few percent of that of the plate when perpendicular to flow (Table 5.2). With the craft stationary in the water, the paddles have a water-speed of 2 m s⁻¹ on both strokes; if the craft moves forward at 1 m s⁻¹, then the power stroke has a water speed of 1 m s⁻¹ and the recovery stroke a speed of 3 m s⁻¹; if the craft moves at 2 m s⁻¹, no thrust can be produced since the power stroke has no motion relative to the water. So the system has an absolute upper speed limit, even if no parasite drag is imposed. How

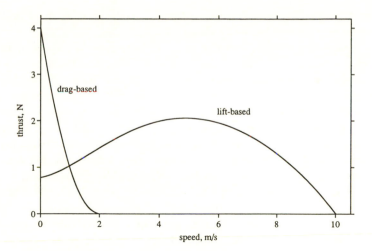

FIGURE 12.11. The results of our simple models of lift- and drag-based swimming, looking at the time-averaged thrust produced as a function of swimming speed.

does its thrust vary with craft speed? Figure 12.11 shows that it's terrific when the craft is stationary but drops off rapidly toward its limit. In fact, it might do even better at the start if we didn't do such an unrelievedly steady-state analysis. For the first power stroke perhaps we ought to drop the duty-cycle discount and thus double our calculation of thrust, and the acceleration reaction (Chapter 16) will give a substantial bonus.

Case (b): Lift-based Thrust

The plates go up and down, keeping the angle of attack adjusted to maximize the lift-to-drag ratio. We assume a speed of movement, relative to the craft, of 1 m s^{-1}; the duty cycle is 100% (no glide phase or reversal delay); the lift-to-drag ratio is assumed 10:1 with a lift coefficient of 0.8 at that L/D. With the craft stationary in the water, the resultant force tilts maximally forward, which is good; but the relative speed of water encountering paddles is only 1 m s^{-1}, which is not so good. If the craft moves, say, 1 m s^{-1}, then the resultant has a relatively smaller forward component, not so good; but the relative water speed has increased to 1.4 m s^{-1}, and the magnitude of that resultant has thus doubled, certainly good. So one thing offsets another, and thrust continues to quite a high speed (or advance ratio, if you prefer). This gives a more complex relationship between the craft speed and the thrust available, as you can see on the graph. Thrust is distinctly submaximal at zero speed (the problem of hovering, again), ascends to a maximum, and then drops (even as parasite drag must be increasing rapidly) at still higher craft speeds.

Interesting comparison! Even without considering power requirements, the drag-based system is very much better when the craft is stationary but the lift-based system is clearly superior at any decent forward speed. Of course we might have picked different conditions, but a very high flapping rate would be needed to make the lift-based system do as well at zero speed or a very high paddling rate to make the drag-based system better at high speeds. I think the comparison tells us a lot about who ought to use one or the other. If initial acceleration is what matters, then a drag-based system is clearly advantageous. A lot of animals are wait-and-lunge predators who might well prefer high acceleration to high steady speed. If speed matters, then flapping is much better than paddling. But two further considerations ought to be mentioned. The first is that no rule disallows combining both systems, even using the same appendages. An animal just has to use a little more complicated stroke, and to vary the stroke parameters with speed in a fancier manner. (And such mixed strokes appear to be used by many noncetacean swimming mammals, according to Fish 1992.) The second goes back to the material on streamlining and the shape of airfoils. The lift-based system is considerably less forgiving with respect to the cross-sectional shape of the plates if a decent L/D is to be achieved. Getting a high ratio of normal drag to parallel drag can be done with cruder equipment. Thus multifunctional appendages ought to be more likely to use the drag-based system.

So what happens in the real world? Penguins clearly flap their wings up and down, propelling themselves by developing lift (Hui 1988a). Sea lions (Otariidae) get thrust solely with pectoral flippers (by contrast with true seals, the Phocidae, which use their tails) with a lift-based system (Feldcamp 1987). Sea turtles, as well, use lift for swimming (Davenport et al. 1984). Most fish propel themselves with their tails, as described earlier, but a few use paired pectorals as lift-based principal propulsors. For instance, the shiner seaperch, *Cymatogaster aggregata*, normally swims with pectoral fins only, using a lift-based system (Webb 1973). The ratfish (so called because of its rat-tail-like tail) *Hydrolagus colliei*, and most rays (Myliobatoidae), both cartilagenous fishes but otherwise distant, do likewise. Far distant, phylogenetically, from all of these is a pteropod mollusk, the sea butterfly, *Clione limacina*, which beats its "wings" in an up-and-down, lift based system (Satterlie et al. 1985). And it appears that both portunid crabs (extant) and euripterids (extinct) use lift as well (Plotnick 1985).

Lift-based systems may be widespread, but they're not universal. Ducks paddle with their legs, using a drag-based arrangement (Prange and Schmidt-Nielsen 1970; Baudinette and Gill 1985), as do muskrats (Fish 1984), freshwater turtles (Davenport et al. 1984), nereidiform polychaetes

(Clark and Tritton 1970), and large water beetles (Nachtigall 1980); further examples can be found in Blake (1981), Braun and Reif (1985), and Fish (1992). At least one fish uses a drag-based system: Blake (1979a) found an angelfish (*Pterophyllum eimekei*) doing it with its pectoral fins when swimming slowly. Overall propulsive efficiency is low—around 0.18—but, as Blake comments (and is consistent with our present argument), the efficiency of lift-based propulsion ought to be even lower at such low speeds.

As expected, many of the drag-based swimmers occur where there's a less than single-minded focus on swimming as a way to get from place to place. Polychaete worms burrow; and ducks, muskrats, and freshwater turtles all walk on the same legs as are used for swimming. By contrast, sea turtles walk only minimally and crudely on their flippers.

Acceleration as a consideration is a little more complicated since lift-based swimmers often use some non-lift-based assist for starts. But it must be important. Webb (1979b) noted that a lift-based system in a fish needs quite a lot more muscle to get comparable accelerations even though the fish can achieve much higher final speeds. Similarly, Davenport et al. (1984) found that, while sea turtles could swim six times as fast as freshwater turtles, sea turtles produced only twice as much static thrust (the stuff from which acceleration is made) when tethered. Ordinary fish do accelerate very well, but they usually use special, high-amplitude (drag-based) bends—the "C-start," as shown by Webb (1976, 1978), Harper and Blake (1989), and Frith and Blake (1991). A diversity of fish can achieve accelerations of about 40 or 50 m s^{-2}, according to Webb (1975). Some do much better: Harper and Blake (1989) measured a peak acceleration in a pike, *Esox lucius*, using a C-start of 245 m s^{-2}, about twenty-five times gravity!

Perhaps the most extreme development of this initial drag-based body bend is in crayfish and similar crustaceans, which begin their impulsive rearward motion with a powerful flexing of the abdomen. Webb (1979a) measured an acceleration of 51 m s^{-2} in a crayfish (*Orconectes virilis*); Daniel and Meyhöfer (1989) got 100 m s^{-2}—about ten times gravity—in a shrimp (*Pandalus danae*); and Spanier et al. (1991) got up to 5 m s^{-2} in a lobster (*Scyllarides latus*). Since maximum accleration decreases with size for basic reasons of scaling (Vogel 1988a), these are probably comparable performances.

At least one partly drag-based system is used in flight: broad-winged butterflies throw periodic vortices earthward with such a scheme (Ellington 1980b).

FILLIPS AND FLOURISHES IN FLYING AND SWIMMING

A few incidentals, to tie up loose ends of lift-based locomotion as practiced in both air and water . . .

Formation Flight in Birds

A bird flying rapidly sheds a continuous tip vortex from each wing, with air on the sides of the vortices toward the bird going downward and air outboard going upward (Figure 12.7c). So a bird flying just behind another should be at a disadvantage. Conversely, a bird flying behind and to one side of another ought to be able to take advantage of the upwardly mobile air to reduce its cost of flight. A good aerodynamic argument thus favors flying in a diagonal line or a vee-formation in a horizontal plane. (While the leader clearly has to work harder than the followers, the leader need work no harder when followers are present than when they're absent.) Lissaman and Shollenberger (1970) calculated that by flying in formation twenty-five birds could get an increase of 70% in distance traveled for a given expenditure of energy. Badgerow and Hainsworth (1981) figured an energy savings of up to 23% for Canada geese (*Branta canadensis*) flying in formation. But achieving best economy demands precise coordination of position and the frequency and phase of wing beating, and the variation that Hainsworth (1989) found implies that the full, calculated savings in induced power are unlikely to be realized in practice.

Schooling in Fish

If birds can do it, why not fish? Weihs (1973) has described a diagonal diamond pattern with which swimming fish can save energy. The vertical, diagonal diamond amounts to a filled-in vee; fish, of course, need little net lift, so one can find a good site directly behind another as long as it's sufficiently far back. Also, lateral neighbors give additional benefit, as long as they're sufficiently far aside. While fish clearly school for nonhydrodynamic reasons, the appropriate spacings and synchronizations for energetic benefit do seem to occur.

Ground Effect

If an airfoil moves just above a solid surface, its performance is improved by a phenomenon termed "ground effect," mainly as a result of a decrease in induced drag. Withers and Timko (1977) have pointed out that skimmers (*Rhynchops nigra*), by flying close to the surface of calm water, can achieve about a 20% reduction in their requisite power (assuming an average wing-to-water distance of 70 mm). A skimmer, as pointed out back in Chapter 7, flies with its lower mandible partly submerged, but the mandible is well streamlined, and its hydrodynamic drag is negligible compared to the aerodynamic forces on the body. Fishing bats (Fish 1990) probably derive similar assistance from ground effect. And the phenomenon came up in the last chapter in connection with the sea anchor soaring of petrels and prions and the gliding of flying fishes. The scheme sounds like a wonderful way to check ground or water for prey, but I suspect that flying speeds are ordinarily too high to allow enough time for the predator to take action before moving well beyond the prey. Ground effect does somewhat reduce the minimum speed of flight, and that ought to be important for a bird (especially a large one) landing on ground or water without an awkward run or roll.

Ground effect seems to be used by fishes that swim just above the bottom, as many bottom-feeders do. Blake (1979b) calculated that it enables a mandarin fish (*Synchropus picturatus*), a negatively buoyant fish that hovers near the bottom, to reduce its power output 30% to 60%.

Combining Clap-and-Fling and Ground Effect

At the end, a speculative note. Flatfish, whether skates and rays or flounder and plaice, begin swimming upward from the bottom by elevating their leading edges. This might be especially effective in initiating the vortices associated with their production of thrust—using the hydrodyamic advantage of the fling or peel mentioned in connection with flapping wings. Together with the ground effect just described and the mechanical advantage of having a solid substratum to push against (used according to Webb 1981 in the fast start of a flatfish, *Citharchthys stigmaeus*), it could provide considerable assistance in starting. The matter was mentioned as a possibility for rays by Heine (1992) but work on the relevant hydrodyamics hasn't been reported. Another place where something of this sort might work is in penguins swimming just beneath the water's surface. Underwater swimming is better than surface swimming, as we'll see in Chapter 17, and being deeper is better than being barely submerged. But moving flippers downward from near the air-water interface might help start circulation and improve their efficiency. And the same possibility might be of consequence in dolphins and other air-breathers that often swim near the surface.

Flows within Pipes and Other Structures

QUITE A FEW chapters have come and gone since we last encountered situations in which the moving fluid was on the inside and the solid side of the interface was on the outside. Pipes, in particular, were put aside after the discussion of continuity and Bernoulli's principle—with the aid of streamlines, imaginary pipes could fill any field of flow. But organisms are filled with pipes and channels through which fluids flow, so it's worth plumbing the intricacies of how fluids go through such conduits. We'll mainly concern ourselves with fully laminar flow, where at every point both instantaneous and time-averaged velocities are the same. What we'll find is a domain far tidier than those to which we applied crude devices such as boundary layers and force coefficients. The tidiness, though, isn't princi- pally a result of dealing with internal rather than external flows—although that does simplify things a bit. Mainly, it reflects crossing a line defined by the Reynolds number. The external flows of interest so far have been either transitional or turbulent; that is, we had to contend with vortices, separa- tion, and true turbulence. Here we're dropping down into fully laminar flows for the first time, simply because internal flows of biological interest are mostly in that domain.

BASIC RULES FOR LAMINAR FLOW

Consider for a start a portion of a long, straight, unbranched pipe, one with rigid walls and a cross section of circular shape and constant size. Fluid has entered the pipe a long way upstream and flows steadily through it.

Continuity works with a vengeance—the amount flowing through any section is the same as that passing through any other section. What can we say, in quantitative terms, about flow through it?

Velocities across a Pipe

One nice thing to know would be the way in which velocity varies across the pipe—the equivalent of the velocity variation with depth in a boundary layer. An engineer usually uses a velocity distribution as a step toward some other calculation; for a biologist the distribution itself is of no slight inter- est. Obviously the velocity at the walls of the pipe will be zero; concomitantly the velocity along the axis must be maximal. We're a long way down the pipe, far enough from any entrance or other complication so that the no-slip condition will have exerted its influence throughout the cross section. In effect, the pipe will contain nothing but boundary layer, so the notion will be of no utility and can be discarded. At the same time the principle of continuity vetoes any chance that the skin friction of the pipe's walls might continually slow the flow—friction has to reduce the pressure, but it can't slow the flow. Thus we might expect that some velocity distribution across the pipe will, once established, persist farther downstream— like the logarithmic boundary layer on the earth's surface rather than the ever-thickening layer behind the leading edge of a flat plate.

Assume a piece of pipe of length l and radius a, as shown in Figure 13.1. What keeps fluid moving through it? It must be a drop in pressure, Δp, between the ends of the piece, exerted as a uniform cross-sectional push. We needn't worry about variations in pressure across the pipe associated with differences in velocity—Bernoulli, remember, is strictly applicable only along streamlines and assumes no frictional losses, which is exactly what is causing the present velocity variation. Thus pressure, not total head, will be constant across the pipe, just as across a boundary layer. What will resist the motion? The only relevant agency is the skin friction of the walls, dependent on the shear stress at the walls, again as in a boundary layer (and as in the definition of viscosity in Chapter 2). In a steady flow, the push and the resistance must just balance.

Since flow will be axisymmetric for a circular pipe, we can consider the piece of pipe as filled with concentric cylinders, each with a particular radius, r, measured from the center. These cylinders will slide past one another, with the center one going fastest and the outer one not moving at all. The force pushing a cylinder and each cylinder within it will thus be the pressure drop over that length of pipe times the cross-sectional area:

$$F_p = \Delta p(\pi r^2).$$

The force resisting the push will be the shear stress (τ) times the surface area of the side walls of the cylinder, and shear stress is the product of viscosity and velocity gradient:

$$F_r = \tau(2\pi rl) = \mu \frac{dU}{dr} (2\pi rl).$$

These two forces will be equal and opposite, so

$$\Delta p\pi r^2 = -\mu \frac{dU}{dr} (2\pi rl).$$

Canceling and rearranging, we get

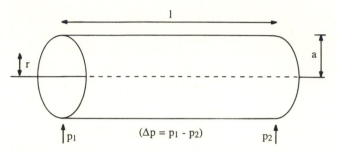

FIGURE 13.1. Conventions and symbols for describing flow through a length of cylindrical pipe.

$$dU = - \frac{\Delta p r \, dr}{2\mu l}$$

And integrating with U and r as the variables,

$$U = - \frac{\Delta p}{2\mu l} \int r \, dr = - \frac{r^2 \Delta p}{4\mu l} + C.$$

The no-slip condition requires that $U = 0$ where $r = a$, so

$$C = \frac{a^2 \Delta p}{4\mu l}.$$

And the whole expression becomes

$$U_r = \frac{\Delta p}{4\mu l} (a^2 - r^2). \tag{13.1}$$

How nice! Equation (13.1) is *exactly* what we sought—a formula for how the velocity varies across the pipe. The distribution, as shown in Figure 13.2, is a particular parabola. At the center ($r = 0$) the velocity is maximal, while at the walls ($r = a$) it's zero. Note that the velocity approaches its zero value at the walls more nearly linearly than asymptotically. The equation says other agreeable things. It asserts that if the pressure drop per unit length ($\Delta p/l$) doesn't change, then the velocity distribution won't either and, of course, vice versa.

Total Flow

Consider the same piece of pipe filled with concentric cylinders of flowing fluid, but now with a face area, dS, on each cylinder in the cross-sectional plane of the pipe. Each of these face areas is

$$dS = 2\pi r \, dr.$$

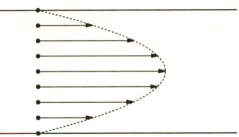

FIGURE 13.2. Velocities at a series of points across the diameter of a long, straight, circular pipe for laminar flow a long distance from the pipe's entrance. Lengths of the arrows are proportional to flow speeds at their bases; the dashed line marks the parabola of both calculation and measurement.

The volume flow (Q, volume per unit time, as opposed to U, distance per unit time or speed of flow) through each face, dQ, will be

$$dQ = U_r \, dS = U_r \, (2\pi r) \, dr.$$

The overall volume flow will then be

$$Q = \int_0^a U_r (2\pi r) \, dr.$$

The integral might look awkward, but we can now have equation (13.1) for U_r and so can plug 'n chug to get

$$Q = \frac{\pi \Delta p a^4}{8\mu l} = \frac{\pi \Delta p d^4}{128 \, \mu l}. \tag{13.2}$$

(The d in the final term is pipe diameter.) This expression is known as "Poiseuille's equation" or the "Hagen-Poiseuille equation" after its independent discoverers. As Prandtl and Tietjens (1934) pointed out, Hagen has precedence by a year (1839 vs. 1840) but he expressed his results in obscure units and for many years missed recognition. (Message from a reviewer of all too many papers: stick with SI.) As for Poiseuille, opinion in the English-speaking world varies on an appropriate mispronunciation.

Equation (13.2) makes the famous statement that volume flow is proportional to the *fourth* power of radius or diameter. A larger pipe of the same length carries *much* more fluid for a given pressure drop than does a smaller one. Thus halving the diameter of a pipe without change in volume flow rate entails a 16-fold greater pressure loss. Remember, though, that two factors co-conspire to give the drastic exponent—the larger pipe has both greater cross-sectional area and less wall area relative to its cross section. So dividing a big pipe into a pair of smaller ones without changing total cross section will, if the pressure drop is unchanged, only halve the overall volume flow rate. Also bear in mind (as some popular literature on coronary arteries does not) that pipes are commonly used in serial arrays, and alteration of the size of one element of an array will almost never change either flow or pressure drop in the way that casual application of equation (13.2) suggests.

Notice, mirabile dictu, the lack of nonsense about arbitrary constants, empirical coefficients, Reynolds numbers, or other bits of deus ex machina. Indeed, the main reason for presenting a derivation of (13.1) and (13.2) was to expose this uncharacteristic absence of chicanery. We did assume a few things—a long, straight pipe, laminar flow, uniform pressure, and viscosity—but we got unique and simple solutions. And these theoretical results correspond very well to reality—so well that (as Massey 1989 points out) the coincidence constitutes one of the better arguments justifying the presumption of a no-slip condition at walls.

Resistance

In laminar flow through a pipe, we've just shown that volume flow is proportional to pressure drop per unit length. The constant of proportionality can be called the resistance (R), as in its analog, Ohm's law, for electrical conduction. Thus

$$R = \frac{\Delta p}{Q} = \frac{8\mu l}{\pi a^4}. \tag{13.3}$$

This quantity, resistance, can be used to characterize some pipe or system of pipes just as the electrical resistance characterizes an ordinary ("ohmic") conductor. It will usually be independent of the particular pressure drop, total flow, or velocity at which it was determined. For laminar flow in a circular pipe beyond the "entrance region" (to be explained shortly), the final term in equation (13.3) can be used to calculate resistance; for resistive elements in general, the middle term applies. For arrays of pipes, the usual (electrical) rules for serial and parallel hookups apply: resistances in series add directly, while for resistances in parallel the sum of the reciprocals of the resistances gives the reciprocal of system resistance. Notice the fourth power of radius in the denominator: resistance, like volume flow, is exceedingly sensitive to small changes in pipe bore. Note also that resistance has peculiar dimensions: force times time divided by length to the fifth power.

Power

Because of the inevitable shear stresses, pushing a fluid through a pipe that has material walls takes work. It's a simple matter to calculate the power required either from the dimensions of the system along with velocity, volume flow, or pressure drop—or power can be obtained from some prior resistance measurement and a datum for volume flow or pressure drop. As with resistance, the case is formally the same as that of electrical conduction, with volume flow and pressure drop taking the place of current and voltage, respectively:

$$P = Q\Delta p = Q^2 R = \frac{(\Delta p)^2}{R}.$$ (13.4)

Power expenditure, as we'll see in the next chapter, is quite a serious consideration for such important arrays of pipes as those of circulatory and filtration systems.

Average and Maximum Velocities

Still other useful relationships fall into our laps from equations (13.1) and (13.2). The average velocity, $\bar{U}$, of flow in a pipe is the total flow, Q, divided by the cross-sectional area, S. From (13.2), then,

$$\bar{U} = \frac{\Delta p\, a^2}{8\,\mu l}.$$ (13.5)

The maximum velocity, occurring along the axis of the pipe, can be obtained from (13.1) by setting $r = 0$:

$$U_{max} = \frac{\Delta p\, a^2}{4\,\mu l}$$ (13.6)

These are truly splendid results! They tell us that the maximum velocity is precisely twice the average velocity, something exceedingly useful. If you mount a calibrated flow probe on the axis of a pipe you can obtain average velocity, total flow, resistance, and power—all from a single datum without the bother of a lateral traverse, integration, or iffy approximation. Conversely, if you can measure total flow, perhaps by catching the output of the pipe over a measured time, you can calibrate a flow probe located along its axis, or you can tell the maximum rate at which some item, in solution or suspended, can possibly be transported.

On several occasions, I've used equation (13.6) to ease awkward problems of measurement. In a transparent tube of liquid, one can readily follow the progress of an advancing front of dye and thus obtain a datum for axial velocity. Combined with the dimensions of the tube and the viscosity of the liquid, it gives the pressure drop in the tube. The setup, then, is a kind of pseudomanometer, useful for measuring very low pressure differences in liquid systems. One need only connect two points in a flow by an appropriate tube and inject a bit of dye. As an example, consider a water-filled tube of 2 mm internal diameter and 20 cm length with a measured axial flow of 2 mm s^{-1}. The corresponding pressure drop is 1.6 Pa or 16 microatmospheres, a really low pressure. But it's the difference between static and dynamic pressures in a water current of 57 mm s^{-1}, a speed of flow that's really in the biological mainstream. As a general rule when doing such pseudomanometry, the Reynolds number, based on the diameter of the tube, should be kept under about 50 in order to avoid bias from entrance phenomena. Also, the flow in the tube should be less than about a tenth of the external current being measured, or else flow through the tube may appreciably relieve the pressure difference being measured. The latter requirement makes little trouble since one can always slow the flow by using a longer tube.

Roughness

The relative irregularity of the inner wall of a pipe has only a small effect on the volume flow or pressure drop—at least where flow is laminar. The same kind of criteria for tolerable roughness apply that were cited for flow over a flat plate (Chapter 8)—the critical thing is the Reynolds number of the heights of projections. Above 30, for pointed projections, and above 50, for rounded projections, roughness appreciably affects the shearing stress and character of the flow. The permissible heights (ϵ) of roughness elements, according to Goldstein (1938), are then

$$\frac{\epsilon}{a} \leq 4\, Re^{-1/2} \text{ (pointed)} \tag{13.7}$$

$$\frac{\epsilon}{a} \leq 5\, Re^{-1/2} \text{ (rounded).} \tag{13.8}$$

(The Reynolds numbers in these formulas are based on the diameters of the pipes and not on the heights of the projections.) We see that the permissible roughness heights relative to pipe radius decrease as the Reynolds number increases—the flow becomes more sensitive to disturbances such as might be caused by bumps and bristles. The limits are very generous, though; even at a Reynolds number as high as 1000 (as with a water flow of 0.1 m s^{-1} through a 10 mm pipe) a pointed protrusion can be up to an eighth of the pipe's radius without making much difference. And for lower Reynolds numbers considerably larger protrusions are tolerable. The main effects, then, of protrusions are reduction in the overall cross-sectional area of a pipe and alteration of the circular cross-sectional shape.

The "Entrance Region"

All the previous formulas assume that a steady-state parabolic velocity distribution has been established somewhere upstream. We turn now to

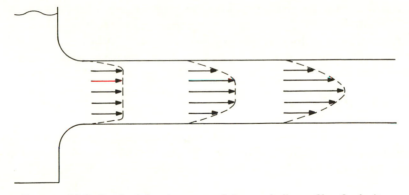

FIGURE 13.3. Gradual development of the parabolic profile of velocity in the entrance region of a cylindrical pipe.

what happens at that "somewhere upstream." As fluid enters a pipe from, say, a reservoir, its speed will be nearly uniform across the pipe; the velocity profile is called "plug flow" or "slug flow." As shown in Figure 13.3, a gradient region gradually develops, beginning at the walls of the pipe and thickening downstream as does a boundary layer until the final parabola is achieved; the flow is then said to be "fully developed." In this entrance region, mass and momentum are transferred from the periphery toward the axis of the pipe—flow in the middle has to speed up as flow near the walls diminishes. The kinetic energy of the fluid increases with distance from the entrance, and the pressure drop per unit length is therefore greater than that predicted by the Hagen-Poiseuille equation (13.2).

The immediate question about the entrance region concerns its length. Here we have to resort to an approximation—seeking the end of an asymptotic curve is a little like looking for the end of a rainbow. The usual approximation, a relatively stringent criterion, calls the flow fully developed when it achieves 99% of its final axial velocity. Just as the thickness relative to downstream distance of a boundary layer depended only on the local Reynolds number, so does the "entrance length" (L') expressed in units of pipe diameters (d):

$$\frac{L'}{d} = 0.058 \, Re = 0.058 \, \frac{\rho U d}{\mu}. \tag{13.9}$$

(Some sources give slightly different numerical constants.) For many purposes it's unnecessary to get so close to full development, and a constant about half that of equation (13.9) may be ample. In fact, the Hagen-Poiseuille equation gives a decent estimate of the pressure drop almost to the very entrance. Conversely, whatever the outcome of applying (13.9), it's best to allow a length of at least one pipe diameter downstream from an entrance for the parabolic profile to develop, no matter how low the Reynolds number.

Another way of looking at development in the entrance region is in terms of the time needed for the process. Since length is average velocity times time,

$$t = 0.058 \frac{\rho d^2}{\mu}. \tag{13.10}$$

The time needed for a flow to become fully developed is independent of its velocity; it varies inversely with either dynamic or kinematic viscosity of the fluid and varies directly with the cross-sectional area of the pipe.

The Limits of Laminarity

The classic experiments of Osborne Reynolds (1883), described in Chapter 5, are again of direct relevance. Reynolds established a value of around $Re = 2000$ for the transition from laminar to turbulent flow in a circular pipe (using as characteristic length the pipe's diameter). Below 2000, a disturbance will not persist but will damp out through viscous action. Above 2000, a disturbance, once started, propagates throughout the fluid as it travels down the pipe. While the transition is commonly sudden—that is, the transitional Reynolds number is well defined—transition doesn't necessarily occur at that exact value. Much depends on the smoothness of entry upstream, the distance from entry or upstream branch point, the straightness of the pipe and smoothness of its walls, and so forth. With meticulous care, transition can be postponed to Reynolds numbers as much as 10-fold higher.

If you want a quick indication of the limit of laminar flow in a pipe, using $Re = 2000$ as the criterion, just remember that for a 10 mm pipe it's about 0.2 m s^{-1} in water and about 3 m s^{-1} in air. Only rarely within organisms are these limits exceeded—the flow through most of our internal pipes is decently laminar, even if many organisms find themselves in situations where turbulent pipe flow is relevant.

Laminar Flow between Parallel Plates

The general tidiness of the equations for laminar flow through circular pipes reflects the laminarity of flow rather than the specific and simple geometry of the pipes. One can do equivalent derivations to obtain equations that apply to other geometries—perhaps the most useful of such equations is a set that applies to a channel between a pair of flat plates (Figure 13.4). By assuming that the plates are very much closer to each other than the channel is wide, we grant ourselves license to ignore the

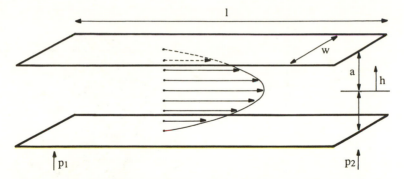

FIGURE 13.4. Flow between a pair of wide but closely spaced parallel plates—conventions and another parabolic profile.

edges of the plates, and we can again look at flow far enough downstream to presume full development. To keep matters as similar as possible to the case developed for pipes, let a now represent half the distance between the plates and h define locations from the center ($h = 0$) to either surface ($h = a$). The analog of equation (13.1) is the following:

$$U_h = \frac{\Delta p (a^2 - h^2)}{2\mu l} \tag{13.11}$$

Things look very much the same! But not exactly—the change from four to two in the denominator means that, while the profile remains parabolic, the parabola is a slightly different one.

A similarly obtained equation for volume flow has the same near but not complete identity with the one for circular pipes. We just treat the flow as a stack of parallel fluid plates bounded by the solid ones. Using w for the width of the channel and d for the distance between plates ($= 2a$), we get an analog of the Hagen-Poiseuille equation (13.2):

$$Q = \frac{2\Delta p w a^3}{3\mu l} = \frac{\Delta p w d^3}{12 \mu l}. \tag{13.12}$$

Note that the fourth power of the radius of the pipe is replaced by the third power of the shorter cross-sectional dimension of the channel times the first power of the longer cross-sectional dimension. The former, of course, provides all (in our derivation) the shear stress that resists the flow.

While at it, we can extend the parallel between circular pipes and parallel plates by giving equations for average and maximum velocities:

$$\bar{U} = \frac{\Delta p a^2}{3\mu l} \tag{13.13}$$

299

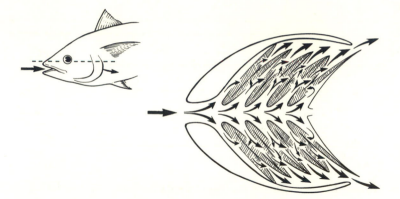

FIGURE 13.5. The sieve units of the gill of a tuna, usefully idealized as a set of closely spaced parallel plates. The view is a frontal section, in the plane indicated by the dotted line in the insert at left.

$$U_{max} = \frac{\Delta p a^2}{2\mu l}.$$ (13.14)

The maximum velocity, instead of being twice the average, is now one and a half times the average. Again, though, the ratio comes out to an engagingly round number.

Quite a few biological situations involve closely spaced and parallel flat plates—the gills of fish and many invertebrates, the book lungs of spiders, various nasal passages, and so forth. It should make little difference if, as in nasal passages, the channels and its walls are rolled as long as channel depth is roughly constant. Equation (13.11), for instance, was used by Stevens and Lightfoot (1986) and (13.12) by Hughes (1966) in work on flow through the gills of fish. A gill of an ordinary bony fish has four slots between its arches (Figure 13.5); in tuna (where they're relatively narrow) a slot (called a "sieve unit") is 127 μm by 20 μm in cross section and about 1.5 mm in the direction of flow. The Reynolds number is under 100 (Stevens and Lightfoot 1986).

Basic Rules for Turbulent Flow

Turbulent flow is less straightforward than laminar flow, both physically and mathematically. In turbulent flow, momentum is continually being transported across a pipe, so the notion of a set of concentric sliding cylinders applies only to a time average of the instantaneously jumbled motion. The precise parabola disappears; and the velocity distribution, even as an average, resists simple description. The speed of flow is still zero at the walls, but flow along the axis is less than twice the average speed. And the roughness of the inside wall of a pipe, of little effect in laminar flow, becomes a major determinant of the pipe's resistance.

Friction Factor

To deal with turbulent pipe flow it's the practice to heap all eccentricities onto a dimensionless variable analogous to the drag coefficient. C_d, you'll recall, could be defined as the ratio of drag per unit surface to the dynamic pressure. Similarly, a "friction factor," f, can be defined as the ratio of the shear stress to the dynamic pressure. The former is what impedes flow in pipes; the latter is what promotes it. Alternatively, the friction factor can be viewed as the ratio of the pressure drop per unit length times pipe diameter to the dynamic pressure:[1]

$$f = \frac{2\Delta pa}{l} \bigg/ \frac{\rho \bar{U}^2}{2} = \frac{4\Delta pa}{\rho l \bar{U}^2}.$$ (13.15)

Just as the drag coefficient is a function of the Reynolds number and the shape of an object, the fraction factor is a function of the Reynolds number and the roughness of the lining of the pipe. And in just the same way, we can plot the friction factor as a function of the Reynolds number, as in Figure 13.6.

This graph of friction factor against Reynolds number has several noteworthy features. The left-hand line refers, of course, to laminar flow, where the friction factor is simply $64/Re$, a result following directly from equations (13.2) and (13.15). (In practice one ignores both friction factor and Reynolds number for laminar flow and just uses equation 13.2.) The right-hand lines are a little more complicated. They begin well above nearest values for laminar flow, indicating that the transition to turbulence involves an abrupt rise in the resistance of a pipe to flow. For smooth pipes the friction factor decreases slowly and steadily with increases in the Reynolds number over quite a range. For rougher pipes that decrease ceases sooner; in all cases the friction factor doesn't change much at yet higher Reynolds numbers.

---

[1] Taken literally, defining friction as shear stress over dynamic pressure generates a coefficient with a numerical constant of 1 rather than 4. Some sources use the constant of 4, as here, but others don't. If it isn't obvious which is intended, find the coefficient that applies to laminar flow. If the latter is $16/Re$, then the constant is 1; if it's $64/Re$, the constant is 4. The names given the dimensionless index vary sufficiently to lack reliability in the matter—pressure drop coefficient, pipe resistance coefficient, Fanning friction factor, and so on. Most recent American sources use the 4 and call the result the friction factor.

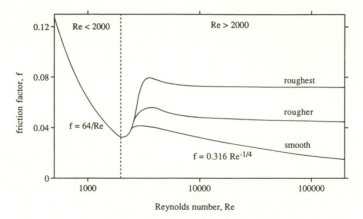

FIGURE 13.6. A plot of friction factor, f, versus Reynolds number for flow through cylindrical pipes of varying degrees of wall roughness. Note the transition at $Re = 2000$ and the fact that roughness matters only above the transition.

Velocity Distribution

As mentioned, the onset of turbulence in a pipe is accompanied by a drastic alteration of the distribution of velocities across the pipe. Not only does the regular and consistent parabolic distribution characteristic of laminar flow disappear, but, even worse, the actual distribution for turbulent flow depends on both wall roughness and Reynolds number. For smooth pipes, the ratio of maximum to average velocity diminishes slowly with increases in the Reynolds number, eventually reaching an asymptote of about 1.25 (Prandtl and Tietjens 1934). For rough pipes, this ratio is nearly constant above the transition range, with the value depending on the roughness. The subject of velocity distributions and friction factors is somewhat more complex than this and is of great technological (if not biological) consequence; it's considered in detail in most textbooks of fluid mechanics.

Entrance Length

The entrance length for turbulent flow is less than that for laminar flow. This results from the lateral transport of momentum in eddies and from a steady-state velocity profile closer to the slug flow of the entrance. Caro et al. (1978) give the following equation for estimating entrance length:

$$\frac{L'}{d} = 0.693 \, Re^{-1/4}. \tag{13.16}$$

As with the formula (13.9) for laminar flow, the criterion is a fairly stringent one and might be relaxed substantially for many purposes.

Flow through Circular Apertures

Two geometrically similar but physically disparate situations share this designation—first, flow of a liquid out a bunghole into a gas and, second, flow of a fluid through what is essentially a pipe of negligible length from one reservoir into another. We'll postpone any consideration of the former to Chapter 17 since it involves fluid-fluid interfaces, focusing here on the latter, sometimes called "flow through a submerged aperture." Flow through such orifices isn't uncommon in biological systems—consider pits and stomata in plants, perforate vessel walls in various places such as the glomeruli of kidneys, and various other filtration devices such as the dermis of sponges. Still, in all such cases one must carefully distinguish between situations in which net transport is osmotically or diffusively driven and those involving true bulk flow forced by hydrostatic pressure differences.[2]

For a single aperture, reasonably far from other apertures, a formula analogous to the Hagen-Poiseuille equation (13.2) is quoted by Happel and Brenner (1965):

$$Q = \frac{a^3 \Delta p}{3\mu} . \tag{13.17}$$

The practical difficulty with equation (13.17) is that it's dependable only up to a Reynolds number of about 3. Above that, edge effects complicate matters, the shape and sharpness of the lip become important, and vortices are generated as fluid leaves the aperture.

For higher Reynolds numbers, another and more commonly cited formula has to be used:

$$Q = C_0 \pi a^2 \sqrt{\frac{2\Delta p}{\rho}} . \tag{13.18}$$

In (13.18) a dimensionless "orifice coefficient," C_0, is needed. The value of the coefficient varies with the Reynolds number but may be taken as 0.61 above about $Re = 30,000$ as a reasonable approximation. Do not, as I once inadvertently did in a lecture, refer to the orifice coefficient as "just another bugger factor." This orifice coefficient has to be distinguished from the so-called coefficient of discharge, which more commonly (usage is inconsistent) applies to nonsubmerged apertures—but sometimes the same

[2] Even if the hydrostatic pressure difference has ultimately been generated by some osmotic potential.

formula applies. Also, whatever the Reynolds number, bear in mind that these formulas presume flows that are hydrostatically driven without any push from preexisting momentum. In practice that means that flow within the orifice must be at least several times faster than the flow upstream of it.

FORMULAS FOR FLOWS AT LOW VERSUS HIGH REYNOLDS NUMBERS

Both for fully developed flow in circular pipes and for flow through sharp-edged circular apertures, we've needed two kinds of formulas. One applies when flow is not only laminar but vortex-free as well; the other is used for other situations—laminar flow with discrete vortices and fully turbulent flow. The contrast between the two kinds of formula is a central and important one in fluid mechanics. It's at least as important to biological applications of fluid mechanics as anywhere in technology because living systems produce and encounter flows on both sides of the transitions. The contrasting character of equations applicable to the two sides of the transition—beyond the geometric details—is worth some comment.

Table 13.1 draws attention to the features that mark this great divide by expressing the relevant equations in equivalent form. Four of the equations come from this chapter—(13.2), (13.15), (13.17), and (13.18). A fifth is equation (5.4), rearranged so the drag on a body (we'll presume a sphere and use frontal area) is expressed as a rearward pressure. The sixth is the analog of the fifth for low Reynolds numbers, Stokes' law, which will appear later as equation (15.1). Looking at the differences between the two columns . . .

1. Pressure drop depends on viscosity rather than on density at low Reynolds numbers; it depends mainly (one can't make a stronger statement with one of those coefficients present) on density at high Reynolds numbers.
2. While pressure drop varies directly with velocity at low Reynolds numbers, it varies essentially (again hedging a little because of the coefficients) on the second power of velocity at high Reynolds numbers.
3. If all terms are multiplied by area, one sees that force varies directly with some linear dimension at low Reynolds numbers but with the square of a linear dimension at high Reynolds numbers.
4. All of what we're calling "high Reynolds number" versions have the form of some dimensionless coefficient times dynamic pressure. That kind of formula, of course, is fully general because of the inclusion of a universal variable constant—a dimensionless coefficient providing a scapegoat for all irregularities. But it's something

TABLE 13.1 EQUATIONS FOR PRESSURE DROP, ΔP, (A) THROUGH A
SECTION OF CIRCULAR PIPE WITH FULLY DEVELOPED FLOW; (B) THROUGH
A SHARP-EDGED CIRCULAR SUBMERGED APERTURE; AND (C) AVERAGED
OVER THE FACE AREA FOR UPSTREAM VERSUS DOWNSTREAM SIDES OF A
SPHERE. IN (C) THE AVERAGE VELOCITY REFERS TO FREE STREAM FLOW.

	Low Re	*High Re*		*Transition Re*
(a) Circular pipe	$\dfrac{8\mu l \bar{U}}{a^2}$	$\dfrac{fl}{2a}$	$\dfrac{\rho \bar{U}^2}{2}$	2000
(b) Circular aperture	$\dfrac{3\pi\mu\bar{U}}{a}$	$\dfrac{1}{C_o^2}$	$\dfrac{\rho \bar{U}^2}{2}$	3
(c) Sphere	$\dfrac{6\mu\bar{U}}{a}$	C_d	$\dfrac{\rho \bar{U}^2}{2}$	0.5

of a last resort, as emphasized back in Chapter 5. By contrast, the low Reynolds number versions are blessedly unambiguous.

5. The transition points are the maximal Reynolds numbers to which the "low" versions can be safely applied. By engineering standards, even what we're calling "high" are really very low Reynolds numbers. But biology lives with a curious asymmetry with respect to this general divide. For apertures and spheres, the divide corresponds fairly neatly to what we'd ordinarily regard as microscopic and macroscopic worlds. For circular pipes, both macroscopic and microscopic systems are mostly on the low Reynolds number side, and we encounter the other domain mainly in either field work or in the design of apparatus such as flow tanks. The message, to repeat, isn't that pipe flows aren't intrinsically tidy and external flows naturally messy (by analogy perhaps with lab and field science) but rather that we're ordinarily on the side of the angels in one case and too often bedeviled in the other.

FLOW THROUGH POROUS MEDIA

Whether flow goes through a pipe or between parallel plates, if it's laminar the average velocity varies directly with the pressure drop per unit length. That's what Gotthilf Hagen and Jean Louis Poiseuille determined in 1839 and 1840, respectively. In 1856, the same basic relationship was shown for what might appear quite a different situation—flow through porous media—by Henri Darcy, as what's now called "Darcy's law" (Rouse and Ince 1957). Its applications are wide, including seepage underneath dams, flows beneath beaches and streams, and filtration by flow through packed columns. On one hand, flows are inevitably slow and passages narrow, so laminarity is certain. On the other hand, the passageways are complicated and irregular, so one can talk only in global or statistical terms about flow and pressure—velocity profiles of the kind we derived earlier can't be easily specified.

Flow though a porous medium made up of packed particles is equivalent to flow through a statistical spaghetti of pipes with ramifying and anastomosing passages. Clearly the constant of proportionality between pressure drop and flow must depend on the average size, the distribution of sizes, and the arrangement of the solid particles. Despite the complications the so-called Kozeny-Carman equation, embodying Darcy's law, turns out to work quite well in practice. Its problem is inclusion of several variables of variable degrees of awkwardness. (Massey 1989 gives a straightforward derivation of the equation and its constituent variables, and Leyton 1975 discusses its biological relevance.) Assume a channel of constant overall cross section filled with particles through which fluid flows under the influence of a specific pressure drop per unit length. One needs to know (1) the volume of the solid particles, V_s, and (2) the voidage or porosity, ϵ, the fraction of the total volume of the pathway made up of the fluid phase. One also needs (3) the total surface area, S, of the particles. All can be determined or approximated from fairly simple measurements. All other relevant variables are lumped into the Kozeny function, k—the tortuosity of the passages, the effective surface area of the particles since some solid surfaces contact others, and so forth. Specifically, then,

$$ U = \frac{\Delta p}{l} \frac{\epsilon^3}{(1 - \epsilon)^2} \frac{1}{\mu k (S/V_s)^2} . \tag{13.19} $$

In practice, according to Massey, packed granular materials of uniform particle size have values of ϵ between 0.3 and 0.5. As particle size becomes more heterogeneous, porosity drops sharply—voids between large particles are increasingly filled with small ones. Immediately adjacent to walls, porosity is greater—a wall prevents contacting particles from interdigitating as effectively. Thus in a narrow packed column or one with a lot of wall relative to cross section, flow may be somewhat greater than calculated. That's rather the opposite of what happens in open columns, but in packed columns the particles rather than the walls constitute the dominant resistance.

Kozeny's function, k, normally ranges between 4.0 and 6.0; and a value of 5.0 is commonly assumed for calculations with equation (13.19).

Considering the abundant and diverse fauna living in porous sediments (as well as the use of separation columns in the laboratory), the Kozeny-Carman equation ought to see more biological application than I've been able to uncover. For macrofauna, penetration of dissolved oxygen will depend strongly on even slight local flows; and for local flow around interstitial microfauna, overall bed resistance is clearly crucial. On a larger scale, Riedl (1971b) and Riedl and Machan (1972) have found that the net flow of water into beaches driven by the gravitational pressure head of the swash has a typical speed of 0.5 mm s^{-1}, and that the resulting flow into the ocean is greater than that of all the world's rivers combined.

Just as the forces of flow affected the shapes of flexible objects, which in turn affected those forces, flow through a porous medium may affect the porosity of the medium, which can then alter the flow. In particular, if water flows upward through a bed of particles at a sufficient rate (which may still be very slow), the particles may "unpack" to some extent. In effect they're supported to some extent by fluid motion as well as by each other; as a result of this "fluidization" the properties of the bed can change dramatically. The most common phenomenon that can be blamed on such upward flow and slight unpacking is so-called quicksand, a sandy bed that looks solid but behaves more like a fairly viscous liquid. I wouldn't be at all surprised to learn that some infaunal organism makes quicksand either incidental to respiratory pumping, as part of a burrowing scheme, or as a feeding device. But at this point I can't cite specific investigations.

CHAPTER 14

Internal Flows in Organisms

THE PAST CHAPTER considered little more than the physical characteristics of internal flows and the various equations that have proven
useful in dealing with them. While choices of scale and regime were made
with living systems in mind, the specifics of these systems were deferred.
These biological systems are both diverse and complex; a whole book
rather than just a chapter could profitably be devoted to the subject. In any
case, I'll use the title of the book to justify an episodic rather than a comprehensive look. And these systems are pretty fancy. Most circulatory systems
of animals involve pulsatile flow of non-Newtonian fluids in pipes of time-
varying cross-sectional areas and shapes; here we have space only for a stab
at the high points. The vessels through which sap ascends in vascular plants
are tiny, interrupted by barriers, and filled with water at wildly subzero (not
just subatmospheric) pressures—their fluid statics have proven sufficiently
enigmatic to loom larger in the literature than their fluid dynamics.

CIRCUMVENTING THE PARABOLIC PROFILE

The velocity distribution across a circular pipe in fully developed laminar flow is parabolic, with an axial velocity precisely twice the average
velocity. In many biological situations either material or heat is exchanged
across the walls of pipes; a moment's consideration should persuade you
that a parabolic profile isn't exactly ideal for exchange. Too much of the
material or heat passes down the middle of the pipe and too little moves
along the edges, near the site of exchange. This places a heavy burden on
molecular diffusion or conduction, weak reeds in a macroscopic, non-
metallic world. And nature has repeatedly found it prudent to arrange
something better than a parabolic profile.

How Far Is Flow from a Wall?

Before revelling in adaptive tricks, it's handy to have an expression for
the distance of the average particle of flowing fluid from the walls of a pipe
or channel. A convenient device is a dimensionless, size-independent index that compares the average distance of the flow from a wall with the
radius of a pipe or the depth of a channel—we might call it the "distance
index," Di.

Consider flow through an axisymmetrical circular pipe. To get the mean distance, the pipe again needs to be viewed as a series of concentric cylinders, each with an annular face. The area of each annulus ($2\pi r\, dr$) must be multiplied by the local velocity (U_r) and the distance from the wall ($a - r$) and divided by the total flow ($\pi a^2 \bar{U}$). The result is then integrated across the radius to get the mean distance, which is divided by the radius of the pipe. Written out as an integral, the procedure can be summarized as

$$Di = \frac{2}{a^3 \bar{U}} \int_0^a U_r (ar - r^2)\, dr = \frac{2\pi}{Qa} \int_0^a U_r (ar - r^2)\, dr. \qquad (14.1)$$

Either form of this expression can be solved by either of two approaches. If an explicit expression is available for the velocity distribution, that is, for U_r as a function of r, the expression can be substituted for U_r and the integral evaluated by the usual mathematics. Alternatively, an experimentally determined set of measurements of velocity at different radii can be used. One by one, each is multiplied by its appropriate dr (now a Δr) and by ($ar - r^2$); the sum of the results is then multiplied by the factors outside the integral sign.[1]

For a uniform velocity (plug or slug flow) across the pipe (Figure 14.1a), $\bar{U} = U_r$. The remainder of the integral is easily evaluated, and everything cancels out after integration except a numerical coefficient. The latter is the desired distance index, 0.333. Flow is, on the average, a third of the way from wall to axis.

For the parabolic profile of fully developed laminar flow (Figure 14.1b), U_r is given by equation (13.1), so one just has to insert that along with equation (13.2) for Q into (14.1). After we do a little more bookkeeping, a numerical coefficient again emerges; this time the distance index is 7/15, or 0.467. Flow, on the average, is 40% closer to the axis than in the preceding case; for exchange of material across walls, parabolic flow is substantially worse than slug flow.

What if the parabolic profile were reversed? That's about what should happen if what is forcing the flow rather than what is resisting the flow is located along the walls. The distance index one calculates depends on the conditions assumed; consider for an example a case in which axial flow is exactly zero, wall flow is maximal (assuming that the gradient region enforced by the no-slip condition is thin enough to ignore), and the parabola is the inverse of the previous one (Figure 14.1c). The distance index comes out to 0.20—flow is 40% closer to the wall than even for slug flow.

To get some examples of the use of equation (14.1) with empirical data, I

---

[1] Note that this procedure isn't universally applicable. The Romberg method, conceptually a little more complex, would require fewer measurements for a given level of precision; for details, see Pennington (1965).

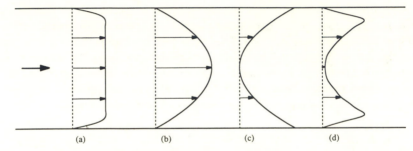

FIGURE 14.1. Velocity profiles across a cylindrical pipe: (a) plug or slug flow, with uniform speed across the pipe; (b) our familiar parabolic profile; (c) a reversed parabolic profile, with zero speed on the axis and highest speeds almost at the walls; (d) an experimentally determined profile for a pipe containing an array of diametric wires.

determined the velocities across a real pipe (about 75 mm in diameter), using it in several arrangements. For an unobstructed pipe, I got a distance index of 0.447—the flow was not quite fully developed ($Di = 0.467$). When I filled the same pipe with wool fibers, the distance index dropped to 0.387—much closer to slug flow figure of 0.333. The effect of the filling is to distribute the resistance to flow across the entire volume of the pipe and thus to reduce the steepness of the parabola. But doing still better was easy. I threaded a set of wires across the pipe, with each wire passing diametrically and thus crossing the axis normally (Figure 14.1d). In effect, the scheme put an axial plug in the pipe. As a result, the distance index dropped to 0.311—better than slug flow. And the pressure drop caused by the wires was substantially less than that of the wool stuffing. It's uncertain whether any organism grows hairs that, by extending to or beyond the axis of a pipe, push the average flow toward the walls and increase exchange between fluid and walls. It's also uncertain, to me at least, whether anyone has looked for such a phenomenon. Don't sniff at the notion—who knows what nasal hairs might do!

The general ways to circumvent parabolic flow, to reduce the distance index, and to improve intimacy of flow with walls are worth a little exploration.

Using Noncircular Cross Sections

From the point of view of exchange, a circular section is the worst geometry, whether we consider slug flow or parabolic flow, whatever its compensating advantages in cheapness of construction, mechanical robustness in the face of pressure differences across the walls, and minimum pressure drop per unit of total flow. Flow between parallel flat plates is better. An analog of equation (14.1) for this latter geometry (using the half channel from center to one wall) is easy to contrive, and the correct parabolic flow is described by equation (13.11). The result is a distance index of 0.625 for parabolic flow and 0.50 for slug flow—slug flow is again better. While both figures sound high, that just reflects the different geometry and a slightly inappropriate comparison. One does a bit better by taking the squares of these figures to shift from a one-dimensional to a two-dimensional channel, getting 0.391 and 0.25 (to compare to the 0.467 and 0.333 earlier).

Large pipes through which exchange takes place are, in fact, often closer in cross section to parallel plates than to circles. A few examples (some illustrated in Figure 14.2) are the internal gills of fish (mentioned in the last chapter) and many other animals; nasal passages, especially those working as countercurrent heat exchange systems (Schmidt-Nielsen 1972); the intestine of an earthworm, with its large dorsal invagination, the typhlosole; the "spiral valve," a spirally-coiled fold that fills most of the small intestine of a shark; and the passages of the open parts of the circulatory systems of the bivalve and gastropod mollusks and of the arthropods. But while most pipes are nicely circular for reasons well rooted in a firm combination of solid mechanics and geometry, parallel plates are only approximately parallel, constrained merely by the present considerations of fluid mechanics and exchange processes. So specific analyses can rest less easily on the equations in the last chapter. Nasal passages, for instance, may be closer to parallel plates than to circular pipes, but a look at Morgan et al. (1991) should persuade anyone that the flows within them are bogglingly complex.

Making the Pipes Very Small

Strictly speaking, this doesn't reduce the distance index since the latter is size independent, but it certainly will improve exchange. Indeed, the use of very small pipes must be the most common way to improve exchange, evident in capillary beds, the parabronchi of bird lungs, renal and Malpighian tubules, and numerous other cases in which exchange occurs as fluid moves through internal conduits. Inevitably (as mentioned in Chapter 3) the total cross-sectional area of arrays of small pipes is huge and the velocities in them are therefore low, providing time for exchange as well as area across which exchange can take place. These low flow speeds also keep the cost of pumping fluid through small pipes from becoming inordinately high, something to which we'll return shortly.

Eddies and Turbulence

Turbulent flow has a flatter velocity profile than does laminar flow and thus is associated with lower distance indices. But in the presence of lateral bulk transport of fluid, the distance index either becomes irrelevant or

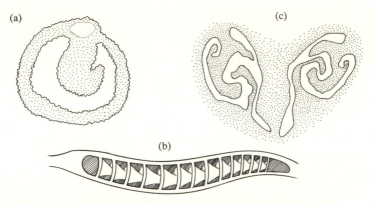

FIGURE 14.2. Examples of pipes with noncircular cross sections: (a) the intestine of an earthworm, with the large dorsal typhlosole protruding downward; (b) the "spiral valve," in cutaway view, which runs the length of a shark's intestine; (c) a cross section of the nasal passages of a kangaroo rat.

takes on quite a different meaning—turbulence will greatly augment exchange. A minor concomitant is the greater cost of producing turbulent flow associated with the greater pressure drop per unit length of pipe. A more important limitation is the impracticality, and often impossibility, of getting turbulent flow at the low Reynolds numbers set by the sizes of most biological pipes. Eddies can be generated by protrusions in a pipe, and these will improve exchange, but the Reynolds number still has to be reasonably high—at least above about 30. The nasal passages of mammals look like reasonable candidates for such deliberate generation of turbulence; the latter would then improve olfaction and heat exchange.

Periodic Boluses

If a solid plug is passed through a pipe, then flow immediately fore and aft of the plug should be more nearly linear than parabolic in profile. We send red blood cells through capillaries at a great rate, and the cells have to deform somewhat to squeeze through. Thus exchange between plasma and capillary walls might well be improved as a purely physical consequence of the presence of red blood cells. In fact, putting boluses through a pipe as close together as are the red cells at a concentration ("hematocrit") of over 40% ought to generate an interesting pattern of flow between them. As Caro et al. (1978) and most other standard sources point out, flow should be toroidal in the spaces between the cells; what resistance is present is still along the walls, so the flow of plasma (relative to the average flow, not the walls) should be forward along the axis and rearward along the walls

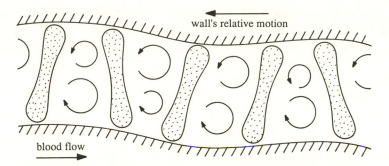

wall's relative motion

blood flow

FIGURE 14.3. Red blood cells moving through a capillary. To view the toroidal flow shown between cells one needs an unusual frame of reference—imagine that the red cells are stationary and the capillary is moving past them from right to left.

(Figure 14.3). Either slug flow or such toroidal flow is obviously better for exchange than is parabolic flow, as suggested by Prothero and Burton (1961).

But there's a problem—one of significance well beyond the particulars of red blood corpuscles in capillaries. When looking at augmentation of diffusion, what's important is the magnitude of the proposed mechanism of augmentation relative to that of diffusion—just as with many problems involving several physical agencies. For this one, where the agencies are convective bulk flow and diffusion, the relevant ratio is the so-called Péclet number, $Pé$,

$$Pé = \frac{lU}{D}, \tag{14.2}$$

where D is the diffusion coefficient of the dissolved substance of interest. A low Péclet number means less importance for convection; for oxygen in blood in a capillary it should be about unity, which is pretty low. In effect, the time for circulatory motion between red cells is long (low U) relative to the time involved in purely diffusive exchange (high D and low l), so the model underlying the textbook bolus effect isn't really appropriate (Duda and Vrentas 1971; Middleman 1972). On the general issue of convection versus diffusion, I suggest consulting Lightfoot (1977).

You may recall that the issue of diffusion versus convection arose earlier (Chapter 8) in connection with diffusive boundary layers. It's always a matter of concern for systems of cellular dimensions. At least in animals,[2]

[2] Plant cells are typically larger than animal cells and commonly have considerable internal bulk flow ("cyclosis"), so the coincidence of physical and biological shifts isn't so tidy.

the boundary between cellular and supercellular domains constitutes an approximate boundary between two physical worlds. In the small world of cells, diffusion can provide the main agency of molecular transport; in the larger supercellular world, augmentation by bulk flow is the nearly universal way to get around the slowness of diffusion over longer distances. I've dwelled on the point elsewhere (Vogel 1988a, 1992a) so I won't make more noise about it here.

Pumping at the Wall

The derivation of the Hagen-Poiseuille equation assumed that the walls provide the resistance to flow, and this assumption generated a parabolic profile. If, by contrast, ciliated walls provide the pumping and the resistance comes from elsewhere in the system and is expressed across the pipe, then the profile will be quite different. A distance index of 0.2 was calculated earlier for the case of an exactly reversed parabolic profile. Of course the no-slip condition must still apply, but here the gradient from zero speed at the wall to the maximum speed of flow will be very steep, occurring over a tiny distance adjacent to the basal part of the the ciliary layer. In fact, a bit of backflow occurs just above the surface, presumably as a result of the return strokes of the cilia (see, for a profile of flow, Figure 15.2a). A reversed parabolic profile is a real phenomenon in at least one biological situation—Liron and Meyer (1980) calculated that it should occur in a layer of fluid above a ciliated surface, and they verified its presence adjacent to the mucociliary membrane of the upper palate of a frog.

Cilia or flagella form propulsive coatings in the flagellated chambers of sponges, the gastroderm of many coelenterates, the gills of many mollusks, the excretory tubules of some flatworms, the oviducts of various animals, and other places. (Curiously, motile cilia are apparently absent in nematodes and arthropods.) In general, the arrangement seems to be more common where exchange processes are taking place and less common where simple propulsion of fluid is the primary mission. While the very steep velocity gradients inevitably associated with ciliary propulsion in tubes certainly improve the efficacy of exchange, those same steep gradients must mean that the cost of operating ciliated pipes as pumps is relatively high—a point made earlier in Chapter 3, and to which we'll return shortly.

With ciliated walls, fluid can be made to flow in a single-ended circular pipe; pushing it one way along the walls guarantees flow in the other along the axis. So the notion of a situation in which the flow profile is even more deviant from the basic parabola than a full inversion isn't, one might say, entirely off the wall.

BERNOULLI VS. HAGEN-POISEUILLE

Textbooks of physiology commonly and commendably begin their treatments of circulatory systems by giving a short introduction to fluid mechanics. However, all too often the first thing presented is Bernoulli's equation, which is then never invoked for any specific application to circulation. The problem, alluded to in Chapter 4, is the implied assumption of frictionless flow. Still, Bernoulli is sometimes useful for internal flows—Venturi meters, for instance, are thoroughly trustworthy devices. In a sense, the problem is that taken straight, Bernoulli's principle and the Hagen-Poiseuille equation make diametrically opposite predictions.[3] If faster flow is associated with a decrease in pressure (Bernoulli), then a flexible pipe will tend to collapse as speed increases. If faster flow requires a greater pressure difference (Hagen-Poiseuille), then the pipe ought to expand. Blow through a cylindrical balloon with its far end cut off and it collapses, with noisy instability. Let a heart push blood faster through an artery, and the artery swells—in taking your pulse you're counting systolic and not diastolic phases.

Comparison between Bernoulli's equation in a simple form (Equation 4.7) and a version of the Hagen-Poiseuille equation (13.5) is useful in deciding how far to trust the former—at least for laminar flows. Giving Bernoulli the benefit of the doubt we set $U_l = 0$ in (4.7) and make an index (called Pi_l, for "tendency to pinch, laminar flow") with the pressure drop due to Bernoulli as the numerator and that due to Hagen-Poiseuille as the denominator (assuming $U =$ for $\bar{U}$ simplicity):

$$Pi_1 = \frac{\frac{1}{2} \rho U_2{}^2}{8 \, \mu l \bar{U}/a^2} = \frac{\rho \bar{U} d^2}{16 \, \mu l}. \tag{14.3}$$

But for the numerical coefficient and the extra length factor, we've managed to reinvent the Reynolds number! What we see is that an increase in speed is associated with a drop in pressure ($Pi_l > 1$) at highish Reynolds numbers (short of the turbulent transition, of course), and for pipes that are wide (large d) and short (small l). For a piece of vein, excised from a circulatory system, with a diameter of 2 mm, a length of 100 mm, and a flow speed of blood of 0.1 m s^{-1}, Pi_l is only about 0.1—so the vein will bulge, not pinch, on account of flow. In fact, all normal vessels of the circulatory system will bulge rather than pinch—even a large artery faces the down-

---

[3] It's possible to combine Bernoulli and Hagen-Poiseuille equations, using the latter as a term for loss of momentum or energy as a result of pipe resistance. Synolakis and Badeer (1989) urge this for physicists; a specific application of the combination is proved by Kingsolver and Daniel (1979).

stream resistance of smaller vessels. About the only time pinching might happen would be in a very local constriction of a large vessel, where d is high and l is low—a stenosis or coarctation—and there it's probably not very much. Hence Engvall et al. (1991) quite reasonably found a pressure drop in a narrowing of the aorta and just as reasonably only slight recovery of the pressure just downstream from the constricted region.

An analogous situation ought to occur in turbulent flow. Using equation (13.15) instead of (13.5) and assuming a value of 0.03 for the friction factor gives a "tendency to pinch for turbulent flow," Pi_t:

$$Pi_t = \frac{2a}{fl} = \frac{67a}{l}.$$

(14.4)

Again, the short fat pipe will tend to collapse while the long, thin pipe will expand—automobile carburetors and laboratory faucet aspirators are wide and not especially long. Once flow becomes turbulent, alteration of the Reynolds number should have little effect. Probably the instabilities and odd effects of downstream conditions found with rubber tubes of flowing water by Ohba et al. (1984; as reported by Matsuzaki 1986) derive from this transition—I calculate an Pi_t value of about 3.0 for a pipe where they encountered instability, compared with 0.5 and 0.4 for the two that gave none.

Of course, a pipe leading to an orifice must at some point get close enough to the orifice so the remaining length is low enough for Pi_l or Pi_t to rise above unity. Hugh Crenshaw has suggested to me that a urethra ought to fulfill this condition. But the pulsating instability of the balloon with its end cut off seems in practice not to afflict urethras—perhaps the most distal region is sufficiently braced and damped against collapse and flutter. On the other hand, flow limitations appear in forced micturition as well as exhalation (Matsuzaki 1986)—perhaps we're arranged to stay normally just on the safe side of Bernoulli-induced mischief. I do wonder if organisms ever make use of a high Pi for deliberate pulsation, spraying, sound production or something. Burton (1972) suggested that the instability is what's involved in snoring; I wonder about purring in cats.

At least one case of vessel shrinkage associated with faster flow clearly can't be blamed on the value of our pinchiness index. Faster ascent of sap in a tree is associated with slight shrinkage of the trunk (Macdougal 1925)— that's due to greater suction at the top and the consequent reduction of static pressure in the vessels.

EFFICIENT BRANCHING ARRAYS OF PIPES

Amid the obvious complexity and diversity of living systems, the biologist ought to treasure any common feature. And the wider the taxonomic range of organisms that show the feature, the more powerful a generalization its recognition represents. The present book is intended to be read, in part, as an argument that the unavoidable imperatives of the physical world underlie much of biological design. I write these grand words at this point prefatory to introducing one of the nicest generalizations about living systems ever derived from considerations of fluid flow.

Murray's Law

Construction, maintenance, and operation of a circulatory system are part of the price of being a large and active animal. The price isn't trivial—simply keeping your blood moving around accounts for about a sixth of your resting metabolic rate. How might a circulatory system be arranged to minimize the cost? The question is an old one; the classic analysis (but not the first) is that of Murray (1926). He considered two factors—keeping the blood going against the pressure losses consequent to the Hagen-Poiseuille equation, and some additional construction and maintenance cost proportional to the volume of the system. The first is certainly reasonable according to the material in the last chapter. The second is sensible because blood has to be made and replenished and because the walls of larger vessels are proportionately thicker (a consequence of Laplace's law), so wall volume is proportional to contained volume rather than surface area.

Murray derived an expression, now called "Murray's law," for the optimal design of a circulatory system based on minimization of his two cost factors; a clearer and more modern derivation and discussion is given by Sherman (1981). In the simplest form it can be stated as

$$Q = ka^3, \tag{14.5}$$

or, volume flow through a vessel should be proportional to the cube of the radius of the vessel. While this may look like a contradiction of the Hagen-Poiseuille equation, it isn't anything of the sort. The latter declares volume flow proportional to the fourth power of the radius, other things—viscosity and pressure drop per unit length—being equal. Murray's law says that for minimum cost, other things shouldn't be equal. In particular, since viscosity is assumed constant, $\Delta p/l$ should vary inversely with radius. Another way to put the rule is to invoke the principle of continuity and express it in terms of the relative radii of vessels before and after a bifurcation—the cube of the radius of the parental vessel should equal the sum of the cubes of the radii of the daughter vessels. Or in terms of a branching manifold in which we can look at any two generations or ranks in the array,

$$a_0^3 = a_1^3 + a_2^3 + \ldots a_n^3. \tag{14.6}$$

Here a_0 is the radius of the parental vessel, and a_1, a_2, etc., are the radii of the daughters, or great-granddaughters, or whatever, as illustrated diagrammatically in Figure 14.4.

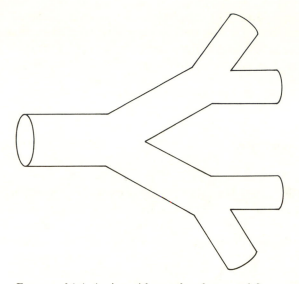

FIGURE 14.4. A pipe with two daughters and four granddaughters drawn so their diameters have the relative sizes specified by Murray's law. The branching angles have no particular significance here. Note that dichotomous branching is incidental—the four granddaughter pipes would have the same size if the original pipe had a 4-fold branch without intermediate daughters.

Murray's law says that when a vessel bifurcates symmetrically into two others, each of the others should have a radius or diameter 79.4% of that of the first. It thereby says that (1) the cross-sectional area of each of the two daughters should be 63.0% of that of the parent—the sums of both radii and areas will be greater for daughters than parents. It also says, by the principle of continuity, that (2) flow will be slower in the daughters by 26%. Good—smaller pipes ought to have slower flow in order not to have inordinate resistance. More specifically it says that (3) the average and axial velocities in any vessel should be directly proportional to the vessel's radius. That follows from the assertion that total flow is proportional to radius cubed and cross section is proportional to radius squared. Furthermore, if velocity is proportional to radius, then (4) a kind of isometry prevails—the same specific parabolic profile characterizes every vessel. And that (recalling the derivation of equation 13.1) means that (5) the velocity gradient at the wall (dU/dr at $r = a$) will be the same throughout the system. Since (for constant viscosity) shear stress is proportional to shear strain or velocity gradient, then (6) all the walls in the system ought to be subjected to the same shear stress.

Do Real Systems Really Follow Murray's Law?

The predictions above are audacious in both specificity and generality. If the size, density, and blood velocity of capillaries are set by the exigencies of diffusion, oxygen transport, and metabolic rates, then the geometry of virtually the entire remainder of a circulatory system should follow in consequence. In fact, the predictions are only a little less powerful than this ideal. Table 14.1 gives the relevant data for the human circulatory system. The largest deviations from Murray's law occur in arterioles and capillaries, the main places where the flow profile isn't really parabolic. But all values are within an order of magnitude, which isn't bad for the cube of a very variable variable's value. For comparison, using the squares of radii leads to values that vary by three orders of magnitude, and using fourth powers of radii gives a variation of four orders. LaBarbera (1990) points out that his analysis is a bit flawed by using the reported averages of radii—the cube of averages isn't quite the same as the average of cubes. He cites less inclusive studies on cats, rats, and hamsters in which nonaveraged measurements of the diameters of specific vessels give almost exactly the predicted exponent of 3.0.

What about systems quite remote from any mammalian circulation? LaBarbera (1990) has made a thorough search for appropriate data against which to test Murray's law. On one hand, very little currently exists; on the other hand, a good set is available for the water passages in sponges, the macroscopic metazoa most distant from mammals. These are consummate water pumpers, processing volumes equal to their own volumes about every five seconds, so cost ought to matter a lot. And the law works about as well for them as for mammalian circulation, with less than an order of

TABLE 14.1 THE AVERAGE RADII AND TOTAL NUMBERS OF THE CONVENTIONAL CATEGORIES OF VESSELS OF THE HUMAN CIRCULATORY SYSTEM ALONG WITH THE SUMS OF THE CUBES OF THEIR RADII.

Vessel	Average Radius (mm)	Number	Σr^3 (mm $\times 10^{-3}$)
aorta	12.5	1	1.95
arteries	2.0	159	1.27
arterioles	0.03	1.4×10^7	0.382
capillaries	0.006	3.9×10^9	0.860
venules	0.02	3.2×10^8	2.55
veins	2.5	200	3.18
vena cava	15.0	1	3.38

SOURCE: LaBarbera 1990.

magnitude variation in the sums of the cubes of the radii for a range of vessel radii of over three orders of magnitude. The exceptions are again the elements in which flow is not parabolic—for sponges the choanocytic chambers, where flagellar pumping takes place, and the short apopyles that lead fluid out from them.

Murray's law fails in several instructives cases that argue that it's not just some geometrical constraint unrelated to flow. Sherman (1981) reminds us of Krogh's (1920) discovery that the branching tracheal system of a larval moth preserved its overall cross-sectional area (or radius squared rather than cubed)—just what's expected in a system relying on diffusion rather than bulk flow. And LaBarbera (1990) mentions systems in which pumping is done by ciliated walls, in particular the gastrovascular transport system in corals (Coelenterata) and the coelomic circulation of crinoids (Echinodermata).

What about all those conduits through which sap ascends in vascular plants? My initial guess was that Murray's law shouldn't work. First, the cells lining the vessels are dead by the time they conduct fluid (a factor whose relevance will be rationalized shortly). Second, because of the very high fluid tensions, large vessels seriously risk cavitation. Finally, little metabolic power is spent in moving water upward: the main pump, an evaporative lifter, uses solar power directly. In fact, the large vessels of tree trunks are smaller than expected from Murray's law. But Canny (1993) has now shown that the law works very well for small vessels in a case that suggests generality. These are vessels with diameters from 4.0 to 25 μm in the veins of sunflower leaves. Here the radii vary by 6-fold, the cubes of the radii vary almost 250-fold, but the sums of the cubes vary by the remarkably low factor of 1.3. To guess again, I'd say that construction rather than operating cost is the key here, made all the more important by the short lives of such leaves.

Self-optimizing Systems

What to me is the most exciting aspect of the renewed appreciation of Murray's law is the way it links functional performance, initial development, and subsequent adjustment. Recall that when the law is obeyed, all the walls in the system will experience the same shear stress. So in order to make a branching array achieve the most efficient arrangement, only a single local command need be given. Endothelial cells lining vessels need only proliferate (along with the underlying tissues) until the shear stress on each in the widening vessel drops to a preestablished set point. A variety of studies (reviewed by Davies 1989, LaBarbera 1990, and Hudlicka 1991) have shown the necessary cellular behavior in systems ranging from entire animals surgically altered to cultured endothelial cells subjected to changes in the flow over them. One needn't invoke some fantastic genetic pre-programming of the details of a circulation; the minimal needs are just some general instructions for building vessels, some chemical released by mildly anoxic tissues to determine where the vessels grow, and the set point of a Murray's law system to determine their relative sizes. And the system can be continuously retuned as the demands (such as the level of aerobic activity) upon it change.

THE ASCENT OF SAP IN TREES

The various vessels used to move sap in vascular plants must add up to the greatest total length of biological conduit through which fluid moves. Almost certainly sap is moved in larger volume—despite the slow speeds—than all animal fluids combined. The pressures involved are more extreme by several orders of magnitude than those in any other multicellular systems. The components of these systems may be unimposing, but these are by any measure nature's prime fluid movers. As an earnest hemlock once noted, the sap also rises. By "sap," incidentally, I mean the liquid that rises in the xylem from roots to leaves, a liquid insignificantly different (physically) from pure water; how fluid is transported in the phloem is yet another story.

What's attracted the most attention and controversy is the origin of these monumental pressures. The tallest trees approach 100 meters; since the hydrostatic pressure against which any sap-lifter works is ρgh, that's about a million pascals—10 atmospheres. To that must be added the pressure drop from flow through vessels whose diameters are inevitably less than half a millimeter. And that sum must be increased by the pressure needed to draw water from the interstices of soil—in nearly dry soils that's another enormous factor. The combined pressures, then, can approach 80 atmospheres. But nowhere in a tree can such high pressures be detected. Roots can generate considerable pressures through an osmotic mechanism, but really high root pressures don't generally occur in tall trees. Nor is staged pumping a viable possibility—the vessels through which sap passes, as mentioned a page or so back, are dead and devoid of moving parts and are therefore completely passive pipes.

What lifts sap is overwhelmingly a pull from above rather than a push from below. Evaporation of water in the interstices of leaves draws water up the pipes—the system is open at the top, but the openings (in the cellulose feltwork of cell walls) are small enough so surface tension proves sufficient to prevent entry of air. Physical scientists react quite negatively upon being told that the vessels contain fluid under millions of pascals of negative pressure. But the matter is as near to settled as anything ever is in biology.

Evaporative pull was suggested a hundred years ago; circumferential shrinkage of tree trunks when sap speed increased was shown in the 1920s; the magnitude of the negative pressures were determined by putting cut twigs in pressure bombs and subjecting them to sufficient pressure to move sap back to the cut end in the 1950s; and by a variety of methods, water has been shown to have much more than adequate tensile strength to permit such pressures. The story is a fine one; it's told by Zimmermann (1983) about as well as I've heard it, but Niklas (1992) or any textbook of plant physiology will give the essentials. I'll skip further detail since it's a tale of fluid statics and this is a book about fluid dynamics. But the basic situation matters here—a gravitational pressure gradient of minus 10 kPa (-0.1 atmosphere) per meter of height, starting at plus 100 kPa ($+1.0$ atmosphere) at the ground. Thus pressure is negative (water is in tension) above a height of, at most, 10 meters.

The xylem elements that carry water upward in trees are small pipes even in tall trees—a 0.25 mm (diameter) vessel in an oak is reckoned large. Individual vessels may be as long as 0.6 m (grapevine, maple) or 3 m (ash). Neither total cross-sectional areas nor vessel diameters change much with height above ground or location in trunk, branches, or twigs. Water moves from vessel to vessel through various pores, pits, and perforated plates, and these may provide substantial resistance to flow. The relative resistance of the pores can be judged by measuring vessel diameter and overall resistance and then subtracting vessel resistance as determined from the Hagen-Poiseuille equation (using equation 13.3). For vines, pore resistance is negligible, and conductivity is fully 100% of that predicted by the Hagen-Poiseuille equation. For a wide variety of trees, including both conifers (with smallish tracheids rather than largish vessels) and hardwoods, conductivity is only around half the calculated value. The figures cited here and elsewhere are a bit rough, though—any calculation based on the Hagen-Poiseuille equation (with its a^4) magnifies uncertainty in measurement of diameter. Thus reduction of vessel diameter by only 16% will by itself reduce calculated conductivity by 50%. And vessels are not at all internally smooth pipes of constant diameter, nor are all vessels in a tree of quite the same length and diameter (Zimmermann 1983). Nor are they all operative all the time.

So, once water is sucked out of soil, two agencies resist its upward flow— the static gravitational gradient of 10 kPa m^{-1}, and the resistance to flow per se. Flow isn't trivial, with speeds of over 0.1 m s^{-1} common in hardwoods such as oaks, which is why trees get detectably skinnier when they're rapidly evaporating water ("transpiring") through their leaves. 0.1 m s^{-1}, 70% of ideal (Hagen-Poiseuille) conductivity, and a diameter of 0.25 mm give a pressure drop of 7.3 kPa m^{-1}, only a little less than the gravitational gradient. Interestingly, just as the gravitational pressure drop is independent of species and both vessel and pore size, so, roughly, is the maximum pressure drop due to flow—but the first is unavoidable physics, while the latter depends on a certain congruence between vessel size and flow speed. Trees that raise water more slowly do so with narrower xylem elements. Of special interest is the similarity of the two figures. Xylem elements seem to be made about as small as nature can get away with without being stuck with a pressure drop due to flow in excess of the drop due to gravity. Or, to put the matter the other way around, little additional flow will be realized by using still larger vessels and reducing resistance further inasmuch as the gravitational pressure drop cannot be altered—a variable resistance is in series with a fixed resistance. The advantages of using many small pipes are probably the safety of numbers and redundancy, greater resistance to implosive collapse (by Laplace's law), and greater resistance to cavitation. A lot of water is being stretched, and embolisms appear to be a significant hazard—according to Tyree and Sperry (1988) vascular plants generally operate quite near the point of catastrophic failure. Good reviews of the present state of the evaporative sap-lifter of plants (with lots of references) are those of Tyree and Sperry (1989) and Tyree and Ewers (1991). Quite another view of how the sap rises will be found in Masters and Johnson (1966).

Low, herbaceous plants never encounter negative pressures, so their higher rates of flow and higher pressure gradients might be taken as an indication of a price paid for extreme height. Their pressure gradients due to flow and gravity, instead of being about 20 kPa m^{-1}, run up to nearly 100 kPa m^{-1} (Begg and Turner 1970), and flow speeds reach not 0.1 m s^{-1} but at least 0.25 m s^{-1} (Passioura 1972).

A PUMP PRIMER

Up to this point, the pressure needed to push fluid was provided by a kind of deus ex machina or, rather more literally than is usual, a power from above. We considered pressure drop, volume flow, and resistance, noting the analogous variables in simple, ohmic, electrical circuits. When talking about thrust production, we were in fact talking about pumps— they may have worked on external rather than internal flows; but momentum flux, actuator disks, and propulsion efficiency were all aspects of the process of expending power to move fluid in ways that the fluid would (on account of inertia or viscosity or both) prefer to avoid. Only putting the thrust producer in a duct is needed to make a recognizable pump, in which power is expended to produce a pressure drop and a volume flow.

Pressure Drop versus Volume Flow

Indeed, that's the basic game for any pump—power output, P, is the product of pressure drop and volume flow, as explained in the last chapter (R is resistance):

$$P = Q\Delta p = Q^2R = \frac{(\Delta P)^2}{R}. \qquad (13.4)$$

What's important here is the implicit trade-off. One pump might produce a high pressure difference from input to output but deal with a relatively low through-put of fluid; another pump might exert the same power but invest it in a high volume flow with only a small increment in pressure. Note carefully that efficiency doesn't enter. We're just looking at how output might be apportioned.

This apportionment of output is rather severely tied to the design or choice of pump. If you use a pump that is a good pressure producer for an application that needs mainly a high volume flow, you won't get the desired result. I say this with the passion peculiar to one who learned the hard way. Quite a few years ago I built a flow tank using a two-horsepower centrifugal pump, the largest such pump that could be operated on the available 115-volt, 15-amp circuit. The pump moved a miserable 2 (U.S.) gallons per second—a maximum flow of a third of a meter per second through a section only 15 cm square. A few years later I wised up and used a marine propeller to push the water around (Vogel and LaBarbera 1978; Vogel 1981). A half-horsepower motor moved 33 gallons per second to give a full meter per second through a section 35 cm square. That's almost two orders of magnitude better with respect to the variable—volume flow—that matters in a flow tank. A centrifugal pump is swell for high-head use such as lifting water a substantial height or making it squirt a long distance through a nozzle. On the other hand, while it imparts only a little pressure, a propeller handles a far higher volume flow.

Nature's pumps are a varied lot—valved hearts, peristaltic tubes, ciliary layers, paddles, evaporative lifters, and others. What isn't sufficiently appreciated is that they operate at widely varying points in this trade-off of pressure drop versus volume flow. For pressure drop, the undisputed champion must be the evaporative sap lifter of tall trees or desert shrubs—Δp's of up to 8 MPa (80 atmospheres) or so. For volume flow, suspension feeders with internal filters, such as sponges, clams, brachiopods, and so forth, must be the best things going. We're in between. A human circulatory system moves about 5 liters per minute at rest, which is about 0.15% of body volume per second, with a pressure drop (left ventricle to right atrium) of around 13 kPa (100 mm Hg).[4] As noted earlier, the cost of our

---

[4] The systemic heart of a resting squid (*Loligo pealei*) does about the same—about 0.18% of body volume per second (Bourne 1987) at about half our blood pressure. While squid have the typically low resting metabolic rate of cold-blooded animals, the oxygen-carrying capacity of their blood is also low, so they have to make a relatively large investment in moving blood around. A further consequence is a poor capacity for increasing aerobic metabolism during activity.

circulation is not insignificant, yet a marine sponge (Foster-Smith 1976; Reiswig 1974) or a mussel (Meyhöfer 1985) may have a volume-specific pumping rate several hundred times ours.[5] Where Q is only a little less than one's body volume per second, one can't afford a high-resistance system that needs a lot of Δp!

Impedance Matching

A productive way to look at pumps is in terms of their "impedance," the general term for the resistance just invoked. A pump that gets a high volume flow from a given investment in power is spoken of as a "low-impedance" pump. The equality of power with Q^2R in equation (13.4) is the key—high Q demands low R. Conversely, one that produces a high pressure from a given power is a "high-impedance" pump; from the equality of power with $(\Delta p^2)/R$ in (13.4) high Δp demands high R. The evaporative sap lifter is obviously an extremely high impedance pump while the mussel's filtration system operates at very low impedance.

The usual technological practice is to divide pumps (only a few don't fit) into positive displacement pumps and dynamic (or fluid dynamic) pumps (see, for instance, French 1988 or Massey 1989). The former enclose fluid in a chamber, and by reducing the volume of the chamber cause the fluid to leave through some prearranged opening. Most often volume is reduced with a piston; the game is really one of fluid statics, not dynamics, which is why no more will be said about their operation here. They're very high impedance devices, with pulsating outputs and an intolerance for suspended particles as their main handicaps. The latter—dynamic pumps—ordinarily operate at lower impedance; they're in turn divided into centrifugal and axial types depending on whether they fling fluid outward or force fluid forward. In general, pumps with centrifugal impellers (such as "squirrel-cage" fans and centrifugal pumps) operate at higher impedance than those with axial impellers (most ventilating fans and propeller pumps). That's the difference in which my nose was rubbed with my first flow tank. (Jet engines get high impedance out of axial fans, but they do so by using a bunch of them in a series, alternating rotor blades on a shaft with stator blades sticking in from the walls.)

Organisms make use of both kinds of pumps as well; for them these matters of types of pumps, operating conditions, and examples are summarized in Table 14.2. The best known positive displacement pumps are valve-and-chamber hearts. These often get especially high overall pressure

[5] Volume-specific pumping rates for freshwater sponges (Frost 1978) and bivalve mollusks (Kryger and Riisgard 1988) are lower than those of marine species, although within the same order of magnitude.

TABLE 14.2. VARIOUS KINDS OF BIOLOGICAL PUMPS.

Type	Category	Impedance	Examples
Evaporative	Positive displ.	Highest	Leaf sapsucker
Osmotic	Positive displ.	Very high	Root sap pusher
Valve/chamber	Positive displ.	High	Heart, bird lungs, squid jet
Peristaltic	Positive displ.	High	Intestine, some hearts
Piston	Positive displ.	Medium	Some tubicolous worms
Valveless/chamber	Positive displ.	Medium	Jellyfish jet, mammalian lung
Drag-based paddles	Fluid dynamic	Medium	Crustaceans in burrows
Lift-based propellers	Fluid dynamic	Low	Hive ventilating honeybees
Ciliary layer	Fluid dynamic	Low	Bivalve gills
Helical/sine wave	Fluid dynamic	Very low	Sponge choanocytes

drops by using serial chambers—contractile vena cavae, atria, ventricles, and various aortic enlargements, for instance. Fish gills (except for the obligate ram ventilators described in Chapter 4) and amphibian bucco-pharyngeal lung inflators also use valve-and-chamber pumps. Another kind of positive displacement pump is close to the piston pumps of our technology. It's used by a variety of polychaete worms to irrigate their burrows—the paradigmatic polychaete *Nereis*, and the really fancy-looking worm *Chaetopterus* (Figure 14.5), both of which have open-ended U-shaped burrows of low resistance. Peristaltic pumps (see Jaffrin and Shapiro 1971; Liron 1976) are also positive displacement devices, and they're quite common in digestive systems and as hearts—peristaltic hearts are quite the usual thing among annelid worms, occasionally supplemented by valves and contractile chambers. The especially ugly lugworm, *Arenicola*, moves water through its burrow—in either direction—by passing peristaltic waves down its body (Wells and Dales 1951). One shaft is filled with sand, and thus the burrow has a substantial resistance; *Arenicola* is a detritus feeder whereas the previously mentioned polychaetes are suspension feeders and require a higher volume flow. Another such pump is the evaporative sap lifter—water is removed by evaporation and can be replaced only by upward flow. As far as I know, all the jet engines of animals are positive displacement devices. Some have valves (squid, for instance); others (such as jellyfish) don't. In general, for best operation the input orifice ought to be bigger than the output since the devices ordinarily make fluid speed up. That can be accomplished with a single, bidirectional orifice by changing its diameter—medusae and jellyfish clearly do so.

The usual kind of fluid dynamic pumps are hard to recognize in organ-

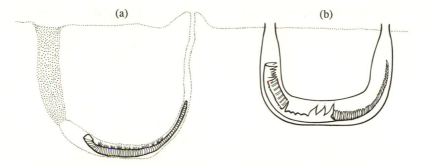

FIGURE 14.5. Polychaete annelids in burrows through which they pump water: (a) the lugworm, *Arenicola*; (b) the parchment worm, *Chaetopterus*.

isms inasmuch as both the axial and centrifugal impellers of our technology inevitably use an "unnatural" wheel-and-axle. That doesn't reflect any scarcity, though. All the lift-based thrusters of Chapter 12 as well as the drag-based paddles of Chapters 7 and 12 are really fluid dynamic pumps. The latter, at least, do service within pipes, for instance in tube-occupying crustaceans such as the amphipod *Corophium*, which pumps water by beating its pleopods (Foster-Smith 1978). Other examples of fluid dynamic pumps are bodies that pass helical or sinusoidal waves backward, and yet others are all the ciliary devices whose operation will be explored in the next chapter.

The impedance range over which either positive displacement or fluid dynamic pumps operate is wide, but the same general distinction made in our technology is apparent—the former are higher impedance devices than the latter. Thus evaporative sap lifters, as already mentioned, achieve pressures of a spectacular 8 MPa. The valve-and-chamber heart of a giraffe can do about 40 kPa (Warren 1974), as can the valve-and-chamber jet of a squid (Trueman 1980), both above the 25 kPa or so that the heart of an exercising human achieves.[6] The peristaltic pump of *Arenicola* achieves around 1200 Pa (Foster-Smith 1978), while the piston pumps of *Nereis* and *Chaetopterus* manage only 80 and 90 Pa, respectively (Riisgard 1989, 1991).

Fluid dynamic pumps encompass a generally lower range of values, although at least one ciliary device gets into a realm more typical of positive displacement pumps—a burrowing echinoderm, *Echinocardium*, achieves about 90 Pa. But it puts a lot of cilia in series rather than in the more common parallel array, limiting (comparatively) the volume flow it can manage (Foster-Smith 1978). Paddles and propellers in ducts are rare in

[6] But jets aren't automatically high-pressure generators—a hydromedusan jellyfish, *Polyorchis*, achieves only 40 Pa (DeMont and Gosline 1988), three orders of magnitude less than a squid. The anal jet of a dragonfly nymph is in between, at 3 kPa (Hughes 1958).

nature, but one can cite the beating pleopods of the amphipod crustacean, *Corophium*, which produce about 40 Pa (Foster-Smith 1978). Such paddles may seem an odd choice; perhaps the best way to view *Corophium* is as a creature constrained by accident of birth into a phylum that doesn't know that cilia can be motile. Nest ventilation by wing beating in stationary honeybees clearly uses a propeller pump, but I don't know of data on pressures achieved. The ciliary and flagellar pumps hold down the low end of the impedance scale. We have quite a lot of data on them, since bivalve mollusks are of both ecological and culinary importance. For such animals as soft-shell clams (*Mya arenaria*) and mussels (*Mytilus edulis*) maximum pressures are about 50 Pa (Foster-Smith 1978; Jørgensen and Riisgard 1988); for marine sponges (Foster-Smith 1976) they run up to about half this value.

A useful way to view the diversity of devices and their operating conditions is as a graph of pressure drop versus volume flow, as in Figure 14.6. In an ideal world, one with no mechanical limitations and perfect (or constant) efficiency, the product of the two variables, power, would be constant; and the graph would come out hyperbolic. In the practical world, the maximum performance of a pump is instead described by quite the opposite kind of curve, with a specific maximum for pressure head even with no flow, and a maximum volume flow even with no pressure head to work against. An organism can use a pump anywhere within the envelope described by that performance line and the axes. The organism, though, can be characterized by a resistance—more flow takes more pressure or vice versa. That resistance generates a line sloping up from the origin—either straight, if the resistance is linear or "ohmic," or slightly curved otherwise. The intersection of the maximum pump performance line and the resistance line marks the best the system can ordinarily do.

Foster-Smith (1978) pointed out a curious feature of this best operating point for at least the feeding pumps of marine invertebrates. They can typically generate pressures quite a bit higher than those they actually use. Thus the 90 Pa of *Chaetopterus* contrasts with the 15 Pa it uses (Riisgard 1989), and the 50 Pa of *Mya* and *Mytilus* with the 1 Pa of normal operation (Jørgensen and Riisgard 1988). That needn't imply serious energetic inefficiency, but one wonders about accidents of ancestry, secondary functions, and so forth. At least wasting potential pressure is more reasonable than wasting potential volume flow, given the function of these pumps. Systems that waste potential volume flow are a lot rarer, enough so that I haven't a handy example.

Measuring either pressure drop or volume flow in systems of very low internal resistance can be quite a tricky business. Not only do the animals not always perform (sponges in captivity are notoriously recalcitrant) but transduction must not impose any significant resistance of its own. For

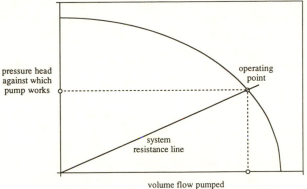

FIGURE 14.6. Pressure drop versus volume flow for a pump. The pump can operate anywhere within the circular curve. The line from the origin—what it can do with a load of fixed resistance—intersects the curve at the maximum output of the pump for that load. The line's slope reflects the impedance of the pump, here fairly low. Notice that in low impedance systems maximal volume flow isn't especially sensitive to the particular pressure drop with which the pump must contend.

pressure drop, old data are scarce; for volume flow, only modern data obtained with tiny flow probes or through calculations from clearance rates should be trusted. To make matters worse, one can't assume a parabolic profile emerging from some excurrent opening and calculate from a single, axial speed, as shown by Charriaud (1982). Further difficulties arise when one tries to determine the relationship between maximum pressure drop and maximum volume flow, the outer envelope in Figure 14.6. Some biological pumps can be impelled to operate under abnormal combinations of the two variables while some cannot. The most direct approach is to adjust the hydrostatic head against which the pump works—for instance, exposing the output side of a U-shaped burrow to either a higher or lower water level than the input side or, as I once did (Vogel 1978b), cannulating an output orifice and connecting it to a device to adjust pressure and measure flow.

Transformers

The ostensible mismatch makes one wonder, as well, about whether nature has come up with anything equivalent to the transformers we use in electrical systems. In a fluid transport system, such a device would allow efficient interconversion of pressure drop and volume flow. Our technology uses quite a range of such devices. Ducted fan-jet engines use their basic jet turbines to move additional air around the engines; they thus get greater flow and less overall pressure drop. As a result of this impedance reduction their Froude propulsion efficiencies are better than early jet engines that lacked a fan and peripheral ducting. A so-called hydraulic ram is a device that raises the impedance of an internal flow and makes water flow uphill—a sudden surge of flow downhill drives a smaller surge to a higher level than that of the original reservoir. For that matter, the combination of the fixed wing and propeller of an ordinary airplane amounts to an impedance reducer—the propeller gives a small stream a high speed and thus a high pressure increase to make its thrust. The wing converts some of that horizontal thrust to an upward force by making a lot of air flow just a little bit downward. In the process it shifts from a high pressure drop, low volume flow to a low pressure drop, high volume flow. After all, the ambient vertical wind is negligible, so a high speed (or pressure) and low volume flow downward would incur an unnecessary cost—something to which a lot of talk was devoted a few chapters back. And these aren't the only examples of impedance-adjusting arrangements.

What seems to me quite curious is the scarcity of such transformers in nature. We do encounter lots of devices that trade off the components of volume flow—cross-sectional area and velocity—using the principle of continuity. But that's not a transformer in the present sense. It's just equivalent to altering the gauge of a wire, trading current density against cross section. Odd omission—unless I'm missing something or simply looking at these systems in some inappropriate way.

But enough about pipes and pumps, lest their efflux develop an effluvial odor; I offer mild apologies for the absence of such items as nice nectar nippers (for instance Kingsolver and Daniel 1979) and bad bloodsuckers (as an example, see Daniel and Kingsolver 1983).

Flow at Very Low Reynolds Numbers

A T THE END of Chapter 13 I made a distinction between flows at moderate or high Reynolds numbers and those at low Reynolds numbers. In the former, drag typically varies with the square of speed, the value of density is of more consequence than that of viscosity, and formulas for drag or pressure drop need a dimensionless coefficient. At low Reynolds numbers, by contrast, drag varies directly with speed, viscosity is the prime variable, and ugly coefficients are unnecessary. We entered this more orderly world in the past two chapters in the context of internal flows; now we turn to their counterparts in which the solid phase is on the inside and the fluid on the outside.

This is the world, as Howard Berg put it, of a person swimming in asphalt on a summer afternoon—a world ruled by viscosity. It's the world of a glacier of particles, the world of flowing glass, of laboriously mixing cold molasses (treacle) and corn (maise) syrup. Of more immediate relevance, it's the everyday world of every microscopic organism that lives in a fluid medium, of fog droplets, of the particulate matter called "marine snow." After all, the Reynolds number can be as well reduced by decreasing size as by decreasing speed or increasing viscosity. "Creeping flow" is the common term in the physical literature; for living systems small size rather than (or as well as) low speed is the more common entry ticket. And it's a counterintuitive—which is to say unfamiliar—world.

At very low Reynolds numbers, flows are typically reversible: a curious temporal symmetry sets in, and flow may move matter around but in doing so doesn't leave much disorder in its wake. Concomitantly, mixing is exceedingly difficult—spreading two miscible fluids amid each other need not actually mix them. Stirring three times clockwise can undo the results of stirring three times counterclockwise—try it by injecting a bit of colored glycerin beneath the surface of a beaker of clear glycerin and stirring with a rod, first in one direction and then in the reverse.

Inertia is negligible compared to drag: when propulsion ceases, motion ceases. Berg (1983) has calculated that if a swimming bacterium ($Re = 10^{-5}$) suddenly stopped rotating its flagellum, it would coast to a stop in a distance much less than the diameter of a hydrogen atom. Propulsion in fluid media is possible, but not by imparting local rearward momentum to the fluid any more than we impart rearward momentum to part of a floor when we walk on it.

Separation behind bluff bodies is unknown—take a look again at the streamlines around a cylinder in Figure 5.5. Separation results from inertia, the tendency of a fluid to continue to move downstream rather than curve around the rear of an obstacle; and where inertia is negligible, fluid oozes around curves and corners with magnificent indifference. Velocity gradients are what the fluid abhors. Streamlining is a fine way to increase drag—the extra surface exposed in the process incurs extra skin friction and, without any separation to be prevented, pressure drag drops little. Shape matters, but to a lesser extent and in a different way than what we've become used to.

Boundary layers are thick because velocity gradients are gentle, and the formal definition of a boundary layer has little or no utility; I urge that the term be avoided for very low Reynolds numbers. Moving with respect to a fluid alters the motion of the fluid a long distance away, and the drag of a body moving through a fluid may be substantially increased by walls around the fluid a hundred or more body-diameters away.

Nor can one create appreciable circulation about an airfoil. Vortices exist only in the upper end of this realm; they're very regular, mostly core, and they dissipate rapidly if their voracious appetite for energy isn't constantly appeased. Turbulence, of course, is unimaginable.

And Galileo must recant even more emphatically than at high Reynolds numbers. An object of density greater than that of the medium still falls, but a large object falls faster than a small one virtually from the time of release—terminal velocity is reached almost immediately. After that, the balance of weight and drag is all that sets speed.

While this queer and counterintuitive range is of some technological interest, its biological importance is enormous. Most often being small means being slow—if for no other reason than that drag is a function of surface and thrust is a function of engine size, or volume. Smallness and slowness lower the Reynolds number in concert; since the vast majority of organisms are tiny, they live in this world of low Reynolds numbers. Flow at very low Reynolds numbers may seem bizarre to us, but the range of flow phenomena with which we commonly contend would undoubtedly seem even stranger to someone whose whole experience was at Reynolds numbers well below unity.

How the tactics of living depend on size and Reynolds number is perhaps best illustrated by some calculations given in a charming essay by Purcell (1977). Consider a bacterium about one micrometer long, swimming through water at about 30 μm s^{-1} and thus at a Reynolds number of about 3×10^{-5}. The bacterium can swim, but should it? Diffusion brings its food to it; to increase its food supply by 10% Purcell calculated that it would have to move at 700 μm s^{-1}[1]. The only reason to swim is to seek a more concentrated patch of food—greener pastures. We've encountered the equivalent of a casual cow who, after eating, just waits for the local grass to regrow.

But while these slow, small-scale flows may seem peculiar, they're orderly (Purcell calls them "majestic") and far more amenable to theoretical treatment than the flows we've previously considered. The most sacred Navier-Stokes equations, obeisance to which was pronounced earlier, contain inertial and viscous terms. A very low Reynolds number implies a preponderance of viscous forces, giving license to ignore the inertial terms. With this simplification, explicit solutions for many cases are possible, and the advent of large computers has permitted approximate solutions to others. In practice, then, we're able to rely more on equations and less on semiempirical graphs and coefficients. The relevant fluid mechanics is well treated in detail by Happel and Brenner (1965), with clarity and brevity by White (1974), and with a more biological perspective by Hutchinson (1967).

DRAG

At high and moderate Reynolds numbers inertial effects were all-important, and we found that drag depended mainly on the product of projecting area and dynamic pressure. A dash of Reynolds number dependence accounted for the vagaries of flow and reflected the residual action of viscosity close to the surface. At Reynolds numbers below 1.0, drag depends mainly on viscous forces and ought to be largely independent of fluid density—since inertial effects are very much less than the effects of shear. The customary drag coefficient, C_{df}, as defined by equation (5.4), is often still used, mainly to maintain consistency with practice at higher Reynolds numbers. Using C_{df} incurs no sin, since it's just the result of a definition and carries no automatic phenomenological implications. But it may be a bit misleading. Thus the drag coefficient of a sphere can be figured as $24/Re$, combining equations (5.4) and (15.1), up to $Re = 1$; as White (1974) points out, this implies a Reynolds number effect where none really exists. Drag just varies with the first rather than the second power of speed. Reynolds number is an unnecessary complication when simple formulas can describe the drag of ordinary objects.

Spheres

Certainly the most useful formula for drag at low Reynolds number is Stokes' law for the drag of a sphere of radius a[2]:

---

[1] Consider feeding on a sugar such as sucrose, whose diffusion coefficient (D) is about 5×10^{-10} m^2 s^{-1}. The Péclet number (Ul/D), used last when we talked about flow and exchange in capillaries, is thus about 0.06; so diffusion is a much more potent agent of exchange than bulk flow.

[2] But for a droplet of gas in a liquid medium, consider using equation (15.9) instead.

$$D = 6\pi\mu a U. \tag{15.1}$$

Stokes' law is trustworthy up to Reynolds numbers of about 1.0—with errors of only a few percent as that upper limit is approached. Above this, various theoretical treatments are available, but the most useful formula I've seen is a curve-fit equation given by White (1974):

$$C_{df} = \frac{24}{Re} + \frac{6}{1 + Re^{1/2}} + 0.4. \tag{15.2}$$

The drag coefficient is the conventional one (5.4) that I just maligned; the characteristic length in Re is the diameter of the sphere. The equation is a useful approximation of the line in Figure 5.4 up to a Reynolds number of about 2×10^5, that is, up to the great drag crisis where the boundary layer becomes turbulent.

Of the drag predicted by Stokes' law, two-thirds comes from skin friction and one-third from a fore-and-aft pressure drop. The negligibility of inertia doesn't imply that flows aren't associated with pressure differences. Persuading a fluid to move from one place to another still takes a difference in pressure. In fact, relatively more pressure is needed at low Reynolds numbers because of the greater retarding effects of viscosity, as one can see in the Hagen-Poiseuille equation (13.2).

Circular Disks

Exact solutions to the basic equations of flow are also available for the drag of circular disks. For a disk of radius a that faces the oncoming flow, one for which flow is parallel to the axis of rotation,

$$D = 16 \ \mu a U. \tag{15.3}$$

For a disk with its faces parallel to the flow and the axis of rotation normal to the flow,

$$D = 10.67 \ \mu a U. \tag{15.4}$$

These last formulas contrast interestingly. They apply to orientations with the most and the least drag, respectively, and they apply to an object that appears about as anisotropic with respect to drag as anything we're likely to devise. First, the difference, though, is only 1.5-fold, the same as we noted for long, nearly flat plates at a Reynolds number of 1.0 (Table 5.2). Second, the values of the drag of a flat plate for the two orientations do not converge further with further decrease in the Reynolds number. Third, while shape (here in the guise of orientation or by comparison with equation 15.1) certainly affects drag, the effect is scarcely overwhelming; indeed, it's counterintuitively modest. Fourth, referred to projected area, a disk broadside to flow has less drag (16 vs. 6π) than a sphere, quite the opposite of what we saw at higher Reynolds numbers; only when referred to wetted area does the broadside disk have more drag. The latter, though, is still a relevant comparison—streamlining may not have its usual meaning, but evidently elongating a body fore and aft can still lower the drag per unit of surface.

Cylinders

As we'll see shortly, the drag of cylinders takes on great importance in explaining propulsion by flagella and cilia. As it turns out, calculating the drag for cylinders is slightly messier than for spheres and disks—cylinders are treated as very long prolate spheroids with axes either parallel or normal to flow. According to Cox (1970), the drag of a cylinder of length l and radius a extending parallel to the flow is

$$D = \frac{2\pi\mu U l}{ln(l/a) - 0.807}. \tag{15.5}$$

Long cylinders parallel to flow—that sounds like silk strands above ballooning spiders. Suter (1991) compared the predictions of (15.5) with careful direct measurements; he found that real strands had about 2.5 times the expected drag. The likely explanations are innocent enough. For one thing, flexible objects may have more drag than rigid objects (recall the leaves in Chapter 6), especially when one is assuming the minimum-drag orientation. For another, these draglines are really a fused pair of cylindrical strands, so exposed surface is much greater than the minimum figured from mass per unit length of a simple cylinder.

The drag of a cylinder with its axis of rotation normal to flow is

$$D = \frac{4\pi\mu U l}{ln(l/a) + 0.193}. \tag{15.6}$$

One occasionally runs across analogous equations in which the numerical constants in the denominators are −0.5 and +0.5, respectively; but, as Brennen and Winet (1977) point out, these are solutions derived for zero Reynolds number, and inertia always makes a small but quite real contribution. And the assertion is occasionally made that a cylinder normal to flow has exactly twice the drag of one parallel to flow. That's nearly right but would really be reached only with a cylinder that's infinitely long. For a length-to-diameter ratio of half a million, the ratio of the drags is still only 1.86. Still, that's significantly above the ratio of 1.5 that we saw for disks in their orientations of greatest and least drag. Oddly enough, at low Reynolds numbers cylinders are more anisotropic than are disks. Or maybe it's not so odd: the steepness of velocity gradients is what matters, and the further downstream, the gentler; a cylinder parallel to flow has about the maximum downstream surface relative to what's upstream.

For a circular cylinder normal to flow, White (1974) offers the following empirical formula for Reynolds numbers from unity to 1×10^5:

$$C_{df} = 1 + 10.0\, Re^{-2/3}. \tag{15.7}$$

It gives about the same results as the line derived from direct measurements in Figure 5.3, and it was found useful by Greene and Johnson (1990a) who looked at the sinking rates of plumed seeds.

Spheroids

Many biological objects may be reasonably approximated by either oblate (pancake or discus) or prolate (cigar or football) spheroids, as mentioned in Chapter 6 in connection with drag at higher Reynolds numbers. The extreme cases are movement with axes of rotation normal or parallel to the free stream; two classes of spheroids and two orientations give four situations of interest. These fill in the gaps between spheres and disks and cylinders. Loosely speaking, a disk is the ultimate result of squeezing a sphere into an oblate spheroid, while a cylinder is the final result of stretching a sphere into a prolate spheroid. Happel and Brenner (1965) give appropriate, if mildly messy, equations for the drag of all these cases; surface areas of spheroids can be calculated from formulas for areas and eccentricity in Beyer (1978).

Visualizing what's happening to drag in these cases is easiest if we look at two kinds of coefficients for drag: drag over the product of speed and viscosity relative to (1) the square root of surface area, and (2) the cube root of volume—with these last replacing the radius as used in equations (15.1), (15.3), and (15.4). Table 15.1 gives these coefficients for a variety of spheroids as well as for the sphere, disk, and cylinder already considered. A number of items are noteworthy. First, the coefficients are not particularly shape-dependent. Again, shape matters, but far less than at high Reynolds numbers. Second, orientation matters, but again not very much—no change in orientation can make even a 2-fold difference in drag. More specifically, relative to surface area, least drag is incurred by a rather long prolate spheroid traveling lengthwise or a flat disk traveling edgewise; a circular cylinder moving crosswise incurs the most drag. Relative to volume, least drag is suffered by a prolate spheroid twice as long as wide and moving lengthwise. Both it and a 1:2-oblate spheroid moving edgewise have less drag relative to their volumes than does a sphere. The difference isn't great, but it's enough to embarrass glib claims that spheres have least drag at low *Re* (Zaret and Kerfoot 1980).

TABLE 15.1 CALCULATED DRAG FACTORS FOR SIMPLE BODIES AT VERY LOW REYNOLDS NUMBERS.

		Flow Parallel of Axis of Rotation			Flow Normal to Axis of Rotation		
	l/d	$S^{1/2}$	$V^{1/3}$	U/U_s	$S^{1/2}$	$V^{1/3}$	U/U_s
Circular disk	1:50	6.383	36.60	0.319	4.255	24.40	0.479
Oblate spheroid	1:4	6.174	16.10	0.726	4.857	12.67	0.923
Oblate spheroid	1:3	6.012	14.82	0.789	4.950	12.14	0.963
Oblate spheroid	1:2	5.794	13.34	0.877	5.074	11.68	1.001
Sphere	1:1	5.317	11.69	1.000	5.317	11.69	1.000
Prolate spheroid	2:1	4.896	11.17	1.047	5.608	12.80	0.914
Prolate spheroid	3:1	4.762	11.39	1.027	5.858	14.01	0.835
Prolate spheroid	4:1	4.734	11.77	0.993	6.093	15.15	0.772
Circular cylinder	50:1	6.598	24.33	0.480	10.448	38.53	0.304

NOTES: The axis of rotation is what a biologist would call the axis of radial symmetry. l/d: length along axis of revolution over maximum width; $S^{1/2}$: drag divided by speed, viscosity, and the square root of surface area; $V^{1/3}$: drag divided by speed, viscosity, and the cube root of volume; U/U_s: terminal velocity relative to that of a sphere of equal volume and weight.

Fluid Spheres

If a sphere of fluid rises or falls through a fluid medium, the passage of the medium will induce a toroidal motion within the sphere, as was shown in Figure 10.6. The result will be lower drag on the sphere, as if the no-slip condition were partially relaxed. The phenomenon depends on the relative viscosities of sphere and medium: for a droplet of water in air the effect is negligible, whereas for a droplet of air in water there is, in effect, perfect slip. Happel and Brenner (1965) give the following formula:

$$D = 6\pi\mu_{ext}aU \ \frac{1 + (^2/_3)(\mu_{ext}/\mu_{int})}{1 + (\mu_{ext}/\mu_{int})} \ , \tag{15.8}$$

where μ_{ext} is the viscosity of the medium and μ_{int} is that of the sphere. For air in water or any gas in a liquid the ratio of the viscosities is essentially infinite, and the equation simplifies to

$$D = 4\pi\mu_{ext}aU, \tag{15.9}$$

which is to say that drag is just two-thirds of that given by equation (15.1) for a rigid sphere. It's not at all uncommon to find this distinction missed. On the other hand, Batchelor (1967) comments that if the interface between a sphere of gas and the surrounding liquid is dirty, surface motion will be much reduced, and equation (15.1) is likely to be closer to reality than is (15.9).

The phenomenon has quite another implication. A reasonable way to develop thrust at low Reynolds numbers is to move a membrane around a sphere or spheroid, extruding it at the front and absorbing it behind. That will happen either if toroidal motion can be created inside, through cyclosis or something similar, or if the membrane can be pulled upon. It's a direct application of the no-slip condition and should work in *any* decently deformable medium. Some such scheme seems to be used by, among others, sporozoan trophozoites (Jahn and Bovee 1969).

Orientation

At intermediate and high Reynolds numbers, a moving body that's symmetrical about each of three mutually perpendicular axes (cylinders, spheroids, rectangular solids, etc.) will either tumble or take up an orientation with the maximum cross-sectional area normal to the direction of motion. Typically this will maximize its drag, for better or worse. A proper arrow requires feathers to maintain its desired orientation. Shot off at a Reynolds number of about 100, a *Pilobolus* sporangium, according to the late Robert Paige, tumbles in flight. By contrast, for symmetrical objects moving through fluids at low Reynolds numbers, any orientation is stable. Released into an unbounded medium, a particle retains its original orientation if it has uniform or at least symmetrically varying density. But as at higher Reynolds numbers, providing protrusions to ensure that an object seeks a preferred orientation following any rotational displacement isn't difficult (McNown and Malaika 1950; Hutchinson 1967).

Wall Effects

Organisms frequently move near solid surfaces or air-water interfaces, and every measurement in a wind tunnel or flow tank involves the imposition of a wall somewhere near the test object. So a lot depends on just how close a wall can be without significantly affecting the flow pattern and the forces on an object. The problem is most acute at low Reynolds numbers, where a body influences the flow a great distance lateral to its outer surface. This lateral influence is, of course, just what makes flow through a pipe develop its parabolic profile nearer an entrance at lower Reynolds numbers (equation 13.9). The lower the Reynolds number, the larger must be the ratio between the distance to a wall and the diameter of an object if interference is to be kept acceptably low. An earlier "gee whiz" example bears repeating. At a Reynolds number of 10^{-4}, the presence of a wall 500 diameters away from a cylinder doubles the effective drag; at $Re = 10^{-3}$, a wall 50 diameters away dominates the determination of drag (White 1946).

Happel and Brenner (1965) give a general treatment of wall effects, but for most experimental purposes a quick index is adequate. Robert Zaret suggests using the following rough-and-ready guide, derived from White (1946), to be reasonably sure that wall effects can be ignored:

$$\frac{y}{l} > \frac{20}{Re}, \tag{15.10}$$

where y is the distance to the nearest wall, and l is a characteristic length of the object. The formula is usable only at Reynolds numbers below unity and is a fairly stringent criterion that could be shaved a bit for nondemanding applications.

The same wall effects will influence the response of hot wire and hot probe flowmeters when these are used near walls, so appropriate caution should be exercised (White 1946). And walls add significantly to the drag experienced by swimming microorganisms (Winet 1973).

Unless an object moves axially in a cylinder or midway between plates, a wall introduces some asymmetry in the drag of even a symmetrical object. A sphere moving parallel to a wall will rotate in the same direction as if it were rolling along the wall. The rotation may, in turn, generate other forces. For instance, constriction of streamlines between moving object and wall will produce an attractive force, a lift as we've defined it earlier, drawing the body toward the wall. These wall effects are pernicious; their presence should be suspected (or even presumed) in any study that doesn't specifically deal with them. As I'll argue shortly, they corrupt quite a lot of data in the biological literature.

TERMINAL VELOCITY

If an object whose density differs from that of the medium is released, it accelerates either upward or downward. As noted, at low Reynolds numbers the period of acceleration is brief and the distance traveled is short; the object very soon reaches (asymptotically, of course) a constant, "terminal" velocity. Not that terminal velocities don't exist at high Reynolds numbers; they're just less commonly of biological relevance and they take more time and distance to achieve. As in gliding, the absence of acceleration implies a balance of forces—weight minus buoyancy (net body force) against drag (resistive force).

Consider a solid sphere of density r falling in a medium of density ρ_0. Gravitational force is gravitational acceleration times the difference between the masses of sphere and displaced fluid, $(\rho - \rho_0)(4/3)(\pi a^3 g)$. Drag is given by Stokes' law (equation 15.1)—$6\pi\mu a U$. If we equate the two and solve for velocity, we get the very well known equation for terminal velocity,

$$U = \frac{2a^2 g(\rho - \rho_0)}{9\mu}.$$ (15.11)

This work for about the same conditions as does Stokes' law—up to a Reynolds number of about 0.5. Notice that terminal velocity is proportional to the square of the radius—a bigger object falls faster, not because it has less drag (it has more), but because for a given density it has more volume and hence a larger net weight. On the other hand, if the investment in mass rather than density is fixed, then a larger sphere will fall more slowly than a smaller one, at least in a medium such as air whose density is much lower than that of any solid sphere. For a given balloon, the greater its inflation, the slower it will fall—although here the Reynolds number is a little high for equation (15.11) to apply strictly.

(Above $Re = 0.5$ a somewhat more complex formula can be derived from equation (15.2) and the definition of the drag coefficient, and at still higher Re's from C_d alone. In effect, the dependence of terminal velocity on radius decreases as the Reynolds number rises, from variation with the square of radius where Stokes' law applies through a direct proportionality just above that, to the square root of radius where the drag coefficient has leveled off (as in Figure 5.4). Bigger things may in general fall faster, but the dependence of the rate of fall on size is less. Thus Brooks and Hutchinson (1950) found for largish narcotized microcrustacean water fleas (*Daphnia*), where drag coefficients ought still to be falling with Reynolds numbers, that sinking rate varied directly with length. A little surprising is what Dodson and Ramacharan (1991) found for *Daphnia*—the sinking rate varied with length to the power 0.58, so something more than passive sinking must have been involved. Incidentally, at constant densities (of both object and medium) proportionality with the square root of length or radius implies that terminal velocity will vary with the sixth root of mass—a much less drastic dependence than one might imagine. "The bigger they are, the harder they fall" is still credible, as long as one recognizes that "harder" alludes more to momentum or kinetic energy than to velocity.)

For objects of other shapes, one can calculate terminal velocities from equations (15.3) through (15.6) or from the data in Table 15.1. For higher Reynolds numbers, some approximation of C_d is needed (such as equations 15.2 or 15.7); since C_d depends on velocity, the nuisance of an iterative solution is hard to avoid.[3] The underlying rule remains that weight minus buoyancy equals drag. For objects of negligible displacement such as disks, or for objects moving in a medium whose density is much less than their own, buoyancy can, of course, be neglected.

---

[3] Don't, repeat *don't*, even think of using Stokes' law and equation (15.11) above $Re = 1.0$. At $Re = 1.6$ the latter overestimates speed by a factor of about 1.2; at $Re = 10$ by 1.8; at 100 by 5-fold; at 1000 over 25-fold.

Stokes' Radius

Free descent or ascent at terminal velocity is a most ordinary phenomenon among tiny biological objects—consider dispersing seeds, spores, baby spiders, gypsy moth caterpillars, and many nonmotile planktonic creatures. The shapes of such small objects are extremely diverse; to separate the effects of shape from those of size and density, a measure of effective size may prove useful. Since most of these objects are either quite small or have elaborate filamentous outgrowths of very small diameter, Reynolds numbers are fairly low. A reasonable measure is the radius of the sphere of the same mass that would ascend or descend at the same rate; let's call it a_s, the "Stokes' radius."[4] To obtain this radius, one just needs to measure the mass and terminal velocity of an object and solve the following equation for a_s:

$$mg - \frac{4}{3}\,\rho_0 g \pi a_s^3 = 6\pi\mu a_s U. \tag{15.12}$$

For solid objects falling through air, the middle term can be safely dropped. The density of the object has been eliminated, so one needn't prejudge the object's effective size. The resulting Stokes' radius can be viewed as a measure of the effectiveness of any protrusions in increasing effective size.

(Several alternatives to the Stokes radius are in use. The most common, generally called the "coefficient of form resistance," is the ratio of the terminal velocity of a sphere of the same density and volume to that of the object or organism at issue [Hutchinson, 1967]. In effect, it's an extra factor hooked on to the right side of equation 15.1.)

Interactions among Falling Objects

What happens if two objects, each going up or down at its terminal velocity, get close to one another? As one might guess from the earlier discussion of wall effects, the behavior of a particle may be substantially altered by another anywhere nearby. A moving particle moves a lot of fluid along with it, and any other particle will be carried along. If two particles are sinking, the one behind will thus tend to fall faster and catch up to the one ahead—even if in isolation the latter might have a slightly lower terminal velocity. Two particles falling in tandem will go faster than either alone, whether they fall one behind the other or side by side. The closer

---

[4] It's been called the "nominal radius" by Kunkel (1948), but I think the name "Stokes' radius" is less ambiguous—Stokes' law and terminal velocities are so commonly considered concurrently.

they travel, the greater the interaction and the faster (by up to 30% or so) they go.

Once one gets used to thinking about these viscosity-dominated systems, intuition becomes a decent guide to quite a few phenomena. For instance, if two spheres sink side by side, each will induce some rotation in the other; and the sides of the particles facing each other will move in the direction of the overall sinking motion. That's reasonable since fluid between the two spheres will to some extent be carried with them, so the retarding effects of passage though fluid will be felt most strongly on the sides farthest from each other. The direction of rotation in this case is opposite that noted for a sphere falling parallel to a wall—here the two solids are moving in the same direction, while in the case of the sphere and wall, the directions are opposite. Again, Happel and Brenner (1965) provide a wealth of additional information in a properly quantitative treatment.

Clearly a group of particles will fall faster than would the individual particles, and such a group will tend to maintain coherence since ones behind will catch up and ones in front will be more strongly retarded. That may explain some puzzling data of Chase (1979), in which aggregated particulates in both fresh and sea water fell at up to ten times the rates predicted by Stokes' law—"a drop of water containing high concentrations of aggregates [was] released into the charging port." At the same time, a cloud of particles that fills a container may fall more slowly than a single particle or a group near the center. If the particles carry fluid downward by viscous interaction, then somewhere an upward flow must compensate to satisfy the principle of continuity. This means trouble for the all-too-common practice of timing descent rates of clouds of organisms to gain some statistical view of sinking and to permit the use of photometric detectors. In short, even the sign of the error introduced by using groups of organisms isn't certain from an ex post facto look at the literature. Furthermore, a descending cloud may tend to concentrate or spread out over time and distance, complicating any correction.

Measuring Terminal Velocity

By this point I hope I've engendered a conviction that existing data must be assumed to be seriously in error unless sufficient procedural detail is given to demonstrate otherwise. What looks at a facile glance to be the simplest of procedures isn't anything of the sort. If walls are far away, convection in the test chamber is quite likely to confound matters, and observations are difficult; if walls are nearby, then they slow the falls. So what is the honest investigator to do? And old but still useful correction is cited by various sources (for instance, Sprackling 1985) for a solid sphere rising or falling in the center of a cylindrical column of fluid:

$$\mu_{true} = \frac{\mu_{apparent}}{(1 + 2.4\, a/R)}, \tag{15.13}$$

where a again is the radius of the sphere and R is the radius of the cylindrical column. The equation can be applied to equation (15.11) either for viscosity determinations or for correcting terminal velocities obtained in columns for unbounded systems ($R = \infty$). A lot of questionable data on sinking rates in the biological literature can be restored to health by therapeutic use of the equation, at least where the author specifies the size of the column. One merely assumes that the viscosity in the denominator of equation (15.11) is the apparent viscosity (as defined by 15.13) and substitutes $[\mu_{true}(1 + 2.4\, a/R)]$ for it.

But that's just a start. The correction is inadequate if an organism moves substantially off-axis in the chamber. Perhaps worst is convection in the chamber—again, the problem gets increasingly severe as chamber size is increased in order to avoid wall effects. Very slight motion is all it takes to contaminate data when, for instance, the sinking rate of a diatom is about 5 μm s^{-1}. That's over three seconds to go a mere millimeter. Even slight irregularity in illumination can make real trouble, as can the residual effects of whatever scheme was used to put the players on stage. Arranging a very gentle thermal gradient in any liquid system to help offset convection is probably a good idea. Booker and Walsby (1979) describe a technique and the appropriate correction formulas for using a sucrose density gradient to control convection.

I'd like to suggest a return to looking at the descent of individual cells or organisms—decontaminating data for unnaturally concentrated assemblages is just too problematic. One might permit an acceptably low density of particles to descend in a thermally jacketed column of generous size, viewing the center of the column with a microscope equipped with optics permitting long working distances. A video camera and recorder can pick up even an occasional item that descends in the right place, illumination can be intermittent or brief, checks for convection can be made under equivalent circumstances with dye or particles of known behavior, and so forth.

When Terminal Velocity Matters

For applications at low Reynolds numbers one thinks immediately of nonflying airborne particles—pollen grains, fungal spores, and the like—for which dispersal distance has a very direct bearing on fitness. By the principle of continuity, as much air must come down as goes up; by the principle of gravity, any passive object denser than air must always be sinking with respect to the air around it. Other things being even vaguely comparable, the smaller the propagule the more effective will be wind dispersal. Ingold (1953) cited, as a record of sorts, 0.5 mm s^{-1} for the rate of fall of the 4.2-μm diameter spore of *Lycoperdon pyriforme*, but he considered 10 mm s^{-1} as a more typical rate for fungal spores. Gregory (1973) provided a wealth of data on size, density, and terminal velocities of spores. Good discussions of the flight of pollen can be found in Burrows (1987) and Niklas (1992); the latter, for instance, calculated a sinking rate of 123 mm s^{-1} for a 64 μm-diameter, 0.137 μg pollen grain of the hemlock, *Tsuga canadensis*. For neither airborne spores nor pollen is density easy to measure (see Figure 15.1a), so it's simpler (neglecting air density, of course) to use a version of equation (15.11) based on mass:

$$U = \frac{mg}{6\pi\mu a_s} \tag{15.14}$$

The interaction of these sinking particles and atmospheric motions (on a variety of temporal and spatial scales) gets beyond the scope of this book; quite a lot is known and can be found under topics such as micrometeorology, biometeorology, and plant (or seed) dispersal.

Certainly the biologists with the greatest interest in terminal velocity are the oceanographers and limnologists. Whether in marine or freshwater systems and whether living, organic but nonliving, or inorganic, most small items are "negatively buoyant"—sinkers. That's quite a spectrum of possibilities, and a large literature concerns itself with sinking speeds—of fecal pellets of microcrustacea (Komar et al. 1981, for instance), of the macroscopic aggregates a few millimeters long of a mixture of living and dead material that's come to be called "marine snow"(Riebesell 1992), of marine bacteria (Pédros-Alío et al. 1989) and fungal spores (Rees 1980), of planktonic ciliates (Jonsson 1989), of various larvae (Bhaud and Cazeux 1990) and embryos (Quetin and Ross 1984), and, in greatest profusion, of phytoplankton such as diatoms.

Living planktonic organisms are almost always denser than their ambient media—even with the use of devices such as gas vacuoles (particularly in freshwater forms), fat droplets, or vacuoles in which ammonia is accumulated and potassium and divalent anions excluded. But they have various ways of reducing sinking rates. Some, such as water fleas (Dodson and Ramacharan 1991) and ciliates (Jonsson 1989), swim upward ("negative geotaxis"). A worm larva, *Lanice*, makes a mucus thread about 15 mm long (Bhaud and Cazeux 1990). Diatoms adjust their composition in response to such factors as light and nutrient concentration (see, for instance, Davey 1988). Some desmids are enveloped in mucilage (Lund 1959). As Hutchinson (1967) points out, even a negatively buoyant coating can reduce sinking rate if the density difference between organism and coating is more than twice the density difference between coating and medium. And the

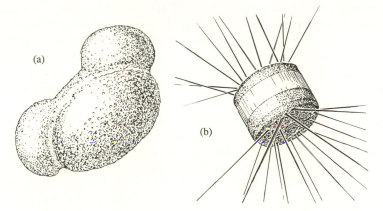

FIGURE 15.1. Small sinkers: (a) a grain of the anemophilous pollen of a pine tree; (b) the diatom, *Thallassiosira*.

irregular shapes of many planktonic organisms are commonly interpreted (at least in part) as evidence that they go to great lengths to sink more slowly.

Unfortunately, the measurements (as obtained) of sinking rates in this literature—especially those on slow sinkers such as diatoms and bacteria— are of very uncertain reliability and applicability to natural circumstances. Indeed, the comments on measurement of sinking rates a page or so back were prompted by my sinking spirits when working through a large number of relevant papers. Some attention has been given to convection in test chambers (Bienfang and Laws 1977) and to wall effects (Quetin and Ross 1984), but little if any to interactions among sinking particles, despite the almost universal practice of measuring terminal velocities of fairly dense suspensions.

Quite a few papers assert an inverse relationship between irregularity (e.g., nonsphericity) in shape and sinking rate; that an increase in relative surface area slows sinking would certainly seem reasonable. But here again, reliable evidence is scanty. The most direct test of the notion was done by Walsby and Xypolyta (1977); they found that removing the long chitan fibers protruding from a diatom, *Thalassiosira fluviatilis* (Figure 15.1b), increased sinking rate from 3.8 to 6.6 μm s^{-1} even though the fibers were much more dense (1495 vs. 1112 kg m^{-3}) than the rest of the cells. Their measurements of sinking rates, though, were made on a turbid layer in which the concentration of cells approached 5%. Some sources give plots showing an inverse relationship between sinking rate and surface-to-volume ratio. The problem with these is that the ratio, S/V, has a residual dimension of inverse length; and thus an axis along which S/V increases may represent decreasing size instead of increasing irregularity—or, for that matter, some combination of the two. I reanalyzed one set of such data for diatoms, using a dimensionless (and hence size independent) surface-to-volume ratio, $S^{1/2}/V^{1/3}$, and found that the ostensible effect of irregularity disappeared entirely. Thus the variation in sinking rates in this particular set of data reflects variation in size rather than shape; correcting for size eliminated the effect.

Since, other things being equal, increased surface area must slow sinking, other things must not be equal. Probably (at least in this one case) the density of the diatoms, the main unmeasured variable, varied systematically with size. Perhaps irregularity is at least in part a device that permits less fastidious control of density in the process of regulating sinking rate. That's at least consistent with the observation of Sournia (1982) that small cells are less irregular than are large cells.

Yet another problem. Why should sinking per se be so ubiquitous even among phytoplankton? Achieving neutral or positive buoyancy should present no special difficulty, even in fresh water—fat droplets and gas vacuoles will always be less dense than water. And sinking takes a photosynthetic organism away from its critical resource, light. The most common argument is that sinking is important for obtaining nutrients, that a small organism that didn't sink would soon deplete the local larder. I'm unpersuaded on account of several counterarguments. At least for actively swimming predators such as ciliates and water fleas, augmenting diffusion with convection in this way is unlikely to matter much. We need to ask how fast sinking would have to be in order to make a difference to diffusional exchange. The diatoms that are the main concern in the literature are very slow sinkers—4 μm s^{-1} is about a foot a day. The cells are around 10 μm in diameter. The Péclet number, used in the last chapter in connection with exchange adjacent to capillary walls and in this one to get a quick view of Berg's (1983) argument against bacterial wandering, again gives an appropriate index of what motion accomplishes. Assuming that carbon dioxide, with a diffusion coefficient in water of 1.5×10^{-9} m^{2}s^{-1}, is what matters for a diatom, the Péclet number (Ul/D) comes out to less than 0.03. That's clearly well below the value of unity (see, for instance, Sherwood et al. 1975) around which the convective augmentation of diffusion might become significant. Put another way, a diatom would have to sink about 30 times faster for sinking to have a significant effect. The argument isn't a new one; in looking for a reason for sinking, Munk and Riley (1952) did some calculations that for such small cells ruled out obtaining nutrients.[5]

---

[5] But for a ciliate protozoan, *Tetrahymena pyriformis*, with $l = 40$ μm and $U = 450$ μm s^{-1} (data from Plesset et al. 1975), trying to get oxygen, $Pé = 10$. Its movement (whatever its other uses) makes convection distinctly nonnegligible. The argument against moving doesn't extend to the whole microscopic world.

Why, then, sink? I'd like very tentatively to suggest that sinking might be a device to keep away from the air-water interface. We're painfully aware that large organisms can't ordinarily support themselves on an air-water interface (Vogel 1988a)—the Bond number, the ratio of the force of gravity to that of surface tension is too high (or its inverse, the Jesus number, is too low). We're less concerned with the boundary condition on the other side, that affecting small organisms and loosely defined by the Weber number. The latter, about which more in Chapter 17 (and defined by equation 17.6), is the ratio of inertial force to surface tension force; in its numerator is length and the square of velocity. Small, slow things can't get enough lU^2 to get loose, and the interface is a very special habitat for which ordinary plankton must be poorly adapted. So, as eloquently pointed out by D'Arcy Thompson (1942), it can snare as well as sustain: "A water-beetle finds the surface of a pool a matter of life and death, a perilous entanglement or an indispensable support." Positive buoyancy may be perilous—better to sink slowly or to counteract sinking by swimming upward to just short of the interface. And (as is commonly noted), natural bodies of water aren't still, so even continuous sinking needn't mean steady and irreversible movement away from the surface.

Propulsion at Low Reynolds Numbers

Consider, again, the central message of the chapter—in no biologically relevant area is the counterintuitive character of fluid mechanics more evident than at low Reynolds numbers. The swimming of spermatozoan and eel may have a certain superficial similarity, but the underlying physical mechanisms turn out not to be at all the same. Put a bit crudely, the eel is imparting local rearward momentum to the fluid, making use of the fluid's inertia. By contrast, the spermatozoan (assume a free-swimming aquatic one, not the sort produced by male readers) is pushing against the fluid more in the manner of a snake slithering on a solid but slightly slippery substratum, using the fluid's viscosity to permit purchase. It's a radically different world, and nowhere is its special character more evident than in the business of propulsion.

The Peculiar Problems and Possibilities

Back in Chapter 11 and early in this one I noted that circulation-based lift isn't worth much at low Reynolds numbers. Thom and Swart (1940), the only explorers of this unpromising territory of whom I know, got a best lift-to-drag ratio of 0.43 at $Re = 10$ and of 0.18 at $Re = 1.0$. Thus tiny organisms insistent on swimming must resort to some drag-based scheme. As it turns out, drag-based propulsive arrangements encounter special problems of their own at low Reynolds numbers.

To begin with, the reversibility of these flows presents a subtle but very real impediment. Consider a simple propulsive system in which append-ages are repeatedly moved fore and aft. They move steadily forward in a slow recovery stroke and steadily rearward in a rapid power stroke. The appendages (perhaps elongated cylinders) present the same shape and area to the fluid in either stroke, and the body from which they protrude doesn't move through the fluid. Does the reciprocation generate a net thrust? Clearly the faster power stroke generates more force than the slower recovery. But what matters in the overall balance sheet is the product of the force and the time over which it is exerted (strictly, force integrated over time)—what's called the *impulse*. The change in momentum of a body will equal the impulse of the force on it.

At high Reynolds numbers drag will be roughly proportional to the square of velocity; since the velocity of the stroke is inversely proportional to its duration, the impulse of a stroke will be proportional to the speed of movement of the appendages. Thus the thrust of the fast power stroke will more than offset the drag of the slow recovery, and a net thrust will be produced, even if inefficiently.

By contrast, at low Reynolds numbers drag is directly proportional to speed. With $F \propto U$ and $t \propto 1/U$, Ft will be constant whatever the speed of the appendage during the stroke. In short, no juggling of speeds of power against recovery strokes will enable them to do anything other than cancel each other's efforts. Not only does low Re imply high drag, but the reduced dependence of drag on speed is a special drawback.

Still, at least two approaches to drag-based locomotion remain possible. Formulas for drag at low Reynolds numbers such as Stokes' law (equation 15.1) have three usable variables, speed (U), shape (for instance, 6π), and area (in the guise of a); and we've ruled out only interstroke manipulation of the first. Organisms, in fact, use both changes in the orientation (effec-tive shape) and changes in the effective area of moving appendages be-tween power and recovery strokes.

Changing the orientation of a reciprocating appendage looks almost as inefficient as using circulation-based lift. The change in drag is minimal for even a most extreme case, a long flat plate shifted from perpendicular to parallel to flow; at $Re = 1.0$ the decrease is only about a third (Table 5.2). That doesn't bode well for a healthy recovery stroke! In fact, as we'll see shortly, propulsion based on the difference in drag between two orienta-tions is widespread; but it's based on a scheme other than the kind of rowing and feathering implied here. For that matter, I can think of no system in which low Reynolds number propulsion is based purely on orien-tation changes of a long, reciprocating paddle.

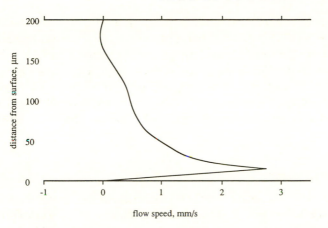

FIGURE 15.2. (a) The velocity gradients near a ciliated surface. 15 μm-long gill cilia of a mussel, *Mytilus edulis*, are propelling water through a 200 μm-wide channel (from Nielsen et al. 1993).

Yet another problem. At low Reynolds numbers velocity gradients are wide and gentle, so a body towed through or falling through a fluid carries a lot of fluid with it, and free-stream velocity is approached only quite far away from the body. But to propel oneself forward one must push against the free stream. Thus an effectively propulsive appendage must protrude through and beyond the gradient region. If such an appendage is short (as, for instance, a cilium), then the gradient must be short and steep, as in Figure 15.2. That means a lot of work must be done against the viscous bugbear of skin friction, and some large portion of an organism's surface ought to be equipped with such appendages—any surface not so equipped will severely offset their action.

The common use of extensive bands of cilia as well as fully ciliated surfaces is thus entirely reasonable. At the same time it raises some awkward problems for the investigator. Say one is interested in the pattern of flow around a ciliated protozoan such as *Paramecium*. You can easily ballast a *Paramecium*-shaped object so it falls at the right Reynolds number in a very viscous liquid. But you most certainly won't get a pattern equivalent to that of a swimming animal—the velocity gradients will be far too gentle for real cilia (allowing for scale) to protrude through them. And you'll vastly overestimate the lateral extent of the disturbance caused by passage of a swimming animal. Queer as it sounds, a swimming *Paramecium* will disturb the surrounding fluid less than one passively sinking at the same speed. The difference is strikingly evident in Wu's (1977) photographs.

Nor can you conclude much about propulsion from looking at the flow field around a tethered creature in still water. Emlet (1990) found, not unexpectedly, that attached molluscan larvae moved less water but disturbed the flow field farther away than ones around which water moved at speeds typical of swimming. Furthermore, the problem we met earlier of using drag data to predict thrust in fish is even more pernicious. One can say almost nothing about the force or power needed for propulsion from measurements of the drag of a model of, say, a ciliated invertebrate larva. Being assured that thrust balances drag gives no easy clue as to how to measure either.

Changing the Area of Appendages: Paddles and Bristles

The real specialists here are small arthropods, mainly aquatic ones but some of the tiniest fliers as well. The general scheme involves bristles (or in larger creatures, oarlets) that extend during a power stroke and fold back during recovery. Nachtigall (1980) has described very clearly how such a combined system works in the swimming appendages of aquatic beetles. The outer segments of the propulsive legs are little more than tubes to which bristles attach. These are spread by the thrusting stroke of the legs until they lie in a plane; if the leg is a flattened one, they lie in the plane of flattening. Two-thirds to three quarters of the thrust is attributable to the bristles; even at these not-so-very-low Reynolds numbers, they produce up to 54% as much thrust as would an equally broad solid surface. During the recovery stroke, the hairs rotate back against each other and against the leg segments. In addition, the leg joints are flexed during the recovery stroke. At least one beetle, *Gyrinus*, has flattened blades in place of cylindrical bristles; spread during the thrust stroke, these overlap like drawn Venetian blinds, giving about 90% of the thrust of a fully solid surface (Figure 7.6). Thus the propulsive legs can be drawn forward with minimal exposure to the free stream both by minimizing effective surface and by keeping them close to the body.

In at least three orders of insects, small ones have converged toward wings similar to these beetle legs (Figure 15.3), as noted by D'Arcy Thompson (1942), who felt that this structure allowed the insects to "row" through the highly viscous air. Ellington (1980a) has shown that the wing fringes of the most common of these bristle-winged groups, the thrips (Thysanoptera), can be actively locked in either an "open" or a "closed" position by forces greater than the aerodynamic and inertial forces on the beating wings.

The scheme will work, of course, at all Reynolds numbers. Its rarity in the really macroscopic world must reflect mainly the existence of better or easier alternatives such as circulation-based lift and recovery strokes in which orientation alone is altered. Indeed, if we don't restrict ourselves to unwebbed bristles, the general scheme is better at higher Reynolds num-

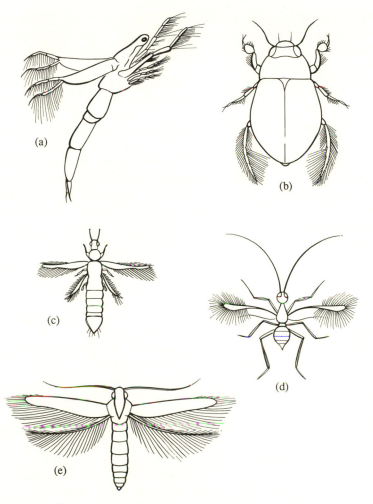

FIGURE 15.3. Animals with propulsive appendages that lack a continuous membrane: (a) a cladoceran crustacea; (b) a diving beetle;
(c) a thrips; (d) a mymarid wasp; (e) a moth. All but the beetle
are very small. The first two swim; the latter three fly.

bers, emphasizing the special difficulties of being slow and small. Williams
(1991) has looked at changes in swimming pattern over the course of development of brine shrimp (*Artemia*) larvae (those crustaceans whose eggs are
sold in pet stores as fish food). The earliest stage ($Re = 2$) moves with a
ratcheting motion using a single pair of appendages—forward with a little
backslip during recovery. By $Re = 5$ a bit of a glide follows the power stroke;

by $Re = 13$ the glide is more pronounced and the backslip is about gone. (Still larger larvae, with Re's between 20 and 40, shift to sequential or "metachronal" strokes of a series of appendages, complicating the comparison.)

What does change with Reynolds number is the nature of the propulsive "surface." In swimming beetles we noted that bristles form dense sheets and may be flattened and overlapped to make a plane surface. By contrast, in the smaller and slower world of a cladoceran crustacean, hairs are spread farther apart; velocity gradients are gentler, and fewer bristles suffice. In a model devised by Zaret and Kerfoot (1980), the propulsive action of the second antenna of the cladoceran, *Bosmina*, turns out to be rather insensitive to the precise position and gap between the bristles. Further information on how small, swimming arthropods use such bristle-paddles can be found in Nachtigall (1974), Hessler (1985), Morris et al. (1985), and Blake (1986).

Changing the Orientation of Appendages: Cilia and Flagella

As mentioned already, the scheme is widely used despite the fact that the drag of a flat plate parallel to flow isn't much less than the drag of one normal to flow. The devices used aren't flat plates but, oddly enough, cylinders; the appendages most often involved are cilia and flagella.

Nothing fundamental distinguishes cilia from flagella in eukaryotic cells, although cilia are usually shorter (5 to 10 μm in length) and more numerous on each cell. A cell rarely has more than a few flagella, but these may be up to 150 μm long (Jahn and Votta 1972). Often "flagellar action" refers to helical or undulatory wave propagation while "ciliary action" refers to a reciprocating beat, with no implication of any other difference. Both cilia and flagella are about 200 nm in diameter and are musclelike devices with motile proteins inside. Bacterial flagella, however, are something quite different; they're composed of a single, nonmotile filament, and they're only 20 nm in diameter (Berg et al. 1982).

The arrangements and actions of these organelles are diverse. An organism (often a single cell) may be covered with cilia or have distinct bands of them (the latter are especially common among marine invertebrate larvae; see Chia et al. 1984). It may have one or more flagella on the anterior that pull, one or more on the posterior that push, or even a flagellum extending laterally that moves the creature in a helical path.[6] The motion of a flagellum may take the form of planar, undulating waves either from base to tip or tip to base; a flagellum may propagate helical

---

[6] Swimming along a helical path is, in fact, very common among microorganisms of just about every group that goes in for active locomotion. For consideration of both its mechanism and functional significance, see Crenshaw (1993).

| power stroke | recovery stroke |

FIGURE 15.4. The motion of a cilium, shown as a left-to-right sequence, pushing water from right to left. It's extended during the power stroke (*left*) and flexed down near the surface for recovery (*right*). The cilium thus moves more distant from the substratum and normal to its long axis during the power stroke and closer to the substratum and parallel to its long axis during recovery. Both differences contribute to net thrust generation.

waves from base to tip; or it may produce spiral waves. By contrast, ciliary coats or bands beat more like our description of the swimming appendages of beetles or microcrustaceans—an extended thrusting stroke is followed by a flexed recovery stroke in which the cylinder moves lengthwise close to the surface of the organism, as in Figure 15.4. Groups of cilia are usually coordinated, and a host of patterns of coordinated beating have been described (and named). Bacterial (prokaryotic) flagella, rigidly helical, are driven in a circular path around the long axis of the helix by proper rotary motors to which they attach; they're the only truly rotational (wheel-and-axle) machines known in organisms (Berg and Anderson 1973). Eukaryotic flagella with helical motion translate in circles.

A common physical mechanism for propulsion really does underlie all this diversity. The basic setup is best introduced by considering what happens if a long circular cylinder is pulled through a viscous medium (Figure 15.5a). If it's pulled lengthwise, equation (15.5) applies; if it's pulled broadside, equation (15.6) applies. For a cylinder 50 times as long as it is wide ($l/a = 100$) the ratio of the two drags is 1.75; for one 150 times as long as wide ($l/a = 300$) the ratio is 1.80—either figure is better than the 1.5 ratio for a long flat plate parallel versus normal to flow. Cilia, beating rather than undulating or rotating, take direct and obvious use of the difference in drags, moving almost broadside on a power stroke and moving almost lengthwise for recovery. And once the ciliated organism is in motion, drag during the recovery stroke is kept down by the additional device of moving the cilia close to the surface of the creature. Even so, such recovery strokes are far more costly than their high-Reynolds number equivalents. The motile algal cell, *Chlamydomonas*, swims in a kind of breast stroke with two similar flagella; during a recovery stroke it loses about a third of the distance gained in a power stroke (Rüffer and Nultsch 1985).

What undulating and rotating systems do is considerably more subtle. The fact that the drags are substantially different means that a rod that's

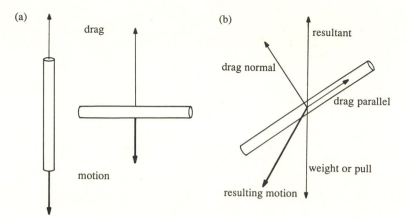

FIGURE 15.5. (a) Pulling a cylinder through a viscous medium; drag is almost twice as high when it's pulled crosswise. (b) An obliquely towed cylinder wants to slew off to one side, much like a boat towed by a rope fixed to one side a little back from the bow.

pulled not quite broadside (obliquely) will slew a bit sideways—it moves lengthwise relatively more easily than it moves broadside. In effect, the difference in drag seen in the two extreme orientations generates a resultant force that's not quite opposite the imposed force trying to move it obliquely (Figure 15.5b). On a long tether, the cylinder will slew off much like a water skier with skis askew or a wedge that persists in going lengthwise even if hit by a glancing blow. You can see this poor sort of gliding in action by dropping an obliquely oriented fine wire in a tank of water. According to Happel and Brenner (1965), the best orientation for slewing is with the long axis of the cylinder about 55° to the direction of pull; it will then slew about 20° off the direction of the pulling force. That glide angle of 70° isn't impressive, and the angle of attack of 35° may seem extreme, but this is something quite different from circulation-based lift.

Sir G. I. Taylor was, I believe, the first to point out (1951) that the phenomenon can be used for propulsion. It works in just the same way that thrust is generated from the lift of an airfoil moving with respect to a craft (Chapter 12). What matters isn't the origin of the lift but simply that some component of a force is developed in a direction opposite the motion of the craft. Move an appendage back and forth, changing its orientation appropriately, and a pulsing component of force will push an organism unidirectionally; the other components will cancel, causing at worst some side-to-side wobble, as shown in Figure 15.6a. If we expand our view from such a paddlelike short segment of a cylinder to a long, axially rotating helix, we can see how rotation of the latter will give steady forward propulsion

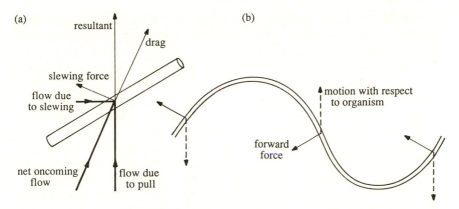

FIGURE 15.6. (a) Moving a oblique portion of an appendage sideways generates a force normal to the motion, which may be used for propulsion ("slewing," locally). If so, the oncoming flow is deflected a bit, as with a poor glider (recall 11.11). (b) A turning helix will generate just such a normal force over its entire length.

(Figure 15.6b). Each element of the cylinder is moving obliquely sideways and thus contributes thrust. That the organism will counterrotate as a result of the remaining force component is either tolerated or offset by another such rotating helix. The game is much like screwing into a cork except that some slip occurs; elements of the helix do not quite advance lengthwise but (in the case described earlier) move $55° - 20° = 35°$ to the long axis of each. It makes little or no difference whether the helix truly rotates (bacterial flagella) or just translates around its long axis (eukaryotic flagella).

Nor does it make much difference if the undulation is in a plane. Pieces of cylinder are still oriented obliquely and move laterally. In planar undulation, though, not all of a flagellum can contribute all at once because some of its length is "wasted" at the upper and lower ends of each wave. On the other hand, a propagated planar undulation doesn't throw the rest of the organism into rotation in the opposite direction.

Neither ciliary nor flagellar propulsion can really be described as efficient. A ciliary tip velocity of 4 mm s^{-1} produces an average flow past the cilium of only about 1 mm s^{-1} (Sleigh 1978). Still, propulsion seems to take only a small part of metabolism of these animals, and any competitor must pay the same price. Ciliary propulsion, in particular, may have fringe benefits. As noted earlier, a swimming *Paramecium* is surrounded by a steeper velocity gradient than even an equivalent passively sinking body. So it's minimally "noisy" in a sense that we'll consider shortly. Furthermore, swimming (by protozoa but not bacteria) can augment diffusion by raising the Péclet number decently above unity; swimming with cilia ought to be especially effective in improving exchange on account of the especially steep velocity gradients.

Usually animals using ciliated surfaces for locomotion are both larger and faster than those using a few flagella. Where sizes overlap, Sleigh and Blake (1977) point out that the ciliated forms go about ten times faster than the flagellated ones. But, as they also note, propulsive efficiency drops with increasing size for either, and no large swimmer in a hurry uses either scheme—muscles have the competitive advantage of another order-of-magnitude increase in speed. The largest creatures using cilia are ctenophores ("comb-jellies"), typically around a centimeter or so long; but they use them in platelike bundles of a kind that occur nowhere else (Tamm 1983). The largest animals swimming with simple cilia appear to be rotifers, an order of magnitude smaller, and they seem notably inefficient (Epp and Lewis 1984).

The diversity of ways in which cilia and flagella are used is only hinted at here. They're commonly used to run pumps involving both water and mucus, as mentioned in Chapter 14. Some flagella have pinnately sequential appendages on them (mastigonemes) and propagate waves in the direction in which the organism is going rather than in the normal, retrograde direction (Brennen 1976)—just as do swimming polychaetes with parapodia sticking out laterally (Chapter 7). The density of cilia implies substantial interaction and some degree of artificiality to any analysis of single organelles. Useful sources range from purely descriptive ones emphasizing biological diversity such as Jahn and Votta (1972), to ones of real mathematical splendor such as Wu (1977) and include Brennen and Winet (1977), Roberts (1981), and the various papers in Wu et al. (1975).

FILTRATION

Feeding by using a filter of some sort ("suspension feeding" usually) almost always involves flow at low Reynolds numbers; edible particles suitable for capture are rarely over a millimeter in diameter, and the capture elements likewise are small. Flow speeds at the filter are most often slow, partly, one may guess, because fine filters may not be mechanically robust and partly because a large filtration surface will, by the principle of continuity, be associated with low speeds. The only high Reynolds number filterers may be the baleen whales, and little seems to be known about their hydrodynamics (Sanderson and Wassersug 1990).

A Few Problems

The preeminent fluid mechanical issues in filtering are those of ensuring contact between edible particles and filtration surfaces and of keeping the resistance of the filter low enough so fluid goes through it. These requirements are substantially antithetical—the finer the mesh of the filter, the greater the resistance to flow. When building a rake, it's all too easy to end up with a paddle, as Cheer and Koehl (1988) put the difficulty. The viscous character of low Reynolds number flow makes matters worse—if you can bail with a sieve, you can't easily filter with it. And it's remarkable just how low the Reynolds numbers of filter-feeding devices can get. Both the tiny microvilli that make up the collars of sponge choanocytes and their intervillar gaps are 0.1 to 0.2 μm across, and water passes through them at about 3 μm s^{-1} (Reiswig 1975b); this gives a Reynolds number of 4×10^{-7}!

The basic problem can be illustrated in what may seem quite a different system—flow through the large pinnate antennae that male saturniid moths use to detect the odorant ("pheromone") molecules of females who may be kilometers away, truly an olfactory sensation. In this context the problem is that exposing more receptors to the flow increases the fraction of the flow that will evade the antenna entirely. I found (Vogel 1983) that only 8% to 18% of the air directly upstream went through; the rest went around. No one should be surprised (by this point in the chapter) that the transmitted fraction increased steadily with free-stream speed. Nor that viscosity proves such a serious impediment to transmission—by contrast with air, 43% of a light beam penetrated an antenna. Calculations of the efficiency of antennae as pheromone detectors should incorporate a correction for viscosity (for a general discussion, see Futrelle 1984).

The problem of flow evading a filter entirely is one with little parallel in conventional technology in which pumps and ducts ensure 100% transmissivity and what matters is the power required to do the job. But evasion is common among suspension feeders. It's been documented in such diverse cases as barnacles (Trager et al. 1990), zooplankton (Hansen and Tiselius 1992), caddisfly nets (Loudon 1990), and black fly larvae (Lacoursière and Craig 1993). In each case the fraction passing through increased dramatically with Reynolds number. To make matters a little more complicated, the filtering structures usually adjusted their configurations as flow speed increased, spreading wider apart at higher flows. The effect is quite the opposite of what would compensate for viscous effects. Forces and optimization of capture in a world of size-varying particles seem more relevant than keeping a constant transmitted fraction. This general issue of transmissivity has been dealt with by Cheer and Koehl (1988).

Mechanisms

The simplest and most familiar kind of filter for suspended particles is a sieve, and the filters of suspension feeders have classically been viewed as sieving devices—particles of diameter greater than mesh size or interfiber distance got caught. As it happens, sieves are relatively unusual biological filters; and most suspension feeders mainly catch particles smaller than could be sieved out by their filters. The turning point in how we viewed the process came with a paper by Rubenstein and Koehl (1977), who adopted and adapted analytical models that had proven useful to engineers dealing with aerosols. Besides simple sieving, they identified five ways in which particles might be captured. Each of these latter particles can be smaller than mesh or interfiber distance, and they approached the problem by looking at the interaction of a particle and a fiber. One of the five, electrostatic attraction, appears unlikely to work in seawater and hasn't yet been shown in any living system, so perhaps it may be put aside. And then there were four, shown diagrammatically in Figure 15.7:

1. In "direct interception" a particle passes within a particle radius of the collecting surface; only in this mode need a particle not deviate from a streamline.
2. In "inertial impaction" flow curves off away from the collector, but a particle keeps going because of its excess density.
3. In "gravitational deposition" a particle similarly deviates from streamlines on account of excess density, but the direction of deviation is always vertical.[7]
4. In "diffusional deposition" or "motile particle deposition" a particle encounters a collector as a result of its Brownian motion or its (assumed random) swimming activity.

Koehl and Rubenstein provided indices for the efficiencies of each of these mechanisms as functions of such things as mesh or fiber size, particle size, flow speed, and viscosity. Recently, their approach has been extended in an impressively clear, penetrating, and iconoclastic analysis by Shimeta and Jumars (1991). They point out that encounter rate is as important as encounter efficiency and that the two don't change with particle size or ambient velocity in the same manner. Taking a few liberties with their quantification, I've summarized that part of the Shimeta-Jumars analysis in Table 15.2.

Shimeta and Jumars also emphasize the distinction between encounter —touching—and retention. Gut content analyses of suspension feeders, commonly carried out and certainly of ecological interest, give the results of both processes, which can't easily be teased apart from the data. Retention depends on such matters as the excess density, size, and drag of particles, which determine whether they'll bounce off or be swept away, and thus on some of the same factors that determined encounter. But it depends on them in different ways and depends as well on the adhesion-determining chemical and physical characteristics of the surface.

---

[7] The distinction between inertial mass and gravitational mass, a fundamental and historic problem in basic physics (see Einstein and Infeld 1966), has thus emerged in this rather odd context!

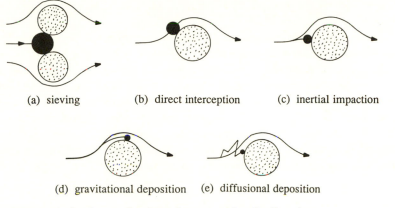

FIGURE 15.7. Five mechanisms for suspension feeding shown diagrammatically.

Other Matters Left Suspended

Anyone thinking about saying anything about the mechanisms of suspension feeding ought to read the last three papers cited here and perhaps also Silvester (1983) and LaBarbera (1984). And don't lose sight of the way the enormous diversity in collector geometry has been swept under the rug in this broad-brush view. Despite a huge literature, we're still a long way from a full appreciation of how animals tune their filters to particular ends such as specialization on a range of prey size and even farther from understanding the schemes by which potential prey can minimize their chance of capture. Some animals suspension-feed from upstream or downstream vortices, as described in Chapter 10. And flows over suspension feeders commonly oscillate back and forth, with substantial consequences as yet

TABLE 15.2 FOUR MECHANISMS FOR PARTICLE CAPTURE BY SUSPENSION FEEDERS AND HOW THEIR EFFECTIVENESS (BOTH RATE AND EFFICIENCY) VARIES WITH PARTICLE SIZE AND SPEED.

Mechanism	Particle Size versus Encounter		Ambient Velocity versus Encounter	
	Rate	Efficiency	Rate	Efficiency
Direct interception	+	+	+	o
Inertial impaction	+	+	+	+
Gravitational deposition	+	+	o	−
Diffusional deposition	−	−	(+)	−

NOTE: "+" indicates a direct relationship, "−" an inverse relationship, and "o" no relationship.

barely touched upon (see, for instance, Hunter 1989 and Trager et al. 1990).

Finally, an important portion of what have commonly been considered suspension feeders are somewhere between that and active predators. Where prey are capable of active evasion and predators grapple and then either accept or reject prey, one encounters a world of encounters for which the preceding discussion approaches irrelevance. This world has been most clearly revealed by the work on copepods of Rudi Strickler and his collaborators; a good bridge to it from ordinary suspension feeding is provided by Koehl and Strickler (1981) and Strickler (1985). But it's really worth special consideration on its own terms.

INFORMATION TRANSFER BY DISTURBING THE ENVIRONMENT

A recurring theme in this chapter has been the gentleness of velocity gradients in a viscosity-dominated domain and the consequent distance over which motion of a solid within a fluid disturbed the fluid. At the very least, any observations on confined animals (as under cover-slips) should obviously be viewed with skepticism. So far, though, attention has been limited to single organisms or, at most, assemblages of like-minded ones. What about interspecific interactions? What's quite clear is that disturbing the flow by moving through it constitutes a signal that can be detected by another organism. What's almost as sure is that a moving organism can detect a nearby nonmoving one by the resulting asymmetry of forces and flows as it swims. Or, in less fancy terms, swimming (or even sinking) both announces one's own presence and provides part of the machinery to detect someone else's presence (Zaret 1980).

What can we say about this peculiar world? First, swimming with a coating of cilia ought to be "quieter" (in a nonacoustic sense) even than simple sinking inasmuch as it involves (as noted earlier) steeper and thus more spatially restricted velocity gradients. Second, swimming steadily ought to be "quieter" than the jerky swimming associated with swinging a single pair of appendages in the manner of cladocerans such as *Bosmina* and *Daphnia* (Figure 2.5). Thus the predatory dipteran larva, *Chaoborus*, about 15 mm long, attacks at a certain threshold of mechanical disturbance. For *Daphnia*, that occurs at a distance of about 3.1 mm, but *Diaptomus*, a steadily swimming copepod (similar to the one in Figure 3.8), can get a lot closer undetected (Kirk 1985). And the steadily swimming copepod, *Epischura*, swims up the wake of a prey animal and then attacks from a short distance above it; by contrast a pulsing swimmer, *Cyclops*, makes a longer-distance strike with a high acceleration (Strickler and Twombly 1975).

In that high acceleration, estimated by Strickler (1975) at 12 m s^{-2}, lurks a partial evasion of this peculiar world. A predator slowly and steadily pursuing its prey may not only alert the prey but will to some extent push the prey forward before actual contact; a burst of speed can briefly drive the Reynolds number up and reduce this awkward effect of viscosity. One copepod briefly manages to get up to $Re = 500$, reaching several hundred body lengths per second in what may be a record for any swimming creature (Strickler 1977). Similarly, moving away from a predator will tend to draw the predator along. A sessile protozoan, *Zoothamnium* (a colonial version of *Vorticella*, shown in Figure 10.4), seems to minimize the problem by contracting its stalk with extreme rapidity and thus achieving a respectably high Reynolds number. Contraction of the "spasmoneme" in the stalk turns out to be the most rapid shortening of any contractile element in any animal (Weis-Fogh and Amos 1972). Whether for predator or prey, these low-Reynolds-number environments aren't exactly the easiest of all possible worlds. While "creeping" may be inappropriate for life within them, "creepy" has some appeal.

Unsteady Flows

$\mathbf{B}$ACK IN THE second chapter we deliberately limited our purview to "steady flows"—ones where speed didn't change over time at any location fixed with respect to an unmoving solid. (Speed was permitted, of course, to change from place to place, so fluid could speed up or slow down as it progressed.) That limit was only occasionally breached, as when talking about the "clap-and-fling" of beating wings in flight, when considering the use of drag-based propulsors for rapid acceleration, or when mentioning the problem of suspension-feeding in bidirectional flows. Other places where the limitation was constraining were either quietly passed over or explicitly deferred. In reality, unsteady flows prove to be of major biological consequence in quite a few situations—appreciation of their relevance has simply been limited and recent.

ADDED MASS AND THE ACCELERATION REACTION

Drag, by definition, is a force that resists motion. Strictly speaking (as we ought), drag resists acceleration only to the extent that acceleration involves motion. The drag of a body accelerating from rest (as in the initial stage of free fall) begins at zero and increases continuously. Drag as a result of motion, though, isn't the only kind of resistance that a fluid imposes on an accelerating body—another is a consequence of acceleration per se. The phenomenon may be introduced with an extreme example, adapted from Birkhoff (1960).

Consider a spherical bubble of air, 1 mm³ in volume, in water at atmospheric pressure, much like a bubble of CO_2 at the bottom of a glass of beer. It is suddenly released and accelerates upward because of its buoyancy—gravitational acceleration times the difference between the masses of displaced water, 1×10^{-6} kg, and bubble, 1.2×10^{-9} kg. Since the bubble's weight is negligible, the buoyant force is 9.8×10^{-6} N. The *initial* acceleration of the bubble can be calculated as that buoyant force divided by the mass of the bubble—8200 m s^{-2} or all of 830 times that of gravity. That's disturbingly high.

But for the bubble to advance upward, some water must move downward. Paths of water and gas are different, and the water movement involves velocity gradients and no distinct volume, so the analytical problem is far from trivial. But at least for some regular shapes such as the present sphere and assuming an ideal fluid, the details have been worked out (see, for instance, Batchelor 1967). To the mass of the bubble must be added an "added mass" (sometimes called "apparent additional mass") that, for linear acceleration of a sphere, equals half the mass of the displaced fluid. We can now recompute the acceleration of the bubble, taking its effective mass as 0.5×10^{-6} instead of 1.2×10^{-9} kg. The result is a value only twice that of gravity, something a lot more credible.

(In what follows, I'm mainly following the explanations given by Daniel 1984 and Denny 1988.) We can now recognize that accelerating a body in a fluid involves a force with three additive components. The first is our old friend drag. The second is the force needed to accelerate the mass of the body forward. And the third is the force needed to accelerate the added mass of fluid backward. How much mass must be added depends on the volume and shape of the object; the shape-dependent part is expressed as the "added mass coefficient," C_a:

$$F = \frac{1}{2} C_d \rho S U^2 + ma + C_a \rho Va. \tag{16.1}$$

(V is, of course, the volume of the body; density in the third term refers to the density of the surrounding fluid.) The force expressed by the last term is commonly called the "acceleration reaction"; the mass of the body plus the added mass ($m + C_a \rho V$) is often called the "virtual mass."

Sometimes virtual mass is used synonymously with added mass, but the meaning is usually clear from the context.

While the added mass coefficient looks rather like the drag coefficient, it proves to be considerably less troublesome—mainly because volume, with which it's multiplied, isn't subject to as many interpretations as is area. Lamb (1932) gives the numbers that everyone else uses. They're calculated for an ideal fluid, but the assumption of an ideal fluid usually isn't too bad. One typically works with the initial stage of movement from rest (where a is high and U is low), and many of the queer correlates of viscous flow such as separation and circulation just haven't had time to get going. And when drag and velocity are high, a is low, and the correction for added mass is small.

For a sphere (as just mentioned), the coefficient is 0.5. For a circular cylinder moving normal to flow it's 1.0. In general, bodies more slender in the streamwise direction have lower added mass coefficients than fatter bodies; Figure 16.1 (from Daniel 1984) summarizes the situation. Thus shapes that minimize drag are also reasonable choices if the acceleration reaction is a substantial concern. In one way, though, the graph is a little misleading—it suggests that a circular flat plate broadside to flow has an

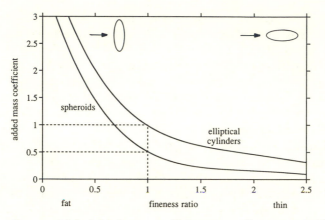

FIGURE 16.1. Added mass coefficient versus fineness ratio (flow-wise over cross-flow lengths, as in 5.7) for spheroids and elliptical cylinders. Spheroids with fineness ratios less than 1.0 are oblate, while ones with ratios above 1.0 are prolate. Spheres and circular cylinders have ratios of just 1.0.

infinite added mass coefficient. That's mainly an accident of definition—the flat plate has zero volume, and the effective product of volume and added mass coefficient (C_aV) is $^8/_3$ r^3 (Batchelor 1967). For an elongate rectangular flat plate, accelerated normal to its surface, the added mass is equal to the fluid mass of a circular cylinder whose diameter is the width of the plate and whose length is that of the plate (Ellington 1984a): C_aV is $\pi/4$ lw^2. That's the same as the value for a circular cylinder of the same length and width, as in Figure 16.1.

Bear in mind that both velocity and acceleration are vectors; while they act in the same direction when velocity is increasing, they act in opposite directions when velocity is decreasing. Drag always slows a body down; the acceleration reaction (like the body's mass) always opposes any change in speed, up or down. So during (positive) acceleration, the three terms of (16.1) are arithmetically additive, while during deceleration the two acceleration terms oppose the drag term.

The Beginning and End of a Swim

The acceleration reaction demands that a swimming animal work harder to get going, exerting an additional force that's proportional to its volume in order to achieve a given acceleration. For many escape responses and lunging predatory strikes, the matter is not negligible. Perhaps the most extreme case so far uncovered occurs in the escape response of a crayfish, *Orconectes* (Webb 1979a). It flexes tail and abdomen and goes rearward with a maximum acceleration of 51 m s^{-2}. Drag turns out to be only around 10% of the resistance, with 90% caused by the masses of crayfish and water—as is reasonable for a high acceleration to a fairly low final speed. The relative importance of the acceleration reaction isn't strongly correlated with body size for movements. It's been shown significant in a small copepod, *Acanthocyclops* (Morris et al. 1990); in shell closing of hinged brachiopods, small creatures looking superficially like bivalve mollusks (Ackerly 1991); in a water boatman bug, *Cenocorixa* (Blake 1986); in an angelfish, *Pteraphyllum* (Blake 1979a); in a frog, *Hymenochirus* (Gal and Blake 1988); and in pike, *Esox* (Frith and Blake 1991). Drag may be only indirectly associated with body volume per se, but the acceleration reaction is very closely tied to it—Webb (1979b) argues that reduction in mass is an important phenomenon in acceleration specialists.

In stopping, the acceleration reaction keeps an animal moving longer than it otherwise would. I haven't heard of a case in which this has biological relevance; where it does matter, though, is in determination of drag by observation of decelerative gliding. Except where deceleration is very gentle, you'll get a significant underestimate of drag if you neglect the correction, as mentioned in Chapter 7. To correct, one needs some figure for added mass coefficient. Such figures aren't as widely available as would be desirable, and estimation from the data of Figure 16.1 is a reasonable approach.

For systems in which impulsive swimming is the normal mode, the acceleration reaction is of special importance. For jet-propelled creatures accelerating from rest, it represents over half the total resistive force—Daniel (1984) got the result for a salp, a dragonfly larva, several coelenterate medusae, and a squid; the accelerations ranged from about 0.2 to 20 m s^{-2}. Again, we're looking at something that may seem odd but certainly isn't trivial.

Making Paddles and Tails Work

I once had a small wind-up toy fish whose rigid tail swung back and forth relative to its body. If one does a crude analysis of the forces acting on the tail fin, assuming symmetrical motion and so forth, one comes to the conclusion (and not just at very low Reynolds numbers) that all forces should balance and the fish should get nowhere. As Figure 16.2 shows, forces are forward and lateral on half-strokes moving toward the midline, with the lateral ones canceling between such half-strokes. Forces are rearward and lateral on the half-strokes moving away from the midline, with the lateral ones again canceling. Unfortunately, the forward and rearward ones cancel as well. That's the result of a "quasi-steady" analysis, ignoring the accel-

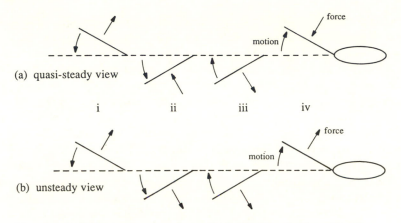

FIGURE 16.2. Forces generated as a rigid tail swings back and forth: (a) the forces anticipated in a quasi-steady analysis; note that all ultimately cancel so no net thrust is generated; (b) the forces predicted by considering the acceleration reaction in an unsteady analysis, as explained by Daniel (1984). Here each quarter stroke produces a forward force, while lateral forces again cancel.

eration reaction. As Daniel (1984) points out, a proper "unsteady" analysis comes to a different conclusion. To the forces on the tail previously considered, it appends a forward force from the accelerative movement toward the midline—forward because the acceleration is rearward. This latter isn't offset; instead, decelerative movement away from the midline also gives a forward force—forward because while movement is forward, it's now decelerative. Since the toy fish does swim, legitimization is comforting. That's not entirely how real fish make caudal fins push, but it shows how the acceleration reaction can be put to use to generate thrust.

A similar unsteady analysis shows how a dytiscid beetle can get thrust by swinging its paired hindlimbs rearward and forward without any asymmetry in speed or configuration (Figure 16.3). What's crucial here is that each hindlimb swing back and forth, not about an axis perpendicular to the long axis of the body, but about one tilted backward—in short, that the legs come closer together in the rear than in the front. Both accelerative rearward and decelerative forward leg movement will produce thrust near the forward extreme of the stroke cycle. The equivalent forces near the rearward extreme of the cycle will be mainly lateral—inward, in fact—and thus won't greatly detract from the thrust. Daniel (1984) calculated that the optimal angle about which the legs should swing is 120° back from the body axis, which corresponds very nicely to what Nachtigall (1960) found. Again it isn't the whole story of how thrust is made, but the animals are certainly doing what we think they ought to if acceleration reactions matter.

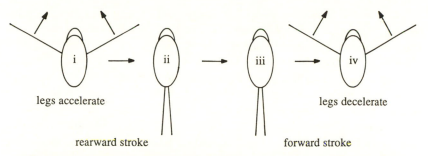

legs accelerate   legs decelerate

rearward stroke   forward stroke

FIGURE 16.3. Forces generated as paddles are swung backward and forward of vertical planes 120° back from the free stream. Both rearward acceleration and forward deceleration near the extreme forward position produce forward force—net thrust. The analogous speed changes near the rearmost position produce mainly lateral forces that cancel.

We should remind ourselves that a drag-based, paddling accelerator encounters the acceleration reaction in two essentially opposing ways. The reaction will mean that for a given stroke the paddles will exert more force than would be the case if they only generated drag. In addition it takes more force to get the body moving than if drag alone counted.

Accelerative Forces on Stationary Objects

Consider what happens when an attached organism is suddenly subjected to a surge of water—perhaps as a result of a nearly breaking wave. Again the acceleration reaction comes into play, but its operation is slightly different from what we saw for impulsive locomotion. As Denny (1988) commented, simply dropping the middle term of equation (16.1) seems reasonable since the object itself no longer is accelerated. As it happens, an equivalent term takes its place. As he explains it, if the object weren't there, a body of water of its volume would be accelerated along with the remaining water; thus the presence of the body is equivalent to accelerating its own volume in the opposite direction. The mass of the body, in that middle term, must consequently be replaced by the mass of water it displaces:

$$F = \frac{1}{2} C_d \rho S U^2 + \rho V a + C_a \rho V a$$

or

$$F = \frac{1}{2} C_d \rho S U^2 + (1 + C_a) \rho V a. \tag{16.2}$$

A really fine argument about the relative relevance of the two terms of (16.2) has been made by Denny et al. (1985). They pointed out that the drag of an attached organism, proportional to projected area, scales with the square of a linear dimension; by contrast, the acceleration reaction, proportional to volume, scales with the cube of a linear dimension. Attachment tenacity ought to scale with attachment area, so, assuming geometric similarity, the large organism is no worse off than the smaller with respect to drag (assuming thin boundary layers and so forth). On the other hand, the large organism is distinctly worse off with respect to the acceleration reaction—with increasing size the latter increases relative to the capacity to hold on. They applied the argument to such creatures as limpets, snails, barnacles, and sea urchins, calculating dislodgement probabilities for various sizes, fluid accelerations, and time durations. In general, the results agree rather nicely with observations on maximum size versus relative exposure in the habitats of the organisms. At least in these extreme environments, nature seems to care very much about the forces specifically caused by the unsteadiness of flow.

Added Mass in Air

Air is eight hundred times less dense than water. Does that mean that added mass is insignificant for creatures that live in air rather than water? After all, density of the medium appears in both terms of equation (16.2), so changes in it shouldn't affect the ratio of either term to their sum. In fact, three considerations minimize the impact of the acceleration reaction in air. First, the speeds of flow are much higher in air, whether we consider average ambient winds, extreme winds (Vogel 1984), or locomotory speeds; and speed squared appears in the first term of (16.1) and (16.2). Second, in all but a few rare instances accelerations aren't especially impressive,[1] whether of organisms taking off or of wind gusts suddenly striking. Finally, the mass of organisms living in air is far higher than the mass of the air they displace, so virtual mass is little different from actual mass. To put this last point another way, mass certainly matters when starting up, but it matters directly and without any aerodynamic chicanery. In short, the last term of equation (16.1) will rarely be of consequence.

But once in a while the acceleration reaction does make some difference. My first publication (Vogel 1962) concerned some variation in the wing-beat frequency of fruit flies. Frequency in many insects is set by the stiffness of the thorax and the moment of inertia of the wings. Changing air density might affect the latter if it alters the effective volume of wings plus boundary layer or the virtual mass (really virtual moment of inertia) of this rapidly oscillating system. In fact, calculations bore out the predictions:

---

[1] Such as the sporangium of the fungus *Pilobolus* when shot from its hypha, flea jumping, and a few other cases; see Vogel (1988a) for data.

added mass is appreciable relative to wing mass (terms 3 and 2 respectively in equation 16.1). Still, added mass is significant only in small flapping fliers with their light wings and high wingbeat frequencies. While it seems always to be less than wing mass, in many insects it's well within the same order of magnitude (Ellington 1984a; Ennos 1989b). To emphasize the lightness of these structures, I note that a kilogram of the wings of *Drosophila virilis* (a biggish fruit fly), laid end to end, would extend from Boston to Washington or London to Glasgow. More relevant is the fact that these insects devote only half of one percent of body mass to wings.

CONSEQUENCES IN THE WAKE OF VORTICES

No rule prohibits steady flows, even through and across rigid objects, from generating unsteady wakes. And a messy wake may come home to haunt you—your troubles have not necessarily been left behind. In particular, vortices shed regularly impose periodically varying forces on the objects that shed them. Such forces may break an object, especially a fairly rigid one; worse, perhaps, such forces may induce motion in the object that increases the shedding and the forces it generates—so-called flutter and galloping instability.

Vortex Shedding: The Von Kármán Trail

In Chapter 5 we saw that rows of vortices were shed behind bluff bodies such as cylinders at Reynolds numbers above about 40. These vortices were left behind in the wake in alternating positions in parallel, streamwise rows; each vortex rotated in the opposite direction of the preceding and succeeding ones (Figure 5.5). Any object in (or crossing) such a wake will be buffeted by these vortices. In an array of closely spaced cylindrical objects exposed to a flow, the foremost ones may be less subject to damage than those farther back.

But even the body that sheds the vortices is affected. Recall the discussion of circulation in Chapter 10. Any circulation created in one place has to be balanced by an equal and opposite circulation somewhere else. Consider any bound vortex around body and the vortices the body has already shed. That bound vortex must be equal in net strength and opposite in net direction to the combined circulation of those shed vortices. The implication is curious: every time a vortex is shed, the circulation around the body must reverse direction! Circulation, of course, produces the transverse force that we've called lift. So the body is shaken normal to flow as a direct result of the shedding of vortices, and the frequency of shaking must be precisely one-half the frequency with which vortices are shed.

Figure 16.4 shows the normal spacing of these vortices that, taken to-

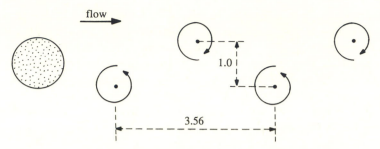

FIGURE 16.4. The spacing of vortices in the Von Kármán trail behind an elongate bluff body such as a cylinder. The ratio of 3.56:1 is stable.

gether, constitute what's called a "Von Kármán trail" or "Von Kármán vortex street." The trails are stable if the distance between successive vortices on each side of the street is 3.56 times the distance between the center lines of the two rows of vortices. This ratio doesn't change with speed or with the way the body produces the wake, although the vortices do diverge a bit well downstream of the body. If the body is free to move transversely to the free stream, the width of the trail will increase in proportion to this increased effective width of the body—but the streamwise spacing between vortices will increase as well, maintaining the 3.56 ratio. Lateral motion of the body will, as well, increase drag, circulation, and lift; and it will reduce somewhat the frequency of vortex shedding (Steinman 1955). Incidentally, the vortices aren't stationary but move in the same (relative) direction as the body at a slower rate. Their speed is proportional to their circulation, so one can calculate the lift and drag of a body from measurements of the dimensions and speed of its trail of vortices, as described by Prandtl and Tietjens (1934).

What determines the frequency with which vortices are shed? As with so many situations in fluid mechanics, general guidance takes the place of a properly precise and universal formula. As mentioned earlier, both the Reynolds number and the drag coefficient can be obtained by dimensional analysis, assuming that four variables (length, velocity, density, and viscosity) are germane. If the same sort of analysis is done with the addition of a fifth variable, frequency (dimensions of T^{-1}), another dimensionless index emerges. In the present application this variable is called the "Strouhal number,"

$$St = \frac{nl}{U}, \qquad (16.3)$$

where n is the frequency of a periodically varying flow; l is a characteristic length of a solid object, usually the diameter or the distance perpendicular to both flow and the long axis of the object; and U is the free-stream velocity. The Strouhal number serves as a dimensionless frequency, just as the drag coefficient works as a dimensionless drag. And, like the drag coefficient, it's a function of shape and Reynolds number and can be conveniently plotted as an ordinate against the Reynolds number; a specific shape gives a characteristic line. The frequency of vortex shedding can easily be obtained from the Strouhal number, the size of an object, and the free-stream speed.

Among biologically interesting shapes, data are available for cylinders and for flat plates normal to flow (Figure 16.5). Fortunately, the Strouhal number varies little with Reynolds number in the range that's most likely to concern us, even though the physical character of the flow changes. Consider, for example, a cylinder. At Reynolds numbers below 40, no vortices are shed, so the Strouhal number is effectively zero. From 40 to 150 the train of vortices is laminar, and the Strouhal number rises smoothly from 0.1 to 0.18. From 150 to 300, the vortices gradually become internally turbulent, and turbulence persists up to about 3×10^5 with a nearly constant Strouhal number of 0.20. Above this drag crisis the wake becomes narrower, the boundary layer is turbulent, and the vortices are disorganized. Above about 3×10^6, a discrete vortex street is reestablished; and it persists up to at least $Re = 10^{10}$. This last figure, by the way, is no wind tunnel determination but comes from a satellite photo of a cloud-marked vortex trail behind the mountain peak of Guadalupe Island (Simiu and Scanlan 1978). Other useful references on Strouhal number are Bishop and Hassan (1964), Goldburg and Florsheim (1966), Blevins (1977), and Van Atta and Gharib (1987).

Von Kármán trails are everyday occurrences. For example, the way suspended electrical wires sung aeolean tones in the wind was what drew Strouhal's attention to vortex shedding in 1878 (Massey 1989). But little in the way of biological mischief has been laid at their door. Not that vortex shedding doesn't happen: in a larch plantation in which trunk diameters were about 0.4 m and wind about 0.6 m s^{-1} the predominant frequency of turbulent eddies was, as expected, about 0.3 Hz (Grace 1977).

Perhaps it's of some importance that organisms such as trees don't shed vortices at the wrong frequencies. The shaking of an object induced by vortex shedding might be of considerable consequence when the object is either flexible or flexibly mounted and the rate of shedding is close to coincidence with some natural oscillatory frequency of the object. In a 20-m s^{-1} wind, a 0.2-m tree trunk will shed vortices at 20 Hz, fortunately too high a frequency to have any bearing on sway or wind throw. Holbo et al. (1980) found a sharp compliance peak in 0.35 m Douglas firs at 0.3 Hz, well below such high-wind vortex shedding frequencies but well above typical gust frequencies of about 0.05 Hz.[2] For a given Strouhal number,

---

[2] And that's about what happens with the loblolly pines (*Pinus taeda*) in my front yard during storms.

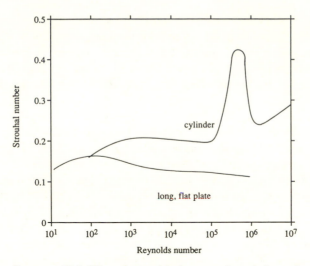

FIGURE 16.5. The relationship between Strouhal number and Reynolds number for flow normal to a cylinder and a long flat plate.

frequency will be proportional to flow speed divided by body diameter. So the tapering of a tree trunk together with any skyward increase in wind speed ought to further reduce the chance of getting shaken up by shedding vortices. Similarly, a cat's whisker of 0.3 mm diameter in a 1-m s^{-1} wind would shed vortices at 1000 Hz. But the whisker is neatly tapered from thick base to fine point, so it isn't likely to hum or purr.

Self-excited Oscillators and Aeroelasticity

Vortex shedding at a frequency determined by the Strouhal number generates a purely aerodynamic or hydrodynamic forced vibration; that is, the periodic force driving the vibration of a structure exists whether or not the structure actually moves. To this mechanism must be added another of somewhat different origin, but which may act in concert with it. This second mechanism also releases a trail of alternating vortices, but the frequency at which they're shed isn't such a simple function of the free-stream velocity.

Steinman (1955) describes the mechanism of these self-excited oscillators in the following way. Consider a half-cylinder with the flat face upstream, as in Figure 16.6a. If the half-cylinder is moving laterally as a wind strikes it, the net or relative wind will approach obliquely. The forward stagnation point will be offset from the center of the face, and more of the flow will move around the leading edge of the cylinder. This difference in flow will produce a circulation around the object, and that in turn will

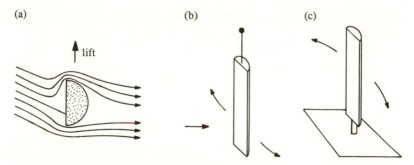

FIGURE 16.6. Self-excited oscillators: (a) circulation and lift on a half-cylinder moving across a flow; the lift will augment the preexisting movement; (b) a pendulum will swing back and forth in a flow as a result of such circulation; (c) a spring mount can substitute for gravity in providing the restoring force necessary for such an oscillator.

generate a lifting (cross-stream) force that will tend to keep the half-cylinder moving in the same direction. In other words, the flow will amplify any initial perturbation from zero cross-stream speed.

All that's needed now to get a proper oscillation is some restoring force that increases with distance traveled laterally from the original position. Steinman describes a version he calls a "Steinman pendulum" (Figure 16.6b) in which the aerodynamic force is eventually offset by gravity, only to reappear in the next half-cycle and keep the pendulum swinging. I use the device for a demonstration of both the fluid-mechanically driven pendulum per se and of circulation and vortex shedding—every time the direction reverses a vortex is shed into the wake and a new bound vortex of the opposite spin begins, just as in flapping flight. A rigid flat plate works as well as a half-cylinder broadside to flow. Mine is about 3 by 20 cm, suspended beneath a bearing (it must be prevented from downstream deflection), and gets half-immersed in a flow tank. It does nothing until tweaked but then oscillates quite persistently, shedding vortices that are easy to mark with dye in the water or sawdust on the surface. Steinman points out that a spring might alternatively provide the restoring force (Figure 16.6c); the scheme then takes on a more distinctly biological odor. I've watched white poplar (congeneric with quaking aspen) leaves flutter both on the tree and in a wind tunnel (Vogel 1989, 1992c). They move side to side on a flattened petiole (Figure 16.6c), and it certainly appears as if they're playing some version of the present game.

The most famous case of self-excited oscillations must be the failure of the Tacoma Narrows suspension bridge in Washington State in 1940 after less than half a year of service. It was designed for static wind loads of 2400 Pa, but in a mild gale that produced a static load of only around 240 Pa it developed spectacular torsional oscillations and, while spectators watched and movies were made, the roadway broke from its cables and dropped into Tacoma Bay. Such oscillation (excluding the rigid Steinman pendulum) depends as much on the character of the solid body as on that of the moving fluid. In particular, it depends on the lack of stiffness of the solid— how it behaves as the spring to provide the restoring force. The subject is usually called "aeroelasticity," and it's of great importance to people who design airplanes and large buildings as well as bridges. (The original explanation of the collapse involved only Von Kármán vortex shedding; it now appears that aeroelastic coupling was the principal culprit—Petroski 1991 gives a nice account of the controversy. On aeroelasticity in general, one might look at Dowell et al. 1989.)

So two things matter—stiffness and shape. The former is a matter both of stiffness in the strict sense (especially the torsional modulus of elasticity) and of damping and resilience. As for shape, an entire cylinder (broadside to flow) is a relatively stable profile, whereas a half-cylinder is stable when the flat side faces downstream, not upstream. As mentioned, a flat plate normal to flow is unstable. A beam shaped like a "T" in cross section is stable if the lower arm faces upstream but unstable if the lower arm faces downstream. Thus a bridge with a cantilevered external sidewalk (the lower arm of the T) is more stable than one with a solid upright truss outboard of the sidewalk.

The cylindrical profiles of trees, large sea anemones, and large algal stipes (as in the sea palm, *Postelsia*) probably limit accidental application of the mechanism, but I'd expect nature to have put it to use here and there, as perhaps she does in arranging for the leaves of aspen to quake. A likely place for this kind of flow-induced, self-excited oscillation is in a scheme by which seeds or spores are preferentially released when a wind is blowing. But at this point I know of no specific case that's been the subject of investigation.

Self-excited oscillation associated with aeroelasticity isn't limited to external flows. Cut the end off a cylindrical balloon and blow through it. The balloon will fill and empty irregularly while making socially unacceptable noises. What seems to be happening here is that increasing flow decreases the pressure, as described by Bernoulli's equation (4.1); low pressure then permits the balloon to collapse inward, increasing its resistance and decreasing flow, which in turn permits expansion (the basis for the phenomenon was described in Chapter 14). It happens in nature, indeed inside some of us. Jones and Fronek (1988) found that a constriction in a flexible pipe generated vibrations in the flow, and these in turn lowered the Reynolds number at which turbulence set in. They used conditions approximating pathological stenoses (narrowings) in the human circulation. And flutter in collapsible tubes provides a persuasive model for the generation of respiratory wheezes (Gavriely et al. 1989).

Still, flexibility isn't always a hazard, imposing risks of aeroelastic flutter and such. Koehl et al. (1991) describe how flexibility can provide a way to escape drag under quite reasonable circumstances. Imagine a symmetrically bidirectional, oscillating flow. A really flexible sessile organism will follow the flow and experience mainly tensile forces. If the organism is longer than the distance the fluid travels in a cycle, some distal portion of the organism will never be pulled upon! The criterion, then, for the point at which drag evasion begins is the ratio of the length of the organism to half the wave period times the average current. For long, marine macroalgae the point is commonly reached—they can get up to around 10 meters long. In fact, long algal blades can actually experience less force than shorter ones The same just won't happen on land: with wind gusts 10 or 20 seconds apart and speeds of 10 or 20 m s^{-1}, a leaf would have to be a hundred or so meters long to benefit.

MORE WAYS TO GET UNSTEADY FLOW

We've seen several ways in which an initially steady flow could interact with some solid structure to produce oscillation—vortex shedding behind a rigid object and self-excited (aeroelastic) oscillation of a flexible object. These certainly don't exhaust the possibilities, and some others should be kept in mind:

1. Flow through a smooth pipe faces a sudden increase in resistance when turbulence sets in, as one can see from Figure 13.6. Assume flow from a constant-head source at a Reynolds number of around 2000. Faster flow will generate turbulence, increasing resistance and slowing itself; flow will then become laminar, decreasing resistance and speeding up. What results is the production of a series of turbulent "plugs" that appear as spurtings from the end of the pipe. Massey (1989) gives a little more detail if you want to contrive a demonstration.

2. The great drag crisis at Reynolds numbers around 100,000 can be made to produce an analogous oscillation. Consider a sphere suspended as a pendulum in a wind so it can swing back and forth (streamwise) at about that value. It can be pushed backward by the higher drag of laminar flow at the lower relative wind concomitant with its backward motion. As it slows down the relative wind will increase, and it can move forward with the lower drag of turbulent flow; as it starts to swing down and back, its drag will again increase.

3. Aircraft can follow an oscillating path. Speeding up due to descent or a local wind gust generates more lift, which stops the descent and slows the craft, which reduces the lift, and so forth. The wavelike or looping path is called a "phugoid" oscillation; it's all too easy to obtain with model airplanes and gliders (Von Kármán 1954; Sutton 1955). This phenomenon and the one below depend on the pitching moment and thus the location of the lift vector as well as on the magnitude of lift—I'm ignoring some complications.

4. Another lift-based oscillation, delayed stall, was dismissed in Chapter 12 as of little consequence for animal flight. If the angle of attack of an airfoil is raised near the stall point, the increase in lift doesn't follow immediately. But when it does appear, it elevates the craft and increases the angle of attack beyond stall. That in turn decreases lift and causes the craft to descend, decreasing the angle of attack back beneath the stall point. In effect, the separation point is moving fore and aft on the top of the airfoil. This (or something closely analogous to it) seems to be what happens when leaves flutter up and down in a mild wind (Perrier et al. 1973).

WHEN IS FLOW UNSTEADY ENOUGH TO MATTER?

Obviously no flow is ever perfectly steady just as no solid is ever perfectly rigid. Unsteadiness is clearly important for animal flight, especially for small fliers and during hovering (as we talked about in Chapter 12). Other areas have received much less attention. Some rules of thumb can help decide when one might safely ignore unsteady effects—when one might get away with averages over time or with a quasi-steady analysis based on a sequence of steady-flow situations. Lighthill (1975) gives a dimensionless criterion, the "aerodynamic frequency parameter" (sometimes the "reduced frequency") for oscillating systems such as beating wings:

$$f_a = \frac{2\pi nc}{U} \tag{16.4}$$

(n is wingbeat frequency and c is wing chord.) It amounts to a a ratio of chordwise flow speed to free stream speed. (Some sources omit the factor of 2π.) Other things being equal, it will be highest during hovering; it will be higher for short, broad wings than for long, thin wings and when wingbeat frequencies are high. If the parameter exceeds 0.5, unsteady effects are likely to be significant. In full forward flight values usually come out to about that number—a locust forewing operates at about 0.25 and a hindwing at about 0.5; a fruit fly wing works at about 0.5 also. So they're awkwardly close to that critical value, and the value is very much higher in slow flight and hovering.

A slightly different test for steadiness is used for pulsating flow in circulatory systems (Womersley 1955). The dimensionless test parameter for pipes is called the "Womersley number," Wo:[3]

$$Wo = a \ \sqrt{\frac{2\pi n \rho}{\mu}} \ . \tag{16.5}$$

Here n is the frequency of a sinusoidally applied pressure gradient. If Wo is less than unity, viscosity predominates and the profile of flow across a pipe is essentially parabolic; flow is then said to be quasi steady. If Wo is above unity (favored by large pipes and frequent pressure cycles), then inertial forces distort the profile and the Hagen-Poiseuille equation cannot be accorded full faith. It's another test, along with Reynolds number and entrance length, that a system must pass for the assumption of proper parabolic flow to be justified. In our aorta, Wo is 10 or more, but that's about as high as we ever get, so the Hagen-Poiseuille equation isn't in very serious trouble in most of our arterial system on that account. For comparison, Wo is around 0.001 in the capillaries (Caro et al. 1978). The Womersley criterion also finds use in the analysis of flow in respiratory airways (see, for instance, Moslehi et al. 1989).

Finally, I should reemphasize the importance of unsteady flows. Crops wave in the wind, with effects on water loss and gas exchange (Finnigan and Mulhearn 1978). In storms trees sway before they fall, and falling results more from the dynamic forces associated with gusts and swaying than from the static loading of steady winds on rigid objects (Grace 1977). Waves striking shores impose unsteady flows on every attached organism there. And, of course, sound production, whether by humans, other animals, or musical instruments, always involves unsteady flow; but that's something about which this book will remain silent. One simply can't do everything.

[3] It might be of some interest to note that the square of the Womersley number equals $\pi/2$ times the product of Reynolds and Strouhal numbers.

Flow at Fluid-Fluid Interfaces

I N   T H I S   C H A P T E R we will relax another of our initial assumptions. Except for the briefest of allusions, a fluid has been permitted to make interfaces only with a solid—even if the solid hasn't always been incorrigibly stiff and unyielding. Let's talk now about situations in which two fluids make contact without mixing—an interfluid interface—and where one fluid moves across the other.

What's required to permit such contact without mixing is that the intermolecular *cohesion* of at least one of the fluids be greater than the *adhesion* of the molecules of one fluid with those of the other. Since intermolecular cohesion in gases is negligible, we don't find proper interfaces between gases. What we do get are gas-liquid and liquid-liquid interfaces, where the condition that cohesion exceeds adhesion can be met. If it's met, work is needed to create surface area—a substance that's free to flow will spontaneously minimize the area it presents to the other substance. If we have two substances, a and b, we can represent the energy per unit area that we'd have to supply to create surface as γ_a and γ_b—these are the two "works of cohesion." Similarly, we can represent the intermolecular adhesion as γ_{ab}. If γ_{ab} exceeds $(\gamma_a + \gamma_b)$, then the fluids mix and no interface occurs. If γ_{ab} is less than $(\gamma_a + \gamma_b)$, then a fluid-fluid interface will form. If so, and if $\gamma_a > \gamma_b$, then substance b will preferentially surround a.

SURFACE TENSION

For our purposes, one fluid, say a, will almost always be water and the other, say b, will be air; that means that $\gamma_b = 0$ and $\gamma_{ab} = 0$. We then only have to worry about the value of γ_a. γ_a generates an additional material property, one that's relevant to a fluid-fluid interface. It is, of course, *surface tension*. How do we get the latter from work (or energy) per unit area? In fact, we have it already—work per area is both dimensionally and practically the same as force per distance, and γ_a, henceforth just called γ, is the surface tension. It's a property of a liquid, more particularly of a liquid making an interface with another liquid or with a gas; its dimensions are force per distance (or MT^{-2}), and its units are newtons per meter.

A note about the peculiar dimensions. We're more accustomed to "tensile force," with dimensions of force, and to "tensile stress," as force per area of cross section than to "tension," as force per distance. If you pull on, say, a sheet of rubber, tensile stress is a good measure of what you're doing—it's the force you apply divided by the product of the width and thickness of the sheet. But when you pull on an air-water interface, making more surface, you're pulling on something that really has no thickness in the normal sense. Or, to put it another way, as you extend the interface, more water molecules move out to the interface, creating more of it—the interface isn't a material, and it doesn't stretch out and get thinner as a result of the pull. So force per distance rather than per distance squared is entirely reasonable for surface tension.

Again, the interface we care about is that between water and air. As might be expected from its basis in intermolecular attraction, the value of surface tension decreases as temperature increases. At $0°$ C, it's 0.756 N m^{-1}; at $10°$ C, it's 0.742; at $20°$ C, it's 0.728; at $30°$ C it's 0.712. Adding inorganic salts increases the surface tension a little; thus for seawater (at a salinity of 35 $°/oo$), the value is about 0.78 at $20°$ C.

In addition to density and viscosity, we now have a third property of fluids relevant to flows of biological importance, this one applicable to liquids only and not to gases. I don't mean to dwell on all the biological phenomena to which surface tension is relevant—plastron respiration in aquatic insects, the prevention of air embolisms in the xylem of leaves, surfactants in lungs, and so forth; I touched on many of these in an earlier book (Vogel 1988a). Here we're concerned with fluids in motion and are ignoring problems of fluid statics—waves and locomotion are what matter.

WAVES

A truly smooth air-water interface is fairly unusual in anything other than small bodies of liquids (hence, I suppose, the expression "tempest in a teacup"); more common are disturbances, and the most common of disturbances are periodic waves. These surface waves are such ordinary things it's easy to forget just what a queer business they are. For one thing, all three fluid properties that we've talked about conspire to prevent them. The density of water under the urging of gravity opposes any nonhorizontal surface, surface tension opposes any surface of nonminimal area, and viscosity opposes the kind of shearing internal motion that waves inevitably involve. For another thing, while waves inevitably move, what travels across the interface is no net volume of liquid but rather a kind of ghost—the form of a disturbance. Disturb the interface in one place by compression (pushing down on it) or by shear (blowing across it), and the disturbance moves laterally for quite a remarkable distance in the form of a traveling elevation difference, a surface wave.

What's going on in a wave is an orbital oscillation of water beneath the surface (Figure 17.1). Water travels in the direction of propagation of the

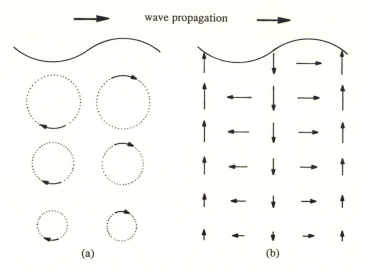

FIGURE 17.1. Two views of what happens as a wave moves across a deep body of water: (a) the orbits of several particles of fluid as a wave passes; note that these are not vortices shearing where adjacent—bits of water are just translating in circles in synchrony with ones above and below; (b) an instantaneous view of the water velocities beneath a passing wave to emphasize the point made above.

wave when it's near the crest and in the opposite direction when it's near the trough. Thus as a wave passes, a particle of water travels in a circle. The radius of the circle reflects the amplitude of the wave, and the period needed to make the circle reflects (inversely) the speed of movement of the wave across the surface. With increasing depth beneath the surface, the radii of these orbits decrease. It's easy to imagine from a diagram showing a few such orbits that a series of vortices exits beneath each wave, but that's not really the case. Instead a whole vertical sheet of water is moving upward, forward, downward, and backward. Adjacent sheets, in the direction of wave propagation, do likewise but at phases that increasingly lag in time. Thus adjacent sheets shear across each other during their upward and downward motion. In shallow water (defined arbitrarily as depths of less than half the wavelength) the circles become increasingly elliptical with depth—right near the bottom water goes only back and forth. Bascom (1980) gives a good general introduction to these phenomena.

We need to define a variable for this motion of a train of disturbances. Velocity as we've used it is a little misleading since almost no net fluid movement is involved. Instead what's used is a variant of velocity called the "celerity," defined as distance per time, but where the distance is that between adjacent wave peaks and the time is that needed for one peak to replace the previous one at a given point—thus celerity is wavelength (λ) over period:

$$c = \frac{\lambda}{t}.$$ (17.1)

As it happens, the celerity depends only negligibly on wave height but quite substantially on wavelength. What sets the relationship between celerity and wavelength are the agencies tending to restore the flat and horizontal equilibrium interface—gravity and surface tension. That an increase in either will make a wave move faster isn't too hard on the intuition (think of a stiffer spring), but somewhat queerer is the fact that the two interact with wavelength in opposite ways. Without troubling with a proper derivation (see, for instance, Denny 1988 if you wish), the relationship among gravity, wavelength, and celerity is

$$c = \sqrt{\frac{g\lambda}{2\pi}}.$$ (17.2)

And that among surface tension, wavelength, and celerity is

$$c = \sqrt{\frac{2\pi\gamma}{\lambda\rho}}.$$ (17.3)

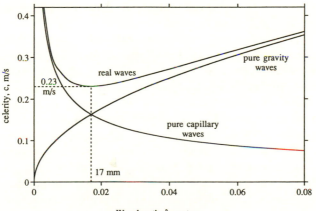

FIGURE 17.2. The relationship between celerity and wavelength for real waves, as well as for gravity waves (were surface tension zero) and capillary waves (were gravity zero). Notice that for pure water on earth real waves have a particular minimum celerity and that the minimum celerity corresponds to a specific wavelength.

Since gravity and surface tension work concomitantly, the celerity is really set by their combination:

$$c = \sqrt{\frac{g\lambda}{2\pi} + \frac{2\pi\gamma}{\lambda\rho}}.$$ (17.4)

This combination generates the peculiar relationship between wavelength and celerity shown in Figure 17.2. It's no surprise that equation (17.3) gives a good approximation of reality for very short wavelengths ("capillary waves") and (17.2) does as well for very long ones (" gravity waves")—which is why (17.3) is relegated to footnote status in books on ocean waves. What is a little startling are the implications that (1) no real train of waves of water on this earth can have a celerity of less than 0.23 m s^{-1}, and (2) the minimum celerity corresponds to a wave of the decently finite and bioportentous wavelength of 17 mm.

SURFACE SHIPS

I think the main thing our analysis of locomotion at the air-water interface has to explain is why it's so uncommon in nature. We swim, albeit poorly, at the air-water interface; and we've been building superbly successful surface-swimming ships for a few thousand years. But nature prefers submarines—despite the facts that the densities of organisms are well matched to operation at the surface and that many of the submarine swimmers are air breathers. Nonetheless a few animals, ranging from whirligig beetles to ducks, do get around at the surface, so no absolute prohibition is enforced.

One might expect that a ship would gain efficiency by traveling at a surface simply because a large part of both surface area and frontal area faces air, with a resistance almost three orders of magnitude lower than that of water. But this gain is offset by an additional component of resistance. A ship moving on the surface generates waves, which means increasing the surface area of the interface and lifting water above its equilibrium height. So generation of surface waves means that work has been done. Thus to skin friction and pressure drag must be added wave-making resistance, and the latter may be no small matter. A moving ship usually produces two waves, one at the bow and one at the stern. Thus the waterline length of the ship ("hull length") almost fully determines the wavelength of the system of spreading waves.

Wavelength, though, determines celerity. For a large ship, where gravity waves are the main thing, celerity is proportional to the square root of wavelength (equation 7.2). And wave celerity proves to have a fairly direct connection to the practical speed of a ship, as illustrated in Figure 17.3. For a ship moving more slowly than these waves, the bow wave moves out of the

FIGURE 17.3. The surface waves associated with the passage of an ordinary ship with a displacement hull. This particular ship is a "rubber ducky" ("Ernie's Genuine Playschool") towed in a flow tank, just slower than hull speed.

way in front of it, at least to some extent, and the stern wave may even elevate the rear. Consequently the ship faces a level or perhaps a slightly downhill course. For a ship moving faster than these waves, things are not so nice. The ship faces an increasingly steep bow wave, while it outruns its stern wave. As a result, forward progression is an uphill battle in the most literal sense—the ship is ever trying to climb its own bow wave. If, instead, it plows through, then it's caught by an alternative problem. With a high bow wave, more water must be shouldered aside in front. Pushing water normal to the hull means that some component is being counterproductively pushed against the ship's motion, even with a fairly sharp prow and narrow beam. At the same time, less water pushes back in on the converging stern to counteract the work done by the ship at the bow.

Among other matters, this problem of bow waves leads to a substantial divergence in form between surface and underfluid craft—compare the bulbous shape of modern submarines, of blimps, and of whales with the sharp-prowed narrow profiles of fast ships. Even a strut that penetrates the air-water interface does best when it has a sharp upstream edge. Which is what Fish et al. (1991) found for the gaffing feet of fish-catching bats and what Withers and Timko (1977) found for the part of the lower bill of a skimmer that's exposed to the interface. In both cases the shape seems to be determined more by interfacial exigencies than by those of the air above or water beneath—the interface is the serious problem. (The phenomenon was noted in connection with streamlined struts in Chapter 7.)

For a ship, then, the cost of propulsion increases severely because of wave-making resistance when it exceeds a critical speed that corresponds to the celerity of a wave of its hull length. Herein lies the problem. This speed will be proportional to the square root of hull length, making it difficult for a short ship to travel rapidly. How short and how fast? We can make a rough-and-ready estimate from equation (17.2) by substituting hull length for wavelength and speed for celerity. We get a prediction that the ship should hit the wall, so to speak, when speed squared equals 1.56 times hull length—on earth and with SI units. As we'll see, that turns out to be about right.

THE FROUDE NUMBER

In Chapter 5 the most useful of all the dimensionless indices, the Reynolds number, was introduced as the ratio between inertial forces and viscous forces. Since then, we've touched on other indices that represented other ratios. We now encounter what is probably the second most useful of these indices, the Froude number, named after William Froude (1810–1878), a British naval engineer. Froude was interested in hull design for surface ships and the possibility of using models for testing. If gravity waves are what matter most, then a reasonable ratio is that between inertial force (ma) and gravitational force (mg). The ratio, designated Fr, is thus

$$Fr = \frac{U^2}{gl}. \tag{17.5}$$

(Sometimes the square root of the expression on the right is used as the Froude number.) In fact, the scaling scheme, while useful, is imperfect. The difficulty isn't the neglect of surface tension—practical technology is well above its scale. Rather it's the neglect of viscosity, and the continued relevance of the Reynolds number. As a little manipulation will convince you, making a scale model that maintains both correct Reynolds and Froude numbers just doesn't work. So while the Froude number is used, models are as large and used as close to normal speed as is practical.

The speed limit alluded to just above can be expressed as a value of the Froude number by inserting our $U^2 = 1.56l$ into equation (17.5)—the result is $Fr = 0.16$. And that turns out to be a reasonable value for real waves and real hulls. A ship that's ten meters long ought to be able to go about 4 m s^{-1}, or 7.7 knots. A hundred-meter ship should go 12.5 m s^{-1}, or 24 knots, as in Figure 17.4. What's biologically interesting is what happens down at the lower left of Figure 17.4b. A 0.3 m hull ought to be able to do only around 0.7 m s^{-1} before the cost begins to rise rapidly. That's a pretty poor speed by the standards of vertebrate swimmers. (A 0.1 m hull should go about 0.4 m s^{-1}, which I once verified by towing a rubber duck in a trough.)

Some data are now available on surface swimmers. Prange and Schmidt-Nielsen (1970) found that a duck with a hull length of 0.33 m will go no faster than 0.7 m s^{-1}, as expected. Its metabolic rate is at that point no-

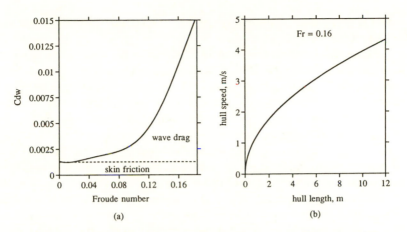

FIGURE 17.4. (a) The resistance (expressed as drag coefficient based on wetted area beneath the water line) versus Froude number for a displacement hull of fairly ordinary shape. (b) "Hull speed," or $Fr = 0.16$, as a function of hull length—the practical speed limit for ordinary surface ships.

where near what it can reach in flight; apparently the investment in leg muscle needed to go much faster hasn't proven evolutionarily cost effective. Fish (1982, 1984) looked at muskrats on a pond and found much the same thing. Few of them ever exceeded $Fr = 0.16$, which for them corresponds to 0.63 m s^{-1}.

While data for the drag of swimming animals gained from towing tests have their problems (Chapter 7), comparison of data from surface and submerged tows is still of some interest. Williams (1983) towed minks and got figures for drag seven to ten times greater at the surface than submerged. With Humboldt penguins, Hui (1988a) found that parasite drag increased much more rapidly with speed (velocity exponent of 2.55) for carcasses towed on the surface than for ones towed submerged (exponent of 2.0), although no sharp break at a particular Froude number was evident. Stephenson et al. (1989) calculated aerobic efficiencies three times greater during diving than while surface swimming for ducks (lesser scaup). From these data, together with a lot more that are summarized by Videler and Nolet (1990), clearly (1) surface swimming is more costly than submerged swimming, and (2) the cost of surface swimming increases with speed more drastically than does the cost of submerged swimming. Put another way, speed-specific drag, D/U^2 (Chapter 6), increases with speed for surface swimming while it decreases with speed for both aerial and aquatic, noninterfacial systems. Or yet another way—Hui (1988b) found that the cost of transport (mass moved per unit distance) was about the

385

same for penguins at the surface and submerged, but the submerged ones went a lot faster.

Admittedly, these comparisons may overstate the case for underwater swimming at least a little. The best shapes for keeping drag down when submerged aren't the same as those for keeping it down when swimming at the surface. If an animal does both kinds, any natural advantage to underwater swimming ought to prompt specialization of body shape for that mode at (inevitably) the expense of surface swimming efficacy.

In at least one instance, near-surface swimming may have an energetic advantage; not surprisingly it involves very large creatures and works best (by calculation) for the largest of these. In a force-5 sea (one beneath a 10 m s^{-1} wind) a fin whale (*Balaenoptera physalus*) ought to be able to get as much as 25% of its power for propulsion from waves when facing them and up to 33% when going with them, according to Bose and Lien (1990).

And our technology has inadvertently provided an analogous opportunity —whether it ever occurs without human contrivances is unknown. That's the bow-wave riding practiced by dolphins. Without any obvious locomotory movements, they "ride" for long distances in a position just in front of ships and just beneath the surface. While they're certainly taking advantage of some inhomogeneity in the local flow, the exact mechanism has been a subject of considerable contention; see, for instance, Fejer and Backus (1960).

MORE SURFACE TRANSPORTATION

Most ships and everything thus far considered use what are called "displacement hulls"—they're Archimedean floaters. These don't exhaust the ways of getting around while keeping one's head above water. I know of several more of some biological relevance; still others may exist. But just what's going on is pretty speculative at this point—which mean we're talking about things ripe for investigation. Thus . . .

Planing

With an appropriate hull shape—a bow that's a bit flattened below and slopes gently downward in the aft direction—and sufficient power, a boat can quite literally climb out of the water and skim along on the surface. The weight of the boat is offset in large part by lift, a downward momentum flux of water. Several agencies contribute to the lift, including the acceleration reaction that we encountered in the last chapter. For a given speed, planing incurs less drag and thus requires less energy expenditure than does propulsion of a hull displacing its own weight of water. But it doesn't work at low speeds, and a boat must ordinarily achieve planing speed as a displacement device.

Planing is put to only limited use by organisms, and it doesn't appear associated with very obvious hydrodynamic specializations. Aquatic birds quite clearly plane as they land. Since flight speeds are almost inevitably higher than swimming speeds and since hovering is costly, one would be surprised if they didn't plane. I've heard that by planing for short distances as they become airborne some swimming birds can manage to evade the hull-speed limit. That certainly seems to be what's happening in movies that I've watched—takeoff is no easy matter for large waterfowl since the water surface severely limits initial wingbeat amplitude. At least one seabird is reported (Klages and Cooper 1992) to plane steadily—a broad-billed prion, *Pachyptila vittata*, extends its wings to get lift and paddles into the wind with its body only barely touching the water. Meanwhile it sticks its head under water and either seizes or filters food, as mentioned earlier.

Slapping

Recall from the last chapter that a flat disk accelerating broadside to flow has a substantial added mass-volume product, $8/3\ r^3$. As Batchelor (1967) pointed out, that's what's at work when you hit the surface of a body of water with hand or hammer. Water is massy stuff, and fights back when asked to accelerate at a high rate. Thus upward force is exerted. In at least one instance the phenomenon is routinely used for locomotion. A genus of iguanid lizards of the new world tropics, *Basiliscus*, popularly known as "Jesus Christ lizards," have large, webbed hind legs and commonly run across streams and other small bodies of water with entirely airborne torsos. The legs enter the water rapidly, moving downward and then backward; they almost certainly make substantial use of the acceleration reaction and little if any of surface tension. An Asian species of agamid lizard, *Hydrosaurus pustulatus*, does very much the same thing. At this writing, the scheme is under active investigation by James Glasheen (Glasheen and McMahon 1992); I'm reporting mainly what I've heard from him.

Using Surface Tension to Move

Neither planing nor slapping can provide static support in the way floating does. But floating is only one of two systems that can provide static support. Just as waves can be dependent on either weight (gravity waves) or surface tension (capillary waves), static support can use either system. And some creatures (Figure 17.5) press surface tension into service for propulsion.

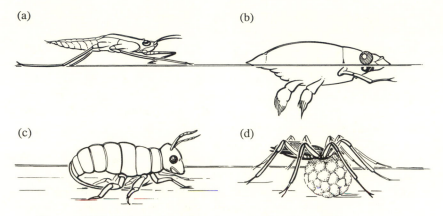

(a)

(b)

(c)

(d)

FIGURE 17.5. Animals that take advantage of surface tension for support and (sometimes) propulsion: (a) a water strider, *Gerris*; (b) a whirligig beetle, *Dineutus*; (c) a springtail, *Podura*; (d) a fishing spider, *Dolomedes*.

Consider, referring to Figure 17.6a, how the foot of an animal such as a water strider (a gerrid bug) presses against the surface of a pond or stream. Forces must balance, which means that the upward force due to surface tension must equal the downward force of weight. The force exerted by the interface will be equal to the surface tension of water times the wetted perimeter of the foot. That force will have a line of action tangent to the water's surface, so its upward component will be its magnitude times the cosine of the angle between its line of action and the vertical.

How can the system be used for locomotion? What a leg has to do is push backward against the water (Figure 17.6b). A leg presses downward and backward, making a dimple in the water's surface. The dimple moves backward, resisted by forces from the water's surface tension, inertia, and viscosity, while the animal accelerates and moves forward. Thus the dimple becomes asymmetrical. And how can we tell what's going on? That asymmetry records the forward force on the leg. Since the surface tension of water is known, the animal is in effect walking on a continuous force-monitoring platform. Not that we're talking about something easy—neither the analyses nor the technical problems of continuous three-dimensional recording of the surface contour are trivial matters.

Moving around on the surface of water has other interesting aspects. Nothing rules out using a combination of displacement and surface tension for support—that seems to be what adult whirligig beetles (Gyrinidae) do. *Dineutus*, about which we have some information, is supported about half by buoyancy and half by surface tension (Tucker 1969). Nor are underwater oars and paddles ruled out as propulsion devices, since beneath the

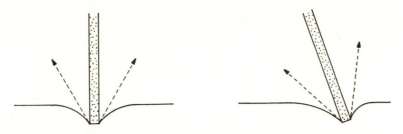

FIGURE 17.6. *Left*: Forces involved when a leg stands on the water's surface. A heavier load pushes the leg in farther, which makes the line of action (tangent to the surface) more nearly vertical, which compensates for
the heavier load. *Right*: Pushing rearward as well as downward against
the water's surface. The surface dimple is asymmetrical, with a net forward component that offsets (and thus permits) the rearward push. Tilt
of the leg at right is incidental.

contact line of the air-water interface the animal is submerged in every
functional sense. A nudibranch mollusk (a shell-less snail) can move
around while hanging from the surface at least in a laboratory sea table; the
contact line extends around a considerable perimeter. They seem to be in
some way using mucus threads for locomotion, but I don't know of any
specific investigation of what, mechanically, is going on.

The size range of animals that can use surface tension for support and
locomotion is severely limited—the water striders, whirligigs, and spring-
tails whose habits are nicely described by Milne and Milne (1978) range
from only about a millimeter to a few centimeters long. I think the problem
is that two different bits of mechanics set top and bottom of what proves to
be a small window of opportunity. If the downward force of weight scales
with length cubed while the upward force of surface tension scales directly
with length, then an upper size limit will certainly be imposed. The central
issue if you're large is static support.

A lower limit alluded to (in a different context) in Chapter 15 may be
imposed by the functional stickiness provided by surface tension. Surface
tension is a little like viscosity—it opposes motion. A body on a surface is
pulled in all horizontal directions equally, so whichever way it moves it will
be opposed by some tensile force. Consider the ratio of inertial force to the
force of surface tension, called the "Weber number":

$$We = \frac{\rho l U^2}{\gamma}.\tag{17.6}$$

A small, slow body will find it difficult to generate enough force to do much
moving around—to offset surface tension—as we can see from the length

and velocity factors in the numerator. Staying up may be easy, but getting around is awkward.

Swimming on the surface is quite an odd thing if you're small. The whirligig beetle mentioned just above makes no bow wave when swimming slowly—real waves must go at least 0.23 m s^{-1} (Figure 17.2). The hull speed barrier for a centimeter-long beetle comes out to 0.16 m s^{-1}, using equation (17.5) and $Fr = 0.16$; that's not very different from the corresponding point for gravity waves in Figure 17.2. But the limitation doesn't apply, and such beetles have been observed to hit 0.4 m s^{-1}. The summed line in the figure is what's relevant, and a beetle is swimming in the range where wavelength is mainly determined by surface tension. At its size, surface swimming isn't quite such a bad thing! Nevertheless, the range is narrow; and adult North American whirligigs range from about 3 to 15 mm in length (Merritt and Cummins 1984)—not exceeding the wavelength corresponding to minimum celerity. What may be the oddest feature is that, since wavelength sets hull speed, the smaller beetle should have the higher speed limit.[1]

If you have a hill, you might arrange to slide down it. Which is what the sport of surfing is all about. Using such a potential difference has a curious analog in the world of surface tension. Small staphylinid beetles of the genera *Dianous* and *Stenus* normally walk slowly on the interface. But they can move rapidly shoreward under duress by ejecting a secretion from the tips of their abdomens that locally reduces the surface tension. Higher surface tension in front then pulls them forward, and they skim across the surface, at 0.6 to 0.7 m s^{-1} (Jenkins 1960). They're using a version of a common demonstration in which a tiny paper boat is propelled by a speck of soap on its stern. Veliid bugs (such as *Velia*) can do the trick as well, moving rapidly when alarmed by discharging surfactant saliva from their beaks (Linsenmair and Jander 1963). Hynes (1970) has a nice summary of all these adaptations.

Sailing

Organisms that support themselves by either buoyancy or surface tension ought to be in a position to do a bit of sailing—using air movements to move about. And sailing can be either lift based, drag based or some combination. With this choice of mechanisms, what's a little surprising is how little use is made of it. Perhaps it's another of those things that, like going about in displacement hulls, work best for a size range larger than that of all but a few organisms. Here one can point to the association of wind and

---

[1] It's too easy to forget that these are adult insects with a complete metamorphosis in their life cycle. Thus they've completed their growth, and one needn't worry about any size-based geriatric slowdown.

waves and suggest that with a really useful wind waves would be awkwardly large.

Still, sailing creatures exist. The most notorious is the Portuguese-man-of-war, *Physalia*, a colonial coelenterate with a large and beautiful float above ferociously armed tentacles. With an bulgy float about 20 or 30 cm long, it's probably a purely drag-based sailor. Another coelenterate, the by-the-wind-sailor, *Velella*, was mentioned in Chapter 11; Francis (1991) found that its sail got a reasonable lift and noted reports that it could sail as much as 63° off the direction of the wind. So it's at least to some extent a lift-based sailor. With a wide skirt on the surface of the water beneath the sail, it may make very slight use of surface tension. But the skirt is more likely to oppose tipping over as a result of its static displacement and virtual mass—a little like the slapping of basilisk lizards except that it's never free of the surface. At least one surface-tension supported animal is a sailor. On windy days, a fishing spider, *Dolomedes*, has been reported (Deshafy 1981) to lift its second pair of legs from the surface and be carried across the water even against currents beneath.

Another form of sailing seems not to have been recognized in any formal sense, yet it's clearly a case of wind-forced movement at the air-water interface. Wind certainly blows the surface of water around—just drop sawdust on the water in any basin and blow across it to see how responsive the surface is. Any organism hanging beneath the surface by local surface tension will certainly be carried along. Such organisms include more than just spores and pollen. *Hydra*, the common freshwater coelenterate, often hangs down from the surface. If it's carried along with a wind-blown surface it will experience a velocity gradient along its length; the latter may expose *Hydra's* tentacles, extending downward, to prey, just as if it actively pumped water or swam around. Mosquito larvae hang down from the surface as well, so they might be blown about the same way.

COMMUNICATION AT THE SURFACE WITH WAVES

Communication has been given little attention in this book, appearing mainly in connection with the wide disturbances to flow at low Reynolds numbers (Chapter 15). Accustomed as we are to communicating with both acoustic waves and electromagnetic waves, the use of surface waves for communication ought to come as no surprise. To some extent it must be just a matter of capitalizing on the unavoidable. If you make waves when you swim, you've left a trace that no predator should ignore.

For a detector at a fixed point, frequency is probably a more directly relevant variable than wavelength, and people working on communication usually use the latter. The frequency and wavelength of surface waves interconvert in just the same way as acoustic and electromagnetic waves—

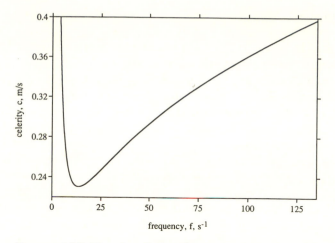

FIGURE 17.7. How the celerity of spreading surface waves varies with the frequency with which a specific place is forced up and down. The graph is basically that of 17.2 with a recalculated abscissa.

frequency is propagation speed (here celerity) divided by wavelength. So the graph in Figure 17.2 can be redrawn with frequency rather than wavelength on the abscissa, as in Figure 17.7.

How can a predator tell what's prey? Falling debris produces trains of concentric waves of short duration, while wind makes nonconcentric trains mainly at frequencies below 10 Hz and at high amplitudes. So high-frequency (10 to 50 Hz), low-amplitude, moderate-duration, concentric wave trains indicate edibles such as struggling insects that have been trapped on the surface. High frequencies attenuate with distance more severely than lower frequencies; at least one fish (*Aplocheilus*) uses relative attenuation to judge distance. Both a back swimmer, *Notonecta*, and the sailing fishing spider, *Dolomedes*, as well as surface-feeding fish are capable of frequency discrimination—the general picture is of highly tuned predators who can tell a lot both about the direction and distance to a signaling food source and about its character. Surface-foraging bats also use surface waves as signals; they detect them through echolocation.

Water striders make by far the fanciest use of surface waves so far uncovered. These waves provide them with the critical cues involved in courtship, copulation, and even postcopulatory behavior. They produce and detect waves of specific frequencies in trains of specific lengths in highly species-specific patterns.

All of this information on communication comes from reviews by Bleckmann (1988) and Wilcox (1988).

Fuzzy Interfaces

At least one other fluid-fluid interface is of considerable consequence to organisms, but it's one that's not anywhere near as familiar. Not that it's at all hard to make in a model system—the phenomenon is just subtle because the same fluid is on both sides of the interface and the boundary is indistinct. A situation of surprisingly great temporal stability may occur when less dense fluid lies above more dense fluid. Thus an "atmospheric inversion" with hot air above cold traps the effluvia of contemporary life in the unpleasant form of haze or smog. And a lake may be severely stratified during the summer, with the surface region depleted in nutrients as material sinks through the so-called thermocline into the cold water beneath. Winds in the atmosphere or blowing across a lake's surface ought to stir things up and put such matters right. Still, inversions often persist for days, and many lakes aren't mixed until they "turn over" as the surface cools in the fall.

The density difference may trace to variation in composition—most commonly salinity—as well as in temperature; that's common in estuaries, where fresh water comes in above salt water.

But flow can disestablish such interfaces. A useful test is available to predict whether these density gradients will remain stable or whether flow-induced instability will produce a layer of mixing vortices and eventual dissipation, as shown in Figure 17.8. The test involves another dimensionless index, the "Richardson number," Ri:

$$Ri = \frac{g(d\rho/dz)}{\bar{\rho}(dU_x/dz)^2}.$$ (17.7)

The expression is less complicated than it may appear. The ratio $(d\rho/dz)$ is the vertical density gradient, which keeps the layering stable; and the ratio (dU_x/dz) is the vertical velocity gradient, which promotes mixing through shear. The two can be obtained experimentally from sets of measurements of temperature or salinity and velocity versus depth. Stability is obviously favored by higher Richardson numbers, and in practice the Richardson number must be below 0.25 for instability to set in. Negative values occur, by contrast with all the other dimensionless numbers we've used so far. Further information on the meaning, use, and (to a limited extent) biological relevance of Richardson numbers may be obtained from Hutchinson (1957) or Mortimer (1974) for lakes, from Fischer (1972) or Dronkers and van Leussen (1988) for estuaries, and from Scorer (1978) or other meteorological texts for the atmosphere.

Wind over water can do other things besides prevent layering—in particular it can give rise to peculiar "windrows," vortices just beneath the surface

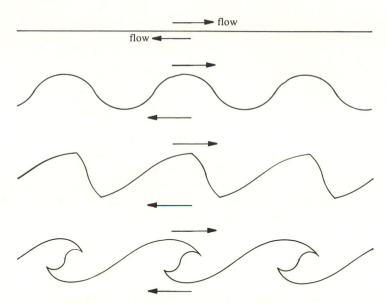

FIGURE 17.8 The relative stability of density gradients subjected to shear. Vortices will form in such stratified systems if the density gradient is sufficiently gentle relative to the velocity gradient. The result will be mixing and a decrease in the steepness of both gradients.

whose long axes run in the direction of the wind. They're often marked by windward streaks of accumulated debris. These vortices are called "Langmuir circulations," and each vortex in the (cross-wind) sequence circulates in the opposite direction from its neighbors (Leibovich 1983 gives a good account of the phenomenon). The distribution patterns of a lot of planktonic organisms are sensitive to such vortices; see, for instance, Hamner and Schneider (1986). At least one, the scyphomedusan coelenterate, *Linuche unguiculata*, seems deliberately to use the patterns to aggregate in the windrows (Larson 1992).

SQUIRTING AND SPRAYING

Textbooks of fluid mechanics usually spend a page or so on the way fluid (usually a liquid) flows through a sharp-edged orifice into another fluid (usually a gas) as a free jet, partly because the case provides a classic application of Bernoulli's equation. The velocity at the place of minimum cross section of the jet (the vena contracta) turns out to be the square root of the product of twice gravitational acceleration and the height of the liquid in the reservoir. The latter, of course, can be expressed as pressure (recall from Chapter 4 that $\Delta p = \rho g z$), which puts the formula in more general terms:

$$U = \sqrt{2gz} = \sqrt{\frac{2\Delta p}{\rho}}. \tag{17.8}$$

For real fluids, with viscosity and surface tension, the velocities are a few percent lower than predicted by the equation. And the vena contracta has a lower cross section than that of the orifice itself—61% to 66% is the ordinary range, although surface tension can raise the value to as much as 72%, according to Massey (1989). Discharge from a reservoir through a sharp-edged orifice isn't a common occurrence in organisms, but it does have relevance to laboratory setups.

More common, if still not exactly ubiquitous, is discharge of liquid into air from a nozzle, a considerably more complicated business in which viscosity and surface tension play major roles. When coming from a small, circular nozzle, a cylinder of liquid soon breaks up into droplets. The primary agency promoting breakup is surface tension—any place where the cylinder is even a little reduced in diameter feels a greater strangling effect of surface tension. What's responsible is Laplace's law, the rule that pressure is proportional to tension divided by radius—tension is most effective in generating pressure (here inward) where the radius is smallest. So the smooth cylinder is unstable.[2]

Breakup of the cylindrical jet happens at an increasing distance from the nozzle as its velocity increases—its persistence time doesn't change much—up to a Reynolds number that's usually between 500 and 3000. Then there's a transition from laminar to turbulent flow, whereupon the breakup length decreases, and ascends again thereafter. This region of turbulent breakup has a different character, though—the stream begins to wander about ("sinuous" breakup) rather than strangling into fairly regular droplets ("varicose" breakup). The ratio of length before breakup to jet diameter is described by several mildly different formulas (cited by Blevins 1984) involving the two dimensionless numbers that we'd expect to matter here, the Weber number and the Reynolds number. A combination, the square root of the Weber number divided by the Reynolds number, anticipates the

---

[2] A very pretty demonstration of this effect of surface tension is quite easy to arrange. Push a pipette through the middle of a 4 to 8 cm sphere of styrofoam or similar light material, and mount it so the sphere rests loosely on the surface of a small, upturned loudspeaker. Connect the pipette to a hose coming from an elevated reservoir of water and the loudspeaker to an audio-frequency oscillator set to a few hundred hertz. An almost inaudible sound is sufficient to shake the pipette enough to regularize droplet formation; the latter can then be viewed as if in slow motion with a repetitive stroboscope. Lord Rayleigh did this (with considerably less convenience) over a hundred years ago.

(a)   (b)

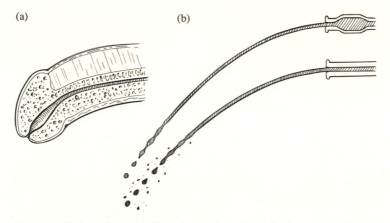

FIGURE 17.9. (a) Longitudinal (sagittal) section of a human penis, showing the anteterminal enlargement called the "navicular fossa." (b) Water
coming out of two glass tubes, one with and one without an anteterminal
enlargement.

character of the broken jet. A low value of this so-called Ohnesorge number indicates that breakup will be varicose (McCarthy and Molloy 1974).

When a circular liquid jet breaks up, it often forms droplets of two widely
different size ranges, the main ones and what are called "satellite droplets";
the phenomenon was first noted by Lord Rayleigh in 1896 (Bogy 1979
gives a nice review). Interest in satellite droplets has increased in recent
years in connection with the development of ink jet printers, for which
they're not at all a good thing. Half of us may be arranged to minimize the
production of satellite droplets. The human penis (and many other mammalian penes) have an enlargement of the urethra just upstream of the
final orifice, the "navicular fossa" (Figure 17.9a). Its function is unclear,
although enough congenital abnormalities are known for its association
with a smooth stream of urine to be fairly certain (Jordan 1987). I compared the stream of water coming from a tube with such a preterminal
enlargement with one that lacked it (Figure 17.9b); the stream from the
latter broke up earlier and produced more satellite droplets.

But any functional significance is at this point pure guesswork. Perhaps
quadrupedal animals that urinate while standing shouldn't spray and thus
mark themselves with odorant, either as predators or as prey. One wonders
about other squirters—such as archer fish (*Toxotes*), which come to the
surface and squirt a jet of water up to a meter long that can knock an insect
into the water. Existing descriptions (such as Waxman and McCleave 1978)
are silent on the geometry of the narrow groove along the roof of the
mouth through which water is forced. And still further, one wonders about

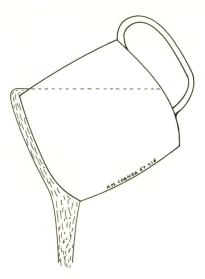

FIGURE 17.10. A jet of fluid is inconveniently drawn along a surface as the so-called Coanda effect—despite gravity and surface tension. A hydrophobic surface recruits the latter and helps a little. Providing a sharp edge that the fluid would have to turn around brings the fluid's inertia into corrective action.

animals such as cats that can either squirt or spray urine—what adjustments are involved?

While micturition might be the most memorable ending for a chapter on interfaces and a book on flow, one more phenomenon, something called the "Coanda effect," is aptly named to serve as coda. If a jet of fluid passes close to a solid object, it tends to adhere and, if the curvature of the solid isn't too severe, to follow the surface. What's happening is that the stream of fluid draws along ("entrains") fluid from around itself, but the supply of such ambient fluid is limited on the side of the jet near the surface. The resulting drop in pressure draws the jet inward, as in Figure 17.10. The phenomenon is put to use in various fluid logic ("fluidic") devices. An analogous attraction occurs with a jet of liquid in a gas; here viscosity and surface tension must also play a role. The effect is a nuisance for teapot spouts, but it can be alleviated to some extent by making the surface hydrophobic—buttering it. The Coanda effect may have relevance in some cardiac pathologies. Of more interest to the biologist is its use, shown by Eisner and Aneshansley (1982), by bombardier beetles. A beetle sprays (usually at ants) a hot jet that it aims by pointing a hind leg, along which the jet of spray is guided.

CHAPTER 18

Do It Yourself

WHEW. I'm winded by all these words and worry that their thrust has left the reader, with head barely above water, more draggy than uplifted. One more topic, though—something of a peroration or admonition, a return to a subject briefly mentioned in the first chapter. With this book, I mean not only to provide an introduction to what has been done in biological fluid mechanics but to what can be done as well. That's an odd intention, given the almost automatically retrospective nature of a fact-laden book in any area of science. After all, while the student goes to the library in hope of finding material on a topic of interest, the intending investigator really prefers to find nothing at all. Most of us wear both hats. We begin with the former and, once launched, discover that the latter fits better. Eventually some of us later revert to the student's hat, finding that, as with so many things, vicarious pleasures are better than none.

At the least, I have tried to promote the ideas that adaptation to fluid flow underlies much biological design and that such adaptation is of relevance to limnologists, marine biologists, natural historians, paleontologists, ecologists of diverse persuasions, comparative and environmental physiologists, and, indeed, all biologists who study organisms that either contain or are exposed to fluid flow. My intention hasn't been to convert unsuspecting biologists into biofluidmechanicians. Rather, it has been provision of an easy introduction to the subject for those who suspect that it might bear relevance to their projects but whose time and vigor is unequal to immersion in a lot of engineering and applied mathematics simply to test that suspicion. Put another way, it's intended as a bit of consciousness-raising for people in other fields who will remain wedded to those fields, but who might perhaps come to regard flow as fascinating and relevant instead of fearsome and peripheral.

But my larger intention has been instigational. I'm trying to lure people into actually doing a little biological fluid mechanics for themselves, to persuade people that one can usefully get one's feet wet without getting in over one's head, without taking a full-time, full-immersion bath. Showing relevance is only one part, the easier or at least the more traditional task.

The other is showing tractability, showing that, once one is familiar with fluid behavior and can envision matters from the viewpoint of the fluid, one can without difficulty generate testable hypotheses. What, then, ought to be said about *doing* biological fluid mechanics rather than just reading about it?

First, much of the subject is substantially counterintuitive. That counterintuitive character means that only by knowing a little about fluid mechanics can you realize what your organisms must contend with and recognize the opportunities for them to be the cleverly adapted rascals in whom we take delight. Equipped with notions such as Reynolds number, continuity, separation, velocity gradient, circulation, advance ratio, propulsion efficiency, and acceleration reaction, you've got the activation energy to make it over that hurdle. But bear in mind that quite a lot of the biological literature has been produced by people still mired in misconception and downright heresy: with a properly jaundiced eye you'll at least be amused; with luck, you'll gain some of that self-confident audacity that permits your creativity to flourish. I keep a small file of conceptually flawed papers for use in teaching—a wonderful sight is an early-stage student newly empowered by discovering the ability to deconstruct the published word.[1]

Second, the idea that fluid mechanics and complex mathematics must walk hand in hand is, I believe, a great misapprehension. Admittedly, a lot of fancy mathematics underlies the simple equations you've seen. And, so I'm informed, the more mathematics one can manage, the more problems one can solve. But I do believe that up to this point our subject has made more progress through experimental work than through purely mathematical analyses and that experimental work needn't require great mathematical sophistication. The real *sine qua non* for the biologist in this game is insight into the operative physical processes. While this is a complex area in which we're second-guessing nature, complex systems are a biologist's stock in trade—the simplest of our systems makes the most byzantine type of flow look like child's play. As a rule of thumb, the more complex the system, the greater the likelihood that the assumptions needed to make equations manageable will also render them inapplicably unrealistic.

Third, no magic formula points to success in unscrambling adaptations to flow. Ideally one lets the question determine the approach, although in practice one's own experience and investigative prejudices play major parts. The investigator is almost inevitably faced with two gradients of opposite sign. Both extend from work on real organisms in their normal habitats, through work on real organisms under controlled circumstances, and through work on physical models in flow tanks or wind tunnels (or, rarely, in the field), to numerical simulation or equations. But along one gradient, running from reality to abstraction, relevance gradually decreases, while along the other, running from abstraction to reality, such

---

[1] But I'd really prefer to keep the list private. Embarrassing people is certainly not the intention and quite as certainly would be the result of dissemination.

complication gradually accumulates that tractability is lost. Where some crucial insight will emerge is predictable, if at all, only on a case-by-case basis. The ideal is a multifaceted study, hitting the problem from enough directions so a solution just can't stay hidden.

Finally, don't be intimidated by an unfamiliar literature, whether your background is in biology, engineering, or perhaps geology, physics, or even medicine. If you insist on limiting your questions to areas in which you're fully equipped and ready to go, you'll not go far. And speaking of literature, the biological one is especially scattered, making trouble for anyone with the temerity to try to keep widely current or even to detect a mainstream. Going through the present bibliography, I find that no journal represents even 10 percent of the citations, and only one has over 3 percent. To give specific numbers,

Journal of Experimental Biology	67
Limnology and Oceanography	21
Canadian Journal of Zoology	18
Journal of Experimental Marine Biology and Ecology	14
Biological Bulletin	11
Hydrobiologia	9
Journal of Theoretical Biology	9

Again, let me emphasize that this book should be considered merely as a starting point by anyone who intends to pursue a flow-related investigation. In illustrating the diversity of points at which fluids and organisms come into proximity, it has of necessity done full justice to neither. I've suggested sources of additional information in the text, to the extent that I'm familiar with them; but most investigations will probably need to move beyond these at a fairly early stage.

For biologists such as me, the engineers themselves are enormously useful. The fact that fluid flow is more in the province of engineers than physicists has put the subject into a tradition with a strongly practical bent, one with a pragmatic willingness to deal with complexity one way or another. That's a great boon to the baffled biologist—even if our problems are often viewed as some kind of comic relief. The biologist, though, shouldn't assume that someone doing physical fluid dynamics has any idea that a lot of really good work has been done on biological problems since the two literatures tend to be quite distinct. Furthermore, the bemused engineer needs a little introduction to our approach to design. Scion of a practical tradition, the engineer designs things or devises rules by which efficient devices may be designed. By contrast we start with the assumption that our organism is well designed (bearing in mind, of course, the constraints intrinsic to natural selection), and we try to figure out just why its design is a good one.

This book began with an exhortation, and I mean for it to end with the same. Biologists seem afflicted with a great faddishness in the choice of items for investigation. One might argue the merits of the situation—that breakthroughs are made by concentrating one's troops. But I'd assert that the history of science indicates the opposite—that the major conceptual advances usually came before their areas became particularly well populated. We're told that the rate of scientific progress has never been so great; we're also told that never have so many been actively doing science. Is it just possible that the rate of progress *per investigator* might be at a low point, with faddishness a contributing factor? Devil's advocate and curmudgeon that I am, I frequently mention the matter to undergraduates awed by the attention given to the most highly funded areas of biology. What matters is one's chance of doing something that really makes a difference to how people view something, and a lot of people chasing a few problems ought to be taken as a counterindication.

Here we have an area in which people needn't trip over each other, where problems are more abundant than investigators. To appropriate the ecologist's jargon, this is an r-selected rather than a K-selected field. Certainly much work has been done in biological fluid mechanics; at the moment I'm in danger of drowning in references. But I remain impressed with the diversity of interesting and accessible questions yet unplumbed, with great gaping lacunae where the hand of an investigator has not yet set foot. Don't wait to see which way the wind is blowing—get in the swim.

List of Symbols

a	acceleration	c_b	concentration in bulk solution
a	radius, half of width (elongate slot)	c_m	concentration at membrane
a_s	Stokes' radius	D	drag
AR	aspect ratio	D	diffusion coefficient
b	span (of wing)	d	diameter, of pipe or disk swept by propeller
C	arbitrary constant	d	zero-plane displacement (of logarithmic boundary layer)
C_a	added mass coefficient	Di	distance index (internal flows)
C_d	drag coefficient	E	exponent for variation of D/U^2 with U.
C_{df}	drag coefficient referred to frontal area	F	force
C_{dl}	local drag coefficient (for point on surface)	f	friction factor, Fanning friction factor
C_{dp}	drag coefficient referred to plan form area	f_a	aerodynamic (or reduced) frequency parameter
C_{dv}	drag coefficient referred to volume$^{2/3}$	Fr	Froude number
C_{dw}	drag coefficient referred to wetted area	G	shear modulus
C_l	lift coefficient	g	gravitational acceleration
C_o	orifice coefficient	H	total head of pressure
C_p	pressure coefficient	h	location from center of slot toward either plate
c	chord (of wing)	J	advance ratio
c	celerity (of surface wave)	k	constant of unspecified value

k	Kozeny function (porous media)	Re_x	Local Reynolds number (at place on surface)
L	the dimension *length*	Ri	Richardson number
L	lift	S	surface area
L'	entrance length (for pipe)	S_d	area of disk swept by beating wings
l	length, characteristic length		
M	the dimension *mass*	S_f	frontal area
m	mass	S_p	plan form area
n	oscillation frequency, revolution rate of propeller	S_v	two-thirds power of volume
		S_w	wetted area
P	power	St	Strouhal number
P_i	induced power (propeller)	T	the dimension *time*
Pi_l	tendency to pinch, laminar flow	T	thrust
		t	time, period
Pi_t	tendency to pinch for turbulent flow	U	speed, velocity, free-stream velocity
p	pressure	U_*	shear velocity or friction velocity
$Pé$	Péclet number		
Q	volume flow rate	$\bar{U}$	average (mean) velocity
R	wing length	U_d	wind component due to autogyro descent
R	resultant force	U_f	wind on blade element due to free stream flow
R	resistance to flow (in pipe)		
r	radial location outward from center of pipe	U_h	velocity at particular place between plates
Re	Reynolds number	U_i	wind induced by propeller action
Re_r	boundary roughness Reynolds number	U_{max}	peak velocity (in pipe or channel)

U_r	wind on blade element due to its rotation		α	angle of attack
U_r	velocity at particular radial position in pipe		Γ	circulation
U_t	tangential velocity (in vortex)		γ	tension, surface tension, surface energy per area
U_w	net wind impinging on blade element		Δ	prefix for difference between two values of variable
U_x	local velocity (component in x-direction)		δ	boundary layer thickness
U_z	local velocity (component in z-direction)		δ_u	thickness of unstirred layer
V	volume		ϵ	height above surface of protrusion or roughness element
V_s	volume of particles (porous media)		ϵ	voidage or porosity (porous media)
W	weight		η_f	Froude propulsion efficiency
w	width of channel or slot		θ	angle of deformation
We	Weber number		θ	glide angle
Wo	Womersley number		κ	Von Kármán's constant
x	distance or direction along surface and with flow		λ	wavelength
y	shortest distance to nearest wall		μ	viscosity, dynamic viscosity
z	distance or direction normal to surface and flow		ν	kinematic viscosity
z_0	roughness parameter or roughness length		ρ	density
			τ	stress, shear.stress
			ϕ	amplitude of wingbeat (stroke angle)

Bibliography and Index of Citations

Abbott, E. A. (1885) *Flatland: A Romance of Many Dimensions.* Reprint. Princeton, NJ: Princeton University Press, 1991. (*208*)

Ackerly, S. C. (1991) Hydrodynamics of rapid shell closure in articulate brachiopods. *J. Exp. Biol. 156*: 287–314. (*365*)

Adair, R. K. (1990) *The Physics of Baseball.* New York: Harper and Row. 110 pp. (*227*)

Alexander, D. E. (1990) Drag coefficients of swimming animals: Effects of using different reference areas. *Biol. Bull. 179*: 186–90. (*91*)

Alexander, D. E., and T. Chen (1990) Comparison of swimming speed and hydrodynamic drag in two species of *Idotea* (Isopoda). *J. Crustacean Biol. 10*: 406–12. (*143, 145*)

Alexander, D. E., and J. Ghiold (1980) The functional significance of the lunules in the sand dollar, *Mellita quinquiesperforata. Biol. Bull. 159*: 561–70. (*65*)

Alexander, R. M. (1971) *Size and Shape.* London: Edward Arnold. (*121, 244*)

Alexander, R. M. (1977) Swimming. In R. M. Alexander and G. Goldspink, eds., *Mechanics and Energetics of Animal Locomotion*, pp. 222–48. London: Chapman and Hall. (*80*)

Aleyev, Y. G. (1977) *Nekton.* Boston: Kluwer. (*29, 152*)

Altman, P. L., and D. S. Dittmer, eds. (1971) *Biological Handbooks: Respiration and Circulation.* Bethesda, Md.: Federation of American Societies for Experimental Biology. (*28*)

Ambühl, H. (1959) Die Bedeutung der Strömung als ökologisches Faktor. *Schweiz Z. Hydrol. 21*: 133–264. (*179*)

Armstrong, J., W. Armstrong, and P. M. Beckett (1992) *Phragmites australis*: Venturi- and humidity-induced pressure flow enhance rhizome aeration and rhizosphere oxidation. *New Phytol. 120*: 197–207. (*71*)

Armstrong, S. L. (1989) The behavior in flow of the morphologically variable seaweed *Hedophyllum sessile* C. Ag. Setchell. *Hydrobiologia 183*: 115–22. (*119, 125*)

Arnold, G. P., and D. Weihs (1978) The hydrodynamics of rheotaxis in the plaice (*Pleuronectes platessa* L.). *J. Exp. Biol. 75*: 147–69. (*65, 176*)

Arya, S. P. (1988) *Introduction to Micrometeorology.* San Diego: Academic Press. (*171*)

Au, D., and D. Weihs (1980) At high speeds dolphins save energy by leaping. *Nature 284*: 548–50. (*151*)

Augspurger, C. K. (1986) Morphology and dispersal potential of wind-dispersed diaspores of neotropical trees. *Amer. J. Bot. 73*: 353–63. (*257, 272*)

Augspurger, C. K. (1988) Mass allocation, moisture content, and dispersal capacity of wind-dispersed tropical diaspores. *New Phytol. 108*: 357–68. (*257, 272*)

Aylor, D. E. (1975) Force required to detach conidia of *Helminthosporium maydis. Plant Physiol. 55*: 99–101. (*193*)

Aylor, D. E., and F. J. Ferrandino (1985) Rebound of pollen and spores during deposition on cylinders by inertial impact. *Atmospheric Environ. 19*: 803–806. (*176*)

Azuma, A., and Y. Okuno (1987) Flight of a samara, *Alsomitra macrocarpa. J. Theor. Biol. 129*: 263–74. (*249, 256, 272*)

Azuma, A., and K. Yasuda (1989) Flight performance of rotary seeds. *J. Theor. Biol.* *138*: 23–54. (*249, 272*)

Badgerow, J. P., and F. R. Hainsworth (1981) Energy savings through formation flight? A re-examinaton of the vee formation. *J. Theor. Biol. 93*: 41–52. (*288*)

Bagnold, R. A. (1941) *The Physics of Blown Sand and Desert Dunes*. London: Methuen. (*187*)

Bahadori, M. N. (1978) Passive cooling systems in Iranian architecture. *Sci. Amer. 238 (2)*: 144–54. (*72*)

Baker, P. S., M. Gewecke, and R. J. Cooter (1984) Flight orientation of swarming *Locusta migratoria*. *Physiol. Entomol. 9*: 247–52. (*223*)

Barry, P. H., and J. M. Diamond (1984) Effects of unstirred layers on membrane phenomena. *Physiol. Rev. 64*: 763–872. (*201*)

Bartholomew, G. A., and R. C. Lasiewski (1965) Heating and cooling rates, heart rate, and simulated diving in the Galapagos marine iguana. *Comp. Biochem. Physiol. 16*: 575–82. (*28*)

Bascom, W. (1980) *Waves and Beaches*. 2nd ed. Garden City, N.Y.: Doubleday. (*6*)

Batchelor, G. K. (1967) *An Introduction to Fluid Dynamics*. Cambridge, U.K.: Cambridge University Press. (*25, 337, 363, 364, 387*)

Baudinette, R. V., and P. Gill (1985) The energetics of "flying" and "paddling" in water: Locomotion of penguins and ducks. *J. Comp. Physiol. B155*: 373–80. (*150, 155, 286*)

Begg, J. E., and N. C. Turner (1970) Water potential gradients in field tobacco. *Plant Physiol. 46*: 343–46. (*323*)

Bennet-Clark, H. C., and G. M. Alder (1979) The effect of air resistance on the jumping performance of insects. *J. Exp. Biol. 82*: 105–21. (*142, 143*)

Berg, H. C. (1983) *Random Walks in Biology*. Princeton, N.J.: Princeton University Press. (*331, 346*)

Berg, H. C. and R. A. Anderson (1973) Bacteria swim by rotating their flagellar filaments. *Nature 245*: 380–82. (*353*)

Berg, H. C., M. D. Manson, and M. P. Conley (1982) Dynamics and energetics of flagellar rotation in bacteria. *Symp. Soc. Exp. Biol. 35*: 1–31. (*352*)

Berger, S. A., L. Talbot, and L.-S. Yao (1983) Flow in curved pipes. *Ann. Rev. Fluid Mech. 15*: 461–512. (*215*)

Best, B. A. (1985) An integrative analysis of passive suspension feeding: the sea pen Ptilosarcus gurneyi as a model organism. Ph.D. dissertation, Duke University, Durham, N.C. (*119*)

Best, B. A. (1988) Passive suspension feeding in a sea pen: Effects of ambient flow on volume flow rate and filtering efficiency. *Biol. Bull. 175*: 332–42. (*113*)

Beyer, W. H. (1978) *CRC Standard Mathematical Tables*. West Palm Beach, Fla.: CRC Press. (*336*)

Bhaud, M. R., and C. P. Cazaux (1990) Buoyancy characteristics of *Lanice conchilega* (Pallas) larvae (Terebellidae). *J. Exp. Mar. Biol. Ecol. 141*: 31–45. (*344*)

Bidder, G. P. (1923) The relationship of the form of a sponge to its currents. *Quart. J. Microscop. Soc. 67*: 292–323. (*39, 190*)

Bienfang, P., and E. Laws (1977) Phytoplankton sinking rate determination: Technical and theoretical aspects, an improved methodology. *J. Exp. Mar. Biol. Ecol. 30*: 283–300. (*345*)

Bird, R. B., W. E. Stewart, and N. Lightfoot (1960) *Transport Phenomena*. New York: John Wiley. *(197)*

Birkhoff, G. (1960) *Hydrodynamics*. 2d ed. Princeton, N.J.: Princeton University Press. *(362)*

Bishop, L. (1990) Meteorological aspects of spider ballooning. *Environ. Entomol. 19*: 1381–87. *(12)*

Bishop, R.E.D., and A. Y. Hassan (1964) The lift and drag forces on a circular cylinder in a flowing fluid. *Proc. Roy. Soc. Lond. A277*: 32–50. *(371)*

Bishop, W. (1961) The development of tailless aircraft and flying wings. *J. Roy. Aero. Soc. 65*: 799–806. *(256)*

Blackburn, P., J. A. Petty, and K. F. Miller (1988) An assessment of the static and dynamic factors in windthrow. *Forestry 61*: 29–43. *(121)*

Blake, R. W. (1979a) The mechanics of labriform locomotion. I. Labriform locomotion in the angelfish (*Pterophyllum eimekei*): An analysis of the power stroke. *J. Exp. Biol. 82*: 255–71. *(287, 365)*

Blake, R. W. (1979b) The energetics of hovering in the mandarin fish (*Synchropus picturatus*). *J. Exp. Biol. 82*: 25–33. *(289)*

Blake, R. W. (1981) Mechanics of drag-based mechanisms of propulsion in aquatic vertebrates. *Symp. Zool. Soc. Lond. 48*: 29–52. *(155, 283, 287)*

Blake, R. W. (1985) Crab carapace hydrodynamics. *J. Zool. Lond. 207*: 407–23. *(143, 146, 176)*

Blake, R. W. (1986) Hydrodynamics of swimming in the water boatman, *Cenocorixa bifida*. *Can. J. Zool. 64*: 1606–13. *(154, 352, 365)*

Bleckmann, H. (1988) Prey identification and prey localization in surface-feeding fish and fishing spiders. In J. Atema, R. R. Fay, A. N. Popper, and W. N. Tavolga, eds., *Sensory Biology of Aquatic Animals*, pp. 619–41. New York: Springer-Verlag. *(392)*

Blevins, R. D. (1977) *Flow-Induced Vibration*. New York: Van Nostrand Reinhold Co. *(371)*

Blevins, R. D. (1984) *Applied Fluid Dynamics Handbook*. New York: Van Nostrand Reinhold Co. *(395)*

Blick, E. F., D. Watson, G. Belie, and H. Chu (1975) Bird aerodynamic experiments. In T. Y.-T. Wu, C. J. Brokaw, and C. Brennen, eds., *Swimming and Flying in Nature*, vol. 2, pp. 939–52. New York: Plenum Press. *(242)*

Blokhin, S. A. (1984) Investigations of gray whales taken in the Chukchi coastal waters. In M. L. Jones, S. L. Swartz, and S. Leatherwood, eds., *The Gray Whale*, Eschrichtius robustus, pp. 487–509. Orlando, Fla.: Academic Press. *(180)*

Bogy, D. B. (1979) Drop formation in a circular liquid jet. *Ann. Rev. Fluid Mech. 11*: 207–28. *(396)*

Bold, H. C., and M. J. Wynne (1978) *Introduction to the Algae*. Englewood Cliffs, N.J.: Prentice Hall. *(125)*

Booker, M. J., and A. E. Walsby (1979) The relative form resistance of straight and helical blue-green algal filaments. *Br. Phycol. J. (Phycol. J.)14*: 141–50. *(343)*

Booth, D. T., and M. E. Feder (1991) Formation of hypoxic boundary layers and their biological implications in a skin-breathing aquatic salamander, *Desmognathus quadramaculatus*. *Physiol. Zool. 64*: 1307–21. *(200)*

Bose, N., and J. Lien (1989) Propulsion of a fin whale (*Balaenoptera physalus*): Why the fin whale is a fast swimmer. *Proc. Roy. Soc. Lond. B237*: 175–200. (*151, 282*)

Bose, N., and J. Lien (1990) Energy absorption from ocean waves: A free ride for cetaceans. *Proc. Roy. Soc. Lond. B240*: 591–605. (*386*)

Bourne, G. B. (1987) Hemodynamics in squid. *Experientia 43*: 500–502. (*324*)

Bowerbank, J. S. (1864) *A Monograph of the British Spongiadae*. Vol. 1. London: The Ray Society. (*39*)

Boyajian, G. E., and M. LaBarbera (1987) Biomechanical analysis of passive flow of stromatoporoids—morphologic, paleoecologic, and systematic implications. *Lethaia 20*: 223–29. (*66*)

Braimah, S. A. (1987) Pattern of flow around filter-feeding structures of immature *Simulium bivittatum* Malloch (Diptera: Simuliidae) and *Isonychia campestris* McDunnough (Ephemeroptera: Oligoneuriidae). *Can. J. Zool. 65*: 514–21. (*103*)

Brancazio, P. J. (1984) *Sport Science: Physical Laws and Optimum Performance*. New York: Simon and Schuster. (227)

Braun, J., and W.-E. Reif (1985) A survey of aquatic locomotion in fishes and tetrapods. *N. Jb. Geol. Paläont. Abh.169*: 307–32. (287)

Breder, C. M. (1926) The locomotion of fishes. *Zoologica (N.Y.) 4*: 159–256. (*281*)

Brennen, C. (1976) Locomotion of flagellates with mastigonemes. *J. Mechanochem. Cell Motil. 3*: 207–17. (*356*)

Brennen, C., and H. Winet (1977) Fluid mechanics of propulsion by cilia and flagella. *Ann. Rev. Fluid Mech. 9*: 339–98. (*335, 356*)

Bridgman, P. W. (1931) *Dimensional Analysis*. New Haven, Conn.: Yale University Press. (*9*)

Brodie, H. J., and P. H. Gregory (1953) The action of wind in the dispersal of spores from cup-shaped plant structures. *Can. J. Bot. 31*: 402–10. (*214*)

Brodsky, A. K. (1991) Vortex formation in the tethered flight of the peacock butterfly, *Inachis io* (Lepidoptera: Nymphalidae) and some aspects of insect flight evolution. *J. Exp. Biol. 161*: 77–96. (276)

Brooks, J. L., and G. E. Hutchinson (1950) On the rate of passive sinking of Daphnia. *Proc. Nat. Acad. Sci. USA 36*: 272–77. (*340*)

Burggren, W. W., and W. E. Bemis (1992) Metabolism and ram gill ventilation in juvenile paddlefish, *Polyodon spathula* (Chondrostei: Polyodontidae). *Physiol. Zool. 65*: 515–39. (*189*)

Burkett, C. W. (1989) Reductions in induced drag by the use of aft swept wing tips. *Aeronaut. J. 93*: 400–405. (*239*)

Burrows, F. M. (1987) The aerial motion of seeds, fruits, spores and pollen. In D. Murray, ed., *Seed Dispersal*, pp. 1–47. New York: Academic Press. (*344*)

Burton, A. C. (1972) *Physiology and Biophysics of the Circulation*. 2d ed. Chicago: Year Book Medical Publishers. (*316*)

Bushnell, D. M., and K. J. Moore (1991) Drag reduction in nature. *Ann. Rev. Fluid Mech. 23*: 65–79. (*152, 153*)

Butman, C. A. (1987) Larval settlement of soft-sediment invertebrates: The spatial scales of patterns explained by active habitat selection and the emerging role of hydrodynamical processes. *Oceanogr. Mar. Biol. Ann. Rev. 25*: 113–65. (*184*)

Cahoon, L. B. (1988) Use of a whirling cup rotor to stir benthic chambers. *Hydrobiologia 160*: 193–98. (*112*)

Canny, M. J. (1993) The transpiration stream in the leaf apoplast: Water and solutes. *Phil. Trans. Roy. Soc. Lond. B341*: 87–100. *(320)*

Carey, D. A. (1983) Particle resuspension in the boundary layer induced by flow around polychaete tubes. *Can J. Fish. Aquat. Sci. 40 (Suppl. 1)*: 301–308. *(217)*

Carey, F. G., J. M. Teal, J. W. Kanwisher, and K. D. Lawson (1971) Warm-blooded fish. *Amer. Zool. 11*: 137–45. *(28)*

Carling, P. A. (1992) The nature of the fluid boundary layer and the selection of parameters for benthic ecology. *Freshwater Biol. 28*: 273–84. *(172)*

Caro, C. G., T. J. Pedley, R. C. Schroter, and W. A. Seed (1978) *The Mechanics of the Circulation.* Oxford: Oxford University Press. *(36, 37, 62, 302, 312, 377)*

Carrington, E. (1990) Drag and dislodgment of an intertidal macroalga: Consequences of morphological variation in *Mastocarpus papillatus* Kutzing. *J. Exp. Mar. Biol. Ecol. 139*: 185–200. *(119)*

Carstens, T. (1968) Wave forces on boundaries and submerged bodies. *Sarsia 34*: 37–60. *(115)*

Chamberlain, A. C., J. A. Garland, and A. C. Wells (1984) Transport of gases and particles to surfaces with widely spaced roughness elements. *Boundary Layer Meteorol. 29*: 343–60. *(186)*

Chamberlain, J. A. (1976) Flow patterns and drag coefficients of cephalopod shells. *Paleontology 19*: 539–63. *(110, 143, 147)*

Chance, M. M., and D. A. Craig (1986) Hydrodynamics and behavior of Simuliidae larvae (Diptera). *Can. J. Zool. 64*: 1295–1309. *(217)*

Charriaud, E. (1982) Direct measurement of velocity profiles and fluxes at the cloacal siphon of the ascidian *Ascidiella aspera*. *Mar. Biol. 70*: 35–40. *(329)*

Chase, R.R.P. (1979) Settling behavior of natural aquatic particulates. *Limnol. Oceanogr. 24*: 417–26. *(342)*

Cheer, A.Y.L., and M.A.R. Koehl (1988) Paddles and rakes: Fluid flow through bristled appendages of small organisms. *J. Theor. Biol. 129*: 17–39. *(357)*

Chia, F. S., J. Buckland-Nicks, and C. M. Young (1984) Locomotion of marine invertebrate larvae—a review. *Can. J. Zool. 62*: 1205–22. *(352)*

Chien, S., S. Usami, R. J. Dellenback, and C. A. Bryant (1971) Comparative hemorheology: Hematological implications of species differences in blood viscosity. *Biorheology 8*: 35–57. *(28)*

Choe, J. C., and K. C. Kim (1991) Microhabitat selection and adaptation of feather mites (Acari: Analgoidea) on murres and kittiwakes. *Can. J. Zool. 69*: 817–21. *(181)*

Chopra, M. G. (1975) Lunate-tail swimming propulsion. In T. Y.-T. Wu, C. J. Brokaw, and C. Brennen, eds., *Swimming and Flying in Nature*, vol. 2, pp. 635–50. New York: Plenum Press. *(282)*

Clark, B. D., and W. Bemis (1979) Kinematics of swimming of penguins at the Detroit Zoo. *J. Zool. Lond. 188*: 411–28. *(149)*

Clark, R. B., and C. O. Hermans (1976) Kinetics of swimming in some smooth-bodied polychaetes. *J. Zool. Lond. 178*: 147–59. *(281)*

Clark, R. B., and D. J. Tritton (1970) Swimming in nereidiform polychaetes. *J. Zool. Lond. 161*: 257–271. *(287)*

Cone, C. D. (1962) The soaring flight of birds. *Sci. Amer. 206 (4)*: 130–40. *(221, 259)*

Conway, C. (1969) *The Joy of Soaring.* Los Angeles: The Soaring Society of America. *(259)*

Cook, P. L. (1977) Colony-wide currents in living Bryozoa. *Cah. Biol. Mar.* 18: 31–47. (*191*)

Cox, R. G. (1970) The motion of long, slender bodies in a viscous fluid. Part I. General theory. *J. Fluid Mech.* 44: 791–810. (*335*)

Cranston, P. S. (1984) The taxonomy and ecology of *Orthocladius fuscimanus*, a hygropetric chironomid dipteran. *J. Nat. Hist.* 18: 873–96. (*183*)

Crenshaw, H. C. (1993) Orientation by helical motion. I. Kinematics of the helical motion of organisms with up to six degrees of freedom. *Bull. Math. Biol.* 55: 197–212. (*352*)

Crisp, D. J. (1955) The behaviour of barnacle ciprids in relation to water movement over a surface. *J. Exp. Biol.* 32: 569–90. (*183, 185*)

Crisp, D. T. (1989) Comparison of the physical properties of real and artificial salmon eggs and of their performance when drifting in an experimental stream channel. *Hydrobiol.* 178: 143–54. (*88*)

Cruickshank, A.R.I., P. G. Small, and M. A. Taylor (1991) Dorsal nostrils and hydrodynamically driven underwater olfaction in plesiosaurs. *Nature* 352: 62–64. (*60*)

Csicáky, M. J. (1977) Body gliding in the zebra finch. *Fortschr. Zool.* 24, 2/3: 275–86. (*239*)

Cummins, K. W. (1964) Factors limiting the microdistribution of the caddisflies *Pycnopsyche lepida* (Hagen) and *Pycnopsyche guttifer* (Walker) in a Michigan stream (Trichoptera: Limnephilidae). *Ecol. Monogr.* 34: 271–95. (*180*)

Dadswell, M. J., and D. Weihs (1990) Size-related hydrodyamic characteristics of the giant scallop, *Placopecten magellanicus* (Bivalvia: Pectinidae). *Can. J. Zool.* 68: 778–85. (*77*)

Daniel, T. L. (1981) Fish mucus: *In situ* measurement of polymer draag reduction. *Biol. Bull.* 160: 376–82. (*152*)

Daniel, T. L. (1983) Mechanics and energetics of medusan jet propulsion. *Can. J. Zool.* 61: 1406–20. (*77*)

Daniel, T. L. (1984) Unsteady aspects of aquatic locomotion. *Amer. Zool.* 24: 121–34. (*363, 365, 366*)

Daniel, T. L., and J. G. Kingsolver (1983) Feeding strategy and the mechanics of blood sucking in insects. *J. Theor. Biol.* 105: 661–72. (*330*)

Daniel, T. L., and E. Meyhöfer (1989) Size limits in escape locomotion of carridean shrimp. *J. Exp. Biol.* 143: 245–66. (*287*)

Dasgupta, A., P. Guenard, J. S. Ultman, J. S. Kimbell, and K. T. Morgan (1993) A photographic method for the visualization of mass uptake patterns in aqueous systems. *Int. J. Heat Mass Transfer.* 36: 453–62. (*197*)

Davenport, J. (1992) Wing-loading, stability, and morphometric relationships in flying fish (Exocoetidae) from the north-eastern Atlantic. *J. Mar. Biol. Assoc. UK* 72: 25–39. (*258*)

Davenport, J., S. A. Munks, and P. J. Oxford (1984) A comparison of the swimming of marine and freshwater turtles. *Proc. Roy. Soc. Lond.* B220: 447–475. (*155, 286, 287*)

Davey, M. C. (1988) The effects of nutrient depletion on the sinking velocity and cellular composition of a freshwater diatom. *Arch. Hydrobiol.* 112: 321–24. (*344*)

Davies, P. F. (1989) How do vascular endothelial cells respond to flow? *News in Physiol. Sci.* 4: 22–25. (*320*)

Davis, J. A., and L. A. Barmuta (1989) An ecologically useful classification of mean and near-bed flows in streams and rivers. *Freshwater Biol. 21*: 271–82. *(172)*

DeMont, M. E. and J. M. Gosline (1988) Mechanics of jet propulsion in the hydromedusan jellyfish, *Polyorchis penicillatus*. II. Energetics of the jet cycle. *J. Exp. Biol. 134*: 333–45. *(77)*

Denny, M. W. (1983) A simple device for recording the maximum force exerted on intertidal organisms. *Limnol. Oceanogr. 28*: 1269–74. *(129, 130)*

Denny, M. W. (1985) Wave forces on intertidal organisms: A case study. *Limnol. Oceanogr. 30*: 1171–87. *(108, 130)*

Denny, M. W. (1987a) Lift as a mechanism of patch initiation in mussel beds. *J. Exp. Mar. Biol. Ecol. 113*: 231–45. *(66)*

Denny, M. W. (1987b) Life in the maelstrom: The biomechanics of wave-swept rocky shores. *Trends in Ecol. Evol. 2*: 61–66. *(130)*

Denny, M. W. (1988) *Biology and the Mechanics of the Wave-Swept Environment*. Princeton, N.J.: Princeton University Press. *(6, 108, 130, 170, 363, 367, 381)*

Denny, M. W. (1989) A limpet shell shape that reduces drag: Laboratory demonstration of a hydrodynamic mechanism and an exploration of its effectiveness in nature. *Can. J. Zool. 67*: 2098–2106. *(66, 101, 108, 176)*

Denny, M. W. (1993) *Air and Water: The Biology and Physics of Life's Media*. Princeton, N.J.: Princeton University Press. *(6)*

Denny, M. W., and S. D. Gaines (1990) On the prediction of maximal intertidal wave forces. *Limnol. Oceanogr. 35*: 1–15. *(130)*

Denny, M. W., and M. F. Shibata (1989) Consequences of surf-zone turbulence for settlement and external fertilization. *Amer. Nat. 134*: 859–89. *(168, 185, 186)*

Denny, M. W., T. L. Daniel, and M.A.R. Koehl (1985) Mechanical limits to size in wave-swept organisms. *Ecol. Monogr. 55*: 69–102. *(367)*

de Paula, E. J., and E. C. de Oliveira F. (1982) Wave exposure and ecotypical differentiation in *Sargassum cynosum* (Phaeophyta: Fucales) *Phycologia 21*: 145–53. *(126)*

Deshafy, G. S. (1981) 'Sailing' behaviour in the fishing spider, *Dolomedes triton* (Walckenaer). *Anim. Behav. 29*: 965–66. *(391)*

Dick, J. B. (1950) The fundamentals of natural ventilation of houses. *Inst. Heat. Vent. Engineers J. 18*: 123–34. *(72)*

Dodds, G. S., and F. L. Hisaw (1924) Ecological studies of aquatic insects. I. Adaptations of mayfly nymphs to swift streams. *Ecology 5*: 137–48. *(139, 178, 200)*

Dodds, G. S., and F. L. Hisaw (1925) Ecological studies of aquatic insects. III. Adaptations of caddisfly larvae to swift streams. *Ecology 6*: 123–37. *(180)*

Dodson, S., and C. Ramacharan (1991) Size specific swimming behavior of *Daphnia pulex*. *J. Plankton Res. 13*: 1367–80. *(30, 340, 344)*

Doty, M. S. (1971) Measurement of water movement in reference to benthic algal growth. *Botanica Marina 14*: 32–35. *(129)*

Dowell, E. H., ed. (1989) *A Modern Course in Aeroelasticity*. 2d ed. Boston: Kluwer. *(374)*

Dronkers, J., and W. van Leussen (1988) *Physical Processes in Estuaries*. Berlin: Springer-Verlag. *(393)*

DuBois, A. B., G. A. Cavagna, and R. Fox (1974) Pressure distribution on the body surface of swimming fish. *J. Exp. Biol. 60*: 581–91. *(67)*

Duda, J. L., and J. S. Vrentas (1971) Heat transfer in a cylindrical cavity. *J. Fluid Mech. 45*: 261–79. *(313)*

Dudley, R. (1985) Fluid-dynamic drag of limpet shells. *Veliger 28*: 6–13. *(119, 120)*

Dudley, R., and P. DeVries (1990) Tropical rain forest structure and the geographical distribution of gliding vertebrates. *Biotropica 22*: 432–34. *(258)*

Dudley, R., and C. P. Ellington (1990a) Mechanics of forward flight in bumblebees. I. Kinematics and morphology. *J. Exp. Biol. 148*: 19–52. *(249, 268)*

Dudley, R., and C. P. Ellington (1990b) Mechanics of forward flight in bumblebees. II. Quasi-steady lift and power requirements. *J. Exp. Biol. 148*: 53–88. *(239, 245, 249, 252)*

Dudley, R., V. A. King, and R. J. Wassersug (1991) The implications of shape and metamorphosis for drag forces on a generalized pond tadpole (*Rana catesbeiana*). *Copeia* (1991): 252–57. *(110, 143, 146)*

Durand, W. F. (1936) *Aerodynamic Theory*. Vol. 6. Reprint. New York: Dover Publications, 1963. *(63)*

Dury, G. H. (1969) Hydraulic geometry. In R. J. Chorley, ed., *Introduction to Fluvial Processes*, pp. 146–56. New York: Barnes and Noble. *(40)*

Eckman, J. E. (1983) Hydrodynamic processes affecting benthic recruitment. *Limnol. Oceanogr. 28*: 241–57. *(220)*

Eckman, J. E. (1990) A model of passive settlement by planktonic larvae onto bottoms of differing roughness. *Limnol. Oceanogr. 35*: 887–901. *(186)*

Eckman, J. E., A.R.M. Nowell, and P. A. Jumars (1981) Sediment destabilization by animal tubes. *J. Mar. Res. 39*: 361–74. *(219)*

Einstein, A., and L. Infeld (1966) *The Evolution of Physics.*, Rev. ed. New York: Simon and Schuster. *(358)*

Eisenhart, C. (1968) Expression of the uncertainty of final results. *Science 160*: 1201–1204. *(11)*

Eisner, T., and D. J. Aneshansley (1982) Spray aiming in bombardier beetles. Jet deflection by the Coanda effect. *Science 215*: 83–85. *(397)*

Elder, J. W. (1960) The flow past a plate of finite width. *J. Fluid Mech. 9*: 133–53. *(136)*

Ellers, O.W.J. (1988) Locomotion via swash-riding in the clam *Donax variabilis*. Ph.D. dissertation, Duke University, Durham, N.C. *(181)*

Ellington, C. P. (1978) The aerodynamics of normal hovering flight: Three approaches. In K. Schmidt-Nielsen, L. Bolis, and S.H.P. Maddrell, eds., *Comparative Physiology: Water, Ions and Fluid Mechanics*, pp. 327–45. Cambridge, U.K.: Cambridge University Press. *(276)*

Ellington, C. P. (1980a) Wing mechanics and take-off preparation of *Thrips* (Thysanoptera). *J. Exp. Biol. 85*: 129–36. *(278, 350)*

Ellington, C. P. (1980b) Vortices and hovering flight. In W. Nachtigall, ed., *Instationäre Effekte an schwingenden Tierflügeln*, pp. 64–101. Wiesbaden: Franz Steiner. *(276, 278, 287)*

Ellington, C. P. (1984a) The aerodynamics of hovering insect flight. *Phil. Trans. Roy. Soc. Lond. B305*: 1–181. *(265, 267, 275, 364, 369)*

Ellington, C. P. (1984b) The aerodynamics of flapping animal flight. *Amer. Zool. 24*: 95–105. *(265, 278, 279, 280)*

Ellington, C. P. (1991) Aerodynamics and the origin of insect flight. *Adv. Insect Physiol. 23*: 171–210. *(98, 284)*

Elliott, J. S. (1984) A note on the propulsive systems of fishes and birds with possible application to man-powered flight. *Aeronaut. J. 88*: 296–98. *(280)*

Emerson, S. B., and M.A.R. Koehl (1990) The interaction of behavioral and morphological change in the evolution of a novel locomotor type—flying frogs. *Evolution 44*: 1931–46. *(258)*

Emlet, R. B. (1990) Flow fields around ciliated larvae: Effects of natural and artificial tethers. *Mar. Ecol. Progr. Ser. 63*: 211–26. *(349)*

Engvall, J., P. Ask, D. Loyd, and B. Wranne (1991) Coarctation of the aorta: A theoretical and experimental analysis of the effects of a centrally-located arterial stenosis. *Medical and Biological Engineering and Computing 29*: 291–96. *(62, 316)*

Ennos, A. R. (1988) The importance of torsion in the design of insect wings. *J. Exp. Biol. 140*: 137–60. *(264)*

Ennos, A. R. (1989a) The effect of size on the optimal shapes of gliding insects and seeds. *J. Zool. Lond. 219*: 61–69. *(241)*

Ennos, A. R. (1989b) The kinematics and aerodynamics of the free flight of some diptera. *J. Exp. Biol. 142*: 49–85. *(264, 265, 268, 278, 369)*

Epp, R. W., and W. M. Lewis, Jr. (1984) Cost and speed of locomotion for rotifers. *Oecologia 61*: 289–92. *(356)*

Eymann, M. (1988) Drag on single larvae of the black fly, *Simulium vittatum* (Diptera: Simuliidae) in a thin, growing boundary layer. *J. N. Amer. Benthol. Soc. 7*: 109–16. *(113, 119)*

Eymann, M. (1991) Flow patterns around cocoons and pupae of black flies of the genus *Simulium* (Diptera: Simuliidae). *Hydrobiol. 215*: 223–29. *(218)*

Fauchald, K., and P. A. Jumars (1979) The diet of worms: A study of polychaete feeding guilds. *Oceanogr. Mar. Biol. Ann. Rev. 17*: 193–284. *(191)*

Feder, M. E., and A. W. Pinder (1988) Ventilation and its effect on "infinite pool" exchangers. *Amer. Zool. 28*: 973–83. *(201)*

Fejer, A. A., and R. H. Backus (1960) Porpoises and the bow-riding of ships under way. *Nature 188*: 700–703. *(386)*

Feldcamp, S. D. (1987) Swimming in the California sea lion: Morphometrics, drag, and energetics. *J. Exp. Biol. 131*: 117–35. *(143, 150, 286)*

Finnigan, J. J., and P. J. Mulhearn (1978) Modelling waving crops in a wind tunnel. *Boundary Layer Meteorol. 14*: 253–78. *(377)*

Fischer, H. B. (1972) Mass transport mechanisms in partially stratified estuaries. *J. Fluid Mech. 53*: 672–87. *(393)*

Fish, F. E. (1982) Aerobic energetics of surface swimming in the muskrat, *Ondatra zibethicus*. *Physiol. Zool. 55*: 180–89. *(385)*

Fish, F. E. (1984) Mechanics, power output and efficiency of the swimming muskrat (*Ondatra zibethicus*). *J. Exp. Biol. 110*: 183–201. *(286, 385)*

Fish, F. E. (1987) Kinematics and power output of jet propulsion by the frogfish genus *Antennarius* (Lophiiformes: Antennariidae). *Copeia 1987*: 1046–48. *(77)*

Fish, F. E. (1990) Wing design and scaling of flying fish with regard to flight performance. *J. Zool. Lond. 221*: 391–403. *(259, 288)*

Fish, F. E. (1992) Aquatic locomotion. In T. Tomasi and T. Horton, eds. *Mammalian Energetics: Interdisciplinary Views of Metabolism and Reproduction*, pp. 34–63. Ithaca, N.Y.: Comstock Publ. Assoc. *(150, 154, 286, 287)*

Fish, F. E., and C. Hui (1991) Dolphin swimming—a review. *Mammal Rev. 21*: 181–95. *(150, 151, 152, 153)*

Fish, F. E., B. R. Blood, and B. D. Clark (1991) Hydrodynamics of the feet of fish-catiching bats: Influence of the water surface on drag and morphological design. *J. Exp. Zool. 258*: 164–73. (*141, 383*)

Fletcher, G. L., and R. T. Haedrich (1987) Rheological properties of rainbow trout blood. *Can. J. Zool. 65*: 879–83. (*28*)

Flettner, A. (1926) *The Story of the Rotor.* New York: F. O. Willhofft. (*227*)

Foster-Smith, R. L. (1976) Pressures generated by the pumping mechanisms of some ciliary filter feeders. *J. Exp. Mar. Biol. Ecol. 25*: 199–206. (*325, 328*)

Foster-Smith, R. L. (1978) An analysis of water flow in tube-living animals. *J. Exp. Mar. Biol. Ecol. 34*: 73–95. (*327, 328*)

Fox, S. H., C. S. Ogilvy, and A. B. DuBois (1990) Search for the biological stimulus of the Cushing response in bluefish (*Potatomus saltatrix*). *Biol. Bull. 179*: 233. (*68*)

Francis, L. (1991) Sailing downwind: Aerodynamic performance of the *Velella* sail. *J. Exp. Biol. 158*: 117–32. (*249, 252, 391*)

Fraser, A. I. (1962) Wind tunnel studies of the forces acting on the crowns of small trees. *Rep. Forest Res. (U.K.)* (1962): 178–83. (*121*)

Freadman, M. A. (1979) Swimming energetics of striped bass (*Morone saxatilis*) and bluefish (*Potatomus saltatrix*): Gill ventilation and swimming metabolism. *J. Exp. Biol. 83*: 217–30. (*110*)

Freadman, M. A. (1981) Swimming energetics of striped bass (*Morone saxatilis*) and bluefish (*Potatomus saltatrix*): Hydrodynamic correlates of locomotion and gill ventilation. *J. Exp. Biol. 90*: 143–62. (*110*)

French, M. J. (1988) *Invention and Evolution: Design in Nature and Engineering.* Cambridge, U.K.: Cambridge University Press. (*325*)

Frith, H. R., and R. W. Blake (1991) Mechanics of the startle response in the northern pike, *Esox lucius. Can. J. Zool. 69*: 2831–39. (*287, 365*)

Frost, T. M. (1978) In situ measurements of clearance rates for the freshwater sponge *Spongilla lacustris. Limnol. Oceanogr. 23*: 1034–39. (*325*)

Futrelle, R. P. (1984) How molecules get to their detectors: The physics of diffusion of insect pheromones. *Trends in Neurosci. 7*: 116–20. (*357*)

Gal, J. M., and R. W. Blake (1987) Hydrodynamic drag of two frog species: *Hymenochirus boettgeri* and *Rana pipiens. Can J. Zool. 65*: 1085–90. (*143, 146*)

Gal, J. M., and R. W. Blake (1988) Biomechanics of frog swimming. I. Estimation of the propulsive force generated by *Hymenochirus boettgeri. J. Exp. Biol. 138*: 399–412. (*154, 365*)

Garratt, J. R. (1992) *The Atmospheric Boundary Layer.* Cambridge, U.K.: Cambridge University Press. (*171*)

Gates, D. M. (1962) *Energy Exchange in the Biosphere.* New York: Harper and Row. (*221*)

Gavriely, N., T. R. Shee, D. W. Cugell, and J. B. Grotberg (1989) Flutter in flow-limited collapsible tubes: A mechanism for generation of wheezes. *J. Appl. Physiol. 66*: 2251–61. (*375*)

Genin, A., P. K. Dayton, P. F. Lonsdale, and F. N. Spiess (1986) Corals of seamount peaks provide evidence of current acceleration over deep-sea topography. *Nature 322*: 59–61. (*40*)

Gerard, V. A., and K. H. Mann (1979) Growth and reproduction of *Laminaria longicruris* populations exposed to different intensities of water movement. *J. Phycol. 15*: 33–41. (*126*)

Gewicke, M., and H. G. Heinzel (1980) Aerodynamic and mechanical properties of the antennae as air-current sense organs in *Locusta migratoria*. I. Static characteristics. *J. Comp. Physiol. A139*: 357–66. (*119*)

Gibo, D. L., and M. J. Pallett (1979) Soaring flight of monarch butterflies. *Can. J. Zool. 57*: 1393–1401. (*12, 255*)

Gies, J. (1963) *Bridge and Men*. New York: Doubleday. (*34*)

Glasheen, J. W., and T. A. McMahon (1992) An analysis of aquatic bipedalism in basilisk lizards. *Amer. Zool. 32*: 144a. (*387*)

Goldburg, A., and B. H. Florsheim (1966) Transition and Strouhal number for the incompressible wake of various bodies. *Phys. Fluids 9*: 45–50. (*371*)

Goldman, D. T., and R. J. Bell, eds. (1986) *The International System of Units (SI)*. National Bureau of Standards Special Publ. 330. Washington, D.C.: Government Printing Office. (*9*)

Goldstein, S. (1938) *Modern Developments in Fluid Dynamics*. Reprint. New York: Dover Publications, 1965. (*6, 19, 77, 102, 135, 160, 216, 244, 296*)

Goldstein, S. (1969) Fluid mechanics in the first half of this century. *Ann. Rev. Fluid Mech. 1*: 1–28. (*177*)

Gordon, J. E. (1978) *Structures, or Why Things Don't Fall Down*. Harmondsworth, Middlesex, England: Penguin Books. (*6, 109*)

Grace, J. (1977) *Plant Response to Wind*. London: Academic Press. (*109, 128, 163, 171, 197, 198, 371, 377*)

Grace, J. (1978) The turbulent boundary layer over a flapping *Populus* leaf. *Pl. Cell Environ. 1*: 35–38. (*193*)

Grace, J., and M. A. Collins (1976) Spore liberation from leaves by wind. In C. H. Dickinson and T. F. Preece, eds., *Microbiology of Aerial Plant Surfaces*, pp. 185–98. London: Academic Press. (*192*)

Grace, J., and J. Wilson (1976) The boundary layer over a *Populus* leaf. *J. Exp. Bot. 27*: 231–41. (*163*)

Grant, J., C. W. Emerson, and S. E. Shumway (1993) Orientation, passive transport, and sediment erosion features of the sea scallop *Placopecten magellanicus* in the benthic boundary layer. *Can. J. Zool. 71*: 953–59. (*219*)

Grant, J.W.G., and I.A.E. Bagley (1981) Predator induction of crests in morphs of the *Daphnia carinata* King complex. *Limnol. Oceanogr. 26*: 201–18. (*30*)

Grant, R. E. (1825) Observations of the structure and function of the sponge. *Edinburgh Phil. J. 13*: 94–107, 333–46. (*38*)

Gray, J. (1936) Studies in animal locomotion. VI. The propulsive powers of the dolphin. *J. Exp. Biol. 13*: 192–99. (*150, 152*)

Green, D. S. (1980) The terminal velocity and dispersal of spinning samaras. *Amer. J. Bot. 67*: 1218–24. (*228, 272*)

Greene, D. F., and E. A. Johnson (1989) A model of wind dispersal of winged or plumed seeds. *Ecology 70*: 339–47. (*272*)

Greene, D. F., and E. A. Johnson (1990a) The aerodynamics of plumed seeds. *Func. Ecol. 4*: 117–25. (*336*)

Greene, D. F., and E. A. Johnson (1990b) The dispersal of winged fruits and seeds differing in autorotative behaviour. *Can. J. Bot. 68*: 2693–97. (*228, 272*)

Greenewalt, C. H. (1962) Dimensional relationships for flying animals. *Smithsonian Misc. Coll. 144*(2): 1–46. *(243, 268)*

Greenstone, M. H. (1990) Meteorological determinants of spider ballooning: The roles of thermals versus the vertical windspeed gradient in becoming airborne. *Oecologia 84*: 164–68. *(223)*

Gregory, P. H. (1973) *The Microbiology of the Atmosphere*. 2d ed. Aylesbury, U.K.: Leonard Hill Books. *(186, 195, 344)*

Guard, C. L., and D. E. Murrish (1975) Effects of temperature on the viscous behavior of blood from antarctic birds and mammals. *Comp. Biochem. Physiol. 52A*: 287–90. *(28)*

Guries, R. P., and E. V. Nordheim (1984) Flight characteristics and dispersal potential of maple samaras. *Forest Sci. 30*: 434–40. *(272)*

Hainsworth, F. R. (1989) Wing movements and positioning for aerodynamic benefit by Canada geese flying in formation. *Can. J. Zool. 67*: 585–89. *(288)*

Hairston, N. G. (1957) Observations on the behavior of *Draco volans* in the Philippines. *Copeia* (1957): 262–65. *(258)*

Hamner, W. M., and D. Schneider (1986) Regularly spaced rows of medusae in the Bering Sea: Role of Langmuir circulation. *Limnol. Oceanogr. 31*: 171–77. *(394)*

Hansen, B., and P. Tiselius (1992) Flow through the feeding structures of suspension feeding zooplankton: A physical model approach. *J. Plankton Res. 14*: 821–34. *(357)*

Hansen, R. A., D. D. Hart, and R. A. Merz (1991) Flow mediates predator-prey interactions between triclad flatworms and larval black flies. *Oikos 60*: 187–96. *(179)*

Happel, J., and H. Brenner (1965) *Low Reynolds Number Hydrodynamics*. Englewood Cliffs, N.J.: Prentice Hall. *(303, 333, 336, 337, 338, 342, 354)*

Harper, D. G., and R. W. Blake (1989) Fast start performance of rainbow trout *Salmo gairdneri* and northern pike *Esox lucius* during escapes. *J. Exp. Biol. 150*: 321–42. *(287)*

Hartnoll, R. G. (1971) The occurrence, methods, and significance of swimming in the Brachyura. *Anim. Behav. 19*: 34–50. *(140)*

Harvell, C. D., and M. LaBarbera (1985) Flexibility—a mechanism for control of local velocities in hydroid colonies. *Biol. Bull. 168*: 312–20. *(118, 119)*

Haslam, S. M. (1978) *River Plants: The Macrophytic Vegetation of Watercourses*. Cambridge, U.K.: Cambridge University Press. *(199)*

Havel, J. E., and S. I. Dodson (1987) Reproductive costs of *Chaoborus*-induced polymorphism of *Daphnia pulex*. *Hydrobiol. 150*: 273–81. *(30)*

Havenhand, J. N., and I. Svane (1991) Roles of hydrodynamics and larval behavior in determining spatial aggregation in the tunicate *Ciona intestinalis*. *Mar. Ecol. Prog. Ser. 68*: 271–76. *(184)*

Hayami, I. (1991) Living and fossil scallop shells as airfoils: An experimental study. *Paleobiology 17*: 1–18. *(249, 252)*

Hebert, P.D.N. (1978) The adaptive significance of cyclomorphosis in *Daphnia*: More possibilities. *Freshwater Biol. 8*: 313–20. *(30)*

Heine, C. (1992) Mechanics of flapping fin locomotion in the cownose ray, *Rhinoptera bonasus* (Elasmobrachii: Myliobatidae). Ph.D. dissertation, Duke University, Durham, N.C. *(289)*

Hertel, H. (1966) *Structure, Form and Movement.* New York: Reinhold. (*135*)

Herzog, H. O. (1925) More facts about the Flettner rotor ship. *Sci. Amer. 132*: 82–83. (*227*)

Hess, F., and J. J. Videler (1984) Fast continuous swimming of saithe (*Pollachius virens*): A dynamic analysis of bending moments and muscle power. *J. Exp. Biol. 109*: 229–51. (*143, 148*)

Hessler, R. R. (1985) Swimming in Crustacea. *Trans. Roy. Soc. Edinburgh 76*: 115–22. (*146, 352*)

Hicks, J. W., and H. S. Badeer (1989) Siphon mechanism in collapsible tubes: Application to circulation of the giraffe head. *Amer. J. Physiol. 256*: R567–71. (*55*)

Hinds, T. E., F. G. Hawksworth, and W. J. McGinnies (1963) Seed dispersal in *Arceuthobium*: A photographic study. *Science 140*: 1236–38. (*195*)

Hinman, F., Jr., ed. (1968) *Hydrodynamics of Micturition.* Springfield, Ill.: Thomas. (*88*)

Hirst, J. M., and O. J. Stedman (1971) Patterns of spore dispersal in crops. In T. F. Preece and C. H. Dickinson, eds., *Ecology of Leaf Surface Micro-organisms*, pp. 229–37. London: Academic Press. (*186*)

Hocking, B. (1953) The intrinsic range and speed of flight of insects. *Trans. Roy. Ent. Soc. Lond. 104*: 223–345. (*239*)

Hoerner, S. F. (1965) *Fluid-Dynamic Drag.* Privately published by the author, now deceased. (*6, 98, 109, 123, 133, 134, 135, 138, 164, 165, 228*)

Holbo, H. R., T. C. Corbett, and P. J. Horton (1980) Aeromechanical behavior of selected douglas-fir. *Agric. Meteorol. 21*: 81–91. (*371*)

Holbrook, N. M., M. W. Denny, and M.A.R. Koehl (1991) Intertidal "trees": Consequences of aggregation on the mechanical and photosynthetic properties of sea-palms *Postelsia palmaeformis* Ruprecht. *J. Exp. Mar. Biol. Ecol. 146*: 39–67. (*122*)

Houston, J., B. N. Dancer, and M. A. Learner (1989) Control of sewage flies using *Bacillus thuringiensis var. israelensis.* I. Acute toxicity tests and pilot scale trial. *Water Res. 23*: 369–78. (*183*)

Hoyt, J. W. (1975) Hydrodynamic drag reduction due to fish slimes. In T. Y.-T. Wu, C. J. Brokaw, and C. Brennen, eds. *Swimming and Flying in Nature*, pp. 653–72. New York: Plenum Press. (*152*)

Hudlicka, O. (1991) Role of mechanical factors in angiogenesis under physiological and pathophysiological circumstances. In M. E. Maragoudakis, P. Gullino, and P. I. Lelkes, eds., *Angiogenesis in Health and Disease*, pp. 207–15. *NATO ASI Ser. A: 227.* (*320*)

Hughes, G. M. (1958) The co-ordination of insect movemnts. *J. Exp. Biol. 35*: 567–83. (*77*)

Hughes, G. M. (1966) The dimensions of fish gills in relation to their function. *J. Exp. Biol. 45*: 177–95. (*300*)

Hughes, R. G. (1975) The distribution of epizoites on the hydroid *Nemertesia antennina* (L.). *J. Mar. Biol. Assoc. U.K. 55*: 275–94. (*190*)

Hui, C. A. (1988a) Penguin swimming. I. Hydrodynamics. *Physiol. Zool. 61*: 333–43. (*149, 286, 385*)

Hui, C. A. (1988b) Penguin swimming. II. Energetics and behavior. *Physiol. Zool. 61*: 344–50. (*385*)

Hui, C. A. (1992) Walking of the shore crab *Pachygrapsus crassipes* in its two natural environments. *J. Exp. Biol. 165*: 213–27. (*140*)

Humphrey, J.A.C. (1987) Fluid mechanic constraints on spider ballooning. *Oecologia 73*: 469–77. (*12, 222*)

Hunter, T. (1989) Suspension feeding in oscillating flow: The effect of colony morphology and flow regime on plankton capture by the hydroid *Obelia longissima. Biol. Bull. 176*: 41–49. (*360*)

Hunter, T., and S. Vogel (1986) Spinning embryos enhance diffusion through gelatinous egg masses. *J. Exp. Mar. Biol. Ecol. 96*: 303–308. (*197*)

Hutchinson, G. E. (1957) *A Treatise on Limnology*. Vol. 1. New York: John Wiley. (*47, 393*)

Hutchinson, G. E. (1967) *A Treatise on Limnology*. Vol. 2. New York: John Wiley. (*29, 30, 333, 338, 341, 344*)

Hynes, H.B.N. (1970) *The Ecology of Running Water*. Liverpool, U.K.: Liverpool University Press. (*40, 177, 390*)

Ingold, C. T. (1953) *Dispersal in Fungi*. Oxford: Clarendon Press. (*344*)

Ingold, C. T., and S. A. Hadland (1959) The ballistics of *Sordaria. New Phytol. 58*: 46–57. (*195*)

Ippen, A. T. (1966) *Estuary and Coast Line Hydrodynamics*. New York: McGraw-Hill. (*168, 170*)

Jacobs, M. R. (1954) The effect of wind sway on the form and development of *Pinus radiata* D. Don. *Australian J. Bot. 2*: 35–51. (*127*)

Jaffe, M. J. (1980) Morphogenetic responses of plants to mechanical stimuli or stress. *Bioscience 30*: 239–43. (*128*)

Jaffrin, M. Y., and A. H. Shapiro (1971) Peristaltic pumping. *Ann. Rev. Fluid Mech. 7*: 213–47. (*326*)

Jahn, T. L., and E. C. Bovee (1969) Protoplasmic movements within cells. *Physiol. Rev. 49*: 793–862. (*337*)

Jahn, T. L., and J. J. Votta (1972) Locomotion of protozoa. *Ann. Rev. Fluid Mech. 4*: 93–116. (*337, 352, 356*)

Janour, Z. (1951) Resistance of a plate in parallel flow at low Reynolds numbers. *Nat. Adv. Comm. Aero. Tech. Mem. No. 1316.* (*119*)

Jenkins, M. F. (1960) On the method by which Stenus and Dianous (Coleoptera: Staphylinidae) return to the banks of a pool. *Trans. Roy. Ent. Soc. Lond. 112*: 1–14. (*390*)

Jenkinson, I. R., and T. Wyatt (1992) Selection and control of Deborah numbers in plankton ecology. *J. Plankton Res. 14*: 1697–1721. (*18*)

Jensen, M. (1956) Biology and physics of locust flight. III. The aerodynamics of locust flight. *Phil. Trans. Roy. Soc. Lond. B239*: 511–52. (*239, 249, 255, 278*)

Jobin, W. R., and A. T. Ippen (1964) Ecological design of irrigation canals for snail control. *Science 145*: 1324–26. (*108*)

Johnson, A. S. (1988) Hydrodynamic study of the functional morphology of the benthic suspension feeder *Phoronopsis viridis* (Phoronida). *Mar. Biol. 100*: 117–26. (*217*)

Johnson, A. S. (1990) Flow around phoronids: Consequences of a neighbor to suspension feeders. *Limnol. Oceanogr. 35*: 1395–1401. (*217*)

Jones, S. A., and A. Fronek (1988) Effects of vibration on steady flow downstream of a stenosis. *J. Biomechanics 21*: 903–14. (*374*)

Jones, W. E., and A. Demetropoulos (1968) Exposure to wave action: Measurements of an important ecological parameter on rocky shores on Anglesey. *J. Exp. Mar. Biol. Ecol.* 2: 46–63. (*129, 130*)

Jonsson, P. R. (1989) Vertical distribution of planktonic ciliates: An experimental analysis of swimming behavior. *Mar. Ecol. Progr. Ser.* 52: 39–54. (*344*)

Jordan, G. H. (1987) Reconstruction of the fossa navicularis. *J. Urol.* 138: 102–104. (*396*)

Jørgensen, C. B., and H. U. Riisgård (1988) Gill pump characteristics of the soft clam *Mya arenaria*. *Mar. Biol.* 99: 107–10. (*328*)

Jumars, P. A., and A.R.M. Nowell (1984) Fluid and sediment dynamic effects on marine benthic community structure. *Amer. Zool.* 24: 45–55. (*169, 170, 172*)

Kaufmann, J. (1974) *Streamlined: Let's Read and Find Out.* New York: Thomas Crowell. (*106*)

Kingsolver, J. G., and T. L. Daniel (1979) On the mechanics and energetics of nectar feeding in butterflies. *J. Theor. Biol.* 76: 167–79. (*315, 330*)

Kirk, K. L. (1985) Water flows produced by *Daphnia* and *Diaptomus*: Implications for prey selection by mechanosensory predators. *Limnol. Oceanogr.* 30: 679–86. (*360*)

Klages, N.T.W., and J. Cooper (1992) Bill morphology and diet of a filter-feeding seabird: The broad-billed prion *Pachyptila vittata* at South Atlantic Gough Island. *J. Zool.* 227: 385–404. (*260, 387*)

Koehl, M.A.R. (1977) Effects of sea anemones on the flow forces they encounter. *J. Exp. Biol.* 69: 127–42. (*67, 115*)

Koehl, M.A.R. (1984) How do benthic organisms withstand moving water? *Amer. Zool.* 24: 57–70. (*125*)

Koehl, M.A.R., and R. S. Alberte (1988) Flow, flapping, and photosynthesis of *Nereocystis luetkeana*: A functional comparison of undulate and flat blade morphologies. *Mar. Biol.* 99: 435–44. (*119, 125, 129*)

Koehl, M.A.R., and J. R. Strickler (1981) Copepod feeding currents: Food capture at low Reynolds number. *Limnol. Oceanogr.* 26: 1062–73. (*360*)

Koehl, M.A.R., T. Hunter, and J. Jed (1991) How do body flexibility and length affect hydrodynamic forces on sessile organisms in waves versus in currents? *Amer. Zool.* 31: 60A. (*125, 375*)

Kokshaysky, N. V. (1979) Tracing the wake of a flying bird. *Nature* 279: 146–48. (*43, 276*)

Komar, P. D., A. P. Morse, L. F. Small, and S. W. Fowler (1981) An analysis of sinking rates of natural copepod and euphausid fecal pellets. *Limnol. Oceanogr.* 26: 172–80. (*344*)

Kramer, M. O. (1960) The dolphin's secret. *New Sci.* 7: 1118–20. (*152*)

Kramer, M. O. (1965) Hydrodynamics of the dolphin. *Adv. Hydrosci.* 2: 111–30. (*152*)

Kramer, P. J. (1959) Transpiration and the water economy of plants. In F. C. Steward, ed., *Plant Physiology*, vol. 2, pp. 607–726. New York: Academic Press. (*37*)

Kramer, P. J., and T. T. Koslowski (1960) *Physiology of Trees.* New York: McGraw-Hill. (*36, 37*)

Kristiansen, U. R., and O.K.O. Petterson (1978) Experiments on the noise heard by human beings when exposed to atmospheric winds. *J. Sound Vib. 58*: 285–92. *(95)*

Krogh, A. (1920) Studien über Tracheenrespiration. 2. Über Gasdiffusion in den Tracheen. *Pflugers Archiv Gesamte Physiol. Mensch. Tiere. 179*: 95–112. *(320)*

Krogh, A. (1941) *The Comparative Physiology of Respiratory Mechanisms.* Philadelphia: University of Pennsylvania Press. *(35, 196)*

Kromkamp, J., and A. E. Walsby (1990) A computer model of buoyancy and vertical migration in cyanobacteria. *J. Plankton Res. 12*: 161–84. *(30)*

Kryger, J., and H. U. Riisgard (1988) Filtration rate capacities in six species of European freshwater bivalves. *Oecologia 77*: 34–38. *(325)*

Kunkel, W. B. (1948) Magnitude and character of errors produced by shape factors in Stokes' law estimates of particle radius. *J. Appl. Phys. 19*: 1056–58. *(340)*

LaBarbera, M. (1977) Brachiopod orientation to water conditions. I. Theory, laboratory behavior, and field conditions. *Paleobiol. 3*: 270–87. *(71)*

LaBarbera, M. (1981) Water flow patterns in and around three species of articulate brachiopods. *J. Exp. Mar. Biol. Ecol. 55*: 185–206. *(44, 71)*

LaBarbera, M. (1983) Why the wheels won't go. *Amer. Nat. 121*: 395–408. *(187)*

LaBarbera, M. (1984) Feeding currents and particle capture mechanisms in suspension feeding animals. *Amer. Zool. 24*: 71–84. *(359)*

LaBarbera, M. (1990) Principles of design of fluid transport systems in zoology. *Science 249*: 992–1000. *(319, 320)*

LaBarbera, M., and S. Vogel (1982) The design of fluid transport systems in organisms. *Amer. Sci. 70*: 54–60. *(35, 39, 196)*

Lacoursière, J. O. (1992) A laboratory study of fluid flow and microhabitat selection by larvae of *Simulium vittatum* (Diptera: Simuliidae). *Can. J. Zool. 70*: 582–96. *(184)*

Lacoursière, J. O., and D. A. Craig (1993) Fluid transmission and filtration efficiency of the labral fans of black fly larvae (Diptera: Simuliidae): Hydrodynamic, mophological, and behavioural aspects. *Can. J. Zool. 71*: 148–62. *(357)*

Lamb, H. (1932) *Hydrodynamics.* 6th ed. Reprint. New York: Dover Publications, 1945. *(363)*

Lamotte, M., and J. Lescure (1989) The rheophilous and hygropetric tadpoles of the old and new world. *Ann. Sci. Nat. Zool. Biol. Anim. 10*: 111–22. *(178)*

Lane, D.J.W., A. R. Beaumont, and J. R. Hunter (1985) Byssus drifting and the drifting threads of the young post-larval mussel, *Mytilus edulis. Mar. Biol. 84*: 301–308. *(223)*

Lang, T. G. (1975) Speed, power, and drag measurements of dolphins and porpoises. In T. Y.-T. Wu, C. J. Brokaw, and C. Brennen, eds., *Swimming and Flying in Nature*, pp. 553–72. New York: Plenum Press. *(151)*

Lang, T. G., and K. S. Norris (1966) Swimming speed of a Pacific bottlenose porpoise. *Science 151*: 588–90. *(151)*

Langhaar, H. L. (1951) *Dimensional Analysis and Theory of Models.* New York: John Wiley. *(9)*

Langmuir, I. (1938) The speed of the deer fly. *Science 87*: 233–34. *(21)*

Larson, R. J. (1992) Riding Langmuir circulations and swimming in circles: A novel form of clustering behavior by the scyphozoan *Linuche unguiculata. Mar. Biol. 112*: 229–35. *(394)*

Lawton, R. O. (1982) Wind stress and elphin stature in a montane rain forest tree: An adaptive explanation. *Amer. J. Bot. 69*: 1224–30. *(127)*

Lehman, J. T. (1977) On calculating drag characteristics for decelerating zooplankton. *Limnol. Oceanogr. 22*: 170–72. *(139)*

Leibovich, S. (1983) The form and dynamics of Langmuir circulations. *Ann. Rev. Fluid Mech. 15*: 391–427. *(394)*

Leopold, L. B., and T. Maddock, Jr. (1953) The hydraulic geometry of stream channels and some physiographic implications. *U.S. Geol. Survey Professional Paper 252. (40)*

Levanoni, M. (1977) Study of fluid flow through scaled-up ink jet nozzles. *IBM J. Res. Devel. 21*: 56–68. *(88)*

Leyton, L. (1975) *Fluid Behavior in Biological Systems*. Oxford: Clarendon Press. *(6, 164, 198, 306)*

Lidgard, S. (1981) Water flow, feeding, and colony form in an encrusting cheilostome. In G. P. Larwood and C. Nielsen, eds, *Recent and Fossil Bryozoa*, pp. 135–42. Fredensborg, Denmark: Olsen and Olsen. *(44, 191)*

Lightfoot, E. N. (1977) *Transport Phenomena and Living Systems*. New York: Wiley-Interscience. *(313)*

Lighthill, M. J. (1969) Hydromechanics of aquatic animal propulsion. *Ann. Rev. Fluid Mech. 1*: 413–46. *(281)*

Lighthill, M. J. (1975) Aerodynamic aspects of animal flight. In T. Y.-T. Wu, C. J. Brokaw, and C. Brennen, eds., *Swimming and Flying in Nature*, vol. 2, pp. 423–91. New York: Plenum Press. *(376)*

Lillywhite, H. B. (1987) Circulatory adaptations of snakes to gravity. *Amer. Zool. 27*: 81–95. *(56)*

Linsenmair, K. E., and R. Jander (1963) Das 'Entspannungsschwimmen' von *Velia* und *Stenus. Naturwiss. 6*: 174–5. *(390)*

Liron, N. (1976) On peristaltic flow and its efficiency. *Bull. Math. Biol. 38*: 573–96. *(326)*

Liron, N., and F. A. Meyer (1980) Fluid transport in a thick layer above an active ciliated surface. *Biophys. J. 30*: 463–72. *(314)*

Lissaman, P.B.S. (1983) Low-Reynolds-number airfoils. *Ann. Rev. Fluid Mech. 15*: 223–39. *(244)*

Lissaman, P.B.S., and C. Shollenberger (1970) Formation flight of birds. *Science 168*: 1003–1005. *(288)*

Lochhead, J. H. (1977) Unsolved problems of interest in the locomotion of Crustacea. In T. J. Pedley, ed., *Scale Effects in Animal Locomotion*, pp. 257–68. London: Academic Press. *(146)*

Loudon, C. (1990) Empirical test of filtering theory: Particle capture by rectangular-mesh nets. *Limnol. Oceanogr. 35*: 143–48. *(357)*

Lovvorn, J. R., D. R. Jones, and R. W. Blake (1991) Mechanics of underwater locomotion of diving ducks: Drag, buoyancy, and acceleration in a size gradient of species. *J. Exp. Biol. 159*: 89–108. *(143, 147)*

Lowry, W. P. (1967) *Weather and Life: An Introduction to Biometeorology*. New York: Academic Press. *(171)*

Lugt, H. J. (1983a) *Vortex Flow in Nature and Technology*. New York: Wiley-Interscience. *(204, 216)*

Lugt, H. J. (1983b) Autorotation. *Ann. Rev. Fluid Mech. 15*: 123–47. *(228)*

Lund, J.W.G. (1959) Buoyancy in relation to the ecology of the freshwater phytoplankton. *Br. Phycol. Bull. (Phycol. Bull.)* 7: 1–17. (*344*)

Lundegårdh, H. (1966) *Plant Physiology*. New York: American Elsevier. (*36, 37*)

Lüscher, M. (1961) Air-conditioned termite nests. *Sci. Amer.* 205 (1): 138–45. (*71*)

McCarthy, M. J., and N. A. Molloy (1974) Review of stability of liquid jets and the
influence of nozzle design. *Chem. Eng. J.* 7: 1–20. (*396*)

McCave, I. N., ed. (1976) *The Benthic Boundary Layer*. New York: Plenum Press. (*171*)

McCleave, J. D. (1980) Swimming performance of European eel, *Anguilla anguilla*
(L.) elvers. *J. Fish Biol.* 16: 445–52. (*281*)

McCutchen, C. W. (1977a) The spinning rotation of ash and tulip tree samaras.
Science 197: 691–92. (*228*)

McCutchen, C. W. (1977b) Froude propulsive efficiency of a small fish, measured by
wake visualization. In T. J. Pedley, ed., *Scale Effects in Animal Locomotion*,
pp. 339–63. London: Academic Press. (*257, 283*)

Macdougal, D. T. (1925) Reversible variations in volume, pressure, and movement
of sap in trees. *Carnegie Inst. Wash.* 395: 1–90. (*316*)

McGahan, J. (1973) Gliding flight of the Andean condor in nature. *J. Exp. Biol. 58*:
225–37. (*243*)

McMahon, T. A., and J. T. Bonner (1983) *On Size and Life*. New York: Scientific
American Books. (*9*)

McMasters, J. H. (1986) Reflections of a paleoaerodynamicist. *Persp. Biol. Med.* 29:
331–84. (*236, 243*)

McNown, J. S., and J. Malaika (1950) Effects of particle shape on settling velocity at
low Reynolds numbers. *Trans. Amer. Geophys. Union 31*: 74–82. (*338*)

McShaffrey, O., and W. P. McCafferty (1987) The behavior and form of *Psephenus
herricki* (DeKay) (Coleoptera: Psephenidae) in relation to water flow. *Freshwater
Biol.* 18: 319–24. (*179*)

Madin, L. P. (1990) Aspects of jet propulsion in salps. *Can. J. Zool.* 68: 765–77. (*77,
79, 80*)

Martel, A., and F.-S. Chia (1991) Foot-raising behavior and active participation
during the initial phase of post-metamorphic drifting in the gastropod, Lacuna
spp. *Mar. Ecol. Progr. Ser.* 72: 247–54. (*223*)

Massey, B. S. (1989) *Mechanics of Fluids*. 6th ed. London: Van Nostrand Reinhold. (*6,
47, 82, 294, 306, 325, 371, 375, 395*)

Masters, W. H., and V. E. Johnson (1966) *Human Sexual Response*. Boston: Little,
Brown and Co. (*323*)

Matlack, G. R. (1987) Diaspore size, shape, and fall behavior in wind-dispersed plant
species. *Amer. J. Bot.* 74: 1150–60. (*257*)

Matlack, G. R. (1989) Secondary dispersal of seed across snow in *Betula lenta*, a gap-
colonizing tree species. *J. Ecol.* 77: 853–69. (*194*)

Matsumoto, G. I., and W. M. Hamner (1988) Modes of water manipulation by the
lobate ctenophore *Leucothea* sp. *Mar. Biol.* 97: 551–58. (*77*)

Matsuzaki, Y. (1986) Self-excited oscillation of a collapsible tube conveying fluid. In
G. W. Schmid-Schönbein, S. L.-Y. Woo, and B. W. Zweifach, eds., *Frontiers in
Biomechanics*, pp. 342–50. New York: Springer-Verlag. (*316*)

Maull, D. J., and P. W. Bearman (1964) The measurement of the drag of bluff
bodies by the wake traverse method. *J. Roy. Aero. Soc.* 68: 843. (*77*)

Mayhead, G. J. (1973) Some drag coefficients for British forest trees derived from wind tunnel studies. *Agric. Meteorol.* 12: 123–30. (*119, 121*)

Maynard Smith, J. (1952) The importance of the nervous system in the evolution of animal flight. *Evolution* 6: 127–29. (*256*)

Meidner, H., and T. A. Mansfield (1968) *Physiology of Stomata.* London: McGraw-Hill. (*197, 198*)

Merritt, R. W., and K. W. Cummins, eds. (1984) *An Introduction to the Aquatic Insects of North America.* 2d ed. Dubuque, Iowa: Kendall-Hunt Publishing Co. (*177, 390*)

Meyhöfer, E. (1985) Comparative pumping rates in suspension-feeding bivalves. *Mar. Biol.* 85: 137–42. (*189, 325*)

Middleman, S. (1972) *Transport Phenomena in the Cardiovascular System.* New York: John Wiley. (*313*)

Mill, P. J., and R. S. Pickard (1975) Jet propulsion in isopteran dragonfly larvae. J. Comp. Physiol. A97: 329–38. (*77*)

Miller, K. F., C. P. Quine, and J. Hunt (1987) The assessment of wind exposure for forestry in upland Britain. *Forestry* 60: 179–92. (*128*)

Miller, P. L. (1966) The supply of oxygen to the active flight muscles of some large beetles. *J. Exp. Biol.* 45: 285–304. (*71*)

Millward, A., and M. A. Whyte (1992) The hydrodynamic characteristics of six scallops of the superfamily Pectinacea, class Bivalva. *J. Zool.* 227: 547–66. (*249, 252*)

Milne, L. J., and M. Milne (1978) Insects of the water surface. *Sci. Amer.* 238(4): 134–42. (*389*)

Milnor, W. R. (1990) *Cardiovascular Physiology.* New York: Oxford University Press. (*37, 62*)

Mises, R. von (1945) *Theory of Flight.* Reprint. New York: Dover Publications, 1959. (*6, 133, 236, 242, 267, 275*)

Monteith, J. L., and M. H. Unsworth (1990) *Principles of Environmental Physics.* 2d ed. London: Edward Arnold. (*168, 170, 198*)

Morgan, K. T., J. S. Kimbell, T. M. Monticello, A. L. Patra, and A. Fleishman (1991) Studies of inspiratory airflow patterns in the nasal passages of the F344 rat and rhesus monkey using nasal molds: Relevance to formaldehyde toxicity. *Toxicol. Appl. Pharmacol.* 110: 223–40. (*311*)

Morgan, R. P. II, R. E. Ulanowicz, V. J. Rasin, Jr., L. A. Noe, and G. B. Gray (1976) Effects of shear on eggs and larvae of striped bass, *Morone saxatilis*, and white perch, *M. americana. Trans. Amer. Fish. Soc.* 1976: 149–54. (*187*)

Morris, M. J., G. Gust, and J. J. Tores (1985) Propulsion efficiency and cost of transport for copepods: A hydromechanical model of crustacean swimming. *Mar. Biol.* 86: 283–95. (*352*)

Morris, M. J., K. Kohlhage, and G. Gust (1990) Mechanics and energetics of swimming in the small copepod, *Acanthocyclops robustus* (Cyclopoida). *Mar. Biol.* 107: 83–92. (*365*)

Mortimer, C. H. (1974) Lake hydrodynamics. *Mitt. Internat. Verein. Limnol.* 20: 124–97. (*393*)

Moslehi, F., J. R. Ligas, M. A. Pisani, and M.A.F. Epstein (1989) The unsteady form of the Bernoulli equation for estimating pressure drop in the airways. *Resp. Physiol.* 76: 319–26. (*377*)

Mulhearn, P. J., H. J. Banks, J. J. Finnigan, and P. C. Annis (1976) Wind forces and their influence on gas loss from grain storage structures. *J. Stored Prod. Res. 12*: 129–42. *(83)*

Mullineaux, L. S., and C. A. Butman (1990) Recruitment of encrusting benthic invertebrates in boundary layer flows: A deep-water experiment on cross sea-mount, north Pacific Ocean. *Limnol. Oceanogr. 35*: 409–23. *(184)*

Mullineaux, L. S., and C. A. Butman (1991) Initial contact exploration and attachment of barnacle, *Balanus amphitrite*, cyprids settling in flow. *Mar. Biol. 110*: 93–104. *(185)*

Munk, W. H., and G. A. Riley (1952) Absorption of nutrients by aquatic plants. *J. Mar. Res. 11*: 215–40. *(346)*

Murray, C. D. (1926) The physiological principle of minimum work. I. The vascular system and the cost of flood volume. *Proc. Nat. Acad. Sci. USA 12*: 207–14. *(188, 317)*

Muus, B. J. (1968) A field method for measuring "exposure" by means of plaster balls. *Sarsia 34*: 61–68. *(129)*

Nachtigall, W. (1960) Über Kinematik, Dynamik und Energetik des Schwimmens einheimischer Dytisciden. *Z. Vergl. Physiol. 43*: 48–180. *(366)*

Nachtigall, W. (1964) Zur Aerodynamic des Coleopterenflugs: Wirken die Elytren als Tragflügel? *Verh. Deutsch. Zool. Ges. (Kiel) 58*: 319–26. *(249)*

Nachtigall, W. (1965) Die aerodynamische Funktion der Schmetterlingsschuppen. *Naturwiss. 52*: 216–17. *(245)*

Nachtigall, W. (1974) Locomotion: Mechanics and hydrodynamics of swimming in aquatic insects. In M. Rockstein, ed., *The Physiology of Insects*, 2d ed., vol. 3, pp. 381–432. New York: Academic Press. *(352)*

Nachtigall, W. (1977a) Swimming mechanics and energetics of locomotion of variously sized water beetles—Dytiscidae, body length 2 to 35 mm. In T. J. Pedley, ed., *Scale Effects in Animal Locomotion*, pp. 269–83. London: Academic Press. *(143, 145)*

Nachtigall, W. (1977b) Die aerodynamische Polare des *Tipula*-Flügels und eine Einrichtung zur halbautomatischen Polarenaufnahme. *Fortschr. Zool. 24*, 2/3: 347–52. *(249)*

Nachtigall, W. (1979a) Gleitflug des Flugbeutlers *Petaurus breviceps papuanus* (Thomas). II. Filmanalysen zur Einstellung von Gleitbahn und Rumpf sowie zur Steuerung des Gleitflugs. *J. Comp. Physiol. 133*: 89–95. *(250)*

Nachtigall, W. (1979b) Gleitflug des Flugbeutlers *Petaurus breviceps papuanus* (Thomas). III. Modellmessungen zum Einfluss des Fellbesatzes auf Un-strömung und Luftkrafterzeugung. *J. Comp. Physiol. 133*: 339–49. *(249, 250)*

Nachtigall, W. (1980) Mechanics of swimming in water-beetles. In H. Y. Elder and E. R. Trueman, eds., *Aspects of Animal Movement*, pp. 107–24. Cambridge, U.K.: Cambridge University Press. *(154, 287, 350)*

Nachtigall, W. (1981) Hydromechanics and biology. *Biophys. Struct. Mech. 8*: 1–22. *(154)*

Nachtigall, W., and D. Bilo (1965) Die Strömungsmechanik des *Dytiscus*-Rumpfes. *Z. Vergl. Physiol. 50*: 371–401. *(145)*

Nachtigall, W., and D. Bilo (1975) Hydrodynamics of the body of *Dytiscus marginalis*

(Dytiscidae, Coleoptera). In T. Y.-T. Wu, C. J. Brokaw, and C. Brennen, eds., *Swimming and Flying in Nature*, pp. 585–95. New York: Plenum Press. (*145*)

Nachtigall, W., and D. Bilo (1980) Strömungsanpassung des Pinguins beim Schwimmen unter Wasser. *J. Comp. Physiol. 137*: 17–26. (*143, 149*)

Nachtigall, W., and U. Hanauer-Thieser (1992) Flight of the honeybee. V. Drag and lift coefficients of the bee's body: Implications for flight dynamics. *J. Comp. Physiol. B162*: 267–77. (*12, 239*)

Nachtigall, W., and B. Kempf (1971) Vergleichende Untersuchung zur Flugbiologischen Funktion des Daumenfittich (Alula spuria) bei Vögeln. I. Der Daumenfittich als Hochauftriebserzeuger. *Z. Vergl. Physiol. 71*: 326–41. (*242, 248*)

Nachtigall, W., R. Grosch, and T. Schultze-Westrum (1974) Gleitflug des Flugbeutlers *Petaurus breviceps papuanus* (Thomas). I. Flugverhalten und Flugsteuerung. *J. Comp. Physiol. 92*: 105–15. (*249*)

Newman, B. G., S. B. Savage, and D. Schouella (1977) Model tests on a wing section of an *Aeschna* dragonfly. In T. J. Pedley, ed., *Scale Effects in Animal Locomotion*, pp. 445–77. London: Academic Press. (*245*)

Nielsen, A. (1950) The torrential invertebrate fauna. *Oikos 2*: 176–96. (*177, 180*)

Nielsen, N. F., P. S. Larsen, H. U. Riisgård, and C. B. Jørgensen (1993) Fluid motion and particle retention in the gill of *Mytilus edulis*: Video recordings and numerical modelling. *Mar. Biol. 116*: 61–71. (*349*)

Niklas, K. J. (1985) The aerodyamics of wind pollination. *Bot. Rev. 51*: 328–86. (*43, 187*)

Niklas, K. J. (1992) *Plant Biomechanics*. Chicago: University of Chicago Press. (*43, 109, 121, 128, 322, 344*)

Niklas, K. J., and S. L. Buchmann (1985) Aerodynamics of wind pollination in *Simmondsia chinensis* (Link) Schneider. *Amer. J. Bot. 72*: 530–39. (*187*)

Niklas, K. J., and S. L. Buchmann (1987) The aerodynamics of pollen capture in two sympatric *Ephedra* species. *Evolution 41*: 104–23. (*187*)

Nobel, P. S. (1974) Boundary layers of air adjacent to cylinders: Estimation of effective thickness and measurement on plant material. *Pl. Physiol. 54*: 177–81. (*163*)

Nobel, P. S. (1975) Effective thickness and resistance of the air boundary layer adjacent to spherical plant parts. *J. Exp. Bot. 26*: 120–30. (*163*)

Norberg, R. Å. (1973) Autorotation, self-stability, and structure of single-winged fruits and seeds (samaras) with comparative remarks on animal flight. *Biol. Rev. 48*: 561–96. (*264, 272, 274*)

Norberg, U. M. (1975) Hovering flight of the pied flycatcher (*Ficedula hypoleuca*). In T. Y.-T. Wu, C. J. Brokaw, and C. Brennen, eds., *Swimming and Flying in Nature*. Vol. 2, pp. 869–81. New York: Plenum Press. (*278*)

Norberg, U. M. (1976) Aerodynamics of hovering flight in the long-eared bat *Plecotus auritus*. *J. Exp. Biol. 65*: 459–70. (*278*)

Norberg, U. M. (1990) *Vertebrate Flight: Mechanics, Physiology, Morphology, Ecology and Evolution*. Berlin: Springer-Verlag. (*236, 241, 248, 257, 259, 266, 275*)

Nowell, A.R.M., and P. A. Jumars (1984) Flow environments of aquatic benthos. *Ann. Rev. Ecol. Syst. 15*: 303–28. (*170, 219*)

Odendaal, F. J., P. Turchin, G. Hoy, P. Wilkens, J. Wells, and G. Schroeder (1992) *Bullia digitalis* (Gastropoda) actively pursues moving prey by swash riding. *J. Zool. 228*: 103–13. *(182)*

O'Dor, R. K., and D. M. Webber (1986) The constraints on cephalopods: Why squid aren't fish. *Can J. Zool. 64*: 1591–1605. *(80)*

Oke, T. R. (1978) *Boundary Layer Climates.* London: Methuen and Co. *(169)*

Oliver, H. R., and G. J. Mayhead (1974) Wind measurements in a pine forest during a destructive gale. *Forestry 47*: 185–94. *(167)*

Oliver, J. A. (1951) "Gliding" in amphibians and reptiles, with a remark on arboreal adaptation in the lizard *Anolis carolinensis* Voight. *Amer. Nat. 85*: 171–76. *(257, 258)*

O'Neill, P. L. (1978) Hydrodynamic analysis of feeding in sand dollars. *Oecologia 34*: 157–74. *(246)*

The Open University (1989) *Waves, Tides and Shallow-Water Processes.* Oxford: Pergamon Press. *(172)*

Orton, L. S., and P. F. Brodie (1987) Engulfing mechanics of fin whales. *Can. J. Zool. 65*: 2898–2907. *(69)*

Paine, R. T. (1988) Habitat suitability and local population persistence of the sea palm *Postelsia palmaeformis. Ecology 69*: 1787–94. *(122)*

Palmer, E., and G. Weddell (1964) The relationship between structure, innervation, and function of the skin of the bottlenose dolphin (*Tursiops truncatus*). *Proc. Zool. Soc. Lond. 143*: 553–68. *(29)*

Palumbi, S. R. (1986) How body plans limit acclimation: Responses of a demosponge to wave force. *Ecology 67*: 208–14. *(128)*

Pankhurst, R. C., and D. W. Holder (1952) *Wind-Tunnel Techniques.* London: Sir Isaac Pitman & Sons. *(56, 96)*

Parlange, J.-Y., and P. E. Waggoner (1972) Boundary layer resistance and temperature distribution on still and flapping leaves. II. Field experiments. *Pl. Physiol. 50*: 60–63. *(163)*

Parrish, J. K., and W. K. Kroen (1988) Sloughed mucus and drag reduction in a school of Atlantic silversides, *Menidia menidia. Mar. Biol. 97*: 165–69. *(152)*

Parry, D. A. (1949) The structure of whale blubber, and a discussion of its thermal properties. *Quart. J. Microscop. Sci. 90*: 13–26. *(29)*

Passioura, J. B. (1972) The effect of root geometry on the yield of wheat growing on stored water. *Aust. J. Agric. Res. 23*: 745–52. *(38, 323)*

Pawlik, J. R., C. A. Butman, and V. R. Starczak (1991) Hydrodynamic facilitation of gregarious settlement of a reef-building tube worm. *Science 251*: 421–24. *(185)*

Paw U, K. T. (1983) The rebound of particles from natural surfaces. *J. Colloid Interface Sci. 93*: 442–52. *(186)*

Payne, R., O. Brazier, E. M. Dorsey, J. S. Perkins, V. J. Rowntree, and A. Titus (1983) External features in southern right whales (*Eubalaena australis*) and their use in identifying individuals. In R. Payne, ed. *Communication and Behavior of Whales,* pp. 371–445. Washington, D.C.: American Association for the Advancement of Science. *(181)*

Pearse, J. S., I. Bosch, V. B. Pearse, and L. V. Basch (1991) Bacterivory by bipinnarias: In the antarctic but not in California. *Amer. Zool. 31*: 8A. *(29)*

Pearson, G. A., and L. V. Evans (1990) Settlement and survival of *Polysiphonia lanosa*

(Ceramiales) spores on *Ascophyllum nodosum* and *Fucus vesiculosus* (Fucales). *J. Phycol. 26*: 597–603. (*187*)

Pedgley, D. (1982) *Windborne Pests and Diseases: Meteorology of Airborne Organisms.* Chichester, U.K.: Ellis Horwood. (*221, 223*)

Pedley, T. J., and J. O. Kessler (1992) Hydrodynamic phenomena in suspensions of swimming microorganisms. *Ann. Rev. Fluid Mech. 24*: 313–58. (*224*)

Pedrós-Alío, C., J. Mas, J. M Gasol, and R. Guerrero (1989) Sinking speeds of free-living phototrophic bacteria determined with covered and uncovered traps. *J. Plankton Res. 11*: 887–906. (*344*)

Pennak, R. W. (1978) *Fresh Water Invertebrates of the United States.* 2d ed. New York: John Wiley. (*179, 180*)

Pennington, R. H. (1965) *Introductory Computer Methods and Numerical Analysis.* 2d ed. New York: John Wiley. (*309*)

Pennycuick, C. J. (1960) Gliding flight of the fulmer petrel. *J. Exp. Biol. 37*: 330–38. (*248*)

Pennycuick, C. J. (1968) A wind-tunnel study of gliding flight in the pigeon *Columba livia. J. Exp. Biol. 49*: 509–26. (*133, 148, 243*)

Pennycuick, C. J. (1971) Control of gliding angle in Rüppell's griffon vulture *Gyps rueppellii. J. Exp. Biol. 55*: 39–46. (*148, 248*)

Pennycuick, C. J. (1972) *Animal Flight.* London: Edward Arnold. (*107, 259*)

Pennycuick, C. J. (1975) Mechanics of flight. In D. S. Farner and J. R. King, eds., *Avian Biology*, pp. 1–75. New York: Academic Press. (*259*)

Pennycuick, C. J. (1982) The flight of petrels and albatrosses (Procellariiformes), observed in South Georgia and its vicinity. *Phil. Trans. Roy. Soc. B300*: 73–106. (*242, 259*)

Pennycuick, C. J. (1988) *Conversion Factors: SI Units and Many Others.* Chicago: University of Chicago Press. (*9*)

Pennycuick, C. J. (1989) *Bird Flight Performance: A Practical Calculation Manual.* New York: Oxford University Press. (*138, 221, 236, 241, 248*)

Pennycuick, C. J. (1992) *Newton Rules Biology.* New York: Oxford University Press. (*9*)

Pennycuick, C. J., H. H. Obrecht, and M. R. Fuller (1988) Empirical estimates of body drag of large waterfowl and raptors. *J. Exp. Biol. 135*: 253–64. (*148*)

Perrier, E. R., A. Aston, and G. F. Arkin (1973) Wind flow characteristics of a soybean leaf compared with a leaf model. *Pl. Physiol. 28*: 106–12. (*163, 376*)

Petroski, H. (1991) Still twisting. *Amer. Sci. 79*: 398–401. (*374*)

Pinder, A. W., and M. E. Feder (1990) Effect of boundary layers on cutaneous gas exchange. *J. Exp. Biol. 143*: 67–80. (*200*)

Plesset, M. S., C. G. Whipple, and H. Winet (1975) Analysis of the steady state of the bioconvection in swarms of swimming micro-organisms. In T. Y.-T. Wu, C. J. Brokaw, and C. Brennen, eds., *Swimming and Flying in Nature*, vol. 1, pp. 339–60. New York: Plenum Press. (*346*)

Plotnick, R. E. (1985) Lift based mechanisms for swimming in eurypterids and portunid crabs. *Trans. Roy. Soc. Edinburgh 76*: 325–37. (*140, 286*)

Podolsky, R. D., and R. B. Emlet (1993) Separating the effects of temperature and viscosity on swimming and water movement by sand dollar larvae (*Dendraster excentricus*). *J. Exp. Biol. 176*: 207–21. (*31*)

Powell, M. S., and N.K.H. Slater (1982) Removal rates of bacterial cells from glass surfaces by shear. *Biotech. Bioengin. 24*: 2527–37. *(176)*

Prandtl, L., and O. G. Tietjens (1934) *Applied Hydro- and Aeromechanics*. Reprint. New York: Dover Publications, 1957. *(6, 56, 72, 77, 293, 302, 370)*

Prange, H. D., and K. Schmidt-Nielsen (1970) The metabolic cost of swimming in ducks. *J. Exp. Biol. 53*: 763–77. *(107, 286, 384)*

Prothero, J., and A. C. Burton (1961) The physics of blood flow in the capillaries. II. The nature of the motion. *Biophys. J. 1*: 565–79. *(313)*

Purcell, E. M. (1977) Life at low Reynolds number. *Amer. J. Physics 45*: 3–11. *(332)*

Quetin, L. B., and R. M. Ross (1984) Depth distribution of developing *Euphausia superba* embryos, predicted from sinking rates. *Mar. Biol. 79*: 47–53. *(30, 344, 345)*

Rainey, R. C. (1963) Meteorology and the migration of desert locusts. *Wld. Met. Org. Tech. Note 54*. *(223)*

Ramus, J., B. E. Kenney, and E. J. Shaughnessy (1989) Drag reducing properties of microalgal exopolymers. *Biotech. Bioengin. 33*: 550–57. *(152)*

Rand, R. H. (1983) Fluid mechanics of green plants. *Ann. Rev. Fluid Mech. 15*: 29–45. *(198)*

Randall, D. J., and C. Daxboeck (1984) Oxygen and carbon dioxide rtransfer across fish gill. In W. S. Hoar and D. J. Randall, eds., *Fish Physiology*, vol. 10, pp. A263–314. New York: Academic Press. *(68)*

Ray, A. J., and R. C. Aller (1985) Physical irrigation of relict burrows: Implications for sediment chemistry. *Mar. Geol. 62*: 371–79. *(72)*

Rayner, J.M.V. (1977) The intermittent flight of birds. In T. J. Pedley, ed., *Scale Effects in Animal Locomotion*, pp. 437–43. London: Academic Press. *(240)*

Rayner, J.M.V. (1979) A new approach to animal flight mechanics. *J. Exp. Biol. 80*: 17–54. *(276)*

Rayner, J.M.V. (1985) Bounding and undulating flight in birds *J. Theor. Biol. 117*: 47–77. *(240)*

Rees, C.J.C. (1975) Aerodynamic properties of an insect wing section and a smooth airfoil compared. *Nature 258*: 141–42. *(245)*

Rees, D., and J. Grace (1980a) The effects of wind on the extension growth of *Pinus contorta* Douglas. *Forestry 53*: 145–53. *(128)*

Rees, D., and J. Grace (1980b) The effects of shaking on the extension growth of *Pinus contorta* Douglas. *Forestry 53*: 155–66. *(128)*

Rees, G. (1980) Factors affecting the sedimentation rate of marine fungal spores. *Bot. Mar. 23*: 375–85. *(344)*

Reif, W.-E. (1985) Morphology and hydrodynamic effects of the scales of fast swimming sharks. *Fortschr. Zool. 30*: 483–85. *(153)*

Reiner, M. (1964) The Deborah number. *Physics Today 17*: 62. *(18)*

Reiswig, H. M. (1971) *In situ* pumping activity of tropical demospongiae. *Mar. Biol. 9*: 38–50. *(189)*

Reiswig, H. M. (1974) Water transport, respiration, and energetics of three tropical marine sponges. *J. Exp. Mar. Biol. Ecol. 14*: 231–49. *(39, 325)*

Reiswig, H. M. (1975a) The aquiferous systems of three marine demospongiae. *J. Morph. 145*: 493–502. *(36, 39)*

Reiswig, H. M. (1975b) Bacteria as food for temperate-water marine sponges. *Can J. Zool.* *53*: 582–89. (*189, 357*)

Resh, V. H., and J. O. Solem (1984) Phylogenetic relationships and evolutionary adaptations of aquatic insects. In R. W. Merritt and K. W. Cummins, eds., *An Introduction to the Aquatic Insects of North America*, 2d ed., pp. 66–75. Dubuque, Iowa: Kendall-Hunt Publishing Co. (*178, 180*)

Reynolds, O. (1883) An experimental investigation of the circumstances which determine whether the motion of water shall be direct or sinuous, and the law of resistance in parallel channels. *Trans. Roy. Soc. Lond. 174*: 935–82. (*47, 84, 298*)

Riebesell, U. (1992) The formation of large marine snow and its sustained residence in surface waters. *Limnol. Oceanogr. 37*: 63–76. (*344*)

Riedl, R. J. (1971a) Water movement. In O. Kinne, ed., *Marine Ecology*, vol. 1, pt. 2, pp. 1085–88, 1124–56. London: Wiley-Interscience. (*109*)

Riedl, R. J. (1971b) How much water passes through sandy beaches? *Int. Rev. Geo. Hydrobiol. 56*: 923–46. (*307*)

Riedl, R. J., and R. Machan (1972) Hydrodynamic patterns of lotic sands and their bioclimatological implications. *Mar. Biol. 13*: 179–209. (*307*)

Riisgård, H. U. (1989) Properties and energy cost of the muscular piston pump in the suspension feeding polychaete *Chaetopterus variopedatus*. *Mar. Ecol. Progr. Ser. 56*: 157–68. (*327, 328*)

Riisgård, H. U. (1991) Suspension feeding in the polychaete *Nereis diversicolor*. *Mar. Ecol. Progr. Ser. 70*: 29–37. (*327*)

Riley, J. J., M. Gad-el-Hak, and R. W. Metcalfe (1988) Compliant coatings. *Ann. Rev. Fluid Mech. 20*: 393–420. (*152*)

Roberts, A. M. (1981) Hydrodynamics of protozoan swimming. In M. Levandowsky and S. H. Hutner, eds., *Biochemistry and Physiology of Protozoa*, 2d ed., vol. 2, pp. 5–66. New York: Academic Press. (*356*)

Rose, C. W. (1966) *Agricultural Physics*. Oxford: Pergamon Press. (*167*)

Rosenberg, N. J., B. L. Blad, and S. B. Verma (1983) *Microclimate: The Biological Environment*. 2d ed. New York: Wiley-Interscience. (*168*)

Rouse, H. (1938) *Fluid Mechanics for Hydraulic Engineers*. Reprint. New York: Dover Publications, 1961. (*6, 89, 160*)

Rouse, H. (1946) *Elementary Mechanics of Fluids*. Reprint. New York: Dover Publications, 1978. (*6*)

Rouse, H., and S. Ince (1957) *History of Hydraulics*. Reprint. New York: Dover Publications, 1963. (*305*)

Rubenstein, D. I., and M.A.R. Koehl (1977) The mechanisms of filter feeding: Some theoretical considerations. *Amer. Natur. 111*: 981–94. (*358*)

Rüffer, U., and W. Nultsch (1985) High-speed cinematographic analysis of the movement of *Chlamydomonas*. *Cell Motil. 5*: 251–63. (*353*)

Ruud, J. T. (1965) The ice fish. *Sci. Amer. 213(5)*: 108–14. (*29*)

Sanderson, S. L., and R. Wassersug (1990) Suspension-feeding vertebrates. *Sci. Amer. 262 (3)*: 96–101. (*188, 356*)

Satterlie, R. A., M. LaBarbera, and A. N. Spencer (1985) Swimming in the pteropod mollusc, *Clione limacina*. I. Behavior and morphology. *J. Exp. Biol. 116*: 189–204. (*286*)

Schmidt-Nielsen, K. (1972) *How Animals Work.* Cambridge, U.K.: Cambridge University Press. (*311*)

Schmitz, F. W. (1960) *Aerodynamik des Flugmodells.* Duisberg, Germany: Carl Lange. (*244*)

Scholander, P. F. (1957) The wonderful net. *Sci. Amer. 196(4)*: 96–107. (*28*)

Schumacher, G. J., and H. A. Whitford (1965) Respiration and P³² uptake in various species of freshwater algae as affected by a current. *J. Phycol. 1*: 78–80. (*199*)

Scorer, R. S. (1978) *Environmental Aerodynamics.* Chichester, U.K.: Ellis Horwood. (*393*)

Sculthorpe, C. D. (1967) *The Biology of Aquatic Vascular Plants.* London: Edward Arnold. (*199*)

Seeley, L. E., R. L. Hummel, and J. W. Smith (1975) Experimental velocity profiles in laminar flow around spheres at intermediate Reynolds numbers. *J. Fluid Mech. 68*: 591–608. (*95, 134*)

Sellers, W. D. (1965) *Physical Climatology.* Chicago: University of Chicago Press. (*169*)

Seter, D., and A. Rosen (1992) Study of the vertical autorotation of a single-winged samara. *Biol. Rev. 67*: 175–97. (*272*)

Shapiro, A. H. (1961) *Shape and Flow.* Garden City, N.Y.: Doubleday. (*6, 102*)

Shapiro, A. H. (1962) Bath-tub vortex. *Nature 196*: 1080–81. (*205*)

Shariff, K., and A. Leonard (1992) Vortex rings. *Ann. Rev. Fluid Mech. 24*: 235–79. (*211*)

Shaw, R. (1960) The influence of hole dimensions on static pressure measurements. *J. Fluid Mech. 7*: 550–64. (*72*)

Sheath, R. G., and J. A. Hambrook (1988) Mechanical adaptations to flow in freshwater red algae. *J. Phycol. 24*: 107–11. (*119*)

Sherman, T. F. (1981) On connecting large vessels to small: The meaning of Murray's law. *J. Gen. Physiol. 78*: 431–53. (*320*)

Sherwood, T. K., R. L. Pigford, and C. R. Wilke (1975) *Mass Transfer.* New York: McGraw-Hill. (*346*)

Shimeta, J., and P. A. Jumars (1991) Physical mechanisms and rates of particle capture by suspension feeders. *Oceanogr. Mar. Biol. Ann. Rev. 29*: 191–257. (*192, 358*)

Sigurdsson, J. B., C. W. Titman, and P. A. Davies (1976) The dispersal of young post-larval molluscs by byssus threads. *Nature 262*: 386–87. (*223*)

Silvester, N. R. (1983) Some hydrodynamic aspects of filter feeding with rectangular-mesh nets. *J. Theor. Biol. 103*: 265–86. (*359*)

Silvester, N. R., and M. A. Sleigh (1985) The forces on microorganisms at surfaces in flowing water. *Freshwater Biol. 15*: 433–48. (*187*)

Simiu, E., and R. H. Scanlan (1978) *Wind Effects on Structures.* New York: John Wiley. (*371*)

Sleigh, M. A. (1978) Fluid propulsion by cilia and flagella. In K. Schmidt-Nielsen, L. Bolis, and S.H.P. Maddrell, eds., *Comparative Physiology: Water, Ions, and Fluid Mechanics*, pp. 255–66. Cambridge, U.K.: Cambridge University Press. (*355*)

Sleigh, M. A., and D. I. Barlow (1976) Collection of food by *Vorticella. Trans. Amer. Microscopical Soc. 95*: 482–86. (*209*)

Sleigh, M. A., and D. I. Barlow (1980) Metachronism and control of locomotion in animals with many propulsive structures. In H. Y. Elder and E. R. Trueman, eds., *Aspects of Animal Movement*, pp. 49–70. Cambridge, U.K.: Cambridge University Press. (*154*)

Sleigh, M. A., and J. R. Blake (1977) Methods of ciliary propulsion and their size limitations. In T. J. Pedley, ed., *Scale Effects in Animal Locomotion*, pp. 243–56. London: Academic Press. (*356*)

Smith, J. A. (Davis), and A. J. Dartnall (1980) Boundary layer control by water pennies (Coleoptera: Psephenidae). *Aquat. Insects 2*: 65–72. (*67, 179*)

Soluk, D. A., and D. A. Craig (1988) Vortex feeding from pits in the sand: A unique method of suspension feeding used by a stream invertebrate. *Limnol. Oceanogr. 33*: 638–45. (*220*)

Soluk, D. A., and D. A. Craig (1990) Digging with a vortex: Flow manipulation facilitates prey capture by a predatory stream mayfly. *Limnol. Oceanogr. 35*: 1201–1206. (*221*)

Sotavalta, O. (1953) Recordings of high wing-stroke and thoracic vibration frequency in some midges. *Biol. Bull. 104*: 439–44. (*268*)

Sournia, A. (1982) Form and function in marine phytoplankton. *Biol. Rev. 57*: 347–94. (*346*)

Spanier, E., D. Weihs, and G. Almog-Shtayer (1991) Swimming of the Mediterranean slipper lobster. *J. Exp. Mar. Biol. Ecol. 145*: 15–32. (*287*)

Spedding, G. R. (1986) The wake of a jackdaw (*Corvus monedula*) in slow flight. *J. Exp. Biol. 125*: 287–307. (*276*)

Spedding, G. R. (1987) The wake of a kestrel (*Falco tinnunculus*) in flapping flight. *J. Exp. Biol. 127*: 59–78. (*276*)

Sponaugle, S., and M. LaBarbera (1991) Drag-induced deformation: A functional feeding strategy in two species of gorgonians. *J. Exp. Mar. Biol. Ecol. 148*: 121–34. (*118, 119*)

Sprackling, M. T. (1985) *Liquids and Solids*. London: Routledge and Kegan Paul. (*342*)

Stamp, N. E., and J. R. Lucas (1983) Ecological correlates of explosive seed dispersal. *Oecologia 59*: 272–78. (*194*)

Stanley, E. N., and R. C. Batten (1968) Viscosity of sea water at high pressures and moderate temperatures. *Nav. Ship Res. Dev. Center Rept. No. 2827.* (*24*)

Statzner, B., J. A. Gore, and V. H. Resh (1988) Hydraulic stream ecology: Observed patterns and potential applications. *J. North Amer. Benthol. Soc. 7*: 307–60. (*175*)

Statzner, B., and T. F. Holm (1982) Morphological adaptations of benthic invertebrates to stream flow—an old question studied by means of a new technique (laser-doppler anemometry). *Oecologia 53*: 290–92. (*179*)

Statzner, B., and R. Müller (1989) Standard hemispheres as indicators of flow characteristics in lotic benthos research. *Freshwater Biol. 21*: 445–59. (*170*)

Steffensen, J. F. (1985) The transition between branchial pumping and ram ventilation in fishes: Energetic consequences and dependence on water oxygen tension. *J. Exp. Biol. 114*: 141–150. (*69, 110*)

Steinman, D. B. (1955) Suspension bridges: The aerodynamic problem and its solution. In G. A. Baitsell, ed., *Science in Progress*, 9th Ser., pp. 241–91. New Haven, Conn.: Yale University Press. (*370, 372*)

Steinmann, P. (1907) Die Tierwelt der Gebirgsbäche. *Ann. Biol. Lacust. Bruxelles 2*: 20–150. (*177*)

Stephenson, R., J. R. Lovvorn, M.R.A. Heieis, D. R. Jones, and R. W. Blake (1989) A hydromechanical estimate of the power requirements of diving and surface swimming in lesser scaup, *Aythya affinis. J. Exp. Biol. 147*: 507–18. (*385*)

Stevens, E. D., and E. N. Lightfoot (1986) Hydrodynamics of water flow in front of and through the gills of skipjack tuna. *Comp. Biochem. Physiol. 83A*: 255–59. (*300*)

Stoker, M.G.P. (1973) Role of diffusion boundary layer in contact inhibition of growth. *Nature 246*: 200–202. (*200*)

Streeter, V. L., and E. B. Wylie (1985) *Fluid Mechanics*. 8th ed. New York: McGraw-Hill. (*6*)

Strickler, J. R. (1975) Intra- and interspecific information flow among planktonic copepods: receptors. *Verh. Internat. Verein. Limnol. 19*: 2951–58. (*360*)

Strickler, J. R. (1977) Observation of swimming performances of planktonic copepods. *Limnol. Oceanogr. 22*: 165–70. (*139, 361*)

Strickler, J. R. (1985) Feeding currents in calanoid copepods: Two new hypotheses. *Symp. Soc. Exp. Biol. 39*: 459–85. (*360*)

Strickler, J. R., and S. Twombly (1975) Reynolds number, diapause, and predatory copepods. *Verh. Internat. Verein. Limnol. 19*: 2943–50. (*360*)

Stride, G. O. (1955) On the respiration of an aquatic African beetle, *Potamodytes tuberosus* Hinton. *Ann. Ent. Soc. Amer. 48*: 344–51. (*70*)

Stride, G. O. (1958) The application of a Bernoulli equation to problems of insect respiration. *Proc. 10th Int. Congr. Entomol. 2*: 335–36. (*71*)

Suter, R. B. (1991) Ballooning in spiders: Results of wind tunnel experiments. *Ethol. Ecol. Evol. 3*: 13–25. (*222*)

Sutton, O. G. (1953) *Micrometeorology*. New York: McGraw-Hill. (*167, 168, 169*)

Sutton, O. G. (1955) *The Science of Flight*. Harmondsworth, Middlesex, England: Penguin Books. (*6, 376*)

Sverdrup, H. U., M. W. Johnson, and R. H. Fleming (1942) *The Oceans: Their Physics, Chemistry, and General Biology*. Englewood Cliffs, N.J.: Prentice Hall. (*24, 29, 47*)

Symbols Committee of the Royal Society (1975) *Quantities, Units, and Symbols*. 2d ed. London: The Royal Society. (*9*)

Synolakis, C. E., and H. S. Badeer (1989) On combining the Bernoulli and Poiseuille equations: A plea to authors of college physics texts. *Amer. J. Physics 57*: 1013–19. (*315*)

Tamm, S. L. (1983) Motility and mechanosensitivity of macrocilia in the ctenophore *Beroë. Nature 305*: 430–33. (*356*)

Taylor, G. I. (1951) Analysis of the swimming of microscopic organisms. *Proc. Roy. Soc. Lond. A211*: 225–39. (*354*)

Telewski, F. W., and M. J. Jaffe (1986) Thigmomorphogenesis: Field and laboratory studies of *Abies fraseri* in response to wind or mechanical perturbation. *Physiol. Plant. 66*: 211–18. (*127, 128*)

Telford, M. (1983) An experimental analysis of lunule function in the sand dollar *Mellita quinquiesperforata. Mar. Biol. 76*: 125–34. (*65*)

Telford, M., and R. Mooi (1987) The art of standing still. *New Scientist*, April 16, 30–35. (*65, 176*)

Thom, A., and P. Swart (1940) The forces on an aerofoil at very low speeds. *J. Roy. Aero. Soc. 44*: 761–70. (*98, 245, 347*)

Thompson, D'Arcy W. (1942) *On Growth and Form.* 2d ed. Cambridge, U.K.: Cambridge University Press. (*347, 350*)

Thorington, R. W., Jr., and L. R. Heaney (1981) Body proportions and gliding adaptatioins of flying squirrels (Petauristinae). *J. Mamm. 62*: 101–14. (*257*)

Thorup, J. (1970) The influence of a short-term flood on a springbrook community. *Arch. Hydrobiol. 66*: 447–57. (*40*)

Tolbert, W. W. (1977) Aerial dispersal behavior of two orb weaving spiders. *Psyche 84*: 13–27. (*194, 222*)

Tracy, C. R., and P. R. Sotherland (1979) Boundary layers of bird eggs: Do they ever constitute a significant barrier to water loss? *Physiol. Zool. 52*: 63–66. (*198*)

Trager, G. C., J.-S. Hwang, and J. R. Strickler (1990) Barnacle suspension feeding in variable flow. *Mar. Biol. 105*: 117–27. (*357, 360*)

Trefethen, L. M., R. W. Bilger, P. T. Fink, R. E. Luxton, and R. I. Tanner (1965) The bath-tub vortex in the southern hemisphere. *Nature 207*: 1084. (*205*)

Trueman, E. R. (1980) Swimming by jet propulsion. In H. Y. Elder and E. R. Trueman, eds., *Aspects of Animal Movement*, pp. 93–105. Cambridge, U.K.: Cambridge University Press. (*327*)

Tucker, V. A. (1969) Wave-making by whirligig beetles (Gyrinidae). *Science 166*: 897–99. (*388*)

Tucker, V. A. (1990a) Body drag, feather drag and interference drag of the mounting strut in a peregrine falcon, *Falco peregrinus. J. Exp. Biol. 149*: 449–68. (*138, 143, 148*)

Tucker, V. A. (1990b) Measuring aerodynamic interference drag between a bird body and the mounting strut of a drag balance. *J. Exp. Biol. 154*: 439–61. (*138, 148*)

Tucker, V. A., and C. Heine (1990) Aerodynamics of gliding flight in a Harris' hawk, *Parabuteo unicinctus. J. Exp. Biol. 149*: 469–89. (*148*)

Tucker, V. A., and G. C. Parrott (1970) Aerodynamics of gliding flight in a falcon and other birds. *J. Exp. Biol. 52*: 345–67. (*133, 138, 248, 253, 256*)

Tunnicliffe, V. (1981) Breakage and propagation of the stony coral *Acropora cervicornis. Proc. Nat. Acad. Sci. USA 78*: 2427–31. (*107*)

Turner, E. J., and D. C. Miller (1991) Behavior and growth of *Mercenaria mercenaria* during simulated storm events. *Mar. Biol. 111*: 55–64. (*192*)

Tyree, M. T., and F. W. Ewers (1991) The hydraulic architecture of trees and other woody plants. *New Phytol. 119*: 345–60. (*323*)

Tyree, M. T., and J. S. Sperry (1988) Do woody plants operate near the point of catastrophic xylem disfunction caused by dynamic water stress? Answers from a model. *Plant Physiol. 88*: 574–80. (*38, 323*)

Tyree, M. T., and J. S. Sperry (1989) Vulnerability of xylem to cavitation and embolism. *Ann. Rev. Plant Phys. Mol. Bio. 40*: 19–38. (*323*)

Van Atta, C. W., and M. Gharib (1987) Ordered and chaotic vortex streets behind circular cylinders at low Reynolds numbers. *J. Fluid Mech. 174*: 113–33. (*209, 371*)

Van der Pijl, L. (1972) *Principles of Dispersal in Higher Plants.* 2d ed. New York: Springer-Verlag. (*187*)

Van Dyke, M. (1982) *An Album of Fluid Motion*. Stanford, Calif.: Parabolic Press. (*6, 42*)

Van Tussenbroek, B. I. (1989) Morphological variations of *Macrocystis pyrifera* in the Falkland Islands, South Atlantic Ocean, in relation to environment and season. *Mar. Biol. 102*: 545–56. (*125*)

Videler, J. J., and A. Groenewold (1991) Field measurements of hanging flight aerodynamics in the kestrel *Falco tinnunculus*. *J. Exp. Biol. 155*: 519–30. (*260*)

Videler, J. J., and B. A. Nolet (1990) Costs of swimming measured at optimum speed: Scale effects, differences between swimming styles, taxonomic groups and submerged and surface swimming. *Comp. Biochem. Physiol. 97A*: 91–99. (*385*)

Videler, J. J., and D. Weihs (1982) Energetic advantages of burst-and-coast swimming of fish at high speeds. *J. Exp. Biol. 97*: 169–78. (*280*)

Vogel, S. (1962) A possible role of the boundary layer in insect flight. *Nature 193*: 1201–1202. (*160, 368*)

Vogel, S. (1966) Flight in *Drosophila*. I. Flight performance of tethered flies. *J. Exp. Biol. 44*: 567–78. (*9, 143*)

Vogel, S. (1967a) Flight in *Drosophila*. II. Variation in stroke parameters and wing contour. *J. Exp. Biol. 46*: 383–92. (*264*)

Vogel, S. (1967b) Flight in *Drosophila*. III. Aerodynamic characteristics of fly wings and wing models. *J. Exp. Biol. 46*: 431–43. (*245, 249, 250*)

Vogel, S. (1976) Flows in organisms induced by movements of the external medium. In T. J. Pedley, ed., *Scale Effects in Animal Locomotion*, pp. 285–97. London: Academic Press. (*73, 103*)

Vogel, S. (1977) Current-induced flow through sponges *in situ*. *Proc. Nat. Acad. Sci. USA 74*: 2069–71. (*71, 189*)

Vogel, S. (1978a) Organisms that capture currents. *Sci. Amer. 239 (2)*: 128–39. (*189*)

Vogel, S. (1978b) Evidence for one-way valves in the water-flow system of sponges. *J. Exp. Biol. 76*: 137–48. (*329*)

Vogel, S. (1981) *Life in Moving Fluids*. Princeton, N.J.: Princeton University Press. (*241, 324*)

Vogel, S. (1983) How much air flows through a silkmoth's antenna? *J. Insect Physiol. 29*: 597–602. (*104, 357*)

Vogel, S. (1984) Drag and flexibility in sessile organisms. *Amer. Zool. 24*: 37–44. (*117, 119, 123, 368*)

Vogel, S. (1985) Flow-assisted shell reopening in swimming scallops. *Biol. Bull. 169*: 624–30. (*56*)

Vogel, S. (1987) Flow-assisted mantle cavity refilling in jetting squid. *Biol. Bull. 172*: 61–68. (*69, 77*)

Vogel, S. (1988a) *Life's Devices: The Physical World of Animals and Plants*. Princeton, N.J.: Princeton University Press. (*35, 78, 103, 142, 195, 287, 314, 347, 368, 379*)

Vogel, S. (1988b) How organisms use flow-induced pressures. *Amer. Sci. 76*: 28–34. (*63*)

Vogel, S. (1989) Drag and reconfiguration of broad leaves in high winds. *J. Exp. Bot. 40*: 941–48. (*119, 123, 193, 373*)

Vogel, S. (1992a) *Vital Circuits: On Pumps, Pipes, and the Workings of Circulatory Systems*. New York: Oxford University Press. (*35, 187, 314*)

Vogel, S. (1992b) Copying nature: A biologist's cautionary comments. *Biomaterials 1*: 63–79. (*106, 150*)

Vogel, S. (1992c) Twist-to-bend ratios and cross-sectional shapes of petioles and stems. *J. Exp. Bot. 43*: 1527–32. (*373*)

Vogel, S. (1993) Leaves in a storm. *Nat. Hist.* 102, 9: 58–62. (*123*)

Vogel, S., and W. L. Bretz (1972) Interfacial organisms: Passive ventilation in the velocity gradients near surfaces. *Science 175*: 210–11. (*71*)

Vogel, S., and N. Feder (1966) Visualization of low-speed flow using suspended plastic particles. *Nature 209*: 186–87. (*44, 45*)

Vogel, S., and M. LaBarbera (1978) Simple flow tanks for research and teaching. *Bioscience 28*: 638–43. (*324*)

Vogel, S., C. P. Ellington, Jr., and D. C. Kilgore, Jr. (1973) Wind-induced ventilation of the burrow the the prairie dog, *Cynomys ludovicianus. J. Comp. Physiol. 85*: 1–14. (*71, 103*)

Von Kármán, T. (1954) *Aerodynamics.* New York: McGraw-Hill. (*6, 106, 107, 376*)

Vosburgh, F. (1982) *Acropora reticulata*: Structure, mechanics and ecology of a reef coral. *Proc. Roy. Soc. Lond. B214*: 481–99. (*119*)

Vugts, H. F., and W.K.R.E. van Wingerden (1976) Meteorological aspects of aeronautic behavior in spiders. *Oikos 27*: 433–44. (*223*)

Wainwright, S. A., and J. R. Dillon (1969) On the orientation of sea fans (genus *Gorgonia*). *Biol. Bull. 136*: 130–39. (*107*)

Wainwright, S. A., W. D. Biggs, J. D. Currey, and J. M. Gosline (1976) *Mechanical Design in Organisms.* New York: John Wiley. (*9*)

Wallace, J. B., and R. W. Merritt (1980) Filter-feeding ecology of aquatic insects. *Ann. Rev. Entomol. 25*: 103–32. (*60*)

Wallace, J. B., J. R. Webster, and W. R. Woodall (1977) The role of filter-feeders in flowing waters. *Arch. Hydrobiol. 79*: 506–32. (*180*)

Walne, P. R. (1972) The influence of current speed, body size and water temperature on the filtration rate of five species of bivalves. *J. Mar. Biol. Assoc. U.K. 52*: 345–74. (*192*)

Walsby, A. E., and A. Xypolyta (1977) The form resistance of chitan fibers attached to the cells of *Thalassiosira fluviatilis* Hustedt. *Br. Phycol. J. 12*: 215–23. (*30, 345*)

Walters, V. (1962) Body form and swimming performance of the scombroid fishes. *Amer. Zool. 2*: 143–49. (*29*)

Ward, J. V. (1992) *Aquatic Insect Ecology. 1. Biology and Habitat.* New York: John Wiley. (*177*)

Ward-Smith, A. J. (1984) *Biophysical Aerodynamics and the Natural Environment.* Chichester, U.K.: John Wiley. (*240*)

Ward-Smith, A. J., and D. Clements (1982) Experimental determinations of the aerodynamic characteristics of ski-jumpers. *Aeronaut. J. 86*: 384–91. (6)

Waringer, J. A. (1989) Resistance of a cased caddis larva to accidental entry into the drift: The contribution of active and passive elements. *Freshwater Biol. 21*: 411–20. (*180*)

Warren, J. V. (1974) The physiology of the giraffe. *Sci. Amer. 231* (5): 96–105. (*55, 327*)

Washburn, J. O., and L. Washburn (1984) Active aerial dispersal of minute wingless arthropods: Exploitation of boundary layer velocity gradients. *Science 223*: 1088–89. *(193)*

Waxman, H. M., and J. D. McCleave (1978) Auto-shaping in the archer fish (*Toxotes chatareus*). *Behav. Biol. 22*: 541–44. *(396)*

Webb, J. E., and J. L. Theodor (1972) Wave-induced circulation in submerged sands. *J. Mar. Biol. Assoc. U.K. 52*: 903–14. *(71)*

Webb, P. W. (1973) Kinematics of pectoral fin propulsion in *Cymatogaster aggregata*. *J. Exp. Biol. 59*: 697–710. *(286)*

Webb, P. W. (1975) Hydrodynamics and energetics of fish propulsion. *Bull. Fish. Res. Bd. Canada 190*: 1–158. *(29, 136, 143, 147, 153, 280, 281, 287)*

Webb, P. W. (1976) The effect of size on the fast-start performance of rainbow trout *Salmo gairdneri* and a consideration of piscivorous predator-prey interactions. *J. Exp. Biol. 65*: 157–77. *(287)*

Webb, P. W. (1978) Fast-start performance and body form in seven species of teleost fish. *J. Exp. Biol. 74*: 211–26. *(287)*

Webb, P. W. (1979a) Mechanics of escape responses in crayfish (*Orconectes virilis*). *J. Exp. Biol. 79*: 245–63. *(287, 365)*

Webb, P. W. (1979b) Reduced skin mass: An adaptation for acceleration in some teleost fish. *Can. J. Zool. 57*: 1570–75. *(287, 365)*

Webb, P. W. (1981) The effect of the bottom on the fast start of a flatfish, *Citharchthys stigmaeus*. *Fishery Bull. 79*: 271–76. *(289)*

Webb, P. W. (1988) Simple physical principles and vertebrate aquatic locomotion. *Amer. Zool. 28*: 709–25. *(280, 281)*

Webb, P. W. (1989) Station-holding by three species of benthic fishes. *J. Exp. Biol. 145*: 303–20. *(65, 176)*

Weibel, E. R. (1963) *Morphometry of the Human Lung*. New York: Springer-Verlag. *(36)*

Weihs, D. (1973) Hydrodynamics of fish schooling. *Nature 241*: 290–91. *(288)*

Weihs, D. (1974) Energetic advantages of burst swimming of fish. *J. Theor. Biol. 48*: 215–29. *(106)*

Weihs, D. (1977) Periodic jet propulsion of aquatic creatures. *Fortschr. Zool. 24 (2–3)*: 171–75. *(79)*

Weihs, D. (1980) Energetic significance of changes in swimming modes during growth of larval anchovy, *Engraulis mordax*. *Fish. Bull. 77*: 597–604. *(281)*

Weir, J. S. (1973) Air flow, evaporation and mineral accumulation in mounds of *Macrotermes subhyalinus* (Rambur). *J. Anim. Ecol. 42*: 509–20. *(71)*

Weis-Fogh, T. (1956) Biology and physics of locust flight. II. Flight performance of the desert locust (*Schistocerca gregaria*). *Phil. Trans. Roy. Soc. Lond. B239*: 459–510. *(143, 144)*

Weis-Fogh, T. (1973) Quick estimates of flight fitness in hovering animals, including novel mechanisms for lift prodcution. *J. Exp. Biol. 59*: 169–230. *(278)*

Weis-Fogh, T. (1975) Unusual mechanisms for the generation of lift in flying animals. *Sci. Amer. 233 (5)*: 81–87. *(277, 279)*

Weis-Fogh, T. (1977) Dimensional analysis of hovering flight. In T. J. Pedley, ed., *Scale Effects in Animal Locomotion*, pp. 405–20. London: Academic Press. *(266)*

Weis-Fogh, T., and R. M. Alexander (1977) The sustained power output obtainable from striated muscle. In T. J. Pedley, ed., *Scale Effects in Animal Locomotion*, pp. 511–25. London: Academic Press. *(266)*

Weis-Fogh, T., and W. B. Amos (1972) Evidence for a new mechanism of cell motility. *Nature 236*: 301–304. (*361*)

Weissenberger, J., H.-Ch. Spatz, A. Emanns, and J. Schwoerbel (1991) Measurement of lift and drag forces in the mN range experienced by benthic arthropods at flow velocities below 1.2 m/s. *Freshwater Biol. 25*: 21–31. (*119, 120, 176, 179*)

Wells, G. P., and R. P. Dales (1951) Spontaneous activity patterns in animal behaviour: The irrigation of the burrow in the polychaetes *Chaetopterus variopedatus* Renier and *Nereis diversicolor* O. F. Müller. *J. Mar. Biol. Assoc. U.K. 29*: 661–80. (*326*)

Wells, M. J. (1990) Oxygen extraction and jet propulsion in cephalopods. *Can. J. Zool. 68*: 815–24. (*80*)

Wells, R.M.G., J. A. Macdonald, and G. diPrisco (1990) Thin blooded antarctic fishes: A rheological comparison of the haemoglobin-free icefishes *Chionodraco kathleenae* and *Cryodraco antarcticus* with a red-blooded nototheniid, *Pagothenia bernacchi. J. Fish Biol. 36*: 595–609. (*29*)

Westlake, D. R. (1967) Some effects of low-velocity currents on the metabolism of aquatic macrophytes. *J. Exp. Bot. 18*: 187–205. (*199*)

Wethey, D. S. (1986) Ranking of settlement cues by barnacle larvae: Influence of surface contour. *Bull. Mar. Sci. 39*: 393–400. (*185*)

White, C. M. (1946) The drag of cylinders in fluids at low speeds. *Proc. Roy. Soc. Lond. A186*: 472–79. (*93, 338*)

White, F. M. (1974) *Viscous Fluid Flow*. New York: McGraw-Hill. (*119, 333, 336*)

Wieringa, J. (1993) Representative roughness parameters for homogeneous terrain. *Boundary-Layer Meteorol. 63*: 323–63. (*172*)

Wilcox, R. S. (1988) Surface wave reception in invertebrates and vertebrates. In J. Atema, R. R. Fay, A. N. Popper, and W. N. Tavolga, eds., *Sensory Biology of Aquatic Animals*, pp. 643–63. New York: Springer-Verlag. (*392*)

Wildish, D. J., and M. P. Miyares (1990) Filtration rate of blue mussels as a function of flow velocity: Preliminary experiments. *J. Exp. Mar. Biol. Ecol. 142*: 213–20. (*192*)

Williams, T. A. (1991) A model of rowing propulsion and the ontogeny of locomotion in *Artemia* larvae. *Amer. Zool. 31*: 60A (and oral presentation). (*351*)

Williams, T. M. (1983) Locomotion in the North American mink, a semi-aquatic mammal. I. Swimming energetics and body drag. *J. Exp. Biol. 103*: 155–68. (*154, 385*)

Williams, T. M., and G. L. Kooyman (1985) Swimming performance and hydrodynamic characteristics of harbor seals, *Phoca vitulina. Physiol. Zool. 58*: 576–89. (*143, 150, 151*)

Winet, H. (1973) Wall-drag on free-moving ciliated microorganisms. *J. Exp. Biol. 59*: 753–66. (*338*)

Withers, P. C. (1979) Aerodynamics and hydrodynamics of the 'hovering' flight of Wilson's storm petrel. *J. Exp. Biol. 80*: 83–91. (*260*)

Withers, P. C. (1981) An aerodyamic analysis of bird wings as fixed aerofoils. *J. Exp. Biol. 90*: 143–62. (*249*)

Withers, P. C., and P. L. (O'Neill) Timko (1977) The significance of ground effect to the aerodynamic cost of flight and energetics of the black skimmer (*Rhyncops nigra*). *J. Exp. Biol. 70*: 13–26. (*141, 288, 383*)

Witman, J. D., and T. H. Suchanek (1984) Mussels in flow: Drag and dislodgement by epizoans. *Mar. Ecol. Progr. Ser. 16*: 259–68. *(119)*

Womersley, J. R. (1955) Method for the calculation of velocity, rate of flow, and viscous drag when the pressure gradient is known. *J. Physiol. 127*: 553–63. *(377)*

Woodcock, A. H. (1987) Mountain breathing revisited: The hyperventilation of a volcano cinder cone. *Bull. Amer. Meteorol. Soc. 68*: 125–30. *(72)*

Woodin, S. A. (1986) Settlement of infauna: Larval choice? *Bull. Mar. Sci. 39*: 401–407. *(184)*

Wu, T.Y.-T. (1977) Hydrodynamics of swimming at low Reynolds numbers. *Fortschr. Zool. 24 (2–3)*: 149–69. *(137, 349, 356)*

Wu, T.Y.-T., C. J. Brokaw, and C. Brennen, eds. (1975) *Swimming and Flying in Nature*. Vol. 1. New York: Plenum Press. *(356)*

Yager, P. L., A.R.M. Nowell, and P. A. Jumars (1993) Enhanced deposition to pits: A local food source for benthos. *J. Mar. Res. 51*: 209–36. *(214)*

Young, C. M., and L. F. Braithwaite (1980) Orientation and current-induced flow in the stalked ascidian *Styela montereyensis*. *Biol. Bull. 159*: 428–40. *(60)*

Zanker, J. M., and K. G. Gotz (1990) The wingbeat of *Drosophila melanogaster*. 2. Dynamics. *Phil. Trans. Roy. Soc. Lond. B327*: 19–44. *(249)*

Zaret, R. E. (1980) The animal and its viscous environment. In W. C. Kerfoot, ed., *Evolution and Ecology of Zooplankton Communities*, pp. 133–55. Hanover, N.H.: University Press of New England. *(360)*

Zaret, R. E., and W. C. Kerfoot (1980) The shape and swimming technique of *Bosmina longirostris*. *Limnol. Oceanogr. 25*: 126–33. *(336, 352)*

Zimmermann, M. H. (1983) *Xylem Structure and the Ascent of Sap*. New York: Springer-Verlag. *(37, 38, 322)*

Zusi, R. L. (1962) Structural adaptations of the head and neck in the Black Skimmer, *Rynchops nigra* Linnaeus. *Publ. Nuttal Ornithol. Club 3*: 1–101. *(141)*

Index

(More complete entries are given under common names than under scientific names.)

Abietenaria (hydroid), 120

Acanthocyclops (crayfish), acceleration reaction, 365

acceleration: air bubble, 362; crayfish escape response, 365; crustaceans, 287; to evade viscous effects, 359; fish swimming, 287; freshwater turtle, 287; jetting animals, 365; of gravity, 34; scaling, 287; sea turtle, 287

acceleration reaction, 138, 362–69; in air, 368; and body volume, 365; defined, 108; direction, contra drag, 364; and dislodgement probabilities, 368; fruit fly wings, 368; in jet propelled animals, 365; scaling, 108; scaling contra drag, 368; stopping, 365; swimming with tail or paddles, 366; wave surge on fixed object, 367. *See also* added mass

accuracy, 11–12

Acer (maple), 228, 272

acetone, properties, 23 (table)

Acilius (water beetle), 145

actuator disk, 275 (fig.); disk loading, 274; in hovering, 276; induced velocity, 274

added mass, 362–69. *See also* acceleration reaction; swimming

added mass coefficient: and decelerative gliding, 365; defined, 363; vs. fineness, data, 364; use, 363; values, 363

advance ratio: vs. angle of attack, 267; animal flight, 266, 267; defined, 267; propeller, 267; samaras, 272; vs. size, best, 268; vs. top speed, 268; whale tail, 282

aeolean singing of wires, 371

aerodynamic frequency parameter, defined, 376

aeroelasticity, 372, 374–5

aerosol particles, 192

Ailanthus altissima (tree-of-heaven), 228

air: density vs. temperature, 22; incompressibility justified, 54; kinematic viscosity, 25, 88

air-water interface. *See* interface, air-water

airfoil: angle of attack, 234; area adjustment, 243; aspect ratio, 236–39; autogyro blades, 268; biological, 246–52, 249; blade element as, 262; camber, 270; Cayley's profile, 106; chord, 234, 239; components of drag, 242; cross section, 233; drag, 109; drag vs. *Re*, 241; elliptical plan form, 239; vs. Flettner rotor, 233; flow pattern for lift, 231; helicopter blades, 268; infinite wing, 236; leading edge, 233; lift distribution, 233; lift production, 230–34; lift vs. drag, 234; operation of various (figs.), 232, 269; polar diagram, 235 (fig.), 248 (insects); power vs. aspect ratio, 238; propeller, 268; and *Re*, 244–46; vs. *Re*, 142; reference area, 90; reversed, 141; separation, 233; separation and vortices, 213; separation point oscillation, 376; shape, 107 (fig.); stall, 242; surface irregularities, 245; swept back tips, 240 (fig.); terminology, 231 (fig.); thickness-to-chord ratio, 109, 114; tip sweepback, 239; trailing edge, 231, 233; very low *Re*, 245; vortices, 233 (fig.); windmill, 268; wing loading, 243–44. *See also* wings

airship, 63, 68; streamlining, 99

albatross: glide angle, 254; lift-to-drag ratio, 248; slope soaring, 259

algae: Bénard cells, 224; current and photosynthesis, 199; sinking speeds, 185; spore settling, 187

algal cells, as flow markers, 44

algal thallus, modeling, 103

Allogamus auricollis (caddisfly larva), 180

Alloptes (feather mite), 181

441

Alsomitra macrocarpa (Javanese cucumber), 256

alula, as anti-stall device, 242

Ametropus neavei (mayfly larva), 200

amphibia: cutaneous respiration and flow, 200; limbless, anguilliform swimming, 281

amphipod: fluid dynamic burrow pump, 327; pleopods as propeller, 328

Amusium (scallop), 252

anchovy, swimming behavior, 281

Ancylus (snail or freshwater limpet), 40, 179

anemometer, 7, 43, 112

anemophilous plants: exposing pollen to wind, 193; pollen trapping, 187

aneroid barometer, 56

aneurysm, 62

angel, wing loading, 244

angelfish: acceleration reaction, 365; swimming with drag, 287

angle of attack, 231 (fig.); vs. advance ratio, 267; blade element, 263; defined, 234; and delayed stall, 376; and lift, 242; vs. lift and drag, 235; and pitching moment, 234; propeller, 264; and stall, 242

angle of incidence, 263

Anguilla (eel), 281

Anolis carolinensis (lizard), parachuting, 257

antarctic animals, blood and circulation, 28

antenna: E-value, locust, 120; modeling, 105; transmissivity, 356

aorta, 8, 35, 55, 62, 316

Aplocheilus (fish), prey detection with waves, 392

apparent additional mass. *See* added mass

Aptendytes forsteri (emperor penguin), 149

aquatic plants, 199

Arceuthobium (dwarf mistletoe), 195

archer fish, squirting, 396

area of reference. *See* reference area

Arenicola (lugworm), 326

Aristotle, 81, 167

Artemia (brine shrimp): flow markers, 44; swimming, 350

arterioles, 8, 319

arteries, 315

ascidian: *Botryllus,* 191 (fig.); jet propulsion, 77, 80; as Pitot tube, 60; preventing recirculation, 190; *Styela,* 61 (fig.)

ash: autorotating samara, 228; vessel length, 322

aspect ratio, 236–39; 231 (fig.); of autorotating plate, 228; and cost of lift, 237; defined, 236; flying fish, 259; and induced drag, 238; lunate fish tail, 282; phalanger, 250; sand dollar airfoil, 246; scallop, 252; and tip vortices, 236; whale tail, 282

aspen, quaking and reconfiguration, 124

aspirator, 58, 316

atmospheric instability, and thermal soaring, 222

autogyrating: vs. autorotating, 228; vs. gliding, 257

autogyro, 269, 272

autogyro rotor: as airfoil, 269; blade contour, 271; origin of upward force, 271; plane of rotation, 271

autorotating: of Flettner rotors, 228; vs. gliding, 257

bacteria: coasting, 331; flagellar rotation, 206; local drag on walls, 176; *Re*, 87; sinking speeds, marine, 344; special flagella, 351

Baetis (mayfly nymph), 139, 177

Balaenoptera physalus (fin whale), 282, 386

Balanus balanoides (barnacle), 183

ballasted case, caddisfly larva, 180

ballistics of seeds, 194

balloon, 12

ballooning by spiderlings and caterpillars, 12, 222–23

barnacle: filtration, 356; settling, 183, 185

Basiliscus (lizard), slapping, 387

bass, ram ventilation, 110

bat, fishing: echolocation of surface waves, 392; leg cross sections, 141, 383; legs as struts, 140

bat, flying: lift coefficient, 278; power, 241

bathtub vortex, 204, 208

beach: interstitial flow, 307; swash, 307; swash-riding, 182

bee: advance ratio, 268; alleged flight

lawlessness, 278; body lift, 239, 252; flight speed, 13; glide angle, 254; wing loading, 243; wing polar diagram, 245

beetle: air bubble, 70; elmid, 70 (fig.); surfactant propulsion, 390

beetle, bombardier, Coanda effect, 397

beetle, diving, 350 (fig.)

beetle, water, 142 (fig.); drag coefficients, 145; rowing stroke, 155 (fig.); swimming, 137, 155, 287, 350, 366

beetle, water penny, 67, 178, 179

beetle, whirligig, 387 (fig.); hull support, 388; size range, 390; wave making, 390

Bernoulli, Daniel, 52

Bernoulli's equation 5, 156; in boundary layer, 5, 157; and orifice, 394; for Pitot tube, 58

Bernoulli's principle, 21, 52–62, 80, 53 (fig.); and circulatory systems, 62; derivation, 52; and drag, maledictions, 81–82; and internal flows, 314; and lift, 65, 232; limitations and precautions, 60; and pressure-induced flow, 72; and rotating cylinder, 226; use in flow measurements, 57–60; use in pressure measurements, 54; vortex interactions, 210

Betula lenta (birch), 194

Bénard cells, 224 (fig.)

bill, skimmer's as strut, 140–41

bioconvection, 224

biometeorology, 5, 10, 221, 344

bird: blood viscosity in antarctic, 28; body drag, 106, 148–49; body lift, 239; downwind flight, 13; egg boundary layers, 198; flight, 137; formation flight, 288; gliding, 21; interference drag, 148; parasite drag, 248; planing in landing, aquatic, 387; skimmer bill as strut, 140, 141; soaring, 221, 260; stroke plane, 276; swooping flight, 239; tip vortices, 276; vortex wakes, 43, 276; wing aerodynamics, 248; wing area adjustment, 243; wing feathers, 248

bivalve mollusk, preventing recirculation, 190

black fly, advance ratio, 268

black fly larva, 113 (fig.); cephalic fans, 184; drag, 113, 184; feeding from ascending vortex, 217; filter elevation, 191; filtration, 356; settling, 184–85; shape of fans, 113

black fly pupa, gills in vortices, 218

blade element: making thrust, 262; torque component, 264

Blephariceridae (fly), 180

blood: as non-Newtonian fluid, 20; viscosity, 28–29

blood cells, rotation in shear flow, 187

blood flow, cost, 317

blood pressure, 55; dinosaurs, 56; fish, 68; giraffes, 55; horses, 55; mammals, 55; reptilian, 56; systolic, 55

blood vessels, no-slip condition, 19. *See also* arteries, etc.

bluefish: intracranial pressure, 68; pressure distribution, 68; ram ventilation, 110

bluff body, 99, 167; drag, 111; drag in boundary layer, 175; drag vs. speed, 117; splitter plates, 110; vortex shedding, 212

body drag, 138. *See also* parasite drag

body lift: birds, 239; bumblebees, 252; insects, 239; ski jumpers, 240; swimming rays, 239; swooping flight, 239

body of revolution, 133; drag coefficient, 135; thickness-to-chord ratio, 134

Bombus terrestris (bumblebee), 268

Bond number, 347

boomerangs, 101

Bosmina (cladoceran crustacean), 351, 359

Botryllidae (ascidians), colonial jets, 190

bottom. *See* substratum

bound vortex, 232

boundary layer, 5, 19, 156–71 (figs., 157, 159); bird eggs, 198; drag in, 164–67, 175–81; empirical formulas, 163; and evaporation, rate, 163; vs. free stream, 162; and heat transfer, 163; laminar flow, 158–61; laminar sublayer, 161; leaves, 163; logarithmic, 167–71; in nature, 162; origin, 157; and principle of continuity, 159; rotation in, 188 (fig.); and settling, 186; and shear rate, 157; speeds within, 160, 193; and surface roughness, 160; thickness, 157–60, 162; thickness vs. leaf size/shape, 199; turbulent flow, 159, 161–62; unbounded, 167–71; vs. unstirred layer,

boundary layer (*cont.*)
201; as vortex generator, 218; at very
low *Re*, 332. *See also* velocity gradient
boundary layer resistance, 197, 198
bow-wave riding, dolphins, 386
brachiopod: filtration flow patterns, 44;
shell closing acceleration, 365; suspension feeding, 189
branching of pipes, 216–21
Branta canadensis (Canada geese), formation flight, 288
brine shrimp: as flow markers, 44; swimming vs. *Re*, 350
bristles: spacing vs. *Re*, 351; for swimming, 350; use in flight, 350
Brownian motion, in filtration, 357
bryozoa: colonywide currents, 44, 191;
Membranipora, 191 (fig.)
bubble, 8; air, of beetle*, 70; air, of beetle**, 70
bug, surfactant propulsion, 390
bumblebee, polar diagram, 248
buoyancy, 14; air bubble, 362; ascending
vortices, 221
burrows, induced flow through relict, 72
butterfly: lift-to-drag ratio, monarch,
255; monarch, 13; vortices in flight,
276, 287; wing scales, 245
byssus thread, 66, 223

C-start, fish swimming, 287
caddisfly larvae (figs., 61, 178): ballasted
case, 180; catch nets, 180, 356; hygropetric, 182; its Pitot tube, 60
caddisfly pupae, on river bottoms, 40
Callinectes sapidus (crab), 146
Cancer productus (crab), 146
capillarity, 22
capillary, 55, 62; aggregate cross section,
37; blood shear rate, 28; bolus flow,
312; diameter, 28; diffusion in, 313;
flow in, 313 (fig.); lungs, 37; Murray's
law, 319; Péclet number, 313; size and
flow rate, 35; toroidal flow, 312; total
number, 37; transmural exchange,
312; Womersley number, 377
capillary waves, 382
Caranx (jack), 281
carburetor, 58, 316
Carchesium (protozoan), 191
cat, micturition, 397

catch net, caddisfly larvae, 180
caterpillar, thermal soaring, 222
caudal peduncle, 282
Cayley, Sir George, 106
celerity: defined, 380; equation for, 283;
vs. gravity, 381; minimum possible,
382; vs. surface tension, 381; vs. wavelength, 381, 382
cells, viscosity, 18
Cenocorixa (water boatman bug), acceleration reaction, 365
cephalopods; ink vortex, 215; jet propulsion, 77; shell as splitter plate, 110
cetaceans: compliant surfaces, 152; drag,
151; drag coefficients, 151; drag reduction by heat release, 29; porpoising, 151; surface heating, 153; tail
sweepback, 239; wetted surface, 90
Chaetopterus (polychaete worm), 326, 327
Chaoborus (dipteran larva), signal for attack, 359
characteristic length, 111
chemical communication, 6
Chironomidae (midges), 182
Chlamydomonas (algal cell), flagellar swimming, 352
choanocytes, 356
chord: airfoil, 234; strut, 134
Chrysopelea (colubrid snale), gliding, 257
Chthalamus fragilis (barnacle), 185
cilia: arrangements, 351; associated velocity gradient, 348 (fig.); bands and
covered surfaces, 349; of bivalves, 39;
in burrow pump, 327; efficiency, 354;
motion, 352 (fig.); operation, 351–55;
of Paramecium, 137; as wall pumps,
314
ciliates: flow, swimming vs. sinking, 349;
sinking speeds, 344; upward swimming, 344
Ciona intestinalis (ascidian), 184
circular aperture, 303–4; biological examples, 303; laminar flow, 303; orifice
coefficient, 303; transition *Re*, 305; turbulent flow, 303
circular cylinder: across velocity gradient,
216; added mass coefficient, 363;
boundary layer, 163; drag at very low
Re, 335–36, 352, fig., 353; drag coefficient vs. *Re*, 93; drag coefficient, data,
92, 112; drag coefficient, equation,

336; drag vs. orientation, very low *Re*, 335; drag vs. roughness and *Re*, 101; E-value, 120; Flettner rotor, 227; flow around, 83, 93–96, 94 (fig.); lift when rotating, 226; Magnus effect, 226; pressure distribution, 81, 83; pressure drag, 97; protruding from substratum, 216–18; rotating, 230; slewing obliquely, 353; vs. streamlined form, 134; Strouhal numbers, 371; theoretical streamlines, 82; turbulent transition, 94; very low *Re* streamlines, 332; vortices behind, 315; wall effect on drag, low *Re*, 339

circular disk, drag, 112, 334

circulation: dimensions, 225; and lift, 226, 230–31; nonsteady, 277; physical variable, 7; plus translation, 226; reversal in vortex shedding, 369; of self-excited oscillator, 373; starting with clap-and-fling, 279; and translation, 230, 231; and vortex speed, 370; and vortices, 210, 224; and vorticity, 224–25; at very low *Re*, 347; wind driven, 6. *See also* circulatory system

circulatory system, 5, 8, 36 (fig.); antarctic animals, 28; Bernoulli vs. Hagen-Poiseuille, 314; and Bernoulli's principle, 62; in blubber, 29; data, 36, 319; features, 308; iguana, 28; instabilities, 374; Murray's law, 319; output, human, 324; power, 28; pulsating flow, 377; shear flow, 188; temperature effects, 28; vessel expansion, 315; Womersley number, 377

Citharchthys stigmaeus (flatfish), ground effect, 289

cladoceran, 350 (fig.); jerky swimming, 359

Cladonia podetia (lichen), 214

clam: and continuity, 38 (fig.); pump pressure, 328; swash-riding, 181, 182 (fig.)

clap-and-fling mechanism, 278–79, 289

Clione limacina (sea butterfly), swimming with lift, 286

Coanda effect, 397 (fig.); bombardier beetles, 397; mechanism, 397

coarctation, 62, 316

coccid (scale insect), 193

cod, burst-and-coast swimming, 281

coefficient of discharge, 303

coefficient of drag. *See* drag coefficient

coefficient of lift. *See* lift coefficient

coefficient of pressure. *See* pressure coefficient

coelenterate gastroderm, cilia on, 314

coelenterate medus, acceleration reaction, 365

colonywide surface currents, bryozoa, 44

Columba livia (pigeon), 148

communication, chemical, 6

communication with waves, 291–92; frequency vs. wavelength, 391; signal diversity, 392

compliant surfaces, drag reduction, 152

compressibility, 102; vs. speed of flow, 21

compressible flow, 5

condor, lift-to-drag ratio, 248

condor, Andean, wing loading, 243

conifers: autogyrating samaras, 272; cones, airflow patterns, 43

conservation of mass. *See* principle of continuity

continuity, principle of. *See* principle of continuity

continuum assumption, 20

convection, 6; Bénard cells, 224; vs. diffusion, 196, 313; imprecision, 11; and sinking speeds, 343, 345; free tree, 221

copepod: acceleration, 360; acceleration reaction, 365; filtration vs. active predation, 359; jerky vs. smooth swimming, 359

copying nature, 106, 150

coral: drag, 107; E-value, 120; gastrovascular system, 320

coral, black, on seamounts, 40

Coriolis force, 204

corners, vortices in, 213

Corophium (amphipod), 327

corselet, 29

countercurrent exchange systems, 311

crab: drag coefficient, 146; legs as struts, 140; lift, 146, 176; swimming, 146, 286

craspedophilic organisms, 174

crayfish, 142 (fig.); acceleration, 287, 365; escape response, 365

creeping flow, defined, 331

crinoid, coelomic circulation, 320

crustacean, wetted surface, 90. *See also* specific creatures

Cryptolepas (ectoparasitic barnacle), 180
ctenophores (comb-jellies): ciliary plates, 355; jet propulsion, 77
cup: of lichen, 214; vortices in, 214
Cupressaceae, autogyrating samaras, 272
current speed: effects of, 175; and photosynthetic rate, 199; and suspension feeding, 192
cutaneous respiration, amphibians, 200
cuttlefish. *See* cephalopods
Cyamus (whale lice), 180
cyanobacteria, density vs. light intensity, 30
cyclomorphosis, *Daphnia,* 30
Cyclops (copepod), jerky swimming, 359
cyclosis, 5, 196
cylinder. *See* circular cyliner
Cymatogaster aggregata (seaperch), swimming mode, 286
Cynocephalus (flying lemur), gliding, 257
Cypselurus (flying fish), gliding, 258

d'Alembert's paradox, 82, 156
damping, with compliant surfaces, 152
dandelion, 13 (fig.)
Daphnia (water flea), 31 (fig.), 340; cyclomorphosis, 30; sinking rates, 30, 340; swimming, 30, 344, 359
Darcy's law, 305
Deborah number, 18
deer fly, reported flying speed, 21
deformation, shear. *See* shear deformation
Dendraster excentricus (sand dollar), 246
density 7, 10, 11, 22–23, 80; constancy with speed, 21; dimensions and units, 10, 22; in kinematic viscosity, 25; measurement in plankton, 30; regulation by diatoms, 346; sorting in vortices, 216; and wave formation, 379
desmids, mucilage coat, 344
Desmognathus quadramaculatus (salamander), 200
detritus feeding, 189; from ascending vortices, 216–18
Dianous (beetle), surfactant propulsion, 390
Diaptomus (copepod), steady swimming, 359
diatoms: fig., 345; composition adjustment, 344; density regulation, 346; fi-

bers and sinking rate, 345; sinking speeds, 344
diffusion, 47; air vs. water, 198; alveoli vs. capillaries, 199; in capillaries, 313; in cellular domain, 313; coefficient, 170, 313; vs. convection, 196, 313; vs. distance, 35; eddy, 170; exchange while swimming, 355; Fick's law, 47, 197; intracellular transport, 196; molecular, 170; Péclet number, 313; speed of, 196; vs. swimming, very low *Re*, 332; and transmural exchange, 35; in transpiration, 197; and "unstirred layer," 201; and velocity gradients, 196–202; in viscous sublayer, 170; water vapor, 197
diffusional deposition, filtration by, 357
diffusive boundary layer, and unstirred layer, 201
diffusive vs. convective transport, 202
dimensional analysis, 89
dimensional homogeneity, 8, 11
dimensionless coefficients, role at high *Re*, 304
dimensionless numbers, 9. *See also* specific numbers such as Bond, Froude, Péclet, Reynolds, Weber, Womersley; *also* drag coefficient, friction factor, etc.
dimensions, 7–9; in dimensional analysis, 89; fundamental, 8; table, 10
Dineutus (whirligig beetle), hull support, 388
dinoflagellates, seasonal polymorphism, 30
dinosaurs, blood pressure, 56
dipteran larva, signal for attack, 359
direct interception, filtration by, 357
discharge coefficient, 303
discharge, rivers, 39
disk, as speed calibrator, 98
disk loading: defined, 274; and induced velocity, 274
dispersal: ballooning, 222; byssus thread drag, 223; mucous thread, 223; propagules, 344; seed, 4, 6, 12, 245, 257, 272, 344
displacement ships. *See* surface ships
distance index, 308. *See also* pipes
disturbance as signal, very low *Re*, 359
Dolomedes (fishing spider): prey detection with waves, 392; sailing, 391

dolphin: bow-wave riding, 386; "dolphin's secret," 150; drag, 151; Gray's paradox, 152; thunniform swimming, 282

Donax (coquina clam), 182 (fig.); swash-riding, 181

Draco volans (lizard), gliding, 257

drag, 7, 74; in accelerating motion, 363; airfoil, 109; axisymmetrical bodies, 113; on beds of mussels, 66; in bidirectional flow, 114; biological relevance, 106–9; black fly larva, 113, 184; bluff body, 111; in boundary layer, 164–67, 175–81; categories summarized, 241; cetaceans, 151; change by flattening, 134; circular cylinder, 97, 335, 352; circular disk, 112, 334; coccids on leaves, 193; coefficients for low *Re*, 336; coral, 107; crabs, 146; dead or towed fish, 280; dependence on speed, 84; dimensional formula, 81; and drag coefficient, 89; energy cost, 132; equality with thrust, 136; and fairing, 111; falling mice, 108; fish, 147–48; and fitness, 107; flag, 123; flat plate, vs. orientation, 351; and flexibility, 114–27, 335; fluid sphere, very low *Re*, 337; flying deer fly, 21; as force, 14; as friction, 80; frogs, 146; fruit fly, 10; glide length as measure, 138–39; golf ball, 142; and growth, 127–28; hollow half-cylinder, 112; hollow hemisphere, 112; imprecision, 11; indirect measurement, 75; insect, 15; jumping flea, 142; leaves, 14, 120–24; with lift, 226; vs. lift, sessile organisms, 101; vs. lift at interfaces, 176; limpet shell, 66; low *Re* factors, 337 (table); macroalgae, 125; marine mammals, 150–51; measurement, pine tree, 121; motile animals, 132–54; mucous thread, 223; oscillation at transition, 375; as part of thrust, 139; penguins, 149–50; plaice, 65; from pressure data, 63; during ram ventilation, 69; reduction by local heating, 29; reduction by roughness, 99; reduction by streamlining, 98–99; sand dollar, 65; scaling contra acceleration reaction, 368; sea anemone, 115; sea fans, 107; and separation, 96, 99; sessile systems, 106–28; and shape, 96–98, 106–27, 132–54; silk strand of spiderling, 12; silk, thread, 222; skin friction vs. pressure drag, 96–98; snails in irrigation canals, 108; solid hemisphere, 111; sources, 6, 7, 102; vs. speed, very low *Re*, 333, 348; speed-specific drag, 117; sphere, 112, 333; spheroid, 336; spider silk strands, 335; spot on surface, 164, 165; of sting, 138; from streamlines, 43; strut, 109, 134; surface swimming, 385; surface swimming, speed-specific, 385; and suspension feeding, 192; in swimming, 280; tadpoles, 145; for thrust, 283–87; torrential fauna, 176–80; tree, 14, 107; turbulent flow vs. laminar flow, 151; at turbulent transition, 94; unsteady flow, 375; and viscosity, 63; from wake width, 98; wall effect, very low *Re*, 339; from waves, 141. *See also* drag coefficient; E-value; induced drag; parasite drag; pressure drag; skin friction

drag coefficient, 89; vs. angle of attack, 235; area conversion table, 133; body of revolution, 135; cetaceans, 151; change by flattening, 134; circular cylinder, 92 (data), 336 (equation); conversion factors, 132; crabs, 146; desert locust, 144; drogue, 129; emperor penguin, 149; vs. fineness ratio, graph, 110; fishing bat legs, 140; flat plate, 907, 98 (table); flat plate vs. motile animals, graph, 144; frogs and tadpoles, 146; and front fairing, 111; gentoo penguin, 149; from glide angle, 253; harbor seals, 150; Humboldt penguin, 149; isopod, 145; jumping fleas, 142; leaves, 123; mackerel, 147; "Microbus," 113; motile animals, table, 143; *Nautilus* shell, 147; paddles, 154; pigeon, 148; pine tree, 121; power strokes, 154; protrusions, 165, 166; and *Re*, 102, 117; reference areas, 90–91; saithe, 148; sea anemone, 115; sea lion, 150; and separation, 96; and shape, 132; sphere, 92 (data), 334 (equation); streamlined animals, 132; streamlined body, 133; strut, 109; swimming human, 151; and terminal velocity, 340; and thickness-to-chord

drag coefficient (*cont.*)
  ratio, 109, 114; trout, 147; underwater ducks, 147; variation with velocity, 116; various bodies and profiles, 112 (fig.); very low *Re*, 333; water beetles, 145
drag coefficient, local, 164–67
drag maximization, seeds, 246, 257
drag reduction: air deflectors, 114; in boundary layer, 175; boundary layer suction, 179; by compliance, 115; compliant surfaces, 152; ejecting high-velocity fluid, 110; flattened torrential fauna, 177; mucus secretions, 152; overall role, 127; porpoising, 151; reality checking, 153; shark scales, 153; splitter plates, 110, 111; by streamlining, 134; suction, 109; surface heating, 153; surface morphology, 153; swimming, 151–54; vs. thrust production, 136
drag, induced. *See* induced drag
drag, interference. *See* interference drag
drag, profile. *See* profile drag
drag-based locomotion, 146, 154–55
dragonfly: eddies in wing pleats, 245; flying, 86; nymph as jet, 78 (fig.)
dragonfly larva: acceleration reaction, 365; jet propulsion, 77
droplet, vortices inside, 214
Drosophila (fruit fly), 143, 268, 369
duck: flying, 86; hull length, 384; hull shape, 107; lift-to-drag ratio, 248; maximum swimming speed, 384; rubber, 383; surface swimming, 283, 385; swimming with drag, 286; underwater drag coefficient, 147
dye marker, 173
dynamic pressure, 53, 58, 62, 69, 97, 99; cylinder, 83; in drag coefficient, 89; in friction factor, 301
dynamic similarity, 88, 102
dynamic similitude, 89
dynamic viscosity, 18, 23–25; cells, 18; and character of flow, 85; vs. density, 24; dimensions and units, 10, 23, 24; glaciers, 18; glass, 18; vs. kinematic viscosity, 24, 25; symbols used, 24; table, 23; tar, 18; vs. temperature, 24, 27–31, 153; in viscometry, 26. *See also* viscosity
Dytiscus (water beetle), 145

E-value: circular cylinder, 120; extrapolation, 120; flat plate, 120; flexible coral, 120; hydroid, 120; kelp, 125; leaves, 123; locust antenna, 120; low, 120; macroalgae, 120; pine tree, 121 polyethylene sheet, 123; in velocity gradients, 120. *See also* speed-specific drug
earthworm, intestine, as pipe, 311–12
Ecdyonurus (mayfly nymph), 178, 179
Echinocardium (echinoderm), burrow pump pressure, 327
ectoparasites, 180
eddies: attached, 93; and separation point, 95; in turbulence, 87; vertical momentum transport, 167
eddy viscosity, 24, 47
eel, 88; anguilliform swimming, 281
efficiency: cilia and flagella, 354; pheromone detection, antenna, 356
Eiffel, Gustav, 235
elevation, to increase wind exposure, 193
elm, winged seed, 251, 257
elphin forest, 127
embolism, in xylem, 38, 323
Encarsia (wasp), 278
endothelial cells: and Murray's law, 320; and shear stress, 188
energy: conservation, 809; cost of drag, 132; dissipation in boundary layer, 157; dissipation in vortex, 209; dissipation in wake, 97; in jet propulsion, 77; vs. momentum, 73, 77; and surface tension, 378
Engraulis mordax (anchovy), 281
entrance region, 297 (fig.); laminar, 296; turbulent, 302
Epeorus (mayfly nymph), 178
Epischura (copepod), steady swimming, 359
Eristalis tenax (hover fly), 265
erosion, 19; shear flow, 187; shear stress, 188; by vortices, 219
Esox (pike): acceleration reaction, 365; C-start acceleration, 287
Eubalaena australis (whale), 181
Euphausia embryos, sinking rates, 30
euripterid, swimming with lift, 286
evaporation rate, and boundary layer, 163
exhalation, flow limitation, 316
exposure, 128–29

fairing: in boundary layer, 166; and drag, 111

Falco peregrinus (falcon), 1458

falcon: glide angle, 254; interference drag, 138; lift-to-drag ratio, 148; parasite drag, 148

Fanning friction factor. *See* friction factor

feathers, and parasite drag, 148

fecal pellets, sinking speeds, 344

fibroblasts, velocity gradient in medium, 200

Ficedula hypoleuca (flycatcher), 278

Ficik's law, 47, 199

filter feeding: and continuity, 39; pumps, 324; vs. suspension feeding, 188; and temperature, 29; and viscosity, 29. *See also* suspension feeding

filtration: vs. advice predation, 359; collector geometry, 358; direct interception, 357; electrostatic attraction, 357; encounter vs. retention, 357, 358; filter resistance, 356; high *Re*, 355; inertial impaction, 357; mechanisms, 356–58; motile particle deposition, 357; rake vs. paddle, 356; range, 356; sieving, 356; transmissivity, 356; very low *Re*, 355–59

fineness ratio, 135; vs. drag coefficient, graph, 110

fir, thigmomorphogenesis, 128

fish: antarctic "ice fish," 29; aspect ratios of flying fish, 259; drag, 106, 147–48; drag and opercular ejection, 110; flying, 258 (fig.); form vs. pressure, 67; gill ventilation, 68; gills, 29, 68, 300 (fig.), 311; high body temperatures, 28; jet propulsion, 77; lift-to-drag ratios of flying fish, 259; lunate tail sweepback, 239; mucus secretion, 152; ram ventilation,60, 68, 110; schooling, 288; shear flow and eggs, 187; slope soaring of flying fish, 259; streamlining, 106; suspension feeding, 189; swimming, 67, 137, 137–48, 286; swetted surface, 90

flag: drag, 123; tatter, 128

flagella: arrangements, 351; bacterial, 206, 331, 351, 352; of *Chlamydomonas*, 352; efficiency, 354; making thrust, 352–55; mastigonemes, 355; motion,

351, 352; operation, 351–55; as pumps, 314; rotation of bacterial, 206; of sponges, 39

flagging, and tree drag, 127

flapping, 262–64. *See also* flight

flatfish: ground effect, 289; lift, 176

flat plate: added mass coefficient, 364; as airfoil, 244; autorotating, 228; boundary layer, 157–63; drag coefficient, 97–98, 136 (graph), 98 (table); drag for laminar flow, 135; drag for turbulent flow, 135; drag vs. orientation, 351; E-value, 120; flow around, 98; low drag paradigm, 135; separation point, 98; as speed calibrator, 98; splitter plate and drag, 111; Strouhal numbers, 371; turbulent boundary layer, 162 (fig.); turbulent transition, 135

flat plates, parallel 310

flattening: feather mites, 181; lift vs. drag, 177; mayfly nymphs, 178; torrential fauna, 178; turbellarian flatworms, 179; water mites, 179; water penny beetles, 179

flea, drag in jump, 142

Flettner, Anton, 227

Flettner rotor, 227–29, 245; vs. airfoil, 233; tree seeds, 228

Flettner ship, 227

flexibility: deformation vs. reconfiguration, 115; drag and, 114–27, 335, 375; macroalgae, 124–27; moth antennae, 105

flexible organisms, pressure distributions, 67–70

flight, 13; advance ratio, 266; birds, 137; with bristle wings, 350; clap-and-fling mechanism, 278, 279; estimating top speed, 268; evolution, 254; formation, birds, 288; ground effect, 288; hovering, 265, 278; insects, 137; nonsteady effects, 277–80; origin of thrust, 262–64; quasi-steady analysis, 278; samaras, 272–73; swooping, 239; vortex wakes in flapping flight, 276; wing motion, 266 (fig.)

flounder, 66

flow, fully developed, 297

flow markers, various, 44

flow meter, 7

flow regimes. *See* laminar flow; turbulent flow

flow tanks, 7, 13, 14, 62, 98, 103

flow through porous media, 6

flow visualization, 41, 44; liquids vs. gases, 44; water vs. air, 102

flow, compressible. *See* compressible flow

flow-induced forces, maximum, 129

flow-induced pressure, 69, 103. *See also* pressure distribution

flowmeter, 62; and wall effects, 339

flows, maximum. *See* maximum flows

fluid: defined, 16, 17, 80; particle, 21, 47; statics, 5

fluidization, porous media, 307

flutter and separation, leaves, 376

flycatcher, lift coefficient, 278

fly pupa, 178, 180

foraminifera, settling, 184

force: dimensions, 9; indirect measurement, 74–77; inertial, 86; viscous, 86. *See also* drag; lift; thrust

forest. *See* tree

forest canopy, wind within, 171

form resistance coefficient, 341

frame of reference, 12–14, 21; acceleration, 108; rolling vortices, 218; thrust and drag, 262

Fraxinus (ash), 228

free-stream velocity, 20, 76, 79, 156, 157, 158

frequency, vortex shedding, 370

fresh water: hydrostatic pressure, 51; properties, 23 (table)

freshwater limpet, 178 (fig.)

friction factor, 7, 301–2

friction velocity. *See* shear velocity

frog: acceleration reaction, 365; drag, 146; gliding, 257–58; leapfrogging, 211; velocity gradient and oxygen uptake, 200. *See also* tadpole

frogfish, jet propulsion, 77–78

frontal area, defined, 90

Froude, William, 79, 384

Froude number, 284–85; vs. drag, 385 (graph); and maximum speed, 284; use in modeling, 384

Froude propulsion efficiency, 237; bird wings, 237; carangiform swimming, 281; drag-based swimming, 287; ducted fan engines, 330; fixed wing, 237; force vs. power, 237; helicopter blades, 271; and hovering, 266; jet propulsion, 79; with zero momentum wake, 280

fruit fly: acceleration reaction, wings, 368; advance ratio, 268; aerodynamic frequency parameter, 376; clap-and-fling mechanism, 278; drag, 10; glide angle, 254; parasite drag, 143; polar diagram, 245, 248; untwisted wings, 264; wingbeat frequency, 268, 368; wing loading, 243; wing mass, 245, 369; wing stall, 250

fundamental dimensions, 10

fungal spore, sinking speed, 344

fungus, detaching conidia, 194

furrow, flow across, 212

Gadus callarias (cod), 281

Galileo, 142

gas, 16; vs. liquid, 22; rarefied, viscosity, 19

geese, formation flight, 288

Gerridae (water striders), 182

gills: cilia on, 314; fish, 29, 300, 311; positive displacement pump, 326; reduced area in rheophiles, 200; ventilation, fish, 68; in vortices, 218

giraffes, blood pressure, 55

glaciers, viscosity, 18

glass, viscosity, 18

Glaucomys volans (flying squirrel), 250

glide angle, 253–55; and lift-to-drag ratio, 253; vs. size of animal, 254; and speed, 253

glide ploar, 255–56

gliding, 252–62; animals, 257–59; and aspect ratio, 253; vs. autogyrating and autorotating, 257; biological gliders, 258 (figs.); bird, 21; decelerative, 139; flying fish, 258; flying vs. sinking speed, 255; Javanese cucumber seedleaf, 256; to measure drag, 138; minimum speed, 255; vs. parachuting, 257; phalanger, 250; size of animals, 254; speed vs. weight, 254; time aloft, 255; turbulence and seeds, 257

glycerin, properties, 23

golf ball: drag, 142; role of dimples, 100

goose barnacles, filter elevation, 191

gorgonian coral: E-value, 120; on sea-mounts, 40
grapevine, vessel length, 322
gravitational deposition, filtration by, 357
gravity: in Bernoulli's equation, 53; vs. drag in jump, 142; estimating acceleration, 34; and hydrostatic pressure, 51; sinking *Paramecium*, 137; and wave celerity, 381; and wave formation, 379
gravity waves, 382
Gray's paradox, dolphin swimming, 152
ground effect, 260; with clap-and-fling, 289; flight, 288; swimming, 289
growth and drag, 127
gun: to get through velocity gradient, 195; for propagule propulsion, 194; in water, 196
Gyps ruppelli (vulture), 148
gypsy moth, thermal soaring, 222
GHyrinus (water beetle): oarlets, 350; rowing stroke, 155 (fig.)

Hagen-Poiseuille equation, 5, 293, 297, 314, 334; applicability to unsteady flow, 377; and cost of circulation, 317; and sap ascent, 322
half-cylinder, as self-excited oscillator, 372
hawk, lift-to-drag ratio, 248
hawk, Harris', feathers in flight, 148
hearts, 55, 56; maximum pressure, 327; output of human, 35; positive displacement pumps, 325; valve-and-chamber pumps, 327; valves, 62
heartbeat, fish, 68
heat exchanger, 28
heat transfer: and boundary layer, 163; convective, 6
Hedophyllum (macroalga), 126 (fig.); drag and structure, 125
helicopter rotor, 269–71; as airfoil, 269; and hovering flight, 271; origin of vertical force, 271
hematocrit, 312
hemlock, sinking speed, pollen, 344
holly, speed-specific drag, 117
honeybee. *See* bee
Hookean material, 17, 18
hoptree, winged seed, 257
horse, blood pressure, 55

horseshoe vortex, 219; digging, 221; mayfly larva, 220
hover fly: stroke plane, 265; wing corrugations, 245
hovering: animal flight, 265; Froude propulsion efficiency, 266; induced drag, 265; kinds of analyses, 276; lift coefficient, 278; vs. size, 266; stroke, plane, 278; vortex wake, 276; and wing loading, 266; wing motion, 267 (fig.)
hull length: vs. hull speed, 285 (graph); and maximum speed, 384; and wavelength, 382
hull shape, ducks, 107
human: circulation and continuity, 35–37; heart output, 35; instabilities in circulation, 374; micturition, 88; penis as nozzle, 396; pressure drop, circulation, 324; respiratory wheezes, 375; swimming drag coefficient, 151; volume flow, circulation, 324
hummingbird: energy consumption, 237; hovering, 266
Hydra, sailing, 391
Hydracarina (water mites), 179
hydraulic ram, 330
hydraulics, 5
hydroid, colonial: commensal suspension feeders on, 190; E-value, 120; flow around, 190
Hydrolagus colliei (ratfish), swimming mode, 286
Hydrosaurus pustulatus (lizard), slapping, 387
hydrostatic paradox, 51
hydrostatic pressure, 51, 70
hydrostatics, 5
hygropetric organisms, examples, 182
Hyla venulosa (tree frog), parachuting, 258
Hymenochirus (frog), 146; acceleration reaction, 365

ice cream freezer, 17
ice fish, blood, 29
ichthyosaurs, 150; thunniform swimming, 282
ideal fluid, 82, 86; deprecated, 52; lift, 232; vortices, 209
Idotea (isopod), 91
iguana, circulation, 28

impact pressure, 130
impedence: of pumps, 325–29; shifting with transformer, 329
imprecision, 11
impulse, 348
Inachis io (butterfly), 277
incompressibility, 21, 208; and principle of continuity, 33
indirect force measurements, 74–77
induced drag, 137, 237–38, 242; and aspect ratio, 238; in hovering, 265; insect wings, 250; phalanger, 250; on polar diagram, 238; vs. speed, 238
induced power, 149; vs. speed, 238
induced velocity, and disk loading, 274
inertial force, and pressure drag, 97
inertial impaction, filtration by, 357
information transfer, at very low *Re*, 359
insects: aquatic, as protrusions, 165; body lift, 239; drag, 15; flight, 86, 137; flight speed, 15; parasite drag of small, 142; polar diagrams, 248; wetted surface, 90; wingbeat, 15; wing performance, 250
interfaces, 21; lift vs. drag, 176; solid-fluid, 18, 71, 156, 157
interface, air-water, 19, 141, 378–92, 394–97; Bond number, 347; clap-and-fling mechanism, 289; Coanda effect, 397; as hazard, 347; Jesus number, 347; Weber number, 347
interfaces, fuzzy, 393; atmospheric inversions, 393; Richardson number, 393; salinity gradient, 393; temperature gradient, 393
interfaces interfluid, 378–79
interference drag, 138, 242; bird bodies, 148; and drag measurement, 138; falcon, 138; half-streamlined body, 165
internal flows. *See* pipes; pumps
internal fluid transport systems, pipe sizes, 35
interstitial flow, beaches, 307. *See also* porous media
Iron (mayfly nymph), 178
irrotational vortex, 206, 207 (fig.), 225
isopod: drag, 91; drag coefficient, 145; reference areas, 91
isotachs, 44–46

Javanese cucumber seed-leaf, 251 (fig.); airfoil, 252; as glider, 256
jellyfish, as jet, 78 (fig.)
Jesus number, 347
jet: as positive displacement pump, 326; to prevent recirculation, 190; of puffball, 195; of sponges, 39; vortex wake, 215
jet propulsion, 73, 77–80; acceleration reaction, 365; accelerations, 365; cephalopods, 77; ctenophore, 77p dragonfly nymphs, 77; efficiency, 79, 237; energy, power, thrust, 77; frogfish, 77; medusae, 77; momentum in, 77; pulsed jet, 79; scaling, 80; scallops, 77, 251; squid, 69; tunicates, 77, 80
jets, colonial, in suspension feeding, 190
Joukowski, N. E., 231
Juglandaceae, autogyrating samaras, 272

kelp, drag, 125
kestrel, slope soaring, 260
killer bee, 13
kinematic viscosity, 23–25; air, 25, 88; and boundary layer, 158; dimensions and units, 10, 25; vs. dynamic viscosity, 24; local lowering by heat, 29; in modeling, 102; vs. temperature, air, 25; vs. temperature, water, 25; in viscometry, 26; water, 25, 88
Kozeny-Carman equation, 306
Kozeny function, 306
Kutta-Joukowski theorem, 226, 230

Lacuna (gastropod snail), 223
Lagenorhyncus obliquidens (cetacean), 151
lakes, 6
laminar flow, 5, 46–49, 84, 87, 290; circular aperture, 303; drag of flat plate, 135; maintaining, 109; parallel plates, 298–300; pipes, 290–98; transition, 11; vortices in, 209; wall pinching index, 315
laminar sublayer, 161, 185
Laminaria, 126 (fig.)
Lanchester, F. W., 230, 232
Langmuir circulations, 394
Lanice (polychaete worm), 217; mucus thread, 344
Laplace's law: for jet propulsion, 78; surface tension and nozzle flow, 395

larvae: antarctic echinoderm, 29; swimming speeds, 185

larval recruitment. *See* recruitment, larval, 40

leaves: aquatic plants, 199; area, 37; boundary layer resistance, 197; boundary layers, 163; drag, 14, 120–24, 100; E-values, 123; as Flettner rotor, 229; flutter as separation shifting, 376; reconfiguration in wind, 91, 123–24; as self-excited oscillators, 373; size/shape and boundary layer thickness, 199; transpiration, 197

leg of fishing bat as strut, 141

Leguminoseae, autogyrating samaras, 272

lemur, gliding, 257

length: characteristic, 85; dimension, 8

Leucothea (ctenophore), 77

lichen, vortices in cup, 214

lift, 7, 74, 225–27 (figs. 227, 232); and airfoils, 230; and angle of attack, 242; on beds of mussels, 66; and Bernoulli's principle, 65, 233; and circulation, 226, 230–31; cost vs. aspect ratio, 237; crabs, 146, 176; from curved surface, 230, 232; distribution on airfoil, 233; vs. drag at interfaces, 176; vs. drag, sessile organisms, 101; finite wing, 237; flatfish, 176; Flettner rotor, 227; flow pattern over airfoil, 231; ground effect, 260; in ideal fluids, 232; indirect measurement, 76; and induced drag, 137, 238; Kutta-Joukowski theorem, 226, 230; vs. lift coefficient, 234, 244; limpets on surface, 66, 176; making thrust, 137, 263, 283–87; mayfly larvae, 176, 179; negative, 176, 179; nonsteady effects, 277; and orientation, snails, 176; oscillation near stall, 376; and peel failure, 176; of plaice, 65; and pressure distribution, 64–67; from protusions, 65; of ray, 65; rays, 176; reduction by separation, 179; reduction by spoilers, 179; from rotating cylinders, 230; of sand dollar, 65, 176; scaling, 243; in seed dispersal, 246; of self-excited oscillator, 373; vs. separation, 177; streamlined object, 64; unconventional mechanisms, 280; at very low *Re*, 347; as wall

effect, very low *Re*, 339; on water penny beetles, 67

lift coefficient, 234–36; vs. angle of attack, 234; bat, 278; flycatcher, 278; formula, 234; fruit fly wing, 250; from glide angle, 253; for hovering, 278; from lift, 234; reference area, 91, 234; steady vs. non-steady, 278

lift, body. *See* body lift

lift-based locomotion, 155

lift-to-drag ratio, 238, 242, 243, 244, 245, 248, 253; and angle of attack, 264; flying fish, 259; gliders, 255; insect wings, 250; and lengthwise wing twist, 264; ski jumper, 240

limpet: E-values, shell, 120; lift, 176; lift and drag of, 66; local drag, freshwater, 179; separation, freshwater, 179

Linuche unguiculata (coelenterate), aggregation in windrows, 394

liquid, 16, 22

liquid-gas interface. *See* interface, air-water

Liriodendron (tuliptree), 228

Littorina (periwinkle snail), 176

lizard: glide angle of *Draco*, 258; gliding, 257, 258 (fig.); slapping locomotion, 387

local drag: bacteria on walls, 176; freshwater limpets, 179; planarian, 179

local drag coefficient, 158, 164–67; and boundary layer thickness, 159; and transition, 161

locomotion, drag-based, 146, 154–55

locust: aerodynamic frequency parameters, 376; body lift, 239; drag coefficient, 144; E-value, antenna, 120; hindwing polar diagram, 248; lift-to-drag ratio, 255; quasi-steady flight, 278; swarm movements, 223

Locusta migratoria (locust), 223

lodging, 108

logarithmic boundary layer. *See* boundary layer, logarithmic

lugworm, 327 (fig.); burrow pump, 326–27

Luna (moth), 104

lunate tail, 282

lung, amphibian, positive displacement pump, 326

lung, mammalian: capillaries, 37; data, 36; volume flow rate, 37

lunules, of sand dollar, 65
Lycoperdon (puffball fungus), 195, 344
Lycopodium (club moss) spores, 193; powder as marker, 214; settling, 186
Lymantria dispar (gypsy moth), 222

mackerel: drag coefficient, 147; ram ventilation, 68; thunniform swimming, 282
macroalgae, 126 (figs.); bryozoans on, 191; drag and flexibility, 124–27; drag and wave period, 125; drag in unsteady flow, 375; E-values, 120; undulate margins, 125
Macrocystis pyrifera (macroalga), 125
Macronema (caddisfly), 60, 61 (fig.)
Magnus, H. G., 226
Magnus effect, 226, 228
mammals, blood viscosity in antarctic, 28
mandarin fish, ground effect, 289
manometer: Chattock gauge, 56; devices, 55 (fig.); inclined-tube, 56; inexpensive multiplier, 55, 56, 60; for Pitot tube, 58, 60; for Venturi meter, 58
manometric height, 54, 55
manometry, 54–57
maple: samara, autogyrating, 228, 252, 272; vessel length, 322
marine mammals, drag, 150–51
marine snow, 331; sinking speeds, 344
marlin, thunniform swimming, 282
mass: vs. density, 22; dimension and units, 8
mastigonemes, on flagella, 355
maximum flows, 128–30; measurement, 129–30, 130 (fig.); prediction, 130
mayfly larva, 178 and 220 (figs.); E-values, 120; flattening, 178; horseshoe vortex, 221; leg struts, 139; lift, 176; separation, 179; streamlining, 177
medusae, jet propulsion, 77
Mellita quinquiesperforata (sand dollar), 65
Membranipora villosa (bryozoa), 191
mercury, properties, 23 (table)
metachronal rhythm, 154
Metridium, 115, 116 (fig.); filter elevation, 191
microcirculation, 37. *See also* capillary
micrometeorology. *See* biometeorology

micturition, 88; flow limitation, 316; squirting vs. spraying, 396, 397
midges: hygropetric, 182; sewage filter flies, 183; wingbeat frequency, 268
migration, monarch butterfly, 13
mink, drag at surface, 385
mistletoe, dwarf, seed as projectile, 195
mites, feather, specializations, 181
mites, water, flattening, 179
mitosis, 5
mixing, and unstirred layer, 202
modeling: constant force, 103; flow-induced pressure, 103; media shifts, data, 104; moth antenna, 105; prairie dog burrows, 103; and *Re*, 102–5; sea anemone, 115; suspension feeding, 103; using highly viscous media, 102
molecular viscosity, 24
molecules, 19, 20; intermolecular cohesion, 378
mollusks, bivalve: cilia, 39; continuity, 39
momentum, 33, 73–77; angular, 205, 208; conservation, 73; dissipation in boundary layer, 157; vs. energy, 77; flux, 75; flux in hovering, 276; flux of beating wing, 273; flux through actuator disk, 273; and force, 73; and impulse, 348; in jet propulsion, 77; momentum equation, 74; near flat plate, 98; and pressure gradient, 95; and propulsion efficiency, 237; transfer in turbulent flow, 47; at very low *Re*, 331
mosquitoes, wingbeat frequency, 268
moths, 350 (fig.); modeling flow through antenna, 104; pheromone detection, 356; tracheal system, 320
motile particle deposition, filtration by, 357
motion, relative. *See* relative motion
mucous threads: drag and dispersal, 223; use by nudibranch mollusks, 389
mucus, as non-Newtonian fluid, 20
mucus secretion for drag reduction, 152
Murray's law, 188, 317–21; for bifurcation, 318; circulatory systems, 319; coral gastrovascular system, 320; crinoid coelomic circulation, 320; and endothelial cells, 320; leaf vessels, 320; moth tracheal system, 320; sap conduits, 320; and self-optimizing system,

321; and shear stress on walls, 318; speed vs. vessel radius, 318; sponges, 319; velocity gradients, 318

muskrat: Froude number, 385; maximum swimming speed, 385; swimming with drag, 286

mussel: dispersal via byssus thread drag, 223; lift on beds, 66; pumping rate, 189; pump pressure, 328

muzzle speed: mistletoe seed, 195; Sordaria projectile, 195

Mya arenaria (clam), 328

Mytilus edulis (mussel), 223, 328

nasal hairs, 310

nasal passages, 312 (fig.); as parallel plates, 311

nautilus (cephalopod), 77, 80; shell drag coefficient, 147

navicular fossa, 396

Navier-Stokes equations, 156, 158; very low *Re*, 333

negative pressure, 38. *See also* sap conduit

Nemertesia antennina (colonial hydroid), 190

Nereis (polychaete worm), 326

Nereocystis (kelp), 125, 126 (fig.)

Newton's first law, 81

Newton's second law, 52, 73, 74, 86, 138

Newton's third law, 75

Newtonian fluid, 18, 20

no-slip condition, 18–20, 156, 167, 314; and gas droplet, 214; and pipe flow, 292; thrust from shear, 338

Noctilio (fishing bat), 140

non-Newtonian fluids, 5, 20; viscometry, 26

nonsteady effects, 277–80; clap-and-fling, 278; insect flight, 249; stall delay, 278; Wagner effect, 277; wing rotation, 280

nonsteady flows, and suspension feeding, 358. *See also* unsteady flows

Notonecta (back swimmer), prey detection with waves, 392

nozzle, 77, 395–97 (figs., 396); breakup of stream, 395; elevation as, 40; flow and Ohnesorge number, 396; flow and *Re*, 395; flow and Weber number, 395; and Laplace's law, 395; mammalian

penes, 396; minimal size, 190; power loss, 33; and principle of continuity, 33; satellite droplets, 396; of sponges, 39; squid, 79; and surface tension 395

nudibranch mollusk, use of mucus threads, 389

oak: E-values, leaves, 124; sap ascent, 37; vessel size, 322

oarlets, for swimming, 350

ocean currents, 5

octopus. *See* cephalopods

Ohnesorge number, 396

olfaction, 356; flow through burrows, 71; plesiosaurs, 60

Orconectes (crayfish): acceleration, 287; escape response, 365

orifice: circular aperture, 303; and continuity, 33; flow through, 394; and surface tension, 395; vena contracta, 395. *See also* nozzle

orifice coefficient, 303

oscillators, self-excited, 372–75

Ostwald viscometer, 104

owl: quiet flight, 242; wing barbs for anti-stall, 242

Pachyptila vittata (prion), 387

paddles: acceleration reaction, 366; amphipod pleopods, 328; drag coefficients, 154; as fluid dynamic pump, 327

Pandalus danae (shrimp), accelerations, 287

parabolic flow. *See* pipes

parachute, 12

parachuting, 257. *See also* gliding

Paramecium (ciliate protozoan), 349; flow around, 137; sinking, 137; velocity gradients around, 354

parapodia, on polychaetes, 355

parasite drag, 137, 240, 241; birds, 148–49, 248; falcon, 148; and feathers, 148; measurement, 139; small insects, 142–43

particles, aerosol, 192

pascal, 23

pascal second, viscosity unit, 24

pathlines, 41–45

penguins: drag, 139, 149–50; drag coefficients, 149; drag at surface, 385;

penguins (*cont.*)
maximum speeds, 149; swimming, 137, 283, 286; transport cost, surface swimming, 386

Penicillium (fungus), spore elevation, 193

penis, 396 (fig.); as nozzle, 396; orifice, 88

peristaltic pumps, 326

Petaurus breviceps (phalanger), 250

petiole, 123

petrel: sea anchor soaring, 260; slope soaring, 259

Péclet number, 202, 313; capillaries, 313; phytoplankton, 346; swimming protozoa, 355

phalanger, 251 (fig.); gliding, 250, 257; lift-to-drag ratio, 253; polar diagram, 250

pheromone, detection by moth, 356

Phoca vitulina (harbor seals), 150

Phoronopsis viridis (phoronid worm), 217

photographic image and velocity gradient, 198 (fig.)

photosynthesis: cost in drag, 107; vs. current, 199; and drag, 127

Phragmatopoma lapidosa (tube worm), attachment, 185

Phragmites australis (reed), 71

Phrynohyas (flying frog), gliding, 257

phugoid oscillation of aircraft, 376

Physalia (Portuguese-man-of-war), sailing, 391

phytoplankton: Péclet number, 346; sinking speeds, 344; why sink?, 346

Picea abies (spruce), 272

Pieris brassicae (butterfly), 276

pigeon: drag coefficient, 148; drag as spheroid, 133; glide angle, 254; wing area adjustment, 243

pike: acceleration reaction, 365; C-start acceleration, 287

Pilobolus (fungus), tumbling sporangium, 338

Pinaceae, autogyrating samaras, 272

pine: cones as pollen traps, 187; drag coefficient, 121; E-value, tree, 121; pollen, 345 (fig.); speed-specific drag, 117, 118 (fig.); thigmomorphogenesis, 128

pinnipeds, drag and swimming, 150

pipes: avoiding parabolic flow, 308–14; Bernoulli vs. Hagen-Poiseuille, 314–

16; biological applications, 308–23; branching arrays, 33, 316, 321, 318 (fig.); capillaries, 311, 312; with ciliated walls, 314; continuity, 290; distance indices, 308–11; earthworm intestine, 311; entrance region, laminar, 296, 297 (fig.); entrance region, turbulent, 302; flow measurement, 295; flow profiles, 310 (fig.); friction factor 301, 302 (graph); Hagen-Poiseuille equation, 293, 297; laminar flow, 290–98; Laplace's law, 317; Malpighian tubules, 311; mean vs. max flow, turbulent, 301; mean vs. max speeds, laminar, 295–96; nasal hair role, 310; noncircular cross sections, 310, 312 (fig.); nonmaterial, 41; no-slip condition, 19; parabronchi of bird lungs, 311; periodic boluses, 312; plug or slug flow, 297; power, laminar, 294; principle of continuity, 32–33; proximity of flow to wall, 308–11; pseudomanometry, 295; pumping at walls, 314; pumps, 323–29; renal tubules, 311; resistance, laminar, 294; resistance, at transition, 301; shear stress on walls, 291; size vs. pressure drop, laminar, 294; speed vs. x-section, turbulent, 301; of sponges, 38; total flow, 292–94, 295; transition, 290, 298, 305; transmural exchange, 308, 311, 312, 314; turbulent spouting, 375; using very small, 311; velocity vs. x-section, laminar, 290–92, 293 (fig.); velocity vs. x-section, turbulent, 300; vortices in bends, 215; wall roughness, laminar, 296; wall roughness, turbulent, 301; Womersley number, 377

pitch: aircraft, 242; blade element, 262; wings, 242

pitching moment, and angle of attack, 234

Pitot tube, 58–60, 59 (fig.), 72, 83; living, 61 (fig.)

Pizonyx (fishing bat), 140

plaice, 66 (fig.); lift and drag of, 65

planarian, local drag, 179

plan form area, 90

planing, 386–87

plankton: irregular shape, 345; measuring density, 30; shape vs. temperature,

30; sinking rates, 29, 88; size vs. temperature, 30. *See also* phytoplankton, zooplankton

plant: cells, 8; growth and wind, 127; lodging, 109; pressure gradients, 323; sap speeds, 323; thigmomorphogenesis, 128

plates, parallel: biological examples, 300; laminar flow, 298–300, 299 (fig.); mean vs. max speed, 300; nasal passages, 311; speed vs. x-section, 299; total flow, 299

Plecotus auritus (bat), 278

plesiosaurs, 150; olfaction, 60

Pleuronectes platessa (plaice, flatfish), 65

plug flow, 297, 309

poise, 24

Poiseuille's equation. *See* Hagen-Poiseuille equation

polar diagram, 234–36, 235 (fig.); for airfoil, 235; bumblebee wing, 245; fruit fly wing, 245; insect wings, 250; lift-to-drag ratio, 238; phalanger, 250

Pollachius virens (saith), 148, 281

pollen, 345 (fig.); exposure to wind, 193; sinking speeds 88, 344; trapping 187

pollination, 43

polychaetes: anguilliform swimming, 281; parapodia on, 355; swimming with drag, 286

polychaetes, tubicolous, filter elevation, 191

polymers, drag reduction with, 152

poplar, white, leaf reconfiguration, 124

Populus alba (white poplar), 124, 193

Populus tremuloides (aspen), 124

Porifera. *See* sponges

porous media, 305–307; fluidization, 307; Kozeny-Carman equation, 306; Kozeny function, 306; quicksand, 307; voidage or porosity, 306

porous substratum, flow in and out of, 173

porpoise, top speed, 151

porpoising, 151

Portuguese-man-of-war, sailing, 391

Postelsia (sea palm), 122, 374, 126 (fig.)

Potamodytes (beetle), 70, 83, 70 (fig.)

power: finite wing, 237; vs. flying speed, 149, 237, 240, 241 (fig.); loss in nozzle,

33; pipe flow, 294; in pumping, 323; of pumps, 328; to stay aloft, 238

prairie dog: burrow ventilation, 71; modeling burrow flow, 103

Prandtl, Ludwig, 156, 232

precision, 12

pressure, 21, 50, 51 (fig.); across airfoil, 233; atmospheric, 54; in beetle's bubble, 70; in Bernoulli's equation, 53; blood, *see* blood pressure; dimensions and units, 8–10; vs. flow speed, 56; and hydrostatic paradox, 51; impact, 130; manometric height, 55; manometry, 54–57; measurement of low pressure, 56; negative, 38; Pitot tube, 58–60; root, 321; units, 55; Venturi meter, 57–58. *See also* static pressure; dynamic pressure

pressure coefficient, 8, 62–64, 67, 90; near protrusion, 71; and *Re*, 102; streamlined object, 63; whale, 69

pressure distribution, 62–70 (figs., 63, 64, 68); beetle's bubble, 70; circular cylinder, 82, 83; fish, 67–69; grain storage buildings, 83; squid, 79; whales, 69

pressure drag, 97, 109, 111, 133, 242; airfoil, 240; inertial force in, 97; and separation, 97; vs. skin friction, 97, 109; in Stokes' law, 334

pressure-induced flow, 71–73; diagram, 72; prairie dog burrows, 71; reed, 71; relict burrows, 72; ripples of sand, 71; sand dollars, 71; sclerosponges, 71; storage structures, 72; stromatoporoids, 71; suspension-feeding brachiopods, 71; termite mounds, 71; tracheae of flying insects, 71; volcano cones, 72

principle of continuity, 5, 32–40 (figs., 33, 34), 53, 63, 80, 82; and actuator disk, 273; and beating wing, 273; and boundary layer, 159; contracting column of liquid, 34; and filter feeding, 39; human circulation, 35–37; and internal fluid transport systems, 35–39; pipes, 290; rivers rising, 39; and sap movement, 38; and sponges, 38–39; streamlines, 40–41

prion: planing, 387; sea anchor soaring, 260; suspension feeding, 260

profile area, 90
profile drag, 137, 240–241, 242, 244; gliding, 255; insect wings, 250; vs. size, 268
profile power, 149
projectiles, form, 194
propagule: elevation, 193–94; as projectile, 194
propeller, 269 (fig.); advance ratio, 267; as airfoil, 268; airstream contraction, 273; angle of attack, 264; blade contour, 270; blade element analysis, 267; and continuity, 34; disk swept by blades, 267; in ducts as pump, 328; efficiency, 79, 282; in ideal fluid, 82; induced velocity, 274; lengthwise twist, 264; like beating wing, 264; local shear flow, 188; lunate tail as, 282; making thrust, 262; torque, 263; as volume pump, 324; whale tail, 282; as windmill, 270
propulsion: efficiency of cilia and flagella, 354; via effective area change, 348, 350–51; irrelevance of drag data, 350; via orientation change, 348, 351, 355; surface slapping, 387; vs. tethered flow, 349; using cilia and flagella, 351–55; using surface tension, 387–90; at very low Re, 347–55; from waves, whale, 386. See also flight, swimming
propulsion efficiency. See Froude propulsion efficiency
propulsion, jet. See jet propulsion
protozoa: Péclet number, swimming, 355; rapid stalk contraction, 360; stalk for filter elevation, 191
protrusions and protuberances: density and flow, 171; drag coefficients, 165, 166; to get eddies, 312; organisms as, 165; in pipes, 296; pressure coefficients, 71; producing lift, 65; Res of, 165; skimming flow, 220; and Stokes' radius, 341; streamlining, 165; and vortices, 172, 219; zero plane displacement, 168. See also roughness
Psephenidae (water penny beetles), 179
Pseudiron centralis (mayfly larva), 221
pseudomanometry, 295
Pseudopterogorgia (gorgonian coral), 120
psychodid fly larva, hygropetric, 182

Ptelea trifoliata (hoptree), winged seed, 257
Pterphyllum (angelfish): acceleration reaction, 365; swimming with drag, 287
pteropod mollusk, swimming with lift, 286
Ptilosarcus (sea pen), 113 (fig.)
puffball's jet, 195
pulmonary artery, 35
Pulvinariella mesembryanthemi (scale insect), 193
pumping rate: bivalve mollusks, 189; sponge, 189
pumps, 323–29; biological examples, 324, 326 (table); cilia on walls, 314; ciliary and flagellar, 39; evaporative sap lifter, 324; filter-feeders, 324; flagellated chambers, 314; impedance matching*, 325; impedance matching**, 329; vs. local flow, 192; maximum pressures, 328; measuring output, 328; peristaltic, 326; of polychaete worms, 326; power output, 323; pressure drop vs. flow, 329 (graph); resistance, 323; in suspension feeding, 189; transformers, 329–30; valve-and-chamber, 325
pumps, fluid dynamic and positive displacement, 325; biological examples, 326–28; impedance, 325; pressure range, 327
Pygoscelis papua (gentoo penguin), 149
Pyralidae (Lepidoptera), 180

quaking, role in aspen, 124
quasi-steady analyses: of flight, 278; swimming, 365; when adequate, 376
Quercus. See oak
Quercus phellos (willow oak), 229
quicksand, 307

ram ventilation, 60, 68, 110; and suspension feeding, 189
Rana (frog), 146, 200
range, Sordaria projectile, 195
Ranunculus pseudofluitans (aquatic plant), 199
rate of shear. See shear rate
ray: body lift, 239; ground effect, 289; lift, 65, 176; pitching body, 239; swimming, 239, 286

Rayleigh, Lord: in drag, 82; jet breakup, 396

recirculation, preventing, 190–92 (fig., 191)

reconfiguration: E-values, 118; flexible bodies in flows, 116–27; *Hedophyllum*, 125; leaves, 123–24 (fig., 124); and speed-specific drag, 117

recruitment, larval, 40. *See also* settling

red blood cells, 28; as boluses, 312

reduced frequency, 376

reed, pressure-induced flow, 71

reference area: in drag coefficient, 90–91; frontal, 132; in lift coefficient, 234; plan form, 132, 236; in polar diagram, 236; profile, 132, 236; volume to the two-thirds power, 132; wetted, 132

reference frame. *See* frame of reference

relative motion, 12–14 (figs., 13, 14)

reptile, blood pressures, 56

resistance: of pipes, 294; of pumps, 323, 328

respiratory system, Womersley number, 377

Reynolds, Osborne, 47, 84, 298

Reynolds, number, 5, 84–86, 87, 89; acceleration to raise, 360; vs. airfoil drag, 241; and airfoils, 142, 244–46; and airfoil stall, 244; *Baetis* legs, 140; and Bernoulli vs. Hagen-Poiseuille, 315; vs. best aspect ratio, 241; and best thickness-to-chord ratio, 134; biological range, 86; birds, 244; vs. boundary layer thickness, 157; and character of flow, 87; and circulation, 245; desert locust, 144; and drag coefficient, 88–93, 117; dynamic similarity, 88, 102; and E-value, 120; examples, table, 86; and flow pattern, 102; and friction factor, 301; and gliding, 254; high vs. low, 244, 304, 305, 331; imprecision, 11; inertia vs. viscosity, 156; insects, 244; vs. lift coefficient, 234, 244; and modelling, 102–5; and nozzle flow, 395; and optimal shape, 109; and organism's medium, 88; and organism's size, 87; physical meaning, 86–88; pipe roughness elements, 296; and pressure drag, 97; and profile drag, 244; and propulsive bristle spacing, 351; of protrusions, 165; and skin friction, 96; and

streamlining, 99; vs. Strouhal numbers, 371; tadpoles, 146; and terminal velocity, 340; trout, 88, 147; and turbulent transition, 85, 95, 111, 135, 290; useful precision, 85; very low *Re*, *see* Reynolds number, very low; and vortex shedding, 369; and vortices, 209, 216; water beetles, 145

Reynolds number, boundary roughness, 170

Reynolds number, local, 166

Reynolds number, very low, 331–60; absence of separation, 332; biological importance, 332; boundary layers, 332; characteristics, 331; circulation and lift, 347; diffusion vs. swimming speed, 332; drag, 338–38; drag coefficients, 333; drag vs. orientation, 334; drag and shape, 332, 334; drag-based propulsion, 347–48; flow reversibility, 348; inertia vs. drag, 331; information transfer, 359–60; negligible circulation, 332; propulsion, 331, 347–55; stirring and mixing, 331; streamlines, 332; suspension feeding, 355–59; terminal velocity, 332, 339–47; velocity gradients, 332, 348; vortices, 332; wall effects, 332, 338–39

Rhacophorus (flying frog), gliding, 257

rheophilic tadpoles, 179

rheotaxis, water penny beetles, 179

Rhithrogena (mayfly nymph), 178

Rhynchops nigra (skimmer), 141, 288

riblets, drag reduction, 153

Richardson, L. F., 47

Richardson number, 395

ripples, vortices in, 212

river, 5; and continuity, 39

root pressure, 321

rotation: bacterial flagellum, 206, 352; vs. circular translation, 206; falling particle pairs, 342; solid vs. fluid, 204; as wall effect, 339. *See also* vortices

rotifers: seasonal polymorphism, 30; swimming with cilia, 355

roughening, and transition *Re*, 100

roughness: for drag reduction, 99; and friction factor, 301; length, 167, 169, parameter, 167; protrusion heights, 160; *Re* of elements, 160; vs. *Re* and drag coefficient, 101; rocky-coastal

roughness (*cont.*)
organisms, 101; and settling, 186; and
spore liberation, 193; surface, and eddies, 167; tree bark, 100

sailing, 390–91; biological examples,
391; mechanisms, 390; with surface
currents, 391; *Velella*, 252
saithe: burst-and-coast swimming, 281;
drag coefficient, 148
Salmo gairdneri (trout), 147
salps: acceleration reaction, 365; jet propulsion, 78, 80
samara: advance ratio, 272; as autogyro,
272; autorotating 228, 229 (figs.); as
Flettner rotor, 228; flight, 272–73 (fig.,
273); lengthwise twist, 264; sinking
speeds, 273
sand, pressure-induced flow through, 71
sand dollar: airfoil, 246, 247 (fig.); circulation and feeding, 246; effective aspect ratio, 246; lift and drag of 65, 66
(fig.), 176; pressure-induced flow, 71
sand dollar larvae, swimming speeds, 31
sap ascent, 37–38, 321–23; and continuity, 38; flow speeds, 38; measuring, 37;
pumping, 308
sap conduits, 321–23; embolisms, 323;
flow speeds, 322; gravitational gradient, 322; and Hagen-Poiseuille equation, 322; Laplace's law, 323; Murray's
law, 320; negative pressures, 308, 321;
pore resistance, 322; size vs. flow
speed, 322, 323; suction and trunk
shrinkage, 316; vessel lengths, 322
sap lifting pump: positive displacement,
326; pressures, 324
Sargassum (seaweed), form vs. site, 126
satellite drops, 88
Savonius rotor, 112
scale insect: sinking speed, 193; standing
on hind legs, 193
scales, on butterfly wings, 245
scaling: acceleration reaction, 108, 368;
drag, 108; drag of flexible forms, 115;
lift and weight, 243
scallop (figs., 78, 251): as airfoil, 251, aspect ratio, 252; asymmetrical shells,
252; jet propulsion, 77, 251; modeling,
103; pumping rate, 189; swimming,
251; vortices around, 219

Schistocerca gregaria (desert locust), 143,
223
schistosomiasis, 108
schooling, fish, 288
sclerosponge, 71
Scomber scombrus (mackerel), 147
Scyllarides latus (lobster), acceleration, 287
sea anemone, 116 (fig.); drag, 115
seabirds, wing sweepback, 239
sea butterfly, swimming with lift, 286
sea fan, drag and orientation, 107
sea lily, filter elevation, 191
sea lions: drag coefficient, 150; drag and
swimming, 150; swimming with lift,
286
seals: drag and swimming, 150; drag coefficients, 150; swimming with lift, 286
seamounts, as nozzles, 40
sea palm, biomechanics, 122
sea pen, 113 (fig.)
sea turtles. *See* turtles
seasonal polymorphism, 30
seawater: hydrostatic pressure, 51; properties, 23 (table), 25
seed dispersal. *See* dispersal, seed
seeds: autorotating samaras, 228–29;
ballistic mechanisms, 194; drag maximization, 257; sinking rates, 14; sliding on snow, 194. *See also* samara
self-excited oscillator, 372–75; mechanisms, 372, 373 (fig.); roles of stiffness
and shape, 374; seed or spore release,
374; Steinman pendulum, 373; Tacoma Narrows suspension bridge, 373
Semibalanus balanoides (barnacle), 185
separation, 94, 102, 111, 133, 143; airfoil, 233; *Donax,* 182; and drag, 99;
flattened stream fauna, 179; and lift
vs. drag, 177; nonsteady effects, 278;
and pressure drag, 97; and stall, 242;
and vortices, 213
separation point, 95, 96 (fig.); black fly
larval preference, 184; detecting, 95;
flat plate, 98; head and torso, 95 oscillation on airfoil, 376; below transition, 95; at turbulent transition, 95
sessile organisms, settling, 183
setae, thrust from drag, 154
settling: from air, 186; algal spores, 187;
ascidian, 184; barnacles, 185; foraminifera, 184; hydrodynamic factors,

184–85; pollen and spores, 186; rough surfaces, 186; velocity gradient, 183–86

sewage filter fly, 183

shape: and drag, 106–27; and drag coefficient, 132; low Re, 134; of streamlined bodies, 133; vs. skin friction, 134; variation with force, 67

sharks: dermal scales and ridges, 153; spiral valve of intestine, 311; thunniform swimming, 282

shear: deformation, 16, 20; spinning a vortex, 211

shear flow: blood cells rotation, 187; endothelial cells, 188; erosion, 187; fish eggs, 187; rotation, 188 (fig.); tumbleweed rotation, 187

shear modulus, 17

shear rate, 18, 25, 87, 153; for barnacle settling, 183; in boundary layer, 157; in vortex, 207

shear strain, 17

shear stress, 17, 20, 24, 50, 62, 72, 168; direct effects, 187–88; in friction factor, 301; and Murray's law, 318, 320; on pipe wall, 291; removing bacteria, 176; and shear velocity, 169

shear velocity, 168–69

Sherwood number. See Péclet number

ships. See surface ships

shock absorbers, 18

shrimp, acceleration, 287

SI, 9, 10, 22, 23, 25, 163

sieving, filtration by, 356

silk thread: ballooning in vortices, 223; drag, 222; telling wind direction, 223

Simulium vittatum (black fly larva), 184, 217

sinking rate. See terminal velocity

ski jumper: body lift, 240; lift-to-drag ratio, 240

skimmer: bill as strut, 141; ground effect, 288; lower bill shape, 383

skimming flow, 162, 172, 219

skin friction, 96, 99, 153, 156, 162, 164, 242; airfoils, 240, 241; and drag coefficient, 135; vs. optimal shape, 134; in pipes, 291; near propulsive appendages, 349; and Re, 96; in Stokes' law, 334; and viscous force, 97

slapping locomotion, 387

sliding on snow by seeds, 194

slime mold, vortices in cup, 214

slimes, 5

slug flow, 297, 302, 309

slurries, 5

snails: lift and orientation, 176; mucous thread drag, 223; on river bottoms, 40

snake: anguilliform swimming, 281; blood pressures, arboreal, 56; gliding, 257; heart position, arboreal, 56

soaring, 252, 259–61; dynamic, 259, 260; gradient, 261; sea anchor, 260; slope, 259, 260 (fig.); static, 259; thermal, 221, 260

solid, 16, 17, 80

Sordaria (fungus), 195; muzzle speed, 195; range of projectile, 195; trajectories, 196 (fig.)

soredia, of lichens, 214

sound, speed of, 21

sound production, unsteady flow, 377

spasmoneme, of protozoan, 360

specific volume, 22

speed: estimating flight maxima, 268; flying deer fly, 21; and glide angle, 253; porpoise, 151; sinking, of samaras, 273; wing loading vs. flight, 243. See also terminal velocity

speed of flow, measurement, 54

speed of sound, 21

speed-specific drag, 117–19; pine, 118 (fig.); and reconfiguration, 117; vs. speed, 117; trees, 117

spermatozoan: Re, 86, 88; swimming, 347

Spheniscus humboldti (Humboldt penguin), 149

sphere: added mass coefficient, 363; boundary layer, 163; drag, 112, 333; drag coefficient, 92, 112, 334; drag oscillation at transition, 375; flow patterns, 96; interactions in sinking, 342; Magnus effect, 227; Stokes' law, 333; terminal velocity, 339; transition point, 96; transition Re, 305. See also Stokes' law

sphere, fluid: drag at very low Re, 337; induced toroidal motion within, 337

spheroid, drag vs. shape and orientation, low Re, 336

sphygmomanometer, 56

spider: ballooning, 12; drag of silk strand, 335; thermal soaring, 222
spider, fishing, 387 (fig.); prey detection with waves, 392; sailing, 391
spider silk, strength, 10
spiral valve, of shark intestine, 311, 312 (fig.)
Spirogyra (alga), 199
splitter plates and drag, 110–11
spoilers, mayfly nymph, 179
sponges, 38–39; and continuity, 38; filtering choanocytes, 356; flagellated chambers, 314, 320; maintaining laminar flow, 109; Murray's law, 319; pipe sizes and flow speeds, 36, 39; pressure-induced flow, 71; preventing recirculation, 190; pumping rate, 38, 189, 319; pump pressure, 328; sclerosponges, 66; suspension feeding, 189
sporangium, of slime mold, 214
spore, liberating and settling, 186, 192
sporozoan trophozoites, thrust from shear, 338
spot on surface, drag, 165
spraying vs. squirting, 88
springs, 18
springtail, 387 (fig.); use of surface tension, 389
squid: acceleration reaction, 365; cost of transport, 79; as jet, 69, 77, 78 (fig.); modeling, 103; pressure distribution, 68, 69
squirrel, flying, 257
squirting and spraying, 88, 394–97
stagnation point, 82, 95
stall, 242; angle of, 244; anti-stall devices, 242; fruit fly wing, 250; nonsteady effects, 278; vs. *Re*, 244; repeatedly delayed, 276; and separation, 242
starting vortex, 232
static hole, 62
static pressure, 51, 53, 58, 61
steady flow, 21; and streamlines, 42
Steinman pendulum, 373
Stenella attenuata (cetacean), 151
stenoses 62, 316; and turbulent oscillation, 374
Stenus (beetle) surfactant propulsion, 390
Stichopathes (coral), 40
stiff materials, 114

sting, and interference drag, 138
stirring, at very low *Re*, 331
Stokes, Sir George, 25, 218
stokes (unit), 25
Stokes' law, 5, 27, 29, 88, 156, 194, 304; equation, 333; gas droplet, 215; origin of drag, 334; and propulsion, 348; *Re* limit, 334; and terminal velocity, 340
Stokes' radius, 341
stomata, 197
stopping vortex, 232
strain, 9; shear, 17
streaklines, 41–44 (fig., 44)
streamlined bodies, 64 and 100 (figs.); bidirectional, 114; biological, 141–51; drag coefficients, 133; drag vs. roughness and *Re*, 101; lift, 64; pressure coefficients, 63; reversed flow, 114; shape, 133
streamlines, 40–41, 42 (fig.), 61, 63, 80, 81, 99; and airfoil lift, 232; and continuity, 41; and lift origin, 226; locating, 41; pathlines vs. streaklines, 42; source of quantitative information, 42–44
streamlining, 98–99, 101, 109, 114, 120; directional sensitivity, 99; fly pupa, 180; half-streamlined bodies, 165, 166; ideal, 135; mayfly nymph, 177; motile animals, 132–36; organisms, 142 (figs.); pressure coefficients, 63; pressure drag, 133; protrusions, 165; vs. *Re*, 99; struts, 139–41 (fig., 140); as vortex prevention, 212
streams, 5; effects of current, 175; torrential fauna, 139
streamtubes, 41, 61, 74, 75
stress, 9, 50; shear, 17
stroke angle, beating wing, 267
stroke plane: beating wings, 265; hover fly, 265; hovering, 278
stromatoporoid, 66, 71
Strouhal number, 370–71, 372 (graph); circular cylinders, 371; as dimensionless frequency, 370; flat plates, 371; vs. *Re*, 371
struts: across air-water interface, 383; biological, 139–41 (fig., 140); crab legs, 140; drag, 134; fishing bat legs, 140; mayfly nymph legs, 139; skimmer bills, 140; streamlining, 139–41, 140 (fig.); thickness-to-chord ratio, 134

Styela montereyensis (ascidian), 60, 61 (fig.)
submarine, 382, 383
substratum: flow next to, 171–73; flow patterns near, 172 (fig.); shear velocity vs. roughness, 171 (graph)
surface currents, 391
surface heating, drag reduction with, 153
surface ships, 141, 382–387; displacement vs. planing hulls, 386; Flettner, 227; hull drag, 137; hull length, 382; maximum speed, 383, 384; planing, 386
surface spot: drag, 164; local drag coefficient, 164; shear on, 165
surface swimming, 384–86; cost, 385; drag, 385; maximum speed, 390, 390; planing, 387
surface tension, 8, 19, 378–79; biological relevance, 379; dimensions and units, 10, 378; and nozzle, 395; and propulsion, 387–90; size vs. speed limit, 390; and surface currents, 391; useful size range for support, 389, 390; and wave celerity, 381; and wave formation, 379; Weber number, 389
surface waves. *See* waves
surface-to-volume ratio, 9; dimensionless version, 346; and sinking speeds, 345
surfactant propulsion, 390
suspension feeding, 4, 216, 355–59; active vs. passive, 40, 189, 190; colonial jets, 190; and drag, 107, 127; fish, 189; high *Re*, 355; and local flow, 192; modeling, 103; preventing recirculation, 190–92; prion (bird), 260; shape of structures, 113; in turbulent flow, 192; in velocity gradients, 188–90; whales, 189
swash: beaches, 307; velocity gradient in, 181, 219
swash-riding, 181–82; clam, 181; whelk, 182
swift, wing sweepback, 239
swimming, 13, 280–83; acceleration, 287; acceleration reaction with tail or paddles, 366; added mass, 364–67; anguilliform, 281; bacterium coasting, 331; brine shrimp, vs. *Re*, 350; bristles, very low *Re*, 350; burst-and-coast, 280; C-start, 287; carangiform, 281; car-

angiform with lunate tail, 282; cilia vs. flagella vs. muscle, 355; crabs, 146; ctenophores, 355; *Daphnia*, 30; vs. diffusion, very low *Re*, 332; drag-based, 154, via drag production, 286; ducks, 147, 283; eel vs. spermatozoan, 347; fish, 137, 147–48; fish schooling, 288; fish, pressure on, 67; flow disturbance, very low *Re*, 349; frogs, 146; glide to measure drag, 138–39; ground effect, 289; heat production, 29; irregularity as signal source, 359; lift vs. drag based, 283–86; modes, 281, 282 (fig.); paddles, very low *Re*, 350; paddling vs. flapping, 283–86; penguins, 137, 283; quasi-steady analysis, 365; rays, 239; sand dollar larvae, 31; scallops, 251; sea lions, 150; seals, 150; vs. sinking as signal source, 359; speeds, fin whales, 69; speeds, pelagic larvae, 185; squid, 69; subcarangiform, 281; surface, 384–86; surface vs. submerged, 382; tadpoles, 145; thunniform, 282; turtles, 137; unsteady analysis, 366; at very low *Re*, 347–55; vortices in, 281, 282; wall effects, microorganisms, 339; water beetles, 137; zero momentum wake, 280. *See also* surface swimming
swooping flight, birds, 239
Synchropus picturatus (mandarin fish), ground effect, 289
synovial fluid, as non-Newtonian, 20
syphon, 55
systematic error, 11
Système Internationale. *See* SI

tadpole: drag, 145; flattened rheophilic, 179; hygropetric, 182; *Re*, 146; tail as splitter plate, 110
tar, viscosity, 18
tatter flags, 128
Taylor, Sir G. I., 353
technology, 7
temperature, 9; animal bodies, 27; vs. blood viscosity, 28; and convective flow, 6; vs. density, 22; and dynamic viscosity, 24, 27–31, 153; vs. kinematic viscosity, 25
tendency to pinch: laminar flow, 315; turbulent flow, 316
tensile strength, water, 322

terminal velocity, 339; biological significance, 343–47; ciliates, 344; diatoms, 344; and drag coefficient, 340; *Euphausia* embryos, 30; falling object interactions, 341–42; fecal pellets, 344; fungal spore, 344; hemlock pollen, 344; marine algae, 185; marine bacteria, 344; marine snow, 344; measurement pitfalls, 342–43, 345; particle groups, 342; phytoplankton, 344; plankton, 88, 29; pollen, spores, 88, 186; reducing sinking rates, 344; and seasonal polymorphism, 30; vs. shape, 340; and shape irregularity, 345; sphere, 339; Stokes' law, 340; Stokes' radius, 341; and surface-to-volume ratio, 345; trout eggs, 88; very low *Re*, 332, 339–47; wall correction formula, 342; vs. weight, samaras, 273; why sink?, 346

termite mound, ventilation, 71

Thalassiosira fluviatilis (diatom), 345

thermal soaring, 221–23 (fig., 222); and atmospheric instability, 222; spiders and moths, 222

thickness-to-chord ratio, 109, 111; body of revolution, 134; and buckling, 134; strut, 134

thigmomorphogenesis, 128

thrips (Thysanoptera), 350 (fig.); flight, 350

thrust, 14, 74, 262, 263 (fig.); blade element analysis, 262; from drag, 154–55; drag vs. lift based, 283–87; equality with drag, 136; from flapping, 262–64; indirect measurement, 76; jet propulsion, 79; from lift, 137, 149, 263; at low *Re*, 354 (diagram); via no-slip, very low *Re*, 338; production as pumping, 323; production vs. drag reduction, 136; rotating propeller, 262

Thunnus (tuna), 282

time, 8, 9

timelines, 44

tip vortices, 232; and aspect ratio, 236; flying bird, 276

torrential fauna, 139, 178 (figs.); dorsoventral flattening in order, 177; drag in boundary layer, 176, 180

total head, 59, 61, 72, 82; in boundary layer, 157

Toxotes (archer fish), squirting, 396

trajectory, *Sordaria* spore clusters, 195, 196 (fig.)

transformers, biological, 329–30

transition point: as flow quality indicator, 96; laminar to turbulent, 11

transpiration, 38; oak, 37; role of diffusion, 197; vs. wind speed, 197

transport: diffusive vs. convective, 202; phenomena, 197

transport cost: vs. flying speed, 240; squid vs. trout, 79

trapping pollen, 187

tree: autorotating leaves, 229; autorotating seeds, 228; bark roughness, 100; and continuity, 38 (fig.); drag, 14, 107, 117; drag of leaves, 100, 120–24; elphin forest, 127; "flagged," 127; high-altitude forms, 127; leaf area, 37; making free convection, 221; negative pressures, 321; pipe data, 36; sap ascent, 37–38, 321–23; thigmomorphogenesis, 128; trunk shrinkage, 322; turning moments, 1231–22; vortex shedding by trunk, 371; wind-throw, 108

tree-of-heaven, autorotating seed, 228

trench, flow across, 212

trout: Cayley's profile, 106, 107 (fig.); cost of transport, 79; drag coefficient, 147; eggs, sinking, 88; ram ventilation, 68

Tsuga canadensis (hemlock), 344

tuliptree, autorotating samara, 228

tumbleweed, rotation in shear flow, 187

tuna, 142 (fig.); corselet, 29; *Re*, 86; swimming, 282

tunicate. *See* ascidian

turbulent flow, 11, 21, 46–49 (fig., 48), 84, 87, 290; biological relevance, 48; in boundary layer, 159; circular aperture, 303; cross-flow transport, 161; diffusion analogy, 47; distance index, 311; and drag, 111; drag of flat plate, 135; intermittent, in pipe, 375; and local momentum, 96; and momentum transfer, 47; noise, 95; on leaves, 163; oscillation at transition, 375; pipes, 300–303, 375; shear velocity, 168; speed vs. x-sec, 301, 311; and spore liberation, 193; near surface, 95; suspension feed-

ing in, 192; transition, 11, 110; vortex shedding, 371; vortices in, 209; vs. unsteady flow, 47; wakes, 167; wall pinching, 316; in wind tunnels, 193
turbulent intensity, 24
turbulent transition: cylinder, 94; drag change, 134; drag coefficient, 94; drag and shape, 133; flat plate, 135; and *Re*, 85; sphere, 96
Tursiops gilli (porpoise), 151
turtle, swimming, 137, 286
typhlosole, of earthworm intestine, 311

Ulmus (elm), winged seed, 257, 251 (fig.)
undulate margin, algae, 126
units, 9–11
Universal Variable Constant. *See* dimensionless coefficient
unsteady analysis, swimming, 366
unsteady flows, 21, 372–77; acceleration reaction, 362–69; added mass, 362–69; aerodynamic frequency parameter, 376; aeroelasticity, 374–75; drag and flexibility, 375; flutter, 369; periodically delayed stall, 376; reduced frequency, 376; respiratory wheezes, 375; rubber tube, 374; self-excited oscillators, 372–75; sound production, 377; sphere at turbulent transition, 375; Strouhal number, 370; turbulent spouting in pipe, 375; virtual mass, 363; Von Kármán trails, 369, 370–72; vortex shedding, 369–74; waves, 108; Womersley number, 377. *See also* nonsteady flows
unstirred layer, 200–202 (fig. 202); and diffusive boundary layer, 201; and mixing, 202
urethra: navicular fossa, 396; pinching vs. expansion, 316
urination. *See* micturition

vein, flow through, 315
Velella (by-the-wind-sailor), 251 (fig.); airfoil, 252; sailing, 391
Velia (bug), surfactant propulsion, 390
velocity, 8; free stream, 20; tangential, 204, 208, 224
velocity gradient, 5, 12, 19, 20, 25, 61, 72, 80, 156–71, 174; on airfoils, 244; in ciliary pumps, 314; ciliated surface,

348 (fig.); cylinder across, 216; diffusion across, 196–202; as dispersal barrier, 192–96; and E-values, 120; and ectoparasites, 180; fibroblast growth, 200; life in, 174–200; and oxygen uptake, frogs, 200; around *Paramecium*, 354; between plates, 176; propulsion through, 194; and propulsive appendage length, 349; rotation, 188 (fig.); settling, 183–86; from shear measurement, 165; and soaring in vortices, 223; and suspension feeding, 188–90; swash, 181, 219; vs. temperature, 29; the "unstirred layer," 200–202; very low *Re*, 332, 348; vortex induction, 212; and vortices, 208; vorticity, 218–21
vena contracta, 394
vent vs. anus, 146
Venturi meter, 57–58 (fig., 58)
vessel. *See* pipes; sap conduits, capillary; etc.
virtual mass, 138, 149; vs. actual mass, in air, 368
viscoelastic solids, 20
viscometer, 17, 26–27 (fig., 27), 104
viscosity, 5, 7, 18, 21, 46, 50, 80, 156; and airfoils, 244; and antennal transmissivity, 356; and Bernoulli's principle, 61; bloods, 28; and boundary layer thickness, 158; and circulatory systems, 28; and drag, 63; eddy, 24; and falling particle interactions, 342; and filter feeding, 29; gases, measurement, 27; glycerin, 26; hot water, 176; ideal fluid, 52; increase with polymers, 29; measurement, 26; molecular, 24; and Newtonian fluids, 20; and no-slip condition, 20; and pressure-induced flows, 72; rarefied gases, 19; sugar syrups, 26; at very low *Re*, 331; and waves, 379. *See also* dynamic viscosity; kinematic viscosity
viscous entrainment, 72, 103
viscous force, and skin friction, 97
viscous sublayer, 170. *See also* laminar sublayer
visualization of flow, 41, 44
volcano cone, 72
volume, specific, 22
volume flow rate, 33, 293

Von Kármán, T., 216

Von Kármán's constant, 167

Von Kármán trail, 93, 95, 96, 216, 369–72 (fig., 370), 374; shedding frequency, 370; vortex spacing, 370

vortex ring. *See* vortices: rings

vortex shedding: cat's whisker, 372; and circulation reversal, 369; frequency, 370; and *Re*, 369; shaking body, 369; tree trunks, 371

Vorticella (protozoan), 191; vortex, 209, 210 (fig.)

vortices, 11, 12, 204–27; of aircraft, 233 (fig.); at apertures, 303; ascending pairs, 216; and bed erosion, 219; Bénard cells, 224; bird wing-tips, 276; behind black fly larva, 217 (fig.); bound, 232; behind butterflies, 277; butterfly flight, 287; cephalopod ink, 215; circulation, 210, 224; clap-and-fling mechanism, 279; in corners, 213; in cups, 214; behind cylinder, 215, 217 (fig.); density-gradient, 224; inside droplets, 214; ends, 210, 219; examples, 213 (fig.); feeding in, 216–18, 220–21; fictitious, 231; in fish swimming, 281; behind flying bird, 276; and formation flight, 288; horseshoe, 173, 219; in ideal fluid, 209; induction in velocity gradient, 212; interactions, 211; irrotational, 206, 208–12; in jet propulsion, 79; in laminar flow, 209; Langmuir circulations, 394; leapfrogging, 211; nonsteady, 277; pairs, 210; periodic shedding, 93; in pipe bends, 215; around protrusion, 219; and *Re*, 87, 93, 209, 216, 332; respiratory role, 218; rings, 210, 211 (fig.), 232, 276; rolling, 223; around rotating cylinder, 208; rotational, 204, 206, 207; in scallop beds, 219; behind self-excited oscillator, 373; shear driven, 214; shedding, 216, 369–74 (*see also* Von Kármán trail); and splitter plantes, 110; starting, 232; stopping, 232; and streamlining, 212; near surface, 172; in swimming, 282; thermal, 221–24; tip, 232; toroidal, 210; in trenches, 212; in turbulent flow, 209; use in suspension feeding, 358; from velocity gradient, 208; viscosity and energy,

209; vortex rings, 232, 276; and vorticity, 218; in wake, 93; wake of jet, 215; wakes of, 369–74; in wakes of fliers, 275–77 (fig., 277); windrows, 393; in wing pleats, 245

vorticity, 218, 225; and circulation, 224–25; velocity gradients, 218–21; and vortices, 218

vulture: lift-to-drag ratio, 248, 255; thermal soaring, 222

Wagner effect, 277

wake: energy dissipation, 97; turbulent, 167; vortices in, 93, 369–74; width and transition, 111; width as drag indicator, 98; zero momentum, 280

walking underwater, 140

wall effects: attractive force, 339; correction formula, 342; drag asymmetry, 339; falling circular cylinder, 93; induced rotation, 339; quick index to significance, 339; swimming microorganisms, 339; very low *Re*, 332, 338–39

wasp, clap-and-fling mechanism, 278

wasp, mymarid, 350 (fig.)

water: density vs. temperature, 22; kinematic viscosity, 25, 88; properties, table, 23; surface tension, 8; tensile strength, 322

water boatman bug: acceleration reaction, 365; swimming, 154

water flea. See *Daphnia*

water strider, 387 (fig.); communication with waves, 392; hygropetric, 182; propulsion, 387; use of surface tension, 389

wave, waves, 5, 22, 379–86, 390–92; acceleration reaction for surge, 367; amplitude, 380; bow, 382–83; capillary, 382; celerity, 381; celerity vs. frequency, 392 (graph); at density gradient, 394 (fig.); drag, 141; frequency vs. wavelength, 392; gravity, 382; and hull, 383 (fig.); metachronal, 154; as moving disturbance, 379; and mussel beds, 66; as orbital oscillations, 379; particle orbits, 380 (fig.); period, 380; as resistance source, 382; stern, 382; and unsteady flow, 108; use in communication, 391–92; use for propulsion, 386; vs. water depth, 380; wavelength, 381 wave period, and drag of macroalgae, 125

Weber number, 347; defined, 389; nozzles, 395

wetted area, 90

whale (figs., 70, 142); drag, 151; ectoparasites, 180; pressure distribution, 69; propulsion from waves, 386; *Re*, 86, 87; shape, 383; suspension feeding, 189, 355; swimming speed, 69; tail as propeller, 282; thunniform swimming, 282. *See also* cetaceans

whale lice, 180

wheat, flow in root vessels, 38

wheezes, respiratory, 375

whelk, swash riding, 182

white poplar, leaf fluttering, 193

willow oak, leaf as Flettner rotor, 229

wind, 6, 12, 13; and leaf temperature, 198; and plant growth, 127; pollination, 43, 187; propagule dispersal, 344; and transpiration rate, 197; and tree falling, 121

wind-driven circulation, 6

windmill blade; as airfoil, 268; contour, 270; operation, 269 (fig.); origin of torque, 270; quixotic, 271

windrows, 393

wind throw, 108

wind tunnel, 7, 11, 13, 14, 15, 98, 103

wing, 21; amplitude of beating, 267; beating, as actuator disk, 273–75; beating, as propeller blade, 264; birds, performance, 248; blade element analysis, 273; bristles and fringes, 350; finite and propulsion efficiency, 237; as fluid dynamic pump, 327; form of stroke, 264; infinitely long, 236; insects, 15, 250; lengthwise twist of beating, 264; lift production, 230–34; mass, fruit fly, 369; momentum flux of beating, 273–75; pitch, 242; plane of beating, 264; power need of finite, 238; scaling, 243; stroke angle of beating, 265, 267; surface irregularities, 245; thrust, 14, 15; twist vs. speed, 264; unsteadiness in beating, 376; wingbeat frequencies, 268. *See also* airfoils

wingbeat frequency, 268

wing loading, 243–44; Andean condor, 243; bee, 243; vs. flight speed, 243; fruit fly, 243; and hovering, 266; pedalled planes, 243; pedalled planes, 243; as scaling problem, 243; wren, 243

Womersley number, defined, 377

worm, phoronid and terrebellid, tube and feeding, 217

Wormaldia (snail), 40

worms, polychaete: burrow pumps, 326; pressures of burrow pumps, 327; tube and feeding, 217

wren, wing loading, 243

Xenopus laevis (frog), 200

xylem. *See* sap conduits

Young's modulus, 17

Zalophus californianus (sea lion), 150

Zanonia macrocarpa (Javanese cucumber), 256

zebra finch, body lift, 239

zero momentum wake, 280

zero plane displacement, 167–69

zooplankton, filtration, 356

Zoothamnium (sessile protozoan), rapid stalk contraction, 360